SIMULATION TECHNIQUES

SIMULATION TECHNIQUES

MODELS OF COMMUNICATION SIGNALS AND PROCESSES

FLOYD M. GARDNER

JOHN D. BAKER

A Wiley-Interscience Publication

JOHN WILEY & SONS, INC.

New York ■ Chichester ■ Brisbane ■ Toronto ■ Singapore ■ Weinheim

This text is printed on acid-free paper.

Copyright © 1997 by John Wiley & Sons, Inc.

All rights reserved. Published simultaneously in Canada.

Reproduction or translation of any part of this work beyond that permitted by Section 107 or 108 of the 1976 United States Copyright Act without the permission of the copyright owner is unlawful. Requests for permission or further information should be addressed to the Permissions Department, John Wiley & Sons, Inc., 605 Third Avenue, New York, NY 10158-0012.

Library of Congress Cataloging-in-Publication Data

Gardner, Floyd Martin, 1929–
    Simulation techniques : models of communication signals and processes / Floyd M. Gardner, John D. Baker.
      p.   cm.
    Includes index.
    ISBN 0-471-51964-2 (cloth : alk. paper)
    ISBN 0-471-51966-9 (set)
    1. Telecommunication systems–Computer simulation. I. Baker, John D. II. Title.
TK5103.G37 1996
621.382'01'13–dc20   95-38580

Printed in the United States of America

10 9 8 7 6 5 4 3 2

To Samuel F.B. Morse

··· ·— —— ··— — · ·—·· —— ——— ·—· ··· ·

*The first of our profession
who could have made good use of a transmission simulator.*

CONTENTS

PREFACE xix

NOMENCLATURE xxiii

CHAPTER 1 INTRODUCTION 1

    1.1 Role of Simulation, 1
        1.1.1 Simulator Capabilities, 2
        1.1.2 User Background, 4
        1.1.3 Limits of Simulation, 5
    1.2 Transmission Links, 6
        1.2.1 Hierarchy, 7
        1.2.2 Link Elements, 8
    1.3 Pertinent Literature, 15
    1.4 Study Plan, 17

PART 1 FOUNDATIONS

CHAPTER 2 REPRESENTATION OF ANALOG SIGNALS AND NOISE 19

    2.1 Baseband Signal Representations, 19
        2.1.1 Time-Limited Versus Bandlimited Pulses, 20
        2.1.2 Pulse Amplitude Modulation (PAM), 21
        2.1.3 Nyquist Shaping, 27
        2.1.4 Other Pulse Formats, 31

2.2 Passband Signal Representations, 34
    2.2.1 Basic Modulation Formats, 34
    2.2.2 Quadrature Amplitude Modulation (QAM), 36
    2.2.3 Examples of Passband Signal Formats, 40
2.3 Power, Energy, and Spectrum Relations, 47
    2.3.1 Baseband Signals, 47
    2.3.2 Passband Signals and Complex Envelopes, 54
2.4 Noise, 61
    2.4.1 Additive Gaussian Noise, 61
    2.4.2 Counterexamples: Non-AWGN, 65
2.5 Key Points, 66
    2.5.1 Baseband Signals, 66
    2.5.2 Passband Signals, 67
    2.5.3 Power, Energy, and Spectrum, 69
    2.5.4 Noise, 70

CHAPTER 3 REPRESENTATION OF ANALOG FILTERS   71

3.1 Categories of Filter Representation, 71
    3.1.1 Differential Equation Representation, 71
    3.1.2 Rational Transfer Functions, 72
    3.1.3 Nonrational Transfer Functions, 75
    3.1.4 Impulse Response, 75
    3.1.5 Summary of Categories, 77
3.2 Filters for Complex Signals, 77
    3.2.1 Bandpass Filters, 78
    3.2.2 Signal Transmission Through a Bandpass Filter, 79
    3.2.3 Filter Structure, 81
3.3 Conversion to Complex Filters, 82
    3.3.1 Frequency-Domain Conversion, 83
    3.3.2 Time-Domain Conversion, 86
3.4 Key Points, 87
    3.4.1 Filter Categories, 87
    3.4.2 Complex Filters, 88
    3.4.3 Conversion to Complex Filters, 89
    3.4.4 STÆDT Processes, 90

APPENDIX 3A COMPLEX CONVERSION OF TIME-DOMAIN FILTERS   90

CHAPTER 4 REPRESENTATION OF ANALOG SIGNAL PROCESSES   96

4.1 Modulators, 97
    4.1.1 Complex Envelope, 97

4.1.2 Angle Modulations, 97
4.1.3 Analytic Signal, 98
4.2 Demodulators, 99
4.2.1 *I–Q* Demodulation, 99
4.2.2 Demodulator Coherence, 100
4.3 PAM Link Model; Matched Filters, 104
4.3.1 PAM–QAM Link Model, 104
4.3.2 Matched Filters, 109
4.4 Symbol Detection, 112
4.4.1 Symbol-by-Symbol Detection, 112
4.4.2 Decision Rules, 113
4.5 Nonlinear Processes, 116
4.6 Key Points, 119
4.6.1 Modulators, 120
4.6.2 Demodulators, 120
4.6.3 PAM Link Model; Matched Filters, 121
4.6.4 Symbol Detection for PAM–QAM Coherent Reception, 122
4.6.5 Nonlinear Processes, 123
4.6.6 STÆDT Processes, 123

CHAPTER 5 DIGITAL SIMULATION OF ANALOG SIGNALS AND PROCESSES 124

5.1 Digitization, 124
5.1.1 Quantization, 125
5.1.2 Sampling, 125
5.1.3 Aliasing, 126
5.1.4 Other Features, 134
5.1.5 Digital Representation, 134
5.2 Digitized Signals in the Frequency Domain, 135
5.2.1 *z*-Transform, 135
5.2.2 Discrete Fourier Transform (DFT), 136
5.2.3 Time/Frequency Relations, 138
5.3 Structures and Nomenclature, 140
5.3.1 Simulation Modules, 140
5.3.2 Data Formats, 141
5.3.3 Properties of the Data Formats, 143
5.4 Energy and Power Relations, 145
5.4.1 Energy Formulas, 146
5.4.2 Energy of Analog Pulse Trains, 147
5.4.3 Power, 149
5.4.4 Rationale, 149
5.5 Key Points, 150
5.5.1 Digitization, 150

5.5.2 Discrete Fourier Transform, 150
5.5.3 Structures and Formats, 151
5.5.4 Energy and Power, 152
5.5.5 STÆDT Processes, 153

APPENDIX **5A** ALIASING ENERGY 153

APPENDIX **5B** UPPER BOUND ON MAGNITUDE OF ALIASING ERROR 157

APPENDIX **5C** ANALOG/DIGITAL ENERGY RELATIONS 159

PART 2 MODELING OF SIGNALS AND PROCESSES

CHAPTER 6 COMPUTER MODELING OF ANALOG FILTERS 162

6.1 Filter Algorithms, 163
    6.1.1 Frequency-Domain Filters, 163
    6.1.2 Time-Domain Convolution with Unit-Sample Response, 166
    6.1.3 Time-Domain Recursive Difference Equations, 168
    6.1.4 Complex Coefficients, 170
    6.1.5 Linear Blocks from Frequency-Domain Filters, 173
6.2 Analog-to-Digital Filter Transformation, 174
    6.2.1 Frequency-Domain Transformations, 175
    6.2.2 Time-Domain Transformations of IIR Filters, 185
    6.2.3 Time-Domain Transformations of FIR Filters, 194
6.3 Key Points, 195
    6.3.1 On-Line Filter Processes, 195
    6.3.2 Filter Preparation, 196
    6.3.3 STÆDT Processes, 197

CHAPTER 7 GENERATION OF DATA SIGNALS 199

7.1 Message Source, 200
    7.1.1 Binary Shift-Register Sequences, 201
    7.1.2 Representative Sequences, 203
7.2 Symbol Mapping, 205
    7.2.1 Mapping for Linear Modulations, 206
    7.2.2 Differential Encoding, 209
    7.2.3 Mapping for Nonlinear Modulations, 213
7.3 Pulse Shaping, 213

7.3.1 Generating Impulse Trains, 214
7.3.2 Pulse-Shaping Filters, 215
7.4 Modulators, 220
7.4.1 Linear Modulation, 220
7.4.2 Nonlinear Modulation, 222
7.5 Amplifiers, 226
7.5.1 Linear Amplifiers and Attenuators, 226
7.5.2 Nonlinear Amplifiers, 227
7.6 Key Points, 235
7.6.1 Message Source, 235
7.6.2 Symbol Mapping, 236
7.6.3 Pulse Shaping, 237
7.6.4 Modulators, 237
7.6.5 Amplifiers, 239
7.6.6 STÆDT Processes, 240

APPENDIX 7A DIFFERENTIAL ENCODING EXAMPLES   **241**

CHAPTER 8 OVERVIEW OF RECEPTION PROCESSES   **244**

8.1 Simple Channels, 244
8.1.1 Model of the Medium, 245
8.1.2 Processes Required in Receiver, 245
8.1.3 Performance Evaluation, 250
8.2 More-Complicated Channels, 252
8.2.1 Nonwhite, Non-Gaussian, and Nonadditive Disturbances, 252
8.2.2 Frequency-Selective Channels, 255
8.2.3 Time-Varying Channels, 256
8.3 Spread-Spectrum Processes, 258
8.4 Afterword, 259

CHAPTER 9 GENERATION OF NOISE   **261**

9.1 Uniformly Distributed Samples, 261
9.2 Gaussian Noise Generation, 263
9.2.1 Transformation Algorithm, 263
9.2.2 Scaling, 264
9.2.3 Colored Noise, 266
9.2.4 Phase Noise, 267
9.3 Non-Gaussian Variates, 267
9.4 Key Points, 268

9.4.1 Noise Generation, 268
9.4.2 STÆDT Processes, 268

APPENDIX 9A COMPLEX ENVELOPE OF BANDPASS GAUSSIAN NOISE 269

CHAPTER 10 SHIFTERS AND INTERPOLATORS 274

10.1 Phase Shifters, 274
    10.1.1 Phase Shifting of Complex Signals, 274
    10.1.2 Phase Shifting of Real Signals, 276
10.2 Frequency Shifters, 276
    10.2.1 Frequency Shifting of Time-Domain Complex Signals, 277
    10.2.2 Frequency Shifting of Frequency-Domain Signals, 278
    10.2.3 Frequency Shifting of Real Signals, 282
10.3 Time Shifters, 284
    10.3.1 Discrete Time Shifts: Delays, 284
    10.3.2 Continuous Time Shifts via Transforms, 285
10.4 Interpolation, 286
    10.4.1 Interpolator Model, 287
    10.4.2 Characteristics of Interpolating Filters, 288
    10.4.3 Simulation of Interpolation, 290
    10.4.4 Variable Delay by Interpolation, 293
10.5 Key Points, 293
    10.5.1 Phase Shifters, 293
    10.5.2 Frequency Shifters, 293
    10.5.3 Time Shifters, 294
    10.5.4 Interpolation, 295
    10.5.5 STÆDT Processes, 295

CHAPTER 11 SYNCHRONIZATION 296

11.1 Synchronization by Calibration, 297
    11.1.1 Hardwire Synchronization, 297
    11.1.2 Calibration by Recovery Synchronizers, 299
11.2 Recovery Synchronizers, 299
    11.2.1 Synchronizer Categories, 300
    11.2.2 Synchronizer Operations, 302
11.3 Synchronizer Elements, 305
    11.3.1 Correctors, 305
    11.3.2 Timing Controllers (Feedback), 307
    11.3.3 Carrier Regenerators, 310

11.3.4 Clock Regenerators, 315
11.3.5 Parameter Extractors, 319
11.3.6 Error Detectors, 325
11.4 Synchronizer Processes, 331
11.5 Key Points, 331
11.5.1 Recovery Configurations, 332
11.5.2 Synchronizer Elements, 332
11.5.3 STÆDT Processes, 333

APPENDIX **11A SYNCHRONIZATION BY CORRELATION**   **334**

APPENDIX **11B CARRIER REGENERATORS**   **338**

APPENDIX **11C LOCATING ZERO CROSSINGS BY INVERSE LINEAR INTERPOLATION**   **343**

CHAPTER **12 SYMBOL DETECTION AND DEMAPPING**   **345**

12.1 PAM–QAM Decisions, 346
12.1.1 Minimum-Distance Decisions, 347
12.1.2 Quantizer Detection, 348
12.1.3 Differential Detection of MPSK, 350
12.2 Multiple-Strobe Decisions, 353
12.3 Key Points, 353
12.3.1 Minimum-Distance Detection, 354
12.3.2 Quantizer Detection, 354
12.3.3 Differential Detection, 354
12.3.4 Largest-of Detection, 354
12.3.5 Demapping, 354
12.3.6 STÆDT Processes, 355

CHAPTER **13 EVALUATION OF PERFORMANCE**   **356**

13.1 Monte Carlo Error Counting, 357
13.1.1 Monte Carlo Configuration, 357
13.1.2 Monte Carlo Statistics, 361
13.1.3 Importance Sampling, 363
13.2 Quasianalytic Method, 364
13.2.1 Overview of QA Methods, 364
13.2.2 QA Computation of Error Probability, 367
13.2.3 Phantom Noise Calibration, 376
13.2.4 Implications of Time Invariance, 379

13.3 Key Points, 380
    13.3.1 Monte Carlo Method, 380
    13.3.2 QA Method, 381
    13.3.3 Alignment, 381
    13.3.4 Signal and Noise Calibration, 381
    13.3.5 STÆDT Processes, 382

APPENDIX 13A ERROR PROBABILITIES FOR QUASIANALYTICAL EVALUATIONS   382

APPENDIX 13B THEORETICAL ERROR PERFORMANCE   388

APPENDIX 13C EVALUATION OF $Q(x)$ 394

APPENDIX 13D DELAY AND PHASE ALIGNMENT FOR PERFORMANCE EVALUATIONS   395

APPENDIX 13E PHANTOM NOISE CALIBRATION   402

APPENDIX 13F TRUNCATION OF INFINITE IMPULSE RESPONSES   406

CHAPTER 14 FEEDBACK LOOPS AND CONTROL SIGNALS   409

14.1 Simulation of Feedback, 409
    14.1.1 Single-Point Format, 410
    14.1.2 Opened-Loop Simulation, 411
    14.1.3 Multiple Loops, 415
14.2 Control Signals, 418
    14.2.1 Configuration and Properties, 418
    14.2.2 Loop Filters for Control Signals, 420
14.3 Key Points, 423
    14.3.1 Feedback Loops, 423
    14.3.2 Control Signals, 423
    14.3.3 STÆDT Processes, 424

CHAPTER 15 SIMULATION OF DIGITAL PROCESSES   425

15.1 Quantization, 426
    15.1.1 Quantizer Laws, 426
    15.1.2 Simulation of Quantizers, 433
    15.1.3 Quantizer Applications, 437

15.2 Other Digital Issues, 440
    15.2.1 Number Representation, 440
    15.2.2 Overflow and Underflow, 441
    15.2.3 Processing Delays, 443
15.3 Key Points, 443
    15.3.1 Quantization, 444
    15.3.2 Other Digital Issues, 444
    15.3.3 STÆDT Processes, 445

CHAPTER 16 SAMPLE RATE CONVERSION   446

16.1 Block-Mode Conversions, 447
    16.1.1 Resampling Definitions, 447
    16.1.2 Block-Mode Issues, 448
    16.1.3 Reblocking, 450
    16.1.4 Block-Mode Interpolation, 451
16.2 Single-Point Conversions, 452
    16.2.1 Flag-Mediated Control, 453
    16.2.2 Internal Scheduling, 455
    16.2.3 Mixed Scheduling, 457
16.3 Rate-Changing Processes, 460
    16.3.1 Downsampling Processes, 460
    16.3.2 Upsampling Processes, 461
    16.3.3 Block Alterations, 461
16.4 Key Points, 462
    16.4.1 Sample Rate Conversion, 462
    16.4.2 Interpolation, 462
    16.4.3 Block Mode, 463
    16.4.4 Single-Point Mode, 463
    16.4.5 STÆDT Processes, 464

PART 3   UTILITIES AND PROGRAMMING

CHAPTER 17 ENTRY AND BUILDER ROUTINES   465

17.1 Overview, 466
17.2 Entry Routines, 469
    17.2.1 File Formats, 469
    17.2.2 Entry Tools, 469
17.3 Conditioning, 470
    17.3.1 Conditioning of Nonlinear Amplifiers, 470

17.3.2 Frequency Response of Filters, 473
17.3.3 Other Conditioning Tasks, 477
17.4 Nonlinearity Builder Routines, 478
17.4.1 Nonlinearity Overview, 478
17.4.2 Function Fitting, 478
17.4.3 Interpolation, 479
17.5 Filter Builder Routines, 481
17.5.1 Filter Builder Constituents, 481
17.5.2 Standard Filters, 484
17.5.3 Frequency-Domain Builder Routines, 485
17.5.4 Time-Domain Builder Routines, 490
17.5.5 Specimens of Frequency Rescaling, 494
17.6 Key Points, 495
17.6.1 Nonlinearity Builders, 495
17.6.2 Filter Builders, 496
17.6.3 STÆDT Processes, 497

CHAPTER 18 GRAPHICAL DISPLAYS 498

18.1 Types of Plots, 498
18.1.1 Time-Domain Plots, 499
18.1.2 Frequency-Domain Plots, 502
18.1.3 Other Plots, 504
18.2 Units and Scales, 507
18.2.1 Abscissas, 508
18.2.2 Ordinates, 509
18.2.3 Plot Windows, 513
18.3 Data Conditioning, 513
18.3.1 Computation of Delay, 513
18.3.2 Spectrum Conditioning, 515
18.4 Key Points, 519
18.4.1 Types of Plots, 519
18.4.2 Units and Scales, 520
18.4.3 Data Conditioning, 521
18.4.4 STÆDT Processes, 522

CHAPTER 19 PROGRAMMING ISSUES 523

19.1 Simulation Language, 523
19.1.1 Evolution of a Simulation Language, 523
19.1.2 Relation to Programming Language, 525
19.1.3 High Level Versus Low Level, 525

19.2 Architecture and Rules, 526
    19.2.1 User's Interface, 526
    19.2.2 Parameter Entry, 528
    19.2.3 Order of Execution, 528
    19.2.4 Data Connectivity, 529
    19.2.5 Types, 530
    19.2.6 Composite Commands, 531
19.3 Interpreters Versus Compilers, 532
    19.3.1 Interpreters, 532
    19.3.2 Compilers, 533
19.4 Auxiliary Features, 534
19.5 Key Points, 537

REFERENCES   **539**

INDEX   **553**

PREFACE

Communications systems grow more complicated every year. They can no longer be analyzed and evaluated adequately by paper-and-pencil methods. Moreover, laboratory breadboards—a long-established alternative to mathematical analysis—have become too expensive and time consuming to build as part of the system design process. These two approaches were the mainstays of system design in the past, but they are no longer sufficient.

So how is a communications engineer to design a communications link, taking into account all of the many complexities that obstruct the old approaches? Simulation on a computer has emerged as an essential tool. The engineer puts together simulation models of the signals and processes of the communications system, and exercises them in computer runs. Outputs from the simulations provide performance evaluations, waveform and spectrum plots, and indications of how and why a particular design might or might not serve its intended purpose.

A simulation circumvents the obstructions to mathematical analysis and can be prepared much more quickly than breadboards. Simulations have become indispensable for the design of nontrivial communications systems. Every communications engineer needs to learn the elements of simulation and to be able to carry out simulations as a routine part of engineering tasks.

Writing a simulation from scratch is exceedingly laborious; most engineers have neither the time nor the specialized knowledge needed for this chore. Fortunately, a goodly number of simulation programs are available commercially. Engineers can choose from among them and avoid the labor of writing their own—an immense saving of effort.

Any worthwhile simulation program incorporates a large number of comparatively small *modules* or *processes*, each of which simulates a particular operation that may arise in a communications system. Pertinent examples include filters, amplifiers, modulators, signal generators, noise generators, simulated test instruments, and many more. The engineer selects those modules pertinent to the task at hand and assembles them, according to the rules of the particular software, to construct a computer *model* of the system to be investigated. An executive engine in the software then exercises the model.

Program reference documentation (typically, quite formidable) tells the characteristics of each of the many modules available and states the rules for their assembly. Some may also provide examples of typical assemblies. A user quickly finds out that the instruction manuals are not enough; they are essential, but more is needed. Documentation for a simulation program does not tell you how to design the model of your problem; it just offers the pieces, and rules for connecting the pieces. In the same fashion, an art-supply store offers paints, brushes, and canvases—it is up to you to paint the masterpiece.

An engineer must understand the mathematical model underlying each program module, how that local model relates to communications operations, and how the modules have to be interlinked to make up a valid model of the overall communications system. Modeling for simulation is not covered in the broader communications literature, nor is it included in the reference manuals of simulation programs. Yet a thorough understanding of models is vital for skillful construction of simulations.

This volume is devoted to the modeling of signals and the processes operating upon them in the transmission or storage of digital data. Because most new communications systems these days transmit digital data, the book's emphasis has been placed fittingly. In addition to a textbook on modeling (the first volume of this work), we have written an accompanying simulation program STÆDT (on a disk placed in *Simulation Techniques: The STÆDT Program*, the reference manual for the program). The program is of a size and cost appropriate for an engineer's personal computer. Despite its small size and low cost, it is a powerful and flexible tool, capable of running engineering simulations on models of realistic systems. It can be instrumental in the analysis and design of practical communications links. It is also invaluable as a vehicle for learning the elements of modeling and simulation and as an aid for students of communications engineering.

Numerous textbooks exist on engineering subjects; numerous programs exist for engineering and mathematics tasks, each accompanied by its own reference documentation; numerous textbooks on computer programs exist, to supplement the reference documentation. This work consists of a textbook, plus a related computer program, plus documentation—items that have been separate in the past. We think that we have something novel here: a structure that will appear more frequently in engineering textbooks.

The text has been written for practicing engineers who need to apply the principles of modeling and who also need a reference work on the subject. The associated program, or other communications simulator, can be used with the help of the text for analysis and evaluation in design projects. The book and program are further intended for students, not only to teach the principles of simulation but also to serve as a tool while learning the elements of communications engineering. Simulations allow a student to explore quickly many aspects of signal processing in far more detail than can be included in a textbook.

We want to thank our friends and colleagues for helping us prepare this volume;

Lewis Franks, Peter Hew, William Lindsey, Earl McCune, Umberto Mengali, Heinrich Meyr, Marc Moeneclaey, and Richard Sherman, have all reviewed portions. Their comments have greatly improved the final product. We particularly want to acknowledge our grateful indebtedness to Lars Erup, who reviewed the entire work in full detail. This is a favor that we can never hope to return adequately. Changes due to his comments are found throughout the text.

FLOYD M. GARDNER

Palo Alto, California

JOHN D. BAKER

Indiana University of Pennsylvania

July 1996

NOMENCLATURE

A, A_0:	amplitude
AM:	amplitude modulation
Avg$[x]$:	generic average, including statistical and time averages
AWGN:	additive white Gaussian noise
a (subscript):	analog signal or filter
a_i, A_i, b_i, B_i:	filter coefficients
$a_m, a(m)$:	real part of mth symbol
BPSK:	binary phase-shift keying
$b_m, b(m)$:	imaginary part of mth symbol
CPM:	continuous-phase modulation
$c_m, c(m)$:	complex value of mth symbol; $c_m = a_m + jb_m$
DFT:	discrete Fourier transform
DSP:	digital signal processing
d (subscript):	digital signal or filter
E$[x]$:	expectation of quantity x
E_b:	energy per bit
E_s:	energy per symbol
E_x:	energy of quantity x
FFT:	fast Fourier transform
FIR:	finite impulse response
FM:	frequency modulation
FP$[x]$:	fractional part of x
FSK:	frequency-shift keying
f:	frequency
f:	frequency variable of continuous Fourier transform
f_c:	carrier frequency
$G(f)$:	Fourier transform of $g(t)$
$g(t)$:	pulse shape; Impulse response of transmission link
$H(f)$:	Fourier transform of $h(t)$; $H(f) = H(s)|_{s=j2\pi f}$

$H(k)$:	discrete Fourier transform of $h(n)$
$H(s)$:	Laplace transform of $h(t)$
$H(z)$:	z-transform of $h(n)$
$h(n)$:	unit-sample response of digital filter
$h(t)$:	impulse response of an analog filter
$I(t), I(n), I(k)$:	real part of complex (I–Q) quantity
IDFT:	inverse DFT
IIR:	infinite impulse response
Im[x]:	imaginary part of x
IP[x]:	integer part of x
i, k, l, m, n, p:	index numbers (integers)
$i_m, i(m)$:	integer value of mth symbol
j:	$\sqrt{-1}$
k:	frequency-domain sample index (wrapped)
k':	frequency-domain sample index (unwrapped)
M:	number of characters in symbol alphabet; number of points in constellation
M_s:	number of samples per symbol interval
MPSK:	M-ary phase-shift keying
MSK:	minimum shift keying
m:	symbol index
N:	number of samples per block
N_0:	spectral density (one-sided) of white noise
NCO:	number-controlled oscillator
n:	time-domain sample index
P_x:	power of quantity x
PAM:	pulse amplitude modulation
PM:	phase modulation
PPM:	pulse position modulation
PSK:	phase shift keying
Pr{·}:	probability
Pr{ϵ}:	probability of decision error
pr{·}:	probability density
$p(n)$:	phase angle in cycles
$Q(t), Q(n), Q(k)$:	imaginary part of complex (I–Q) quantity
$Q(X)$:	complement of normal probability integral
Re[x]:	real part of x
$R(t), R(n)$:	magnitude of signal; $R^2 = I^2 + Q^2$
$S(f)$:	Fourier transform of $s(t)$
$S^+(f)$:	Fourier transform of $s^+(t)$
s:	complex variable of Laplace transform
$s(t)$:	time-continuous real signal
$s^+(t)$:	time-continuous analytic (complex) signal; $s(t) = \text{Re}[s^+(t)]$; $s^+(t) = z(t)e^{j2\pi f_c t}$
T:	symbol interval

t:	time (continuous)
t_s:	sample interval
$u(m)$:	error-detector output for mth symbol
$u(n)$, $v(n)$:	real and imaginary parts of additive white Gaussian noise
$W(n)$:	control signal
$w(n)$:	complex additive white Gaussian noise; $w(n) = u(n) + jv(n)$
$X(f)$:	continuous Fourier transform of $y(t)$
$X(k)$:	discrete Fourier transform of $y(n)$
$x(t)$, $x(n)$:	real part of complex signal (time domain)
$Y(f)$:	continuous Fourier transform of $x(t)$
$Y(k)$:	discrete Fourier transform of $x(n)$
$y(t)$, $y(n)$:	imaginary part of complex signal (time domain)
$Z(f)$:	continuous Fourier transform of $y(t)$
$Z(k)$:	discrete Fourier transform of $y(n)$
z:	complex variable of z-transform
$z(t)$, $z(n)$:	complex envelope; $z(\cdot) = I(\cdot) + jQ(\cdot)$, or $z(\cdot) = x(\cdot) + jy(\cdot)$, or $z(\cdot) = R(\cdot)e^{j\theta(\cdot)}$

Greek Characters

α:	excess bandwidth factor
γ:	frequency scaling factor
$\gamma(x)$, $\Gamma(x)$:	nonlinear functions of x
$\delta(n)$:	unit sample sequence
$\delta(t)$:	unit impulse; delta function
δ_i:	distance from strobe value or constellation point to decision boundary
θ:	phase angle in radians
σ:	standard deviation
τ:	delay; timing error; time shift
ϕ:	phase angle in radians
$\psi(\cdot)$:	correlation
$\Psi(\cdot)$:	spectral density
ω:	$2\pi f$
$\otimes$:	convolution
$*$:	complex conjugate
$\tilde{x}$:	filtered signal
$\bar{x}$:	average (usually with respect to time)
$\{\cdot\}$:	sequence
$\hat{}$:	estimate, or symbol decision

The characters and abbreviations above are used widely throughout the text. Due to the limited number of characters in the Roman and Greek alphabets, some characters have to be reused with differing meanings. Additional characters, not listed here, are employed where needed in localized portions of the text.

SIMULATION TECHNIQUES

1

INTRODUCTION

This book is a text on techniques for computer simulation of data-signal transmission. A versatile simulation program named STÆDT (Simulation Tools for Analysis and Evaluation of Data-signal Transmission) has been developed as a companion work.* The two works are intended to be used together. When you first approach the subject, read the textbook to learn the principles of computer modeling, and at the same time apply the program to run its accompanying examples, or simulations of your own devising. After the principles have been absorbed, the STÆDT program becomes a valuable tool for analysis of real-world communications links, and the text then serves as a source book for models of signal processes. Because STÆDT is highly modular, you can readily apply it to an enormous variety of practical communications-engineering problems.

Although the two works have been written to support one another, this book also stands by itself for learning the principles of modeling and can be used in conjunction with simulation programs other than STÆDT. In this first chapter we state why simulation is important, provide an overview of the communications elements that are to be simulated, and introduce the material that is covered in the remainder of the book.

1.1 ROLE OF SIMULATION

Why should a communications engineer take an interest in simulations? Communications systems traditionally have been designed with the aid of two complementary tactics: by analysis, and by empirical methods using laboratory breadboards. Because of their increasing size and complexity, present-day transmission links tend to be

---

*J. D. Baker and F. M. Gardner, *Simulation Techniques: The STÆDT Program*, New York: Wiley, 1997.

extremely difficult to analyze in detail, even for someone with highly polished mathematical skills. For the rest of us, thorough analysis becomes virtually impossible.

On the other hand, building a breadboard of a transmission link, plus gathering together the necessary test instrumentation, can itself constitute a massive engineering task, one that is afflicted with all of the usual laboratory distractions that are unrelated to the eventual system to be built. Construction of a breadboard can require months or even years. Once it is constructed, changes are usually unwieldy to incorporate. A breadboard is useful—even necessary—to validate a particular design, but it is extremely cumbersome for evaluating alternative candidates or in arriving at an initial design.

A simulation program serves as a software breadboard. An experienced user with a good model base, working in a friendly environment, can plan and construct a simulation of a transmission link in hours or days and can have it debugged and producing useful data in just a few more hours. Once a particular link has been modeled, parameter changes can be accomplished in minutes. Simulations of completely different links can be prepared in about the same time required for the first model, so that radically different techniques can be compared expeditiously.

Test "instruments" are built into a program's repertory and can be called upon by the user for measurements and evaluations of performance of the simulated link. No laboratory gremlins trouble a simulation (although software bugs always lurk). A simulation program is an intermediate tool between paper-and-pencil analysis and a hardware breadboard. The traditional tools are no longer adequate; anyone seriously designing transmission links or modems today also needs a simulation program.

1.1.1 Simulator Capabilities

Among other features, a simulation program includes a library of modular subprograms, each of which is a model of one element of a transmission link. The program also contains means whereby a user can specify the elements to be used in a simulation, the interconnections between elements, and the parameters that characterize each element. In addition, there is a manager or executive that takes the user specifications and runs the program. A computer runs a simulation by generating models of signals, processing those signals through models of the link elements, and evaluating the signal and data models that are produced. Another part of the program reduces data and displays the results. These run-time activities are all performed by the computer and generally make no demands on the user.

A user might employ only those modules contained in the program library, or can write new modules to supplement the library. A good program will have a library large enough that new modules rarely need to be written. A good program will also have support for incorporating new modules easily, so that task need not be unduly burdensome.

Module Processes

There is a long list of signal-processing tasks performed by modules:

1 Simple arithmetic—addition and multiplication—underlies many models. More specialized modules perform phase rotations, frequency shifts, or time

delays. Others do correlations or convolutions. Fourier transformations are invaluable.

2. Filtering is a major task that must be performed by any simulation program; a variety of filter models should be in the library.

3. Nonlinear elements can be modeled in several different ways and should be well represented in a library.

4. Signal and data generators are needed. Feedback-shift registers generate pseudorandom data sequences, function generators produce impulses, steps, sine waves, and so on, and noise generators furnish random noise with specified statistics.

5. Software "measurement" instruments, such as power meters, spectrum analyzers, and oscilloscopes, are used in the simulation and for displays. Each is typically a special-purpose module.

A user is repeatedly called upon to enter complicated characteristics of elements (e.g., measured frequency response of a filter) into the computer so that they can be incorporated into a simulation. A simulation suite ought to include provisions that minimize the entry labor and include routines that convert raw data into formats compatible with the library modules. Other, more-elaborate modules are needed for calibration and evaluation of the simulation. A good program will allow a user to combine relatively simple modules into a more complicated module. A user then sees only the larger module; the details are hidden from view.

Displays

Part of the value of a simulation program lies in the ability of the user to probe the signals and elements at any point along the link and to examine their properties in familiar patterns. To this end, a program must generate displays for the user. Displays include plots of such items as:

- Waveforms
- Impulse responses of filters
- Frequency responses
- Power spectra
- Correlations
- Eye diagrams
- Signal-space diagrams
- Signal constellations (sampled signal-space diagrams)
- Bit-error performance versus SNR

Examples of displays generated by STÆDT are shown in Chapter 18.

Printed displays include tabulations of any of the above and, crucially, documentation of the simulation performed. Such documentation includes a list of the modules incorporated into the simulation, their interconnections, and the parameters of each block, as well as results produced by a simulation. Adequate documentation permits interpretation of the data by a new user long after the simulation has been performed and the original user has gone on to other activities.

1.1.2 User Background

A skilled user of a simulation program needs a fairly sophisticated level of knowledge. He or she must have a good foundation in communication theory and the mathematics underlying it, particularly for noise and filtering. Also needed is an acquaintance with transmission of digital signals, including signal formats and modulation, and some knowledge of signal-processing fundamentals. Subsequent chapters review salient points. Much of this background must be brought to the simulation task; it cannot be learned quickly on the job.

Yet a simulation program is a marvelous tool for improving one's understanding of these prerequisites or as a learning aid for absorbing them in the first place. A program allows a user to set up fairly complicated examples of particular topics and then lets the computer perform the mind-numbing labor of calculation that human beings find so burdensome. The computer quickly furnishes displays of results. Changes in conditions are easily entered and changed results are seen rapidly. Even more valuable is the fact that arranging an example for simulation compels a user to prepare an exact statement of the problem. A computer does not accept vague formulations. The authors have found that working on simulations is a powerful force to improve understanding of the basic theory.

In addition to the conventional topics listed above, adroit application of simulation requires that the user understand computer modeling in considerable depth. It is often argued that a superior program hides details from a user. But those are *programming* details, not *modeling* details. A user must understand modeling (as opposed to programming) thoroughly to make intelligent use of a program. Experienced users of our acquaintance unanimously concur that modeling competence is crucial for preparation of trustworthy simulations.

In contrast to the basic subjects, modeling is not generally taught in standard courses on communications. Those courses, quite rightly, concentrate on analytical description of communications systems. Modeling needs to be learned by other means. But as vast as it is, the communications literature has included remarkably little information on modeling. Until recently, persons interested in the subject have had to teach it to themselves. One of the purposes of this book—a purpose to which the greater part of the book is devoted—is to explain computer modeling of transmission links and elements.

To use a program intelligently, a prospective user should also have some understanding of simulation programming. If the program is well designed, a casual user does not need much knowledge of program structure. But if a more aggressive user wants to add to or otherwise alter the program, understanding of structure becomes more important. Another purpose of this book is to outline some programming issues at a level appropriate to an engineer-user rather than a programmer.

Like learning to swim, one cannot learn simulation just by reading about it. There is no substitute for hands-on experience. It is to that end that the simulation program STÆDT was developed. STÆDT is a fully operational simulation program. Readers are urged to use it to construct their own simulations and explore the principles further. Beyond its value as a learning aid, STÆDT is intended as a working engineering tool for the design and analysis of real-world communications systems. Its full use just begins once a reader has absorbed the information in this book.

Source code for the STÆDT library modules is stored on the program disks. In addition to reading about modeling principles in the book, a user can print out that code to see exactly how the computer executes the principles.

1.1.3 Limits of Simulation

For various reasons, a simulation is not a complete substitute for a hardware breadboard. Among these reasons are the following:

1. A computer exercises a *mathematical model*, not the actual hardware itself. Program users must continually ask themselves: "Is this model correct? Does it really apply to the problem at hand?"

2. Errors creep in. The wrong model may have been used; there can be program bugs; the wrong parameters might be entered. Experience, critical examination of results, and constant vigilance reduce the incidence of undetected errors, but they never vanish entirely.

3. Computer simulation inevitably requires approximations in the mathematical models, particularly for analog signals and processes. These approximations necessarily introduce *distortions* in the results. With care, the distortions can be held to low levels, but a user must always be aware of their existence and be on guard against misleading results arising from distortions. The nature of the most common distortions is treated in the chapters on modeling.

4. A computer simulation is an *experiment*. If any signals or noises in the experiment are random variables the results are also random variables. If the experiment is repeated with a different realization of the random signals, the results will also be different; the results exhibit *statistical fluctuations*. A simulation must be adequately long, so that the fluctuations become but a small part of the observed results.

5. A simulation runs extremely slowly compared to virtually any hardware implementation of a system. If a long simulation is needed to reduce statistical fluctuations and if the events being studied are comparatively rare, the run time can become intolerable. A user needs to learn to evaluate requirements on run lengths (e.g., number of data bits simulated) before embarking blindly on a course of simulation.

6 ☐ INTRODUCTION

6 Some systems have significant events occurring on widely disparate time scales. An example might be a spread-spectrum system in which the spreading code sequence runs at millions of chips per second, whereas the tracking loop for that code has a bandwidth of just a few Hertz. Collecting meaningful statistics for the tracker might require simulated time intervals of many hundreds or thousands of seconds while still simulating millions of code chips for each second. Situations of this kind impose a severe burden upon any practical simulator. Where possible, a user is well advised to break the problem into more manageable simulation pieces with the aid of analytical tools.

Limitations associated with particular models are identified within the text.

1.2 TRANSMISSION LINKS

This work is specialized to digital transmission links, as illustrated in Fig. 1.1. The link's task is to convey digital numbers from a message source to a message sink. The digital numbers must be sent at a specified rate, with some restriction on power levels, and the received message must suffer no more than some small probability of transmission errors.

Digits are provided to a transmitter, which generates corresponding electrical waveforms (or signals). These are launched onto a transmission (or storage) medium, commonly as electromagnetic waves. Disturbances of many kinds afflict the signals while they traverse the medium. A receiver accepts disturbed signals and processes them into appropriate electrical waveforms. Ultimately, the processed waveforms must be interpreted to estimate the digits that gave rise to them.

Transmitted and received messages consist of discrete digital numbers, by def-

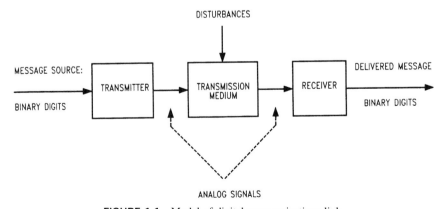

FIGURE 1.1 Model of digital communications link.

inition, but the signals employed for transmitting them invariably are time-continuous, amplitude-continuous, analog waveforms. Thus the transmitter inherently must perform a digital-to-analog conversion, and the receiver inherently must perform an analog-to-digital conversion.

By contrast to digital transmission, there are also analog links (such as those that convey ordinary voice and video) wherein the transmitted waveform itself constitutes the message. Analog transmission systems have criteria of fidelity that are different from those of digital systems. Only digital transmission is treated in these pages. Although analog and digital systems have many common features, specialization to digital is justified by the increasing prevalence of digital transmission in the world's communications systems.

1.2.1 Hierarchy

Networks

Communications systems may be regarded as hierarchical, with *networks* at the top of the hierarchy. A network might be composed of multiple transmission links connecting multiple message sources and sinks. Network analysis is an engineering problem of great importance that often is carried out by computer simulation. Matters investigated in these simulations include the performance of protocols, throughput attainable with a particular network configuration, or the effects of disruptions upon the network.

At the network level, an individual transmission link appears as a simple object described by a few parameters. These parameters may be provided from lower-level simulations of the individual transmission links. The sheer complexity of a network simulation ordinarily prohibits employing any complicated analytic model of the links. Networks are not considered in this book, nor can they be simulated with STÆDT.

Transmission Links

The *transmission link* is at the next level down in the hierarchy and is the only level treated in this book or addressed by STÆDT. A link is composed of various black boxes whose characteristics will receive close attention in the pages to follow. Analysis of a transmission link is needed for design purposes, for performance evaluation, or for comparison of processing techniques. Significant analysis of any but the simplest of links is difficult or impossible by paper-and-pencil methods; practical analysis most often requires the aid of a computer.

Programs for transmission simulation consist of mathematical models of the constituent black boxes; these models generate, process, and evaluate models of the signals. Programs can easily grow to be large and complicated. To restrict the programs to manageable size, the model of each black box is distilled from the much more complicated description of the actual circuits. The model necessarily is a simplification of reality.

Lower Levels

At one level farther down are found the circuits of the black boxes, composed of devices such as transistors, resistors, capacitors, antennas, laser diodes, and so on. Circuit-simulation programs are in widespread use; SPICE is the most prominent example. Below the circuits are the devices; below the devices are materials; below materials are chemistry and physics, where the hierarchy ends.

As a rule, no simulation at one hierarchical level looks up to the next-higher level. A simulation may look down to the next-lower level to obtain the simplified models for its constituents. To carry the full detail of one level of simulation up to the next-higher level typically would impose an insupportable computing burden. Therefore, models at one level are abstractions of the greater detail at lower levels.

This book and the STÆDT program are devoted to the methods of simulation of transmission links. Levels higher or lower in the hierarchy are not considered. Adequate explanations for simulations at other levels would require separate books comparable in size to this one.

1.2.2 Link Elements

Figure 1.2 shows typical elements of a transmission link in greater detail. A brief examination of that block diagram will introduce some of the terminology and subjects to be covered in the pages to follow. The twofold purpose is to establish a skeleton communications link as a base for further discussion, and to indicate the kinds of elements to be simulated. A transmitter is shown at the top of Fig. 1.2, the transmission medium and transducers are in the center, and a receiver is at the bottom. Data originate in the message source, flow through the transmitter, medium, and receiver, and are delivered to the message sink.

Message Source

The source data can be regarded as a collection or stream of binary digits, or bits. These digits from the source are the *information bits* of the system. There are innumerable examples of digital message sources, including computer communications, digitized voice, facsimile, teletype, digitized music, and digital television. The origin of the data does not concern us here, although its intended use can have a strong influence upon requirements on quality of transmission. To simplify analysis, the data are often assumed to be random, chosen by the equivalent of flipping a coin for each bit. Simulation programs typically generate pseudorandom data patterns, by means explained in Chapter 7.

Digital Operations

Message bits might be delivered directly to the transmitter, or they may first be subjected to one or more digital processes, such as encoding, encrypting, scrambling, bit stuffing, punctuation (word markers, frame markers), or interleaving. These, together with their inverse operations in the receiver, are vital functions in many communi-

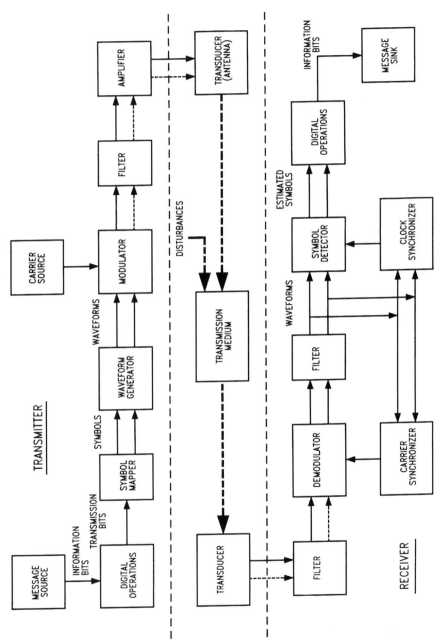

FIGURE 1.2 Elements of transmission link.

cations links. They have been described extensively in the literature and important research is still under way, particularly on coding. Despite the importance of coding and the other digital operations, they are largely omitted from further consideration in this book in favor of more basic topics. Performance of systems employing coding is not readily evaluated analytically; simulations are often needed as a means of experimental evaluation. Simulation methods similar to those described in this book are appropriate to the investigation of encoded signals.

Symbol Mapper

This block takes transmission bits in groups of one or more and maps them into *symbols* in a finite *alphabet*. A symbol can be thought of as a discrete number but no longer necessarily binary. Symbol alphabets are explored in Chapter 4 and symbol mapping is covered in Chapter 7.

In many applications, it is useful to regard the symbols as being two-dimensional, or *complex*. Reasons for that will become apparent in Chapter 2. Complex representation of symbols and signals is a prominent feature in transmission simulations. The two-dimensional nature of complex symbols appears in tangible form in actual hardware; real and imaginary components of the symbols are generated and processed on two physically separate lines. Those two lines are shown in Fig. 1.2 at the outputs of the symbol mapper and subsequent blocks. (Symbols can also be one-dimensional—that is, purely real—in which case the second line in the block diagram—the imaginary component—should be deleted.)

Symbol Examples. A binary alphabet might have symbols 0 and 1, or -1 and $+1$. A ternary alphabet can have symbols -1, 0, and $+1$, and a quaternary alphabet can have symbols -3, -1, $+1$, and $+3$. These are all real alphabets. Some examples of complex alphabets include: $(\pm 1 \pm j1)$ for quaternary; $\exp[j\pi(n+\frac{1}{2})/4]$, $n = 0$ to 7, for 8-ary; and $(a+jb)$, $a = \pm 1, \pm 3$, $b = \pm 1, \pm 3$, for 16-ary. All of these, and many others, are in widespread use in present-day communications systems. A versatile simulation program needs easy provision for modeling many different symbol formats.

Waveform Generator

An electrical *waveform* is generated for each symbol delivered from the symbol mapper. Each symbol from the alphabet has a distinctive waveform assigned to it. If the symbol is complex, its waveform is also complex; that is, two separate waveforms are generated and carried on two separate lines.

Waveforms will be assumed to be generated and transmitted *synchronously*—that is, at uniform time intervals. The overwhelming majority of digital communications links operate synchronously, even those said to be asynchronous. (The latter send waveforms that are uniformly spaced over small groups of symbols, but the groups might begin at uneven times.) All transmission-simulation programs known to the authors work exclusively with synchronous operation. All subsequent discussion is restricted to synchronous transmission.

Typical waveforms will be described later, particularly in Chapters 4 and 7. For now, think of each waveform as an electrical pulse. A sequence of symbol waveforms is called a *signal*.

Modulator

In many (but by no means all) communications systems, the waveforms are impressed onto a periodic *carrier wave*, which is usually sinusoidal. The process is known as *modulation* and the device in which modulation takes place is known as the *modulator*. Several kinds of modulation and modulators are covered in Chapters 4 and 7. The sequence of symbol waveforms provided to the modulator and the processes that generated them are called *baseband* signals and processes. Signals delivered out of the modulator and the processes that operate upon them are called *passband* signals and processes.

Baseband signals often, but not necessarily, have a lowpass spectrum. For a passband signal the carrier frequency often will be large compared to the spectral width of the baseband signal applied to the modulator, in which case the spectrum of the modulated signal will be concentrated in a relatively narrow bandpass region around the carrier frequency. However, that constraint on the carrier frequency is not necessary; some common systems have carrier frequencies that are smaller than the bandwidth of the baseband signal.

A modulated signal is a real signal, even when generated from complex baseband waveforms. Nonetheless, it is convenient, especially in simulations, to represent it in a complex format. The reasons and the format are explained in Chapter 2. The passband signal path in Fig. 1.2 is shown with double connecting lines as a reminder that simulations often treat a passband signal in its complex format. But because the physical passband signal is actually real, one of the lines is shown dashed.

Filter

A *filter* block is shown in the transmitter after the modulator in Fig. 1.2. That is just one possible location; filters could be inserted to operate on any waveform signal at any location in the transmitter or receiver. For example, the waveform generator itself is probably based on a filter. Filtering is a dominating operation in a simulation program. It is not unusual to find simulations in which filter operations constitute the largest single part of the effort. Filter calculations impose much of the computing burden, modeling the filters accounts for a significant portion of the programmer's task, and entering filter characteristics is a large part of a user's labor in setting up a simulation run. Aspects of filters are covered repeatedly in subsequent chapters, first in Chapter 3, then again in Chapters 6 and 7, and yet again in Chapter 17.

Amplifier

An amplifier is shown following the filter. An ideal amplifier is trivial to model; one simply scales all signal values by the amplifier's gain. A real amplifier is more complicated; it has frequency-dependent gain, and an efficient amplifier almost always is significantly nonlinear. Frequency dependence is modeled as a filter effect in a simulation, but nonlinearity requires completely different treatment. Various aspects of modeling nonlinear elements are examined in Chapters 4, 7, and 17. Nonlinear operation is not restricted to amplifiers; any nonlinear element can employ the nonlinear models to be presented.

Transmission Medium

Ignore the transducers temporarily and consider the *transmission medium*. There are many different media employed for signal transmission (or for signal storage—much the same principles apply and much the same simulation techniques are used for storage systems as for transmission systems). Various categories of media can be established, such as:

- Transmission versus storage
- Electrical versus radio versus optical versus magnetic
- Guided waves versus space waves (for electromagnetic-wave media)
- Baseband versus passband

Many examples are extant, including:

- Electrical transmission via wire lines (can be baseband or passband)
- Magnetic storage on tape or disk
- Storage by physical deformations on optical disk
- Radio transmission, such as free space, high-frequency ionospheric reflection, tropospheric scatter, satellite repeaters, terrestrial microwave, and ultrahigh-frequency land mobile (all radio transmission is necessarily passband)
- Lightwave transmission via glass fibers or free space

Disturbances of numerous kinds afflict transmissions; some disturbances are imposed by nature, some are contributed by our fellow humans pursuing their own activities, and others are built into the equipment by ourselves. (The last are not necessarily disturbances in the medium, but they will be treated under that heading anyhow.) Examples include:

- *Attenuation.*
- *Fading* (time-varying attenuation).
- *Additive noise*, including thermal noise in the sky or in lossy circuit elements; shot noise in electronic devices; impulse noise from lightning, ignition systems, relays, or motors; miscellaneous broadband noise from computers, televisions, fluorescent lights, power-line corona, and so on.
- *Bandwidth restrictions* due to multipath in radio channels, gap width in magnetic-recording heads, or resistance, inductance, and capacitance in wire lines. Restricted bandwidth causes smearing and overlapping of signal wave-

forms. Bandwidth restrictions are often described in terms of amplitude distortion and delay or phase distortion.

- Other signals, known as *interference* if inadvertent and *jamming* if deliberate. In wire-line transmission, coupling from one channel to another is known as *crosstalk*. For radio systems, the interference is *cochannel* if it shares the same passband as the desired signal, and *adjacent-channel* if the interference spectrum only partially overlaps the signal spectrum.
- *Time-base variation*, caused by Doppler shift in moving vehicles or mechanical speed variations in moving storage systems. Tape recorders are particularly afflicted by a speed variation called flutter.
- *Quantum noise* in lightwave systems, and granularity irregularities in magnetic media.
- *Phase noise* in oscillators.
- *Timing fluctuations* induced by randomness of data.

Additive white Gaussian noise (AWGN) receives much attention in this book (Chapters 2, 9, and 13) and its associated program because (1) it is a nearly ubiquitous disturbance arising in almost all systems, (2) it has a tractable representation and can be analyzed and simulated readily, and (3) it has received overwhelming attention in the communications literature.

Bandwidth restrictions of media are handled by means of the filter models described in the book and provided in the program. A user will find that interference is easily simulated with the tools introduced in the text and included in the program, even though interference per se receives little attention henceforth. Constant attenuation, of course, is trivial to simulate. Timing fluctuations caused by random data patterns arise indirectly during simulation of timing-recovery elements. Timing recovery is explained in Chapter 11.

References are provided for methods of modeling some of the other disturbances, but most are rather specialized and accordingly are not treated extensively here. The associated program is flexible; a user with moderate programming abilities should be able to incorporate additional disturbance models into the program. It is fair to point out that modeling of some disturbances (e.g., phase noise) might press up against limits of computer capacity or theoretical understanding of the underlying phenomenon.

Transducers

A *transducer* is a device for coupling the transmitter to the medium, or the medium to the receiver. The nature of the transducers depends primarily on the medium. Some transducer examples are antennas for radio media, lasers and photodetectors for lightwave media, magnetic heads for tape recorders, and cable connectors for wire lines. For many simulation applications, transducers can be modeled with the basic tools (e.g., filters, nonlinearities) of any link simulation program. Special applications require development of appropriate special models.

Receiver Filters

A receiver typically contains several filters, placed at several locations. Simulation of ordinary filters in the receiver employs exactly the same types of filter models as used in the transmitter.

In addition, many receivers employ *adaptive filters*. These are structures with feedback control provisions to adjust the filter characteristics according to some performance criterion. Adaptive filters are used: as *equalizers* to correct for bandwidth restrictions in the transmission medium, as *echo cancelers* to permit simultaneous cochannel transmission and reception at a terminal by canceling one's own transmitted signal from the receiver, and with *antenna arrays* to cancel interference coming from a direction other than that of the desired signal.

Adaptive filters are a fascinating subject of great significance, often studied by means of computer simulation. It is possible to apply the methods of this book and the associated STÆDT program to the simulation of adaptive filters, but that is an advanced topic not covered here.

Demodulator

Demodulation is the inverse operation to modulation; it is the recovery of the baseband signal from a passband signal. No demodulator is needed with baseband transmission, where no modulation is performed in the transmitter. Demodulation can be *coherent*, whereby the phase of the incoming passband signal is recovered, or it can be *noncoherent*, whereby the phase of the incoming signal is ignored. Demodulation also can be *differentially coherent*, whereby the change of phase from symbol to symbol is recovered, but the absolute phase is ignored. All three kinds are treated in Chapters 4 and 10.

Coherent demodulation requires a locally generated carrier for its operation. A *carrier synchronizer* (Chapter 11) extracts a properly phased local carrier from the received signal. Differential demodulation and noncoherent demodulation do not include a carrier synchronizer.

Symbol Detector

Demodulated, filtered, disturbed signals are examined in the *symbol detector*, where each symbol waveform is reduced to a digital number, one per symbol. Details are given in Chapters 4 and 12. In a *hard-decision detector*, the number is the receiver's best estimate of the original symbol that gave rise to the particular transmitted waveform. For a *soft-decision detector*, the number is merely a summary description of the disturbed waveform for that symbol; the number is passed on for *decoding* in the digital operations block. This book concentrates upon hard decisions, but soft decisions are easily produced by the associated program.

To perform its operations correctly, the symbol detector must examine each waveform at the correct time instant. A *clock synchronizer* (Chapter 11) extracts the *timing* of those examination instants from the incoming signal. *Timing recovery* is expedited by synchronous transmission of the symbol waveforms; detection would be much more difficult if transmission timing were irregular.

Because of disturbances in the medium, decisions are not always correct. Probability of symbol error, or of bit error, is a vital characteristic of a communications link. Paper-and-pencil evaluation of transmission error statistics is virtually impossible for all but the simplest configurations. A prime reason for simulation is to evaluate error probabilities of realistic transmission systems, by methods covered in Chapter 13.

Receiver Digital Operations

There really should be a *symbol demapper* (Chapter 12) shown after the symbol detector, at least for hard-decision operation. A symbol demapper accepts a sequence of digital numbers representing symbols and converts each number into its assigned group of bits. That function has been relegated to the digital operations block in Fig. 1.2. Other digital operations include deinterleaving, decoding, decryption, destuffing, and frame synchronization based on punctuation. These are all crucial operations in many systems but are omitted from this work. Ultimately, information bits are delivered to the message sink.

Analog/Digital Interfaces

Despite our earlier protestations on the subject, no digital-to-analog or analog-to-digital interface has been identified unequivocally in Fig. 1.2 or the discussion above. That is because locations of the actual interfaces are highly implementation dependent. At one time the transmitters and receivers were almost entirely composed of analog circuits, so the interfaces were at the symbol mapper and symbol detector. As this is being written, the design of modems is caught up in the digital revolution and many formerly analog functions are now being performed by digital processors. Interfaces in both the transmitters and receivers are moving much closer to the transducers.

These digital processors perform such tasks as filtering, modulation, and demodulation; they are quite distinct from the digital operations listed above. Digital processors in some ways are easier than analog operations to simulate on a digital computer, but they have special considerations of their own that require attention. This book concentrates on simulation of analog-implemented processes, but the special features of digital implementation are addressed in Chapter 15.

1.3 PERTINENT LITERATURE

Communications is a topic of intense study and interest; it is favored and burdened with an enormous literature. In this section we list a small portion of the literature that has been found helpful. Others might well produce a different list; inclusions reflect the authors' own libraries as much as any other factor.

Among U.S. journals, the two IEEE publications *Transactions on Communications* and *Journal on Selected Areas in Communications* are preeminent. Between them, they publish some 3000 pages of articles per year, a significant portion of which applies at least in part to digital transmission. Other IEEE publications that often have articles on digital transmission include *Proceedings, Transactions on Information Theory, Transactions on Aerospace and Electronic Systems,* and *Journal of Lightwave Technology.* The *IEEE Transactions on Signal Processing* contains numerous articles on digital signal processing, many of which apply to signal transmission. Past issues of the *Bell System Technical Journal* contained many valuable papers; this journal has ceased publication as a consequence of the breakup of the Bell System.

Two books on communication engineering stand out. Wozencraft and Jacobs [1.1] is a rigorous account of the mathematical basis of communication theory that has been long established as a textbook for graduate courses. Proakis [1.2] is a more recent description of much the same material; it is more up to date on applications. In our experience, the Proakis book is written so that individual topics can be read and understood without the necessity of first learning all of the preceding material. It has been very helpful in our work.

For many years, the preeminent book on data signal transmission has been that by Lucky, Salz, and Weldon [1.3]. It belongs on the shelf of every specialist in the field. More recently, massive books by Lee and Messerschmitt [1.4], Benedetto, Biglieri, and Castellani [1.5], and Gitlin, Hayes, and Weinstein [1.6] have appeared. The first [1.4] is an introductory text (for students who have had the necessary background in noise and circuit analysis), while the second [1.5] is directed toward a more advanced audience. The third [1.6] contains much valuable information on equalizers and echo cancelers, based upon solid engineering experience. Because of their later date and larger size, the latter three are able to include more material than appears in [1.3].

A new book by Simon, Hinedi, and Lindsey [1.7] was published just as our work was nearing completion. It contains succinct accounts of many of the important results and problems in digital communications and should prove to be a valuable reference for communications engineers. In addition, there must easily be three dozen other books on communications in general and on specialized aspects of digital communications. One of note by Bingham [1.8] describes the particular field of voice-frequency telephone-line modems.

Digital signal processing (DSP) is an indispensable ingredient of signal simulation. The standard works on the subject are those by Oppenheim and Schafer [1.9, 1.10] and Rabiner and Gold [1.11]. A recent publication by Proakis and Manolakis [1.12] contains more detail than found in the two classics and may be easier for self-study. New books on DSP seem to appear almost monthly.

Other topics of importance to simulation include circuit theory and statistical analysis of noise. These subjects are a prerequisite to study of communications systems; it is assumed that all readers have adequate background. Innumerable texts exist on these topics and each reader has his/her own favorites. No references will be cited.

In contrast to the rich literature on the background subjects, simulation of signal transmission has received comparatively little published attention until recently. No book on transmission simulation was in existence when we began ours, but a splendid text by Jeruchim, Balaban, and Shanmugam [1.13] appeared after our writing was well advanced. Their contribution treats the same subjects as those covered here, but the emphasis is entirely different. Whereas [1.13] takes a more abstract overview of simulation, with greater emphasis on mathematical foundations, our account is more concerned with the everyday techniques of modeling. The two works are complementary, with remarkably little overlap.

Prior to [1.13], there was only a smattering of simulation papers in the journals. Moreover, a goodly portion of those that appeared consisted of descriptions of capabilities of specific programs and did not address modeling. The largest collection of articles is in two special issues of the *IEEE Journal on Selected Areas in Communications* [1.14, 1.15]. References to most earlier papers may be found in these collections. In subsequent text other papers are cited, where such are pertinent and known, but surprisingly few exist.

1.4 STUDY PLAN

This book is divided into three parts: Part I (Chapters 2 to 5) is called Foundations, Part II (Chapters 6 to 16) is Modeling of Signals and Processes, and Part III (Chapters 17 to 19) is Utilities and Programming. The material has an intimidating bulk. When you first look at it, you will probably conclude that it contains far more than you ever wanted to know about simulation.

Important concepts are summarized at the end of each chapter under the heading *Key Points.** You may find it helpful to read Key Points before, or instead of, the main text. That would convey the essence of the material without the labor of reading it all.

Beyond that, by all means go directly to the STÆDT program itself and its user instructions, or familiarize yourself with another signal-transmission program; several are available commercially. Although the program and book have been prepared together as an integrated whole, this book can be studied separately to learn the fundamentals in conjunction with any other program for simulation of data-signal transmission. The underlying principles are the same for any simulation program.

Any program has certain mechanical details that have to be mastered before the program can be put to use. Once the mechanics have been learned, you will want to begin constructing your own simulations. At that point, you will find it necessary to understand the simulation models in intimate detail. A well-written program manual describes the model of each individual process in its program library. But descriptions of individual processes are not sufficient; you need a broader overview of modeling, to understand what lies behind the processes and how processes are to be put together to construct a simulation. That need for understanding leads right back to this book

---

*Beginning with Chapter 3, the Key Points sections include a subheading entitled *STÆDT Processes*. Under these subheadings are lists of names and purposes of commands in the STÆDT library for accomplishing functions introduced in the text. All commands are described fully in Chapter 5 of *The STÆDT Program*.

(or [1.13]). The alternative is to work out the models for yourself from first principles; this was the only choice prior to the publication of this work or [1.13].

Part I of this book consists mainly of background material on signals, filters, and processes (Chapters 2 to 4) and an introduction (Chapter 5) to digital simulation of communications elements. Chapters 2 to 4 cover material that should already be familiar to most readers, but are written with an emphasis on simulation-related features, an emphasis quite different from that found in conventional texts on communications. A reader already well versed in communications theory and filters could skip Chapters 2 to 4 and start directly with Chapter 5. A less-conversant reader might prefer to study the background material, starting with Chapter 2.

Part II is an exhaustive description of the modeling of different elements encountered in a communications system, the evaluation of performance of a simulated system, and special considerations for simulation of feedback loops, digitally implemented modems, and systems with multiple sampling rates. It is the heart of this book.

Part III considers various utilities needed in a simulation program and gives some attention to issues faced by a programmer (as opposed to communications-engineer users, who are the primary audience for this work).

Although this book and the STÆDT program are each intended to be able to stand alone, we recommend that they be used together to facilitate learning of the subject. A quick simulation is invaluable for illustrating a topic from the text, and reference to the text will often be essential for understanding a model in a simulation. This reliance upon both program and text is helpful while learning but is also likely to be necessary even after the subject has been mastered thoroughly and STÆDT (or another simulation program) is being used as an engineering tool.

Despite appearances to the contrary, we have tried to write this book in as nonmathematical a style as possible. Formal derivations have been avoided. Algorithms are either just stated, or developed informally without proof. Conditions of applicability, needed for rigor, have been omitted. No arcane mathematics have been included to awe the reader.

Nonetheless, a computer model is a *mathematical* model. To describe a model is to write a formula. It is not possible to explain modeling in any but mathematical language, which you will find in profusion throughout the book. Do not be intimidated; there is nothing here that should seem strange to a graduate electrical engineer.

2

REPRESENTATION OF ANALOG SIGNALS AND NOISE

Modeling of analog signals and processes* constitutes a large part of transmission simulation. To understand modeling, it is first necessary to be familiar with analytical representations for the signals and processes; that is the subject matter of this and the next several chapters. The information is intended as review and to establish nomenclature, so derivations are abbreviated and proofs are omitted. Modified representations imposed by computer simulations are introduced in Chapter 5 and pursued intensively in the remainder of the book.

Many modems are implemented by digital methods instead of analog. Representations of digital processes, and the special considerations that arise with them, are deferred until after simulation of analog processes has been explained.

Signals can be categorized as *baseband* or *passband* and are considered in that order. Power and energy relations are stated and then noise is characterized in the same formats as those applied to signals.

2.1 BASEBAND SIGNAL REPRESENTATIONS

A data symbol is a discrete representation of a single number drawn from a finite alphabet of numbers. Symbols are transmitted in a sequence. Associated with each symbol in a sequence is a distinctive waveform employed for the transmission of

---

*Readers will also be familiar with *stochastic processes*: random ensembles whose realizations are time-dependent functions. Throughout this work, *process* will only mean an operation performed on a signal, never a stochastic process.

that symbol. Although the data symbols have discrete numeric (digital) values, the transmission signals are composed of time-continuous, amplitude-continuous (analog) waveforms. Almost invariably, the waveform for a symbol will be a pulse.

For the mth symbol in a sequence there is a pulse whose waveform is denoted $g_m(t)$, where t represents time. Symbol information is carried in some recognizable parameter of the pulse. Typical parameters in common use include:

- Pulse amplitude
- Pulse position
- Pulse width
- Pulse waveform
- Pulse frequency (a subclass of waveform)
- Pulse phase (most applicable to passband signals)

Transmission is assumed to be *synchronous*; pulses are sent at uniform intervals T. A baseband signal may be represented as the sum of time-displaced pulses

$$x(t) = \sum_{m=-\infty}^{\infty} g_m(t - mT) \qquad (2.1)$$

Infinite limits are shown on the summation, but physical signals always have a finite number of pulses.

2.1.1 Time-Limited Versus Bandlimited Pulses

A pulse is said to be *time-limited* if its waveform is identically zero outside a finite-duration time interval:

$$g_m(t) \equiv 0, \qquad t \notin [t_1, t_2] \qquad (2.2)$$

The pulse duration is $t_2 - t_1$.

The Fourier transform of $g_m(t)$ is given by [2.1, 2.2]

$$G_m(f) = \int_{-\infty}^{\infty} g_m(t) e^{-j2\pi ft}\, dt \qquad (2.3)$$

where f denotes frequency. The Fourier transform exists for any pulse of practical interest. Recall that frequency f takes on both positive and negative values.

A baseband pulse is said to be *bandlimited* if its Fourier transform is identically zero outside a finite frequency interval:

$$G_m(f) \equiv 0, \qquad |f| > B \qquad (2.4)$$

The spectrum of a time-limited pulse can be zero only at discrete frequencies.* The waveform of a bandlimited pulse cannot be zero at any but a countable number of discrete times. No physically realizable pulse can be either strictly time-limited or strictly bandlimited. No pulse, not even if it is ideal, can be both time-limited and bandlimited. Engineers commonly work analytically with idealized pulses that are time- or bandlimited, or sometimes they even pretend that a pulse is nearly both. But they recognize that physical pulses depart from the ideal, and they take account of the departures when designing equipment.

Unless pulses are strictly time-limited to a duration of T or less, successive pulses from different symbols will overlap one another. Overlapping can interfere with detection of the pulses for the individual symbols. In the communications community, this overlap is known as *intersymbol interference* (ISI). In the magnetic-recording community, it is known as *pulse crowding*. Pulse overlap is unavoidable in bandwidth-restricted systems. Provisions for countering the potential ill effects of overlap are an important task for an engineer, who could employ simulation to analyze and evaluate the problem.

2.1.2 Pulse Amplitude Modulation (PAM)

The data value of a symbol can be carried in the amplitude of the pulse. This method of transmission is by far the most widespread in communications systems and the one that will receive the most attention in this work.

Definition

A pulse amplitude modulation (PAM) signal employs the same pulse waveform for all symbols; designate that waveform as $g(t)$. Symbol value for the mth pulse is carried in its amplitude, designated a_m. A PAM baseband signal is represented as

$$x(t) = \sum_{m=-\infty}^{\infty} a_m g(t - mT) \tag{2.5}$$

where a_m is a discrete value taken from a finite alphabet of M characters.

Pulse Examples

Countless varieties of pulse shapes can be imagined. A few examples are illustrated in Fig. 2.1. The first two lines of the figure show rectangular pulses. These pulses do not overlap since they are time-limited to the symbol interval T. Nearly rectangular pulses require transmission systems with large bandwidth to preserve their steep transitions.

---

*Statements regarding mathematical properties of Fourier transforms and signals are based on the literature of Fourier transforms (such as [2.1] and [2.2]) or the theory of functions of a complex variable (such as [2.3] and [2.4]).

22 ◻ REPRESENTATION OF ANALOG SIGNALS AND NOISE

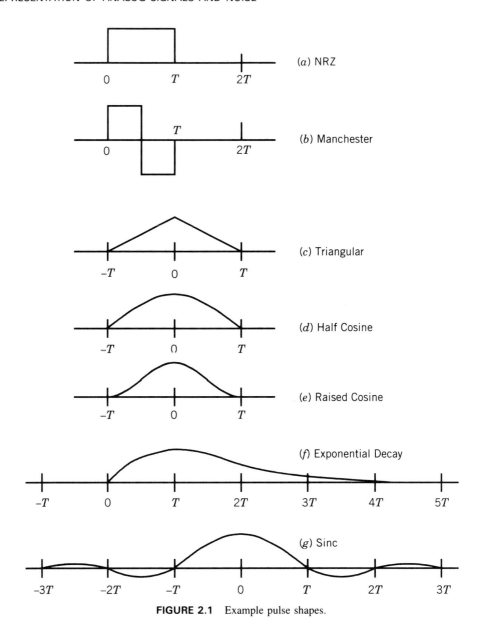

FIGURE 2.1 Example pulse shapes.

Near-rectangularity may be plausible for transmitted pulses in a wideband system, but not after filtering in the receiver. (Refer to the receiver structure in Fig. 1.2.) After optimum matched filtering (introduced in Chapter 4), the rectangular pulse of line a would take on the triangular shape of line c. Lines c, d, and e of Fig. 2.1 show time-limited pulses with durations of exactly 2T. Adjacent pulses of these shapes at spacing T would overlap. Nevertheless, as explained below, pulses with these particular shapes do not interfere with one another, despite the overlap. The half-cosine pulse (line d) of duration $2T$ is associated with pulses in a modulation format known as minimum-shift keying (MSK) [2.5], described in Sec. 2.2.3. A raised-cosine shape, as in line e, is typical of well-designed receiver-output pulses in a system with moderate bandwidth restriction.

Line f sketches a pulse with exponential decay and thus semi-infinite duration. The tail of this pulse interferes with all succeeding pulses. A quasiexponential decay is typical of many bandwidth-restricted systems before corrective pulse shaping has been performed.

Line g illustrates a sinc pulse, where

$$\text{sinc}(t/T) \triangleq \frac{\sin \pi t/T}{\pi t/T}$$

This pulse has infinite time span in both directions. It is strictly bandlimited, a characteristic not possessed by any of the other examples of Fig. 2.1. Also, observe that its tails have nulls equally spaced at intervals T. Despite the fact that any one pulse overlaps all other pulses in a train, there is no interference between pulses, as explained below. This pulse, although nonrealizable, has important theoretical significance.

Some of the pulses in Fig. 2.1 are shown as beginning at time $t = 0$, and they are zero for $t < 0$. These are called *causal* pulses. All physical pulses are causal. Other pulses in the figure are shown as beginning before $t = 0$; these are called *noncausal*. Although a physical pulse can never be noncausal, it is often convenient to perform analysis with noncausal pulses nonetheless. If a noncausal pulse is zero for all time $t < -t_0$ (where t_0 is positive), it can be made causal by delaying it by t_0. If the pulse extends to $t = -\infty$, as in line g, no finite delay can make it causal.

Pulse Spectra

Waveform representation $g(t)$ describes the pulse in the *time domain*. An equivalent *frequency-domain* representation $G(f)$ contains exactly the same information about the pulse, but displays it differently. The two representations $g(t)$ and $G(f)$ are Fourier transform pairs:

$$G(f) = \int_{-\infty}^{\infty} g(t) e^{-j2\pi ft} \, dt$$
$$g(t) = \int_{-\infty}^{\infty} G(f) e^{j2\pi ft} \, df \qquad (2.6)$$

The example pulse shapes of Fig. 2.1 are listed in Table 2.1 along with their respective Fourier transforms. Pulses have been scaled so that each has unity peak amplitude in the time domain. Magnitudes of their Fourier transforms are plotted on

TABLE 2.1: Examples of pulse shapes and spectra

Pulse Name	$g(t)$	$G(f)$
(a) NRZ	$= 1, \quad 0 < t < T$ $= 0, \quad \text{otherwise}$	$Te^{-j\pi fT}\dfrac{\sin \pi fT}{\pi fT}$
(b) Manchester	$= 1, \quad 0 < t < T/2$ $= -1, \quad T/2 < t < T$ $= 0, \quad \text{otherwise}$	$jTe^{-j\pi fT}\dfrac{\sin^2 \pi fT/2}{\pi fT/2}$
(c) Trianglular	$1 - |t|/T, \quad |t| \leq T$ $= 0 \quad |t| > T$	$T\left(\dfrac{\sin \pi fT}{\pi fT}\right)^2$
(d) Half cosine	$\cos \dfrac{\pi t}{2T}, \quad |t| < T$ $= 0, \quad \text{otherwise}$	$\dfrac{\pi T}{(\pi/2)^2 - (2\pi fT)^2}\cos 2\pi fT$
(e) Raised cosine	$\cos^2 \dfrac{\pi t}{2T}, \quad |t| < T$ $= 0, \quad \text{otherwise}$	$\dfrac{\pi^2 T}{\pi^2 - (2\pi fT)^2}\dfrac{\sin 2\pi fT}{2\pi fT}$
(f) Exponential decay	$e \cdot \dfrac{t}{T}e^{-t/T}, \quad t > 0$ $= 0, \quad t < 0$	$\dfrac{eT}{(1 + j2\pi fT)^2}$
(g) Sinc	$\dfrac{\sin \pi t/T}{\pi t/T}$	$= T, \quad |f| < 1/2T$ $= 0, \quad |f| > 1/2T$

a decibel scale in Figs. 2.2 to 2.5. Since the example pulses are all real, $|G(f)|$ is an even function of frequency; consequently, only positive frequencies are shown in the spectrum plots. The figures illustrate the general proposition that as a pulse is made broader and smoother in the time domain, its width in the frequency domain is reduced. All of these graphs are on the same scales, to permit easy comparison.

Figure 2.2 compares the spectra of two widely used rectangular pulses: NRZ (nonreturn to zero) and Manchester. The latter has a spectral null at zero frequency, allowing ready transmission through systems lacking direct-current (dc) response (e.g., magnetic tape and disk, transformer-coupled cables) but with a penalty of widened bandwidth occupancy.

Figure 2.3 shows spectra of three different shapes that might be employed for transmitted pulses. The bandwidth-conservation benefits of pulse shaping are quite evident. Figure 2.4 plots the spectra for three shapes that might be delivered as the output pulses from PAM receivers. The spectrum for the exponential-decay pulse is plotted in Fig. 2.5 along with that for the sinc pulse. An experienced communications engineer would quickly perceive that a system producing the example exponential-

BASEBAND SIGNAL REPRESENTATIONS 25

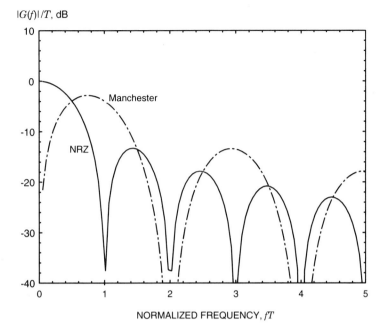

FIGURE 2.2 Comparison of spectra for NRZ and Manchester pulses.

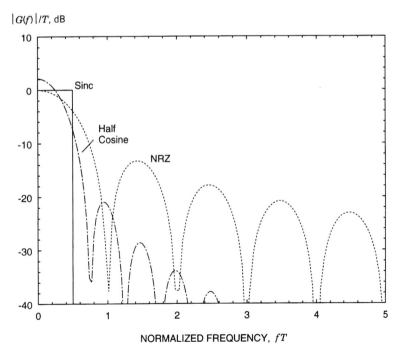

FIGURE 2.3 Spectra for transmitted pulses.

26 ☐ REPRESENTATION OF ANALOG SIGNALS AND NOISE

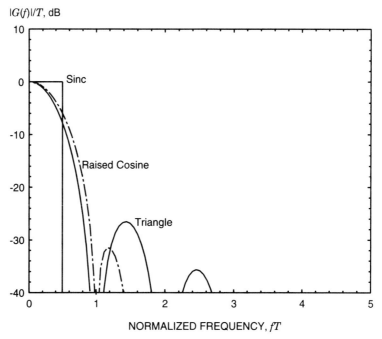

FIGURE 2.4 Spectra for receiver-output pulses.

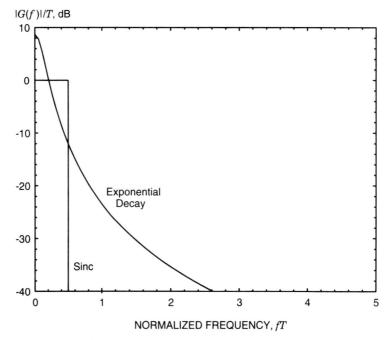

FIGURE 2.5 Spectrum for exponential-decay pulse.

decay pulse has excessive bandwidth restriction and would require appreciable equalization or other corrective measures. The basis for this insight will be described presently.

PAM Link Model

A transmission link based on PAM has a particularly simple model, as shown in Fig. 2.6. Data symbols (digital numbers) are delivered to the transmitter, which encodes the information of each symbol as the amplitude of a transmitted signaling pulse. All transmitted pulses have the same waveshape, differing only in their amplitudes. Pulses are transmitted synchronously, at uniform intervals T.

A wide variety of alphabets (choices of amplitudes) are found in practice. Binary signaling (one bit per symbol) could use an alphabet $\{0, 1\}$ or $\{-1, 1\}$. Quaternary signaling (two bits per symbol) could use $\{-3, -1, 1, 3\}$. Ternary signaling ($\log_2 3 = 1.58$ bits per symbol) might use $\{-1, 0, 1\}$. Most matters considered in this chapter are independent of the alphabet selected.

Assuming for now that the transmission medium is linear, time invariant, and nondispersive, an attenuated, undistorted signal is applied to the receiver along with additive noise. The receiver filters the signal and produces an output pulse train in which all pulses have shape $g(t)$. Filtered noise accompanies the output signal.

Receiver-output pulses are sampled once per symbol interval T; all further operations are performed on the samples. Because the word *sample* has broad application in simulations by digital computer, these symbol-interval samples at the receiver output will be given the special name of *symbol strobes*, or just *strobes*. Satisfactory performance of the receiver requires that the strobes be timed correctly. Strobes are delivered, one at a time, to a decision device (symbol detector) whose task it is to determine the strobe amplitude and thereby estimate the value of the symbol transmitted. This procedure is known as *symbol-by-symbol detection*.

COMMENT: A different detection procedure is used if the symbol stream has been encoded. The strobes are passed to a decoder and symbol decisions are based upon observations on multiple strobes.

2.1.3 Nyquist Shaping

Consider desirable shapes for the receiver-output pulse $g(t)$. Each pulse carries information about only one transmitted symbol. Information retrieval, in symbol-by-sym-

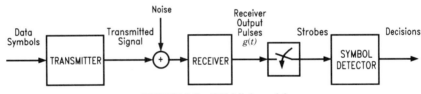

FIGURE 2.6 PAM link model.

bol detection, should be based upon just one pulse at a time. But the examples have demonstrated that receiver-output pulses often overlap. Overlap can extend over many symbol intervals if bandwidth is restricted. How are the pulse amplitudes to be retrieved in isolation when the pulses overlap one another?

The solution lies in the strobing of the receiver-output pulses. Each pulse, although it extends over a nonzero duration, is measured at only a single time instant. If that instant is free of interference from all other pulses, the measurement includes contributions only from the one desired pulse, plus noise. Freedom from intersymbol interference (ISI) will be achieved if all pulses, except the one being strobed, go through zero amplitude at each strobe time.

More formally: Let the optimum strobe point on the pulse shape $g(t)$ occur at $t = t_0$. Then the *Nyquist criterion* [2.6] for freedom from ISI is

$$g(mT + t_0) = 0, \quad m \neq 0 \tag{2.7}$$

Strictly speaking, this is Nyquist's *first* criterion; there are two others that apply to more specialized conditions [2.7, Chap. 5]. Where it is feasible, engineers often strive diligently for Nyquist-1 shaping in designing communications systems.

Nyquist Properties

A time-limited pulse is Nyquist-1 if it is zero at and beyond $t_0 \pm T$. All of the time-limited examples in Fig. 2.1 meet this condition. If the pulse extends beyond $t_0 \pm T$, the only way that the Nyquist criterion can be met will be if the pulse tails—both trailing and leading—pass through zero at the times $t_0 + mT (m \neq 0)$. The sinc pulse of Fig. 2.1 is the epitome of Nyquist-1 shaping.

In the frequency domain, a Nyquist pulse must have a Fourier transform $G(f)$ which, when folded into the frequency range $|f| \leq 1/2T$, has constant amplitude and linear phase [2.8, Chap. 4]. That is, let

$$\overline{G}(f) = \sum_{k=-\infty}^{\infty} G\left(f + \frac{k}{T}\right)$$

Then $G(f)$ is Nyquist if $|\overline{G}(f)|$ is constant and the phase of $\overline{G}(f)$ is linear in f for $|f| \leq 1/2T$.

If the pulse spectrum is bandlimited to $|f| \leq 1/T$, certain symmetry properties apply to $G(f)$, as illustrated in Fig. 2.7 and listed below. For the bandlimited pulse to be Nyquist:

- The phase of $G(f)$ must be linear.
- The amplitude $|G(1/2T)| = G(0)/2$ (i.e., -6 dB).
- The amplitude is zero for $|f| > (1 + \alpha)/2T$. The parameter α, known

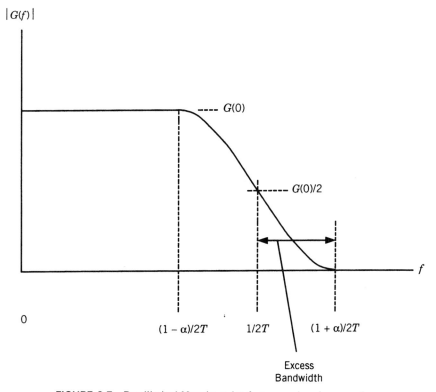

FIGURE 2.7 Bandlimited Nyquist pulse frequency-domain properties.

as the *excess-bandwidth factor*, takes values from 0 to 1. *Excess bandwidth* is defined as $\alpha/2T$.

- The amplitude is flat in the region $|f| < (1 - \alpha)/2T$.
- The amplitude rolls off in the region $(1 - \alpha)/2T \leq |f| \leq (1 + \alpha)/2T$.
- The rolloff has odd symmetry about $|f| = 1/2T$.

If the excess bandwidth is zero ($\alpha = 0$), the pulse shape becomes the sinc pulse of Fig. 2.1, whose Fourier transform is constant and nonzero in the frequency range $|f| < 1/2T$. This is the smallest bandwidth that will support a pulse rate of $1/T$ pulses per second without ISI.

Example. The spectrum of the exponential-decay pulse of Fig. 2.1 was shown in Fig. 2.5. The amplitude of the spectrum is down by about 21 dB from its peak at the frequency $f = 1/2T$. The time and frequency plots both show clearly that the pulse does not meet the Nyquist-1 criterion; corrective measures will be needed to mitigate ISI. Although not strictly bandlimited, the pulse may be regarded as approximately bandlimited for purposes of estimating the magnitude of corrective measures.

30 ☐ REPRESENTATION OF ANALOG SIGNALS AND NOISE

To achieve a spectrum that is only 6 dB down at $f = 1/2T$ requires $21 - 6 = 15$ dB equalization boost.

Cosine-Rolloff Spectra

Even with the Nyquist symmetry conditions listed above, there still could be a boundless number of bandlimited spectra among which to choose. A bandlimited spectrum is an analytical idealization; one might as well select a tractable function for analysis purposes. To that end, the *cosine-rolloff spectra* are widely used [2.8, Chap. 4]. These spectra obey the bandlimited Nyquist-symmetry rules listed above and their rolloff shapes are sections of raised cosines. (For this reason, they are also known as *spectral raised-cosine* pulses, or even *raised-cosine* pulses, which leads to confusion with the time-domain raised-cosine pulse of Fig. 2.1. We prefer the terminology *cosine rolloff* as being less ambiguous.)

The formal definition of a cosine-rolloff spectrum is

$$G(f) = \begin{cases} T, & |f| < \dfrac{1-\alpha}{2T} \\ T\cos^2\left[\dfrac{\pi}{4\alpha}(|2fT| - 1 + \alpha)\right], & \dfrac{1-\alpha}{2T} < |f| < \dfrac{1+\alpha}{2T} \\ 0, & |f| > \dfrac{1+\alpha}{2T} \end{cases} \quad (2.8)$$

and is plotted for several values of α in Fig. 2.8.

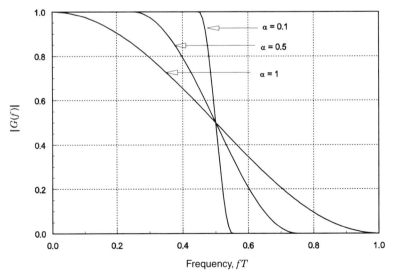

FIGURE 2.8 Cosine-rolloff spectra (data generated by STÆDT).

Take the inverse Fourier transform to find the corresponding time-domain pulse as

$$g(t) = \frac{\sin \pi t/T}{\pi t/T} \frac{\cos \alpha \pi t/T}{1 - (2\alpha t/T)^2} \qquad (2.9)$$

The leading sinc factor assures Nyquist-1 behavior, while the second factor on the right serves as a tapered window (of infinite duration) that reduces the time side lobes. Scaling is such that peak amplitude of $g(t)$ is unity. Examples of $g(t)$ are plotted in Fig. 2.9 for several values of α.

2.1.4 Other Pulse Formats

Although PAM is the most prevalent method of transmission and receives the most attention in this work, other pulse formats are often encountered and must be acknowledged. For the most part, methods of simulation for any other pulse format will be similar to methods employed for PAM.

Pulse Position Modulation

Divide the symbol interval T into an integer number M of subintervals or *slots*, each of duration T/M. To send information, a standard pulse of shape $g_T(t)$ and unit amplitude will be transmitted in just one of the slots, and no pulse will be transmitted in the

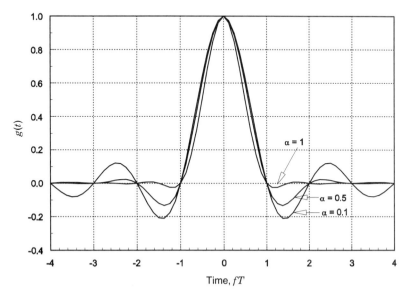

FIGURE 2.9 Pulse waveforms with cosine-rolloff spectra (data generated by STÆDT). Waveforms were computed with eight points per symbol interval. Characteristic irregularities visible in the plots are introduced by the plotting program, which connects adjacent data points with straight lines.

other $M-1$ slots. All pulses transmitted have the same amplitude. A pulse position modulation (PPM) system with M slots can transmit $\log_2 M$ bits per symbol. The task of a PPM receiver is to determine which slot contains the pulse in each symbol interval.

Represent the value of the mth symbol as a_m, taken from the alphabet $\{0, 1, \ldots, M-1\}$. Then the transmitted PPM signal can be represented as the sum of pulses

$$x(t) = \sum_{m=-\infty}^{\infty} g_T\left(t - a_m \frac{T}{M} - mT\right) \tag{2.10}$$

A PPM receiver reshapes the standard pulse to $g(t)$ and then strobes its output M times per symbol interval. Symbol value a_m is decided from identifying the strobe with the largest amplitude.

To avoid interpulse interference within a time slot, the receiver output pulse must obey the Nyquist-1 criterion in the time-slot interval T/M:

$$g(mT/M) = 0 \quad \text{for all } m \neq 0 \tag{2.11}$$

For the same basic pulse shape, a PPM signal requires M times as much bandwidth as an M-level PAM system with the same symbol rate. Pulse position modulation is sometimes employed in lightwave systems, particularly if the system bandwidth is very large compared to the symbol rate.

Pulse Width Modulation

Numerical data can be transmitted by pulses whose width corresponds to the symbol value. Analysis of pulse width modulation (PWM) is related to that of PPM [2.9]. Since PWM is not widely used for data transmission, it will not be considered further.

Pulse Waveform Transmission

Let each symbol take its value a_m from an M-character alphabet. One method of transmission is to assign a different pulse shape for each character of the alphabet and transmit on the mth pulse just the one shape corresponding to the symbol value a_m. The transmitted pulse train is represented as

$$x(t) = \sum_{m=-\infty}^{\infty} g_T[(t - mT); a_m] \tag{2.12}$$

A receiver for M different waveforms requires M separate signal paths, one for each pulse shape employed. All M paths are strobed in parallel, once per symbol interval. The most likely transmitted pulse shape (and consequently, the most likely symbol value) is estimated by selecting the path with the largest strobe amplitude. For further explanation, refer to discussions on orthogonal signaling methods in [1.1] and [1.2].

Example. A pair of waveforms widely used in magnetic recording is illustrated in Fig. 2.10. The waveforms are ordinarily specified in terms of binary encoding rules:

- If $a_m = 1$, record a transition at the center of the mth bit. (*Bit* and *symbol* are synonymous in a binary system.)
- If $a_m = 0$, do not record a transition at the center of the mth bit.

This *FM encoding*, and modifications thereof, underlie many digital magnetic recording systems.

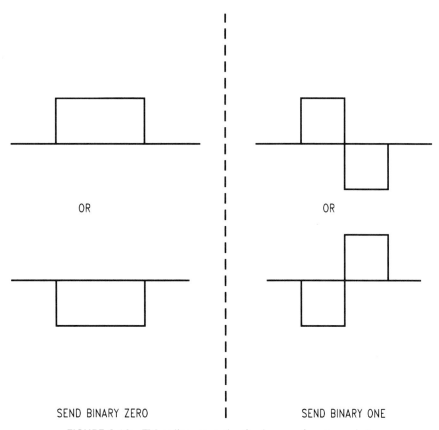

FIGURE 2.10 FM coding: example of pulse waveform transmission.

Pulse Frequency Modulation and Frequency-Shift Keying

One particular form of waveform transmission uses a pulsed sine wave, whose frequency within the pulse is determined by the symbol value a_m. This scheme is variously known as pulse frequency modulation (PFM) or frequency-shift keying (FSK). Strictly speaking, PFM–FSK is a passband signal and ought not be introduced until the next section. However, it is an important example of non-PAM signaling, so it is presented at this point in the narrative instead of later.

With a symbol alphabet of M characters, the transmitted pulse can take on any one of M corresponding frequencies. Most often, the pulse frequencies will be equally spaced by an amount Δf. A PFM–FSK pulse train takes the form

$$x(t) = \sum_{m=-\infty}^{\infty} \rho(t - mT) \cos\left[2\pi(f_0 + a_m \Delta f)(t - mT) + \theta_m\right] \qquad (2.13)$$

where $a_m \in \{0, \ldots, M-1\}$, f_0 is an arbitrary base frequency, and θ_m is the initial phase of the mth pulse. The multiplier $\rho(t)$ represents the pulse envelope, which is often (but not necessarily) rectangular with duration T.

2.2 PASSBAND SIGNAL REPRESENTATIONS

Many transmission media—most notably, radio or optical channels—cannot support baseband signals; they demand signals with bandpass spectra. A passband signal is produced by modulating a baseband signal $x(t)$ onto a sinusoidal carrier at frequency f_c; the spectrum of the resulting passband signal will be concentrated in the vicinity of f_c. Often, but not always, the spectral width of the baseband signal will be small compared to the carrier frequency. In the material to follow, the terms *passband signal* and *modulated signal* are synonymous.

2.2.1 Basic Modulation Formats

There are two fundamental forms of modulation: amplitude modulation (AM), in which the baseband signal $x(t)$ is simply multiplied by the sinusoidal carrier to produce the passband signal

$$s(t) = x(t) \cos 2\pi f_c t \qquad (2.14)$$

or angle modulation, which takes the form

$$s(t) = \cos\left[2\pi f_c t + \theta(t)\right] \qquad (2.15)$$

where the baseband information is conveyed in the phase $\theta(t)$. Angle modulation can be subdivided into phase modulation (PM), where the phase is given by

$$\theta(t) = K_p x(t) \tag{2.16}$$

and frequency modulation (FM), where the phase depends upon the integral of the baseband signal according to

$$\theta(t) = K_f \int_{-\infty}^{t} x(\lambda)\,d\lambda \tag{2.17}$$

Coefficients K_p and K_f represent modulator gain factors.

Elementary Properties of Modulation Formats

Linearity. Let $s_1(t)$ be the modulated signal associated with baseband signal $x_1(t)$, and $s_2(t)$ be associated with $x_2(t)$. Furthermore, let $s_0(t)$ be the modulated signal associated with the baseband signal $x_1(t) + x_2(t)$. Then a modulation format is *linear* if $s_0(t) = s_1(t) + s_2(t)$. It is readily shown that AM is linear according to this definition, but PM and FM are not.

Passband Spectrum. Amplitude modulation produces a passband spectrum that replicates the spectrum of the baseband signal, except translated to $\pm f_c$. Angle modulation also produces a passband spectrum concentrated about the carrier frequency but does not preserve the baseband spectral shape because of inherent nonlinearity. Examples are sketched in Fig. 2.11. If the baseband spectrum is bandlimited, the resulting AM passband spectrum is also bandlimited. However, a nontrivial angle-modulation spectrum is never bandlimited, irrespective of baseband spectrum. The angle-modulation process always spreads the spectrum in some degree.

Envelope Variations. Except under special circumstances, an AM signal exhibits variations in the amplitude of its envelope. By contrast, an angle-modulated signal has a constant-amplitude envelope (barring excessive filtering that induces amplitude variations). Therefore, an AM signal will be distorted by nonlinear amplifiers, whereas a constant-amplitude angle-modulated signal can pass through a nonlinear amplifier without distortion of the modulation.

Residual Carrier. An AM signal includes a discrete spectral line at the carrier frequency only if the baseband signal has a nonzero dc value. If the baseband signal has zero mean, there is no carrier line; the transmission has suppressed the carrier. Residual carrier in an angle-modulated signal depends on the detailed baseband waveforms and the modulation index as well as the mean of the baseband signal. Data signals are commonly (but not always) transmitted by suppressed carrier methods, on the grounds that a residual carrier conveys no message information and therefore wastes power. The modulation formats studied in this book primarily generate suppressed-carrier signals.

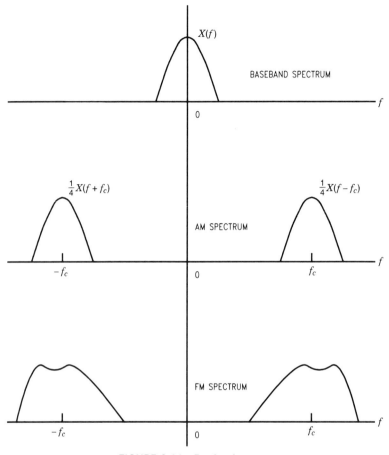

FIGURE 2.11 Passband spectra.

2.2.2 Quadrature Amplitude Modulation (QAM)

The foregoing simplistic modulation formats are all one-dimensional; they support just one real baseband signal $x(t)$. A one-dimensional modulated signal is wasteful of bandwidth because a passband signal can support two baseband signals; a passband signal can be two-dimensional. For example, a signal can contain both amplitude and angle modulation, each produced by independent baseband signals. The two separate modulations can be recovered by a receiver incorporating both amplitude and angle demodulators.

QAM Definition

A more pertinent example is quadrature amplitude modulation (QAM), which uses two carriers of identical frequency, summed in phase quadrature after modulation. If $x(t)$ and $y(t)$ are two baseband signals, a QAM passband signal is represented by

$$s(t) = x(t)\cos 2\pi f_c t - y(t)\sin 2\pi f_c t \qquad (2.18)$$

The two-dimensional QAM signal is the quadrature sum of two one-dimensional AM signals. This representation is of major practical and theoretical importance and will be encountered repeatedly in the sequel. [The term *QAM* is often restricted to signals for which $x(t)$ and $y(t)$ are both PAM baseband signals having the same symbol rate and same pulse shape. No such restriction on terminology applies here; $x(t)$ and $y(t)$ can be entirely independent.]

If $x(t)$ and $y(t)$ have the same spectral widths, the QAM signal of (2.18) consumes no more passband spectrum than either baseband component alone, modulated as a one-dimensional signal onto a simple AM carrier. Both quadrature components of the QAM signal occupy the same passband spectrum; thus QAM offers spectral saving by a factor of 2.

Complex Envelopes and Analytic Signals

Rectangular Format. The QAM signal has a *complex envelope*, defined as

$$z(t) = x(t) + jy(t) \qquad (2.19)$$

Although the baseband signals $x(t)$ and $y(t)$ and the passband signal $s(t)$ are each real, $z(t)$ is a complex function of time. Moreover, since $x(t)$ and $y(t)$ are both baseband signals, $z(t)$ is also a baseband signal with a baseband spectrum, even though it pertains to a passband signal $s(t)$. The complex-envelope representation discards information on carrier frequency. The complex envelope is an embodiment of the *inphase-quadrature (I–Q)* representation of signals, where $x(t)$ is the *I* component and $y(t)$ is the *Q* component.

Simulation of a real passband signal would require calculation of the sinusoidal carrier, a highly computationally intensive operation. By dealing with complex envelopes instead, the computational effort can be greatly reduced. In consequence, almost all simulation programs treat the complex envelope $z(t)$, when possible, rather than the real signal $s(t)$.

Analytic Signal. Along with the complex envelope, there is another complex representation known as the *analytic signal*,

$$s^+(t) = z(t)e^{j2\pi f_c t} \qquad (2.20)$$

which preserves carrier information. The real passband signal is simply

$$s(t) = \text{Re}\,[s^+(t)] \qquad (2.21)$$

where the notation Re [·] means "real part of."

COMMENT: The "+" superscript indicates that positive—that is, counterclockwise–rotation has been applied, causing a positive frequency shift on $z(t)$ to form $s^+(t)$. If the spectrum of $z(t)$ is bandlimited to frequencies $|f| < f_c$ (a condition not necessarily encountered in some important practical systems or their simulations), the spectrum of $s^+(t)$ is confined to positive frequencies. A clockwise rotation would form $s^-(t)$, which has the same properties as noted for $s^+(t)$, except that "negative" should be substituted for "positive."

Polar Coordinates. The complex envelope can also be cast into polar coordinates

$$z(t) = R(t) e^{j\theta(t)} \tag{2.22}$$

where amplitude $R(t)$ and phase $\theta(t)$ are defined as

$$R(t) = \sqrt{x^2(t) + y^2(t)}$$
$$\theta(t) = \tan^{-1} \frac{y(t)}{x(t)} \tag{2.23}$$

Equivalently, the polar representation can be converted to rectangular according to

$$x(t) = R(t) \cos \theta(t)$$
$$y(t) = R(t) \sin \theta(t) \tag{2.24}$$

Simulations employ both rectangular and polar representations of complex envelopes.

The real and analytic signals can be rewritten in a polar format exactly equivalent to the previous definitions in rectangular format:

$$s(t) = R(t) \cos [2\pi f_c t + \theta(t)] \tag{2.18a}$$
$$s^+(t) = R(t) e^{j[2\pi f_c t + \theta(t)]} \tag{2.20a}$$

It now becomes apparent [compare (2.18a) to (2.15)] that combined amplitude/angle modulation, or even straight angle modulation itself, with constant amplitude, can be represented in QAM formalism with the aid of the nonlinear transformations of (2.24). This relation among widely different modulation formats is a valuable tool for devising simulation models.

Although the rectangular components $x(t)$ and $y(t)$ might be strictly bandlimited, the polar components $R(t)$ and $\theta(t)$ of the same bandlimited signal can never be bandlimited. It is as if the nonlinear transformations (2.23) were spreading the spectra of the polar components. Moreover, in angle modulation, even if $\theta(t)$ were bandlimited, the modulation nonlinearity would spread the spectrum of $s(t)$ so that it was not bandlimited.

Alternative Definitions. The reader should be cautioned that the foregoing definition of a complex envelope is not unique. Rice [2.10] puts forward three definitions, each of which has claims to validity in appropriate circumstances. The definition given above corresponds to one of his three. Another definition will be invoked in a later chapter when bandpass filters are to be modeled.

Using (2.18) and (2.19) for the definition of complex envelope, the analytic signal $s^+(t)$ can be shown to have a single-sided spectrum in the sense that $S^+(f) = 0$ for $f < 0$, provided that $X(f)$ and $Y(f)$ are bandlimited to $|f| < f_c$. [Notation: $S^+(f)$, $X(f)$ and $Y(f)$ are the Fourier transforms of $s^+(t)$, $x(t)$, and $y(t)$, respectively.] If $X(f)$ and $Y(f)$ are not bandlimited as stated, $S^+(f)$ will lap over to negative frequencies.

Franks [2.11, 2.12] and Papoulis [2.22, Sec. 10-6] define a subtly different complex envelope, using Hilbert transforms, such that the spectrum of the analytic signal is always single-sided, irrespective of whether or not $X(f)$ and $Y(f)$ are bandlimited. Their definition agrees with (2.20) if the rectangular components are bandlimited appropriately. The Hilbert transform of a nonbandlimited passband signal does not necessarily lead back to the original $x(t)$ and $y(t)$ of the baseband modulating signals. For this reason the Hilbert transform must be applied with care. We do not use Hilbert transforms in this book.

Underlying most use of complex envelopes is the tacit assumption that $X(f)$ and $Y(f)$ are indeed bandlimited to $|f| < f_c$ (equivalent to assuming that the positive- or negative-frequency sidebands of the real passband signal do not overlap zero frequency). If the assumption is not valid in particular circumstances, the model must be reexamined.

QAM Categories

The baseband signals $x(t)$ and $y(t)$ in the QAM representation (2.18) may be completely general; they need not be data signals at all. For example, as will be shown in Sec. 2.4.1, additive bandpass noise can be resolved into quadrature components and modeled in the QAM representation, in much the same manner as the desired signals. Assume that $x(t)$ and $y(t)$ are data signals. They can be entirely independent of one another, with different symbol rates, different alphabets, different pulse shapes, and different amplitudes. The resulting signal format is commonly known as *unbalanced* QAM, particularly when the amplitudes differ.

If the symbol rates of x and y differ, the symbol rates can be mutually synchronous or nonsynchronous. Symbol rates are synchronous if they are derived from a common clock, and nonsynchronous if derived from independent clocks. Nonsynchronous rates may be awkward to simulate, for reasons that emerge in later chapters.

Attention in this book is concentrated primarily on *balanced* QAM, wherein x and y have equal, synchronized pulse rates. The x and y pulses might be time-aligned, or they may be offset from one another by an amount ϵT ($0 \leq \epsilon < 1$). The latter format is known as *offset* QAM (OQAM) or *staggered* QAM. Offset factors $\epsilon = 0.5$ or 0 (i.e., nonoffset) are most prevalent.

2.2.3 Examples of Passband Signal Formats

PAM–QAM Signals

A large and diverse group of signals can all be represented in balanced PAM–QAM format, wherein

$$x(t) = \sum_{m=-\infty}^{\infty} a_m g(t - mT)$$

$$y(t) = \sum_{m=-\infty}^{\infty} b_m g(t - \epsilon T - mT) \qquad (2.25)$$

are PAM pulse streams. Pulse shapes $g(t)$ are assumed to be real and identical for both x and y. Offset, if any, is accommodated by incorporating ϵT into the pulse shape argument for $y(t)$. Data-symbol values a_m and b_m may be independent of each other in some formats and dependent in others. Symbol values might be taken from a wide variety of alphabets.

On–Off Keying (OOK). Let $a_m \in \{0, 1\}$ and $b_m \equiv 0$. Although this scheme is no longer much used in electrical or radio communication, it is widespread in noncoherent optical systems. (*Noncoherent* means that the receiver ignores the phase of the signal; *coherent* means that the signal phase is taken into account. Coherent operation typically is more efficient in both power and bandwidth consumption.)

Binary Phase-Shift Keying (BPSK). Let $a_m \in \{-1, 1\}$ and $b_m \equiv 0$. This is antipodal signaling (one character in the alphabet is the opposite of the other). It is simple, power efficient, and robust, so it has seen extensive service. But it is not bandwidth efficient and so is being displaced by other formats in applications where spectrum conservation is important.

The two preceding formats are one-dimensional and therefore are unbalanced QAM, by strict definition. However, no one ever refers to them in that manner.

Quadrature Phase-Shift Keying (QPSK*). Let $a_m \in \{-1, 1\}$ and also $b_m \in \{-1, 1\}$. In essence, this signal is composed of two BPSK signals that occupy the same frequency range, thereby effecting spectral conservation by a factor of 2. It can be shown that QPSK has the same power efficiency as BPSK, so no power penalty need be paid for the spectral saving. QPSK has been used extensively in satellite digital communications links. QPSK can be nonoffset ($\epsilon = 0$) or offset (typical practice: $\epsilon = 0.5$). If offset, it is designated OQPSK.

---

*To be more rigorous, we probably should use the term *4QAM* instead of *QPSK*, but QPSK is the term that almost everyone uses.

M-ary QAM (MQAM). A graphical depiction, known as a *constellation*, is helpful in visualizing the structure of PAM–QAM signals. The constellation is simply a plotting of all points $a_m + jb_m$ in the complex a, b (or x, y) plane. If there are M such complex points in the combined alphabets of a and b, the PAM–QAM signal is said to have an MQAM format. Some example constellations are plotted in Fig. 2.12.

If the alphabets for a and b consist of equispaced numbers (e.g., $\{\pm 1, \pm 3, \pm 5, \ldots\}$), the points of the constellation will lie on a uniform square grid, which is con-

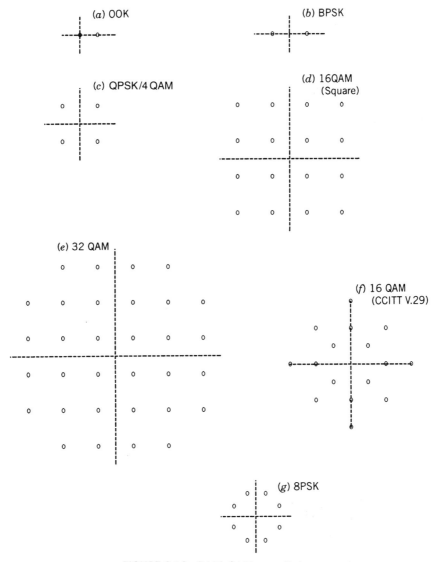

FIGURE 2.12 PAM–QAM constellations.

ducive to simple implementation of symbol detection. If, in addition, a_m and b_m are selected independently from a common alphabet, the constellation itself will be square, and the number M must be the square of an integer L: that is, $M = L^2$.

Figure 2.12d shows a 16-QAM square constellation, which has been employed in many terrestrial microwave digital radios. As this book is being written, 64-QAM systems are in commercial service and 256-QAM systems are under serious consideration. Investigations with 1024-QAM have been reported. The objective of such large constellations is to conserve bandwidth, our most limited radio resource. If $M > 4$, a power penalty is the cost paid for achieving the spectral efficiency.

Also shown, as Fig. 2.12e, is a nonsquare 32-QAM constellation, with all points on a square grid (a 36-point array with the corner points deleted). It is clear that a_m and b_m cannot be selected entirely independently if the constellation is not square.

The points of the constellation need not lie on a uniform, square grid. Figure 2.12f shows a 16-QAM constellation that is nonuniform. It was devised to minimize the effects of phase disturbances; its points occupy only eight distinct phases, as opposed to 12 in the square 16-QAM constellation. Symbol detection is substantially more complicated than for points on a square grid.

If timing of the x and y pulses is not offset, it is meaningful to speak of a *complex symbol*, defined as

$$c_m = a_m + jb_m \qquad (2.26)$$

The complex envelope of such a signal can be written as

$$z(t) = \sum_{m=-\infty}^{\infty} c_m g(t - mT) \qquad (2.27)$$

which may be more compact in some instances than the overtly rectangular form of (2.19) and (2.25).

M-ary Phase-Shift Keying (MPSK). The MQAM formats that have been illustrated must transmit multiple amplitudes (if $M > 4$), which could be severely distorted by any nonlinear device (e.g., a saturating amplifier). Distortion effects are reduced if all points in the constellation have equal amplitudes: that is, if all M points lie on a circle centered at the origin of the complex plane.

In an MPSK format, the points on the circle are all equally spaced: typically, $c_m \in \{e^{\pm j\pi/M}, e^{\pm j3\pi/M}, \ldots\}$. Figure 2.12$g$ illustrates 8PSK, and Fig. 2.12c shows 4PSK (which is exactly the same as QPSK, which was introduced as square-constellation 4QAM; there is often more than one way to classify a signal). Complex-symbol notation is advantageous in MPSK, when a_m and b_m cannot be selected independently. Indeed, since $|c_m|$ = constant, it is required that $a_m^2 + b_m^2$ = constant for MPSK.

For $M > 4$, MPSK incurs a power penalty compared to square-grid MQAM with the same number of points. Large MPSK constellations (e.g., $M > 8$) are not often encountered, at least partly because of this penalty.

Partial-Response Pulses. Embedded in the foregoing is the tacit assumption that the constellation of the signal transmitted is the same as the constellation of the receiver output. That is not necessarily true unless the link has an overall Nyquist pulse shape. In some systems, it is advantageous to employ *partial-response* pulse shapes, in which a known amount of intersymbol interference is introduced deliberately. Partial-response shaping causes an expansion of the receiver constellation to a larger number of points than exist in the complex symbol alphabet at the transmitter.

A decoding operation is required for partial-response reception. The term *partial response* alludes to the fact that all of the information on one pulse is not fully contained in just one receiver strobe (as it would be for Nyquist pulses), but is partially contained in two or more strobes. Partial-response signaling is an important topic for communications engineers [2.13] to [2.16], but is not given much space in this work. It is simulated by techniques virtually identical to those to be described for near-Nyquist pulses.

Complex Pulses

In the preceding paragraphs, the examples all had real pulse shapes and could have complex symbol values (complex amplitudes for the real PAM pulses). A different class of signals employs only real symbol values a_m but allows complex pulse shapes in the passband. Represent the complex pulse shape as

$$\overline{g}(t) = g(t) + j\tilde{g}(t) \qquad (2.28)$$

where the tilde (~) indicates that $\tilde{g}(t)$ was obtained from $g(t)$ by a linear filtering operation. The complex envelope becomes

$$z(t) = \sum_{m=-\infty}^{\infty} a_m \overline{g}(t - mT) \qquad (2.29)$$

With suitably chosen filtering, the real signal $s(t)$ can have a vestigial sideband (VSB) spectrum. If the filter is a Hilbert transformer [i.e., $\tilde{g}(t) = \hat{g}(t)$, where $\hat{g}(t)$ is the Hilbert transform of $g(t)$], the real signal has a single-sideband (SSB) spectrum.

In earlier times, some data transmission links employed VSB or SSB, with the objective of conserving bandwidth. These systems predated the full realization that complex-symbol QAM is as efficient as SSB in spectrum conservation and somewhat easier to implement. More recently, VSB has been put forward for transmission of digital television. Although described here under a different heading, VSB and SSB are PAM–QAM formats. There is little explicit mention of VSB or SSB in the sequel, but they can be simulated by the same methods that apply to other QAM signals.

44 ☐ REPRESENTATION OF ANALOG SIGNALS AND NOISE

Non-PAM Signals

Some non-PAM formats were introduced earlier under the heading of baseband signals. These can be modulated onto carriers to produce passband signals. For example, a PPM baseband stream (2.10) can key an optical transmitter on and off. This combination might be dubbed PPM–OOK.

FSK. Frequency shift keying has already been identified as a passband signal, even though it was presented first in the baseband section. There are different varieties of FSK signals, depending on the choice of characteristics in the defining equation (2.13). Often the pulse envelope $p(t)$ is rectangular, with duration T; a rectangular, NRZ pulse shape provides a constant-amplitude envelope for the passband FSK signal. Most commonly, the signaling alphabet is binary ($M = 2$), but much larger alphabets have been used.

The base frequency f_0 has been defined in (2.13) as the lowest signaling frequency, but it is more common to define a carrier frequency at the center of the set of frequencies. If M is even, the centered carrier frequency is not one of the signaling frequencies.

Phase can be either continuous or discontinuous from one symbol pulse to the next. Continuous-phase FSK changes only its frequency at pulse boundaries; the starting phase for the new pulse is the same as the ending phase for the preceding pulse. Continuous-phase operation is obtained if FSK is generated by switching the frequency of a single oscillator. Discontinuous phase arises if the signaling pulses are generated by switching among M independent oscillators. A random phase jump occurs at each symbol boundary. Discontinuous-phase operation produces a broader RF spectrum than continuous phase, and so generates more adjacent-channel interference.

The *modulation index* or *deviation ratio*, defined as

$$h = \Delta f T$$

has considerable influence on the passband spectrum and on the achievable performance of the link. See [1.1] to [1.5] for more information on FSK systems.

CPM. Continuous-phase modulation [2.17] is a constant-envelope signal format that has received much attention in recent years. It is attractive because of its constant envelope, the ability to control spectral occupancy, and potential enhancement of power efficiency. Each CPM data symbol rotates the phase of the signal by an amount dependent upon the value of the symbol. Message data are recovered from observations of the phase rotations. Phase of a CPM signal always changes smoothly, never discontinuously: thus the name *continuous*-phase modulation.

The real-signal format for CPM is given by

$$s(t) = \cos\left[2\pi f_c t + \theta_0 + \pi \sum_{m=-\infty}^{\infty} a_m h_m p(t - mT_1)\right] \quad (2.30)$$

where f_c is carrier frequency, θ_0 a random initial phase, a_m the symbol value of the mth symbol, h_m the modulation index for the mth symbol, $p(t)$ the *phase pulse*, and T_1 the symbol interval. The reason for using the notation T_1 for the symbol interval, instead of T, is explained below.

The phase pulse is defined as the integral of the *frequency pulse* $q(t)$:

$$p(t) = \int_{-\infty}^{t} q(t') \, dt' \qquad (2.31)$$

and $q(t)$ is subject to the restriction

$$\int_{-\infty}^{\infty} q(t) \, dt = 1 \qquad (2.32)$$

Equivalently, if $Q(f)$ is the Fourier transform of $q(t)$, then (2.32) implies that $Q(0) = 1$. With this restriction, the mth pulse ultimately contributes exactly $a_m h_m \pi$ radians to the total accumulated phase of the signal.

COMMENT: Other authors define terms such that $\int q(t) \, dt = 0.5$, and replace π by 2π in (2.30). Our preference for the version shown here is mainly esthetic. Both versions are correct; they both describe exactly the same signal.

The complex envelope of a CPM signal is $z(t) = e^{j\theta(t)}$, and phase $\theta(t)$ is

$$\theta(t) = \theta_0 + \pi \sum_{m=-\infty}^{\infty} a_m h_m \int_0^t q(t' - mT_1) \, dt'$$

$$= \theta_0 + \pi \int_0^t \sum_{m=-\infty}^{\infty} a_m h_m q(t' - mT_1) \, dt' \qquad (2.30a)$$

where the sum in the second line will be recognized as a PAM baseband data signal with essentially the same format as (2.5). Thus CPM can be generated, in part, by the same techniques used for PAM. If $q(t)$ has no singularities (e.g., impulses), $p(t)$ is continuous over all time, and so is the total phase of the signal.

Some benefit accrues from *multi-h* operation (i.e., imposing a different h_m for each symbol [2.17, Chap. 3]), but simpler implementation is possible if h is a constant. If coherent reception is desired, h is chosen to be a rational number so that the phase trajectory (reduced modulo-2π) is constrained to traverse only a finite number of different paths in each symbol interval. The symbol values a_m are taken from an M-character alphabet; $M = 2$ is commonplace, but larger M can be beneficial.

A CPM signal is said to be *full response* if the duration of the frequency pulse occupies only one symbol interval, and *partial response* if the frequency pulse occupies more than one symbol interval (frequency pulses overlap). Spectral occupancy for a given information rate can be reduced by decreasing h, by increasing the duration of the frequency pulse, by smoothing the frequency pulse, or by increasing the size M of the symbol alphabet. A wide range of trade-offs between spectral economy and power efficiency are theoretically possible [2.18]. However, because CPM is nonlinear (like any angle modulation), it can never achieve the theoretical spectral efficiency offered by a strictly bandlimited linear modulation.

Observe that the instantaneous phase of (2.30a) includes contributions from every preceding symbol. The CPM signal has *data memory* that did not appear in the Nyquist PAM schemes. The memory can be regarded as a form of encoding and can be exploited in the receiver to obtain coding gain, even without explicit coding of the data symbols. Instantaneous phase can accumulate without bound, a feature that requires attention when devising a simulation model.

A CPM receiver can be coherent or can employ differential detection or a frequency discriminator. These alternatives are described further in later chapters.

In a particular form of CPM known as continuous-phase FSK (CPFSK), the frequency pulse $q(t)$ is rectangular so that the phase advances (or recedes) as a piecewise linear function, with constant slopes for finite intervals. The constant-slope interval is T_1 if the duration of $q(t)$ is an integer multiple of T_1.

Minimum-Shift Keying

A signal format known as minimum-shift keying (MSK) [2.5, 2.19] can be described equally well as (1) a particular form of offset QPSK, as (2) continuous-phase FSK, or as (3) full-response CPM. This diversity of allowable representation is most unusual and has made MSK the object of much study.

As OQPSK in the representation of (2.25) and (2.18), MSK has offset factor $\epsilon = \frac{1}{2}$ and transmits the half-cosine pulse shape of Fig. 2.1d. Envelopes of the I and Q pulses overlap in such manner that the real signal has a constant-amplitude envelope, an unusual feature in an amplitude-modulated signal. The interval between adjacent pulses in the I channel (or Q channel) is designated T. Because of the half-symbol offset, a Q-channel pulse follows the preceding I-channel pulse by $T/2$.

The same signal can be regarded as a binary continuous-phase FSK signal with rectangular NRZ pulses $p(t)$ of duration $T_1 = T/2$. The modulation index $h = \Delta f T_1 = \frac{1}{2}$ is rather small for most FSK signals: thus the designation "minimum"-shift keying. Regarded as a CPM signal, each MSK binary symbol causes a phase excursion of exactly $\pm\pi/2$ in each binary-symbol interval of $T_1 = T/2$. The phase excursion is piecewise linear as a function of time. All such phase trajectories, reduced modulo-2π, form a phase trellis.

A reason for using $T_1 = T/2$ as the symbol interval has now been illustrated for MSK. A suitable definition for the symbol interval depends upon the conceptual origin of the signal representation. If MSK is regarded as a variation on QPSK, then T is a plausible choice for symbol interval. If MSK is regarded as FSK or CPM, then T_1 is a natural choice. Heated arguments could ensue over which is the better choice, but it is more important to recognize that ambiguity does exist and to act accordingly. Otherwise, substantial confusion is possible. The same confusion can arise in any offset PAM–QAM signal with $\epsilon = \frac{1}{2}$, as well as CPM.

Further Directions

Although considerable space has been devoted above to non-PAM examples (partly because of their greater complexity), the bulk of the material to follow will concentrate on PAM formats. Only after simulation models for those have been worked out will non-PAM signals receive much attention. Most techniques devised for PAM will carry over nearly unchanged to non-PAM signals.

2.3 POWER, ENERGY, AND SPECTRUM RELATIONS

Performance of a communications system is usually evaluated in terms of the energy or power in the signal and in the noise. This section presents energy, power, and spectrum expressions for the analog data signals that were introduced earlier in the chapter.

2.3.1 Baseband Signals

Energy and Power

The "energy" of a pulse $g(t)$ is defined as

$$E_g = \int_{-\infty}^{\infty} |g(t)|^2 \, dt \tag{2.33}$$

["Energy" is placed in quotation marks because the units of E_g are not really units of energy—that is, joules. If the units of $g(t)$ were volts, the units of E_g would be volt^2-sec, which may be regarded as the energy in joules that would be dissipated in a 1-ohm resistor. Quotation marks will be omitted henceforth, but the abuse of units should be kept in mind. Similar comments apply to "power."]

By Parseval's relation, the energy may also be expressed in the frequency domain by

$$E_g = \int_{-\infty}^{\infty} |G(f)|^2 \, df \tag{2.33a}$$

where $G(f)$ is the Fourier transform of $g(t)$. Any pulse of practical interest has a finite, well-defined energy.

Since a signal composed of an infinite sequence of pulses has infinite energy, it must be described in terms of its power, which will be finite if each pulse has finite energy and the pulses are transmitted at discrete intervals. A real, baseband PAM signal has the form

$$x(t) = A_0 \sum_{m=-\infty}^{\infty} a_m g(t - mT) \tag{2.34}$$

where A_0 is a constant amplitude and a_m is the value of the mth symbol, taken from an M-character alphabet. (Previous signal definitions used $A_0 = 1$.) For simplicity, a_m and $g(t)$ are assumed to be real. Data values a_m are assumed to be selected at random from the alphabet.

The mean-square value of $x(t)$ is the quantity denoted "power." To determine the power, first take the statistical expectation of $x^2(t)$ (the second moment):

$$E[x^2(t)] = A_0^2 E\left[\sum_{m=-\infty}^{\infty} a_m g(t - mT) \sum_{i=-\infty}^{\infty} a_i g(t - iT)\right]$$

$$= A_0^2 \sum_m \sum_i g(t - mT) g(t - iT) E[a_m a_i] \tag{2.35}$$

(The symbol E is used both for energy and for expectation. To minimize confusion, expectation will be written as $E[\cdot]$.)

For a special case that is frequently applicable, assume that the data values are uncorrelated and have zero mean, so that

$$E[a_m a_i] = \begin{cases} E[a_m^2] = \sigma_a^2, & i = m \\ 0, & i \neq m \end{cases} \tag{2.36}$$

For the more general case of correlated data values or nonzero mean, see Bennett [2.20].

With uncorrelated, zero-mean symbols, the second moment reduces to

$$E[x^2(t)] = A_0^2 \sigma_a^2 \sum_{m=-\infty}^{\infty} g^2(t - mT) \tag{2.37}$$

In contrast to the moments of stationary random noise, this second moment is *nonstationary*; its value depends on the time t. If t were replaced by $t' = t + kT$ (k = integer), then (2.37) is unchanged. Therefore, the second moment is periodic in time, with period T. Random signals with periodic statistics are said to be *cyclostationary* [2.20].

A meaningful definition of average power—one consistent with the measurement delivered by a laboratory meter—would be the time average of $\mathrm{E}[x^2(t)]$. Since this quantity is periodic, it is sufficient to average it over a single period of duration T. Therefore, the average power of the real PAM signal $x(t)$ is

$$\overline{P}_x = \frac{1}{T} \int_{-T/2}^{T/2} \mathrm{E}[x^2(t)] \, dt$$

$$= \frac{A_0^2 \sigma_a^2}{T} \sum_{m=-\infty}^{\infty} \int_{-T/2}^{T/2} g^2(t - mT) \, dt$$

$$= \frac{A_0^2 \sigma_a^2}{T} \int_{-\infty}^{\infty} g^2(t - mT) \, dt \qquad (2.38)$$

where the order of summation and integration were interchanged in the second line. Integrals in the second line adjoin one another in time, without overlaps and without gaps; their infinite sum is the same as a single integral over infinite time, as in the third line. The overbar signifies time averaging.

That last integral will be recognized as nothing but the pulse energy E_g, so the time-averaged power in the real PAM signal is given by

$$\overline{P}_x = \frac{A_0^2 \sigma_a^2 E_g}{T} \qquad (2.39)$$

Multiple averaging similar to this example is needed often when dealing with data signals. Henceforth, notation of the form

$$\mathrm{Avg}\,[x^2(t)]$$

is shorthand to indicate that all necessary averages are to be taken, no matter how many there may be.

It is often necessary to determine E_b, the average energy associated with each bit of information being transmitted. If the information bit rate for $x(t)$ is denoted R_b, the energy per bit in a PAM baseband signal is

$$E_b = \frac{\overline{P}_x}{R_b} = \frac{A_0^2 \sigma_a^2 E_g}{R_b T} \qquad (2.40)$$

Spectrum Relations

Periodogram. A single pulse $g(t)$ has a Fourier transform denoted $G(f)$. A finite PAM sequence of L pulses can be written as

$$x_L(t) = A_0 \sum_{m=0}^{L-1} a_m g(t - mT) \qquad (2.41a)$$

Since $x_L(t)$ has finite energy, its Fourier transform exists and is given by

$$X_L(f) = A_0 G(f) \sum_{m=0}^{L-1} a_m e^{-jm2\pi fT} \qquad (2.41b)$$

Each Fourier transform $X_L(f)$ is related uniquely to just one sequence $x_L(t)$, and vice versa.

That Fourier transform depends upon the pulse spectrum $G(f)$, and also upon the particular data pattern comprising the sequence. For an M-character alphabet and a sequence of length L, there can be M^L distinct sequences, and therefore M^L distinct Fourier transforms. Radically different Fourier transforms can result, depending on the data pattern.

The *periodogram* of the finite sequence is defined as $|X_L(f)|^2/T_L$, where T_L is the duration of the sequence. Subtleties arise in the definition of duration T_L. The question of duration is evaded by misusing $|X_L(f)|^2$ as a scaled version of the true periodogram and omitting T_L. This subject is raised again in Chapter 18.

Because phase information has been discarded, numerous sequences generate the same periodogram. Nonetheless, many different periodograms will still be possible from an L-pulse sequence, depending on the specific data pattern. The periodogram, or variations thereon, is the dominant method of computing the spectrum of simulated finite-duration data signals.

Ensemble Spectrum. An infinite PAM sequence does not have a Fourier transform; one deals with the ensemble power spectral density (PSD) instead. Bennett derives the PSD in [2.20]; his procedure is outlined briefly in the following paragraphs.

An infinite-sequence PAM baseband signal is given by

$$x(t) = A_0 \sum_{m=-\infty}^{\infty} a_m g(t - mT)$$

The first step is to find the autocorrelation of the signal. If the data symbols are uncorrelated and zero mean (correlated symbols are treated in [2.20] and [1.7, Chap. 2]), the autocorrelation is

$$E[x(t)x(t - \tau)] = A_0^2 \sigma_a^2 \sum_{m=-\infty}^{\infty} g(t - mT)g(t - \tau - mT) \qquad (2.42)$$

The autocorrelation is cyclostationary, since it depends on time t and since it is periodic in T.

Averaging (2.42) over the period T gives a time-averaged autocorrelation

$$\overline{\psi}_x(\tau) = \frac{A_0^2 \sigma_a^2}{T} \int_{-\infty}^{\infty} g(t)g(t - \tau)\, dt \qquad (2.43)$$

where ψ designates correlation. Taking the Fourier transform of the time-averaged autocorrelation yields the time-averaged power spectral density

$$\overline{\Psi}_x(f) = \frac{A_0^2 \sigma_a^2}{T} |G(f)|^2 \qquad (2.44)$$

For details, see [2.20]. The key finding is that $\overline{\Psi}_x(f)$ is proportional to the squared magnitude of the pulse's Fourier transform.

A laboratory spectrum analyzer would display an approximation to $\overline{\Psi}_x(f)$, provided that the signal duration was long enough and provided that averaging in the spectrum analyzer was adequate. Take careful notice that time averaging conceals features of the cyclostationarity of the signal, features that might be crucial in some circumstances. One example of inapplicability of $\overline{\Psi}_x(f)$ may be found in [2.21].

Equation (2.44) shows only continuous-frequency components. If data symbols do not have zero mean, or if they are correlated, discrete lines might appear in the spectrum; see the references for full detail.

Application to Simulation

The power $\overline{P}_x$ and the spectrum $\overline{\Psi}_x(f)$ are averages of statistical properties of the infinite ensemble of all infinite-duration signals represented by $x(t)$ in (2.34). They are well-defined quantities that, in principle, can be determined by analysis once the signal parameters [i.e., A_0, $g(t)$, T, and the symbol alphabet] are specified. These are the quantities invariably used as "average power" and "spectral density" in theoretical treatments of signal transmission. However, a simulation necessarily is of finite duration. It is not possible to measure the infinite-ensemble quantities $\overline{P}_x$ and $\overline{\Psi}_x(f)$ from a finite segment of a particular realization of $x(t)$. Moreover, it is often excessively difficult or impossible to compute the ensemble quantities analytically from a signal model.

Common practice in simulation is to *estimate* power and spectrum from *measurements* on the finite-duration signal model. Power is estimated by measuring the energy of a signal segment and then dividing by the duration of the segment. Spectrum is represented by the periodogram, or a modification thereof. These measurements and estimates will be explained further in later chapters. Estimates will depend on the particular data sequence appearing in a signal segment. Different sequences can yield widely different estimates, particularly for spectrum. But to be able to compare simulation results to theoretical performance, it is necessary that the estimates closely resemble the ensemble quantities. That suggests either extremely long data sequences (to attempt

52 □ REPRESENTATION OF ANALOG SIGNALS AND NOISE

a brute-force approximation to an infinite sequence) or sequences with special properties (e.g., pseudonoise sequences), or both. Problems with finite-length sequences arise repeatedly in simulation and will receive further attention throughout this work.

Several examples of periodograms of finite sequences of pulses are shown in Fig. 2.13, as generated by STÆDT. All are drawn to the same scale, include the same number of pulses, and have the same pulse shapes. The signals differ only in their data sequences, but their spectra clearly differ greatly from one another. Moreover,

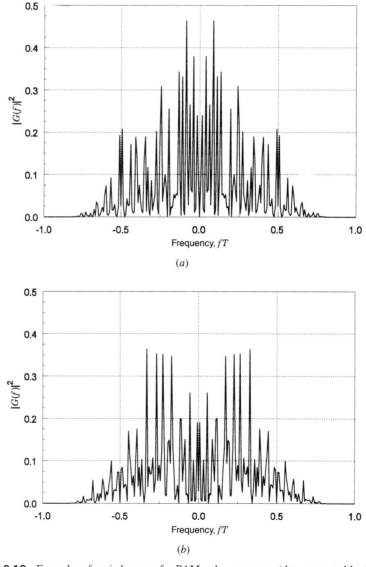

FIGURE 2.13 Examples of periodograms for PAM pulse sequences (data generated by STÆDT.)

POWER, ENERGY, AND SPECTRUM RELATIONS   □   **53**

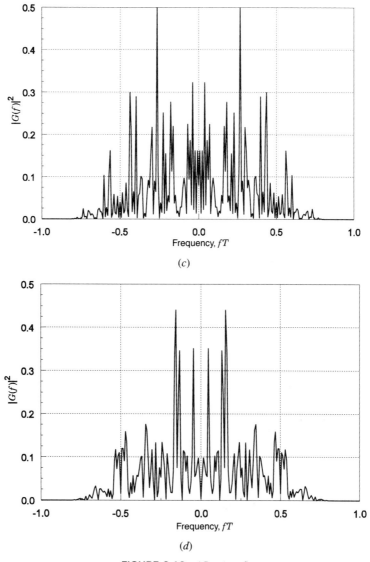

FIGURE 2.13 (*Continued*)

the spectra are exceedingly ragged and irregular. Spectral displays are revisited in Chapter 18, where smoothing to mitigate the irregularities is introduced.

Spectrum Literature

Extensive research over many years has been devoted to spectral analysis. A readable introduction to the subject and further references to the literature can be found in [1.12, Chap. 11].

2.3.2 Passband Signals and Complex Envelopes

Power

Introduce a random phase ϕ, uniformly distributed over 2π, into the real formulation of the passband signal of (2.18) to obtain a slightly modified form,

$$s(t) = x(t) \cos(2\pi f_c t + \phi) - y(t) \sin(2\pi f_c t + \phi) \tag{2.45}$$

Time-averaged power in the passband signal is

$$\overline{P}_s = \text{Avg}\,[s^2(t)] = \tfrac{1}{2}\text{Avg}\,[x^2(t)] + \tfrac{1}{2}\text{Avg}\,[y^2(t)] + \tfrac{1}{2}\text{Avg}\,[(x^2 + y^2) \\ \cdot \cos(4\pi f_c t + 2\phi)] + \text{Avg}\,[xy \sin(4\pi f_c t + 2\phi)] \tag{2.46}$$

If (2.46) is averaged over ϕ (assumed uncorrelated with x or y), the sine and cosine terms will vanish and the signal power is

$$\overline{P}_s = \tfrac{1}{2}\overline{P}_x + \tfrac{1}{2}\overline{P}_y \tag{2.47}$$

If $x(t)$ and $y(t)$ are each doubly infinite sequences of PAM pulses, an expression for $\overline{P}_x$ was derived in (2.35) to (2.39), and $\overline{P}_y$ is defined in the same manner.

The complex envelope of the signal is $z(t) = x(t) + jy(t)$. Time-averaged power in the envelope is

$$\overline{P}_z = \text{Avg}\,[|z(t)|^2] = \text{Avg}\,[x^2(t) + y^2(t)] = \overline{P}_x + \overline{P}_y = 2\overline{P}_s \tag{2.48}$$

By similar means, the analytic signal of (2.20) can be shown to have the same power as the complex envelope. This factor of 2 in power compared to the real signal must be accounted for when dealing with complex envelopes or analytic signals.

The foregoing definitions of power are ensemble averages. They are ideally suited for analytical studies but cannot be applied to physical signals in the laboratory or to simulations. For the latter situations, only a signal segment of duration T_0 is available, and the segment is just one particular realization from the infinite ensemble.

Define the finite-duration passband signal as

$$s_0(t) = x_0(t) \cos 2\pi f_c t - y_0(t) \sin 2\pi f_c t \tag{2.49}$$

Its power can be *estimated* from the time average

$$\begin{aligned}
\hat{P}_{s0} &= \frac{1}{T_0} \int_{T_0} s_0^2(t)\,dt \\
&= \frac{1}{2T_0} \int_{T_0} [x_0^2(t) + y_0^2(t)]\,dt \\
&\quad + \frac{1}{2T_0} \int_{T_0} [x_0^2(t) - y_0^2(t)] \cos 4\pi f_c t\,dt - \frac{1}{T_0} \int_{T_0} x_0(t) y_0(t) \sin 4\pi f_c t\,dt \\
&= \frac{1}{2} \frac{E_{x0}}{T_0} + \frac{1}{2} \frac{E_{y0}}{T_0} + \text{integrals of terms at frequency } 2f_c
\end{aligned} \qquad (2.50)$$

where E_{x0} and E_{y0} are the energies in the baseband signals x_0 and y_0.

If T_0 is sufficiently large and the data sequences are sufficiently random, one expects $\hat{P}_{s0}$ to approach the ensemble average $\overline{P}_s$ of (2.46) and (2.47). But estimates made on different sample functions from the ensemble will produce differing estimates of power. The estimates exhibit *statistical scatter*, a feature that turns up repeatedly in simulation results. What contributes to the scatter?

1. Double-frequency terms contribute ripple.
2. The segment boundaries do not necessarily coincide with symbol boundaries; there will be end effects when pulses are truncated.
3. Different data realizations can lead to widely different segment energies, particularly if the symbol alphabet maps into multiple pulse magnitudes.

Intuition suggests that the integrations of (2.50) will smooth out the double-frequency terms, causing them to vanish if T_0 is sufficiently large and if the carrier frequency is large compared to the symbol rate ($f_c T \gg 1$). Yet it is often necessary to extract estimates (or perform measurements) in comparatively short time intervals, so T_0 is not always large. Also, it is not always true that the carrier frequency is large compared to the symbol rate; there are important applications in which the carrier frequency is actually less than the symbol rate.

Similarly, end effects are likely to recede in importance as the number of symbols in the segment becomes large: that is, as T_0 becomes very large compared to T. But a long-duration signal may be an unaffordable luxury, one whose cost is unacceptable.

The power estimate can be especially sensitive to the particular data sequence in the signal segment. To obtain an estimate that approximates the desired ensemble average, it is necessary that the data sequence be representative of the entire ensemble. These and other problems with estimates arise in most simulations. Methods of dealing with the problems are treated in later chapters.

The complex envelope of the T_0-duration signal segment is

$$z_0(t) = x_0(t) + j y_0(t) \qquad (2.51)$$

Envelope power can be estimated by

$$\hat{P}_{z0} = \frac{1}{T_0} \int_{T_0} |z_0(t)|^2 \, dt$$

$$= \frac{1}{T_0} \int_{T_0} [x_0^2(t) + y_0^2(t)] \, dt$$

$$= \frac{E_{x0}}{T_0} + \frac{E_{y0}}{T_0} \qquad (2.52)$$

Double-frequency terms are absent from the power estimate of the complex envelope. Otherwise, the envelope has twice the power of the real passband signal.

An analytic signal has the same power estimate as that of its complex envelope. Thus, when measuring the power of a complex signal, it is immaterial whether the representation is that of the complex envelope or of the analytic signal.

Correlations and Spectra

In this section we examine the power spectrum of cyclostationary passband data signals. The spectrum of the infinite ensemble is developed conventionally via the Fourier transform of the autocorrelation function. Cross-correlation and cross-spectral terms arise that do not appear in baseband signals. Comparisons are drawn between spectra of real passband signals and of their complex envelopes. The ensemble spectrum cannot be computed for a finite-duration segment of signal; the periodogram is used instead as an approximation.

Ensemble Correlation. Take the statistical expectation $E[s(t)s(t - \tau)]$ of the real passband signal of (2.45) and average it over the random phase ϕ, as was done to find the ensemble average power of (2.47). Omitting details, the interim result is

$$\begin{aligned} \text{Avg}_\phi \, [E[s(t)s(t - \tau)]] \\ = \tfrac{1}{2} [E[x(t)x(t - \tau)] + E[y(t)y(t - \tau)]] \cos 2\pi f_c \tau \\ + \tfrac{1}{2} [E[x(t)y(t - \tau) - x(t - \tau)y(t)]] \sin 2\pi f_c \tau \qquad (2.53) \end{aligned}$$

Often, x and y will be such that their cross correlations (here designated as *cross-channel* correlations, for reasons that will soon become apparent) would vanish for all τ. Cross correlations will be retained in this account so as to point up their potential existence.

The autocorrelations for x and y are the same investigated earlier for baseband; if these signals are PAM, their autocorrelations are cyclostationary. Furthermore, if the x and y each are cyclostationary, their cross correlations are also cyclostationary. Assuming cyclostationarity, the time-averaged autocorrelation of $s(t)$ is

POWER, ENERGY, AND SPECTRUM RELATIONS □ 57

$$\overline{\psi}_s(\tau) = \tfrac{1}{2}\overline{\psi}_x(\tau)\cos 2\pi f_c \tau + \tfrac{1}{2}\overline{\psi}_y(\tau)\cos 2\pi f_c \tau$$
$$+ \tfrac{1}{2}[\overline{\psi}_{xy}(\tau) - \overline{\psi}_{yx}(\tau)]\sin 2\pi f_c \tau \qquad (2.54)$$

(Double subscripts are employed only where essential, such as here for cross correlations. Where possible otherwise, single subscripts will be used to minimize print clutter.)

Ensemble Spectrum. Time-averaged spectral density is the Fourier transform of the time-averaged correlation:

$$\overline{\Psi}_s(f) = \frac{1}{4}\overline{\Psi}_x(f - f_c) + \frac{1}{4}\overline{\Psi}_x(f + f_c) + \frac{1}{4}\overline{\Psi}_y(f - f_c) + \frac{1}{4}\overline{\Psi}_y(f + f_c)$$
$$+ \frac{1}{4j}\left[\overline{\Psi}_{xy}(f - f_c) - \overline{\Psi}_{xy}(f + f_c) - \overline{\Psi}_{yx}(f - f_c) + \overline{\Psi}_{yx}(f + f_c)\right] \quad (2.55)$$

The self-spectrum terms are just baseband spectra of x and y, as derived in (2.42) to (2.44), but translated to $\pm f_c$. Cross-channel spectral terms have no equivalent in real baseband signals. Although the cross spectrum will not be pursued in further detail now, the reader should bear in mind that it does exist and should not automatically assume that it is always zero or negligible.

Envelope Spectrum. To find the (ensemble) spectrum of the complex envelope $z(t) = x(t) + jy(t)$, first take the statistical expectation

$$E[z(t)z^*(t - \tau)] = E[x(t)x(t - \tau)] + E[y(t)y(t - \tau)] + jE[x(t - \tau)y(t)]$$
$$- jE[x(t)y(t - \tau)] \qquad (2.56)$$

where * indicates complex conjugate. If cyclostationarity applies, the time-averaged correlation is

$$\overline{\psi}_z(\tau) = \overline{\psi}_x(\tau) + \overline{\psi}_y(\tau) + j\overline{\psi}_{yx}(\tau) - j\overline{\psi}_{xy}(\tau) \qquad (2.57)$$

Baseband signals x and y have been assumed real, so their auto and cross correlations must also be real, but the correlation function of the complex envelope $z(t)$ will be complex if x and y are correlated. The baseband cross-channel correlations comprise the imaginary part of the time-averaged correlation of the complex envelope $z(t)$. Only if the cross correlation is zero for all τ will the the envelope's correlation function be purely real.

Taking Fourier transforms gives time-averaged spectral density of the envelope as

$$\overline{\Psi}_z(f) = \overline{\Psi}_x(f) + \overline{\Psi}_y(f) + j[\overline{\Psi}_{yx}(f) - \overline{\Psi}_{xy}(f)] \qquad (2.58)$$

where the $\overline{\Psi}$'s have the same meaning as in (2.55). The cross-spectra must satisfy $\overline{\Psi}_{yx}(f) = \overline{\Psi}_{xy}^*(f)$ [2.22, p. 339], so the difference of bracketed terms is pure imaginary, whereupon the spectral density $\overline{\Psi}_z(f)$ is pure real. Furthermore, it must be nonnegative [2.23, p. 105].

The envelope spectrum (2.58) differs from the spectrum (2.55) of the real passband signal in two respects:

1. Envelope-spectrum terms are concentrated in a single peak around $f = 0$ instead of in two peaks around $f = \pm f_c$. That is, the complex envelope has a baseband spectrum rather than a passband spectrum.

2. The area under the curve $\overline{\Psi}_z(f)$ is double that under $\overline{\Psi}_s(f)$, implying that the envelope has twice the power of the real signal. That is consistent with the power calculations of (2.47) and (2.48).

Corresponding to the real signal $s(t)$ there is a positive-frequency analytic signal that can be represented by

$$s^+(t) = z(t) e^{j2\pi f_c t} \qquad (2.59)$$

The correlation function of this analytic signal is

$$E[s^+(t) s^{+*}(t - \tau)] = E[z(t) z^*(t - \tau)] e^{j2\pi f_c \tau} \qquad (2.60)$$

so that its time-averaged correlation is

$$\overline{\Psi}_s^+(\tau) = \overline{\Psi}_z(\tau) e^{j2\pi f_c \tau} \qquad (2.61)$$

which has a time-averaged spectral density of

$$\overline{\Psi}_s^+(f) = \overline{\Psi}_z(f - f_c) \qquad (2.62)$$

In words: The analytic signal has a spectrum identical to that of the complex envelope, but translated to $f = +f_c$ from $f = 0$.

Spectra of Finite-Duration Signals. All of the spectra in (2.55) to (2.62) apply to the infinite ensemble. In a simulation, typically only one signal of finite duration is available. An amplitude-scaled periodogram of that one signal is to be employed as an estimate for the ensemble power spectrum, while recognizing that estimates have statistical scatter.

Power estimation was performed on a signal segment of duration T_0, conceptually snipped out of a signal of infinite duration. While satisfactory for power measurements, drastic rectangular truncation of a signal is likely to have drastic effects upon the spectrum, particularly if the original signal is bandlimited. A less violent expedient is needed.

Techniques for spectral estimation will be treated again in Chapter 18. For now, consider a signal whose baseband components each contain L PAM pulses spaced at intervals T. This model does not truncate pulses and so should have a less drastic effect on the spectrum. Specifically, an L-pulse model is not the same as a T_0-truncated model, not even if $T_0 = LT$, unless each pulse is time limited to T (pulses do not overlap) and the T_0 interval starts and ends exactly on a pulse boundary.

If the pulses are bandlimited, each pulse—and therefore the overall signal—has infinite duration. Methods for simulating infinite duration signals in finite time are an important topic to be covered in subsequent chapters. For now, represent the L-pulse passband signal as

$$s_L(t) = x_L(t) \cos 2\pi f_c t - y_L(t) \sin 2\pi f_c t \quad (2.63)$$

Since only L pulses are present, the signal can be Fourier transformed to

$$S_L(f) = \frac{1}{2} X_L(f - f_c) + \frac{1}{2} X_L(f + f_c) + \frac{j}{2} Y_L(f - f_c) - \frac{j}{2} Y_L(f + f_c) \quad (2.64)$$

Its scaled periodogram is

$$\begin{aligned}
|S_L(f)|^2 = & \frac{1}{4} \left[|X_L(f-f_c)|^2 + |X_L(f+f_c)|^2 + |Y_L(f-f_c)|^2 + |Y_L(f+f_c)|^2 \right] \\
& + \frac{1}{4} \left[X_L(f-f_c) X_L^*(f+f_c) + X_L(f+f_c) X_L^*(f-f_c) \right. \\
& \left. - Y_L(f-f_c) Y_L^*(f+f_c) - Y_L(f+f_c) Y_L^*(f-f_c) \right] \\
& + \frac{j}{4} \left[-X_L(f-f_c) Y_L^*(f-f_c) + X_L^*(f-f_c) Y_L(f-f_c) \right. \\
& \left. + X_L(f+f_c) Y_L^*(f+f_c) - X_L^*(f+f_c) Y_L(f+f_c) \right] \\
& + \frac{j}{4} \left[X_L(f-f_c) Y_L^*(f+f_c) - X_L^*(f-f_c) Y_L(f+f_c) \right. \\
& \left. - X_L(f+f_c) Y_L^*(f-f_c) + X_L^*(f+f_c) Y_L(f-f_c) \right] \quad (2.65)
\end{aligned}$$

The first set of brackets in (2.65) contains periodograms of the baseband signals, translated to $\pm f_c$. They correspond closely to the self-spectrum terms of (2.55). All following brackets contain cross terms; they correspond in some manner to the cross terms of (2.55).

The second set of brackets contains *cross-frequency, co-channel* terms, of the typical form $X_L(f-f_c) X_L^*(f+f_c)$. If X_L and Y_L are bandlimited to $|f| < f_c$, there is no frequency for which both factors of any term in the second set of brackets are simultaneously nonzero. Thus the second-bracketed terms all vanish for an adequately bandlimited signal. By the same argument, all cross-frequency, cross-channel terms in the fourth set of brackets also vanish for a bandlimited signal. Physical pulses are never perfectly bandlimited, but often the greatest portion of their spectrum will be concentrated below the carrier frequency f_c. In that event, the cross-frequency terms would be small compared to the co-frequency terms.

The third set of brackets contain cross-channel, co-frequency terms of typical form $X_L(f+f_c)Y_L^*(f+f_c)$. These two factors occupy the same spectral region, so their product does not vanish everywhere, not even for bandlimited signals. If the terms inside the third set of brackets are expanded and collected, the entire third term reduces to

$$\tfrac{1}{2}\{-\operatorname{Im}[X_L(f-f_c)]\operatorname{Re}[Y_L(f-f_c)]+\operatorname{Re}[X_L(f-f_c)]\operatorname{Im}[Y_L(f-f_c)] \\ +\operatorname{Im}[X_L(f+f_c)]\operatorname{Re}[Y_L(f+f_c)]-\operatorname{Re}[X_L(f+f_c)]\operatorname{Im}[Y_L(f+f_c)]\}$$

[Note that this works out to be pure real, despite the $j/4$ leading coefficient in (2.65).]

Often the ensemble cross-channel correlation is zero for all τ, in which case the ensemble cross-channel spectra vanish: $\overline{\Psi}_{yx}(f) \equiv 0 \equiv \overline{\Psi}_{xy}(f)$ for all f. But zero cross correlation in the ensemble does not assure zero cross correlation in an L-pulse sample function from the ensemble. There can be nonzero cross-channel, co-spectral terms in $|S_L(f)|^2$, even from an uncorrelated ensemble. These terms contribute to the statistical scatter of the periodogram.

Although (2.65) is expanded to show a large number of individual terms, an experimental periodogram exhibits only the sum of terms $|S_L(f)|^2$; the various types of terms cannot be separated out readily.

The complex envelope $z_L(t) = x_L(t) + jy_L(t)$ of the L-pulse signal has a Fourier transform of

$$Z_L(f) = X_L(f) + jY_L(f) \qquad (2.66)$$

Its periodogram is

$$|Z_L(f)|^2 = |X_L(f)|^2 + |Y_L(f)|^2 - jX_L(f)Y_L^*(f) + jX_L^*(f)Y_L(f) \qquad (2.67)$$

No cross-frequency terms appear in (2.67), neither co-channel nor cross-channel, irrespective of pulse bandwidth. Co-frequency, cross-channel terms are closely related to the real-signal co-frequency cross-channel terms in the third set of brackets of (2.65).

The analytic signal is $s_L^+(t) = z_L(t)e^{j2\pi f_c t}$, which has a periodogram

$$|S_L^+(f)|^2 = |Z_L(f-f_c)|^2 \qquad (2.68)$$

that is the same as the periodogram for the complex envelope, but shifted to $f = f_c$.

2.4 NOISE

Many kinds of disturbances can afflict a communications link; examples have been listed in Chapter 1. By far the most ubiquitous disturbance of all is additive Gaussian noise. Its properties are summarized in the following sections. Noises with other characteristics are mentioned briefly in Sec. 2.4.2.

2.4.1 Additive Gaussian Noise

Properties

Probability Density Function. Denote the noise* as $n(t)$. If the noise is Gaussian, it has a probability-density function (pdf) of the form

$$\text{pr}(n) = \frac{1}{\sqrt{2\pi}\sigma_n} \exp\left[-\frac{(n-\mu_n)^2}{2\sigma_n^2}\right] \tag{2.69}$$

where μ_n is the mean value of the noise and σ_n^2 is its variance. For the most part, only zero-mean noise is of interest in later chapters.

[Conventionally, probability density functions and discrete probabilities are assigned the notation $p(\cdot)$ and $P(\cdot)$, respectively. However, these letters have been preempted in this book for other quantities; to avoid confusion, probability densities and probabilities are denoted by pr{·} and Pr{·} instead. Also conventionally, the probability notation carries subscripts indicating the quantity whose probability is shown. Those subscripts will be omitted when no ambiguity is caused thereby, in the interest of reducing clutter.]

Assume that the noise is stationary; its pdf or other statistics do not depend upon time.

Spectral Density. Gaussian noise is fully described by its pdf and its second-order statistics—either its autocorrelation function $\psi_n(\tau)$ or, equivalently, the spectral density $\Psi_n(f)$. (Autocorrelation and spectral density are Fourier transforms of one another.) If $n(t)$ is measured in volts, then $\Psi_n(f)$ has units of (volts)^2/Hz.

It is often convenient to deal with white noise, defined by spectral density $\Psi_n(f) = N_0/2$ for all f (positive and negative), where N_0 is a constant. The autocorrelation function for white noise is $\psi_n(\tau) = \delta(\tau)N_0/2$.

White noise is a mathematical fiction that cannot exist in reality. Engineers consider noise to be effectively white if its spectral density is nearly constant over the bandwidth of interest. Assuming noise to be white, where that approximation is justified, greatly simplifies analytical effort.

---

*More rigorously: the noise random process $n(t)$. But the term *process* is reserved for operations on signals.

Additive Property. Let a receiver be presented with a signal $s(t)$ and additive noise $n(t)$. Then the composite signal-plus-noise $r(t)$ in the receiver is designated by

$$r(t) = s(t) + n(t) \tag{2.70}$$

The noise is almost always considered to be independent of the signal $s(t)$.

AWGN. Additive white Gaussian noise (with implied zero mean and independence of any accompanying signals) has been studied extensively. It is commonly abbreviated as AWGN in the communications literature.

Linear Invariance. Let Gaussian noise $n_i(t)$ be applied to a linear network, which delivers $n_o(t)$ in response. Then $n_o(t)$ is also Gaussian, albeit with altered variance and, perhaps, altered mean if the input mean is not zero.

Bandpass Noise

If the spectrum of Gaussian noise is concentrated about a frequency f_0, the noise wave can be expanded into in-phase and quadrature components according to

$$n(t) = n_c(t) \cos 2\pi f_0 t - n_s(t) \sin 2\pi f_0 t \tag{2.71}$$

which will be recognized as the same two-dimensional format as (2.18) for bandpass signals. If $n(t)$ is Gaussian with variance (i.e., "power") σ_n^2 and zero mean, then n_c and n_s are also Gaussian and each have variance σ_n^2 and zero mean.

Complex Envelope. The real representation of noise (2.71) has a corresponding complex envelope, defined by

$$z_n(t) = n_c(t) + j n_s(t) \tag{2.72}$$

Variance of the envelope is

$$E[|z_n(t)|^2] = E[n_c^2(t)] + E[n_s^2(t)] = 2\sigma_n^2 \tag{2.73}$$

That is, the complex envelope of the noise has double the power of the real representation, just as was found in (2.48) for signals. Therefore, signal-to-noise ratios are the same for real representations and for complex-envelope representations of signal and noise. The factors of 2 cancel in the ratio and so can be ignored by the simulationist when dealing with signal-to-noise ratios.

The power spectrum of the complex envelope can be found, by the same methods that led to (2.58), to be

$$\Psi_{zn}(f) = \Psi_{nc}(f) + \Psi_{ns}(f) + j[\Psi_{nsc}(f) - \Psi_{ncs}(f)] \tag{2.74}$$

where subscript zn applies to the entire complex noise envelope, subscripts nc and ns apply to the two individual quadrature components, and subscripts nsc and ncs apply to cross spectra between the quadrature components. Figure 2.14 illustrates how $\Psi_{zn}(f)$ is developed from $\Psi_n(f)$.

Noise Periodograms

The foregoing noise powers and spectra are averages, applicable only to the infinite ensemble of all realizations of the noise. In a physical setting or in a simulation, only one, finite-duration, particular realization of the noise is available from the ensemble. Designate it as $n_r(t)$. Because of the finite duration, the Fourier transform $N_r(f)$ exists. Fourier transformation is a linear operation, so $N_r(f)$ is also Gaussian and just as random as the time-domain noise representation $n_r(t)$. No two distinct realizations will have the same Fourier transform.

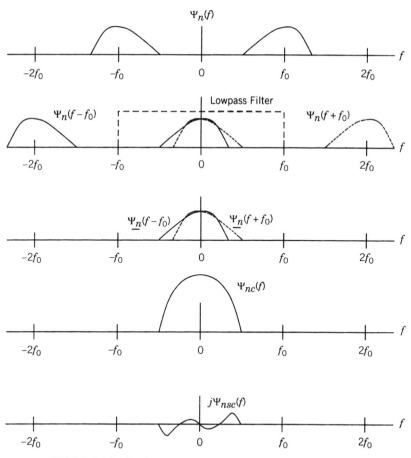

FIGURE 2.14 Quadrature components of bandpass Gaussian noise.

One could compute the noise periodogram $|N_r(f)|^2$, but it will bear little resemblance to the ensemble spectrum $\Psi_n(f)$. Periodograms of random noise are well known [2.23, p. 107] to be unsatisfactory approximations to the underlying power spectrum. In particular, if $n_r(t)$ is drawn from an ensemble with white noise spectrum $\Psi_n(f) = N_0/2$, its periodogram will not be even nearly constant with frequency. Examples are shown in Fig. 2.15. This topic will be reexamined in the light of simulation requirements when generation and use of noise are considered in later chapters.

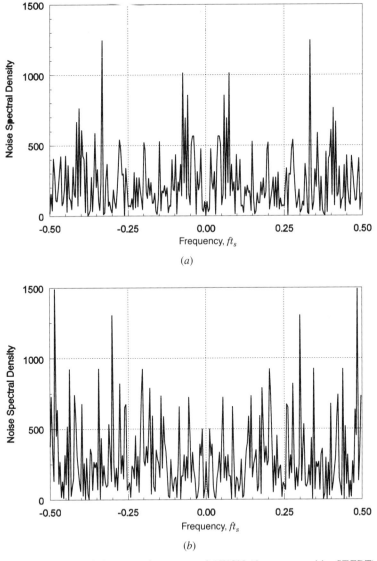

FIGURE 2.15 Periodograms of segments of AWGN (data generated by STÆDT).

2.4.2 Counterexamples: Non-AWGN

Despite the predominance of AWGN in the communications literature and despite its importance in so many real applications, the actual noise afflicting a particular system need be neither additive nor white nor Gaussian (nor zero-mean nor independent of the signal). Several counterexamples are presented briefly.

Processed Noise

If Gaussian noise is applied to a nonlinear device, the resulting output is not Gaussian. As an example, let zero-mean Gaussian noise $n(t)$ with variance σ_n^2 be passed through a square-law nonlinearity; output will be $n^2(t)$. The pdf of $n^2(t)$ is chi-squared, not Gaussian [2.23, Chap. 12]. Although the input has zero mean, the output has a mean value of σ_n^2. If $n(t)$ has a rectangular, lowpass spectrum confined to $|f| < B$, the spectrum of $n^2(t)$ has a triangular shape with a peak at $f = 0$ and extending over $|f| < 2B$.

Impulse Noise

Some communication links are subjected to impulse noise, in addition to AWGN. Transmission in the high-frequency radio band is a prominent example. Distant lightning strokes cause frequent, discrete, high-amplitude disturbances. Also, much human-made noise has an impulsive character. Impulse noise is additive, but it is neither Gaussian nor white. Its spectrum typically falls off with increasing frequency (although the spectrum is virtually constant over the passband of typical signals). Description of the noise in statistical terms is not nearly so compact as for Gaussian noise. Reference [2.24] contains further background on the subject.

Quantum Noise

In optical communication, a receiver essentially counts the individual photons present in a signal pulse. Effective communication is possible with just tens or hundreds of photons per pulse. The number of photons fluctuates randomly from pulse to pulse and the intervals between photon arrival times fluctuate within each pulse. Arrival times have a Poisson distribution. Consequently, photocurrent in a detector exhibits shot noise, an irregular sequence of discrete events. As signal intensity becomes large (large number of photons per pulse), the pdf of the shot noise approaches Gaussian, but it is distinctly non-Gaussian for low optical intensity [2.25]. Shot noise is embedded in the signal; if signal intensity increases, so does the noise. In this sense, shot noise is not additive, and certainly not independent of signal intensity.

Phase Noise

All oscillators exhibit random phase noise [2.26] that interferes with communications. Output of an oscillator with phase noise $\phi(t)$ can be represented by

$$v(t) = A \sin[\omega_c t + \phi(t)]$$

When an oscillator is employed to generate a carrier signal or to translate the frequency of a modulated signal within a transmitter or receiver, the phase noise of the oscillator is impressed upon the signal phase. Impressed noise is multiplicative, not additive. Increasing the signal strength cannot improve the signal-to-phase noise ratio.

Phase noise is nonstationary—its variance grows with time, somewhat in the fashion of a random walk. Strictly speaking, a nonstationary function cannot have a conventional power spectrum. Nonetheless, the "spectrum" of phase noise can be measured in the laboratory by spectrum analyzers; it often has the form

$$\Psi_\phi(f) = \frac{h_\nu}{f^\nu}$$

where h_ν is a scaling factor that depends on the exponent ν. Particular values of ν tend to prevail over comparatively wide frequency ranges. Typical values of the exponent for phase noise are $\nu \simeq 2$ and $\nu \simeq 3$. The former is caused by white noise in the oscillator circuit and the latter by flicker noise. A phase noise spectral density increases without bound toward low frequencies. The theoretical foundations of phase noise have not yet been firmly established and are still a subject of active research [2.27].

Relation to Simulation

Methods for simulation with AWGN are well established and are treated at length in subsequent chapters. Simulations of other kinds of noise often encounter difficulties and are rarely as straightforward as simulations of AWGN. This book concentrates on AWGN and gives only passing notice to non-AWGN.

2.5 KEY POINTS

2.5.1 Baseband Signals

A data signal is most often composed of a sequence of information-bearing *pulses*. The information consists of discrete numeric values of data *symbols*. Only *synchronous* transmission is considered, whereby the pulses are uniformly spaced at the *symbol interval T*. Idealized pulses are often regarded as *time-limited* or *bandlimited*. A time-limited pulse has finite extent in time but infinite extent in frequency. A bandlimited pulse has finite extent in frequency but infinite extent in time. No physical pulse can be strictly time- or bandlimited; these are approximations useful for analytical purposes.

PAM

The most widespread methods of data transmission employ pulse amplitude modulation (PAM) wherein the symbol value a_m is transmitted in the amplitude of the *m*th pulse of the sequence. All pulses in a PAM sequence have the same pulse shape $g(t)$. A PAM baseband signal $x(t)$ is represented by

$$x(t) = \sum_{m=-\infty}^{\infty} a_m g(t - mT)$$

Link Model. A PAM link can be modeled as a pulse transmitter plus a receiver that consists of a filter and a sampler. The sampler extracts one properly timed *strobe* sample per symbol pulse. Each strobe is applied to a *symbol detector* that decides the pulse amplitude and therefore the symbol value a_m associated with the pulse.

Intersymbol Interference. Pulses overlap unless they are strictly time limited to the symbol interval T. Such time limiting is not possible if there are any significant bandwidth restrictions in the channel; pulses almost always overlap after filtering in the receiver.

Uncontrolled overlap causes *intersymbol interference* (ISI) whereby the strobe for the mth symbol contains interference from other nearby symbols. It is possible to avoid ISI at the strobe instant t_0 if the pulse shape is

$$g(t_0 + mT) = 0 \quad \text{for} \quad m \neq 0$$

That is, the pulse tails go through nulls at the strobe times for all symbols except the one being detected. This *Nyquist shaping* is a goal sought in the designs of many communications systems. A cosine-rolloff spectrum is often used as a good analytic approximation for bandlimited Nyquist pulses.

Other Signal Formats

Pulse position modulation (PPM), pulse width modulation (PWM), and pulse waveform modulation (such as frequency-shift keying, FSK) are used instead of PAM in some links. These other formats often have PAM-like features, so that they can be constructed in part through the application of PAM techniques.

2.5.2 Passband Signals

Many transmission media do not support baseband signals. To circumvent this obstacle, it is common practice to modulate the baseband signal onto a sinusoidal carrier at a frequency f_c that is supported by the medium. The modulated signal is also known as a passband signal.

Modulation Formats

Amplitude modulation (AM) and angle modulation—either phase modulation (PM) or frequency modulation (FM)—are the fundamental forms of carrier modulations.

Amplitude modulation treats the baseband signal linearly; angle modulation is a nonlinear operation. Amplitude modulation preserves the spectrum of the baseband signal, just translating it to $\pm f_c$. Angle modulation spreads the spectrum of the baseband, besides translating it to $\pm f_c$.

Amplitude modulation has envelope fluctuations, leading to potential distortion in nonlinear amplifiers. Angle modulation has constant amplitude and so is unaffected by nonlinear amplifiers. Passband signals of interest usually do not have a discrete carrier component. The carrier is suppressed to maximize power efficiency.

QAM

Spectral efficiency of a passband signal is doubled by employing two equal-frequency carriers in phase quadrature, according to

$$s(t) = x(t) \cos 2\pi f_c t - y(t) \sin 2\pi f_c t$$

where $x(t)$ and $y(t)$ are two baseband signals. This quadrature amplitude modulation (QAM) representation of a passband signal has great theoretical and practical importance and is employed throughout the book. The representation can be applied to *all* passband signals, not just those that have been amplitude modulated.

The complex envelope of the QAM representation is

$$z(t) = x(t) + jy(t)$$

Passband signals are usually simulated by their complex envelopes and the carrier sinusoid is omitted, thereby greatly reducing the computation burden. The analytic signal

$$s^+(t) = z(t)e^{j2\pi f_c t}$$

is a frequency-shifted version of the complex envelope. The real signal is simply

$$s(t) = \text{Re}\,[s^+(t)]$$

where Re[·] means "real part of."

The complex envelope can be cast into polar coordinates as

$$z(t) = R(t)e^{j\theta(t)}$$

where

$$R(t) = \sqrt{x^2(t) + y^2(t)}, \qquad \theta(t) = \tan^{-1}\frac{y(t)}{x(t)}$$

and the reverse relations are

$$x(t) = R(t)\cos\theta(t), \qquad y(t) = R(t)\sin\theta(t)$$

It can be seen that an angle modulation signal with constant amplitude [$R(t)$ = constant] is representable in QAM format through these formulas. This property is crucial for simulations, since it allows all passband signals to be processed in much the same manner, even though modulation methods might differ greatly.

Passband Signal Categories

This book concentrates on balanced PAM–QAM signals which can be described geometrically by two-dimensional *constellations*. Each point of a constellation depicts one individual complex-symbol value. Constellations laid out on a square grid and phase-shift constellations disposed uniformly around a circle are widely employed.

Other signals are not PAM–QAM based and cannot be represented by two-dimensional constellations. Significant examples include single-sideband or vestigial-sideband modulation (SSB or VSB), frequency-shift keying (FSK), continuous-phase modulation (CPM), and minimum-shift keying (MSK). Nonetheless, the QAM representation of a passband signal can be applied to all of these examples.

2.5.3 Power, Energy, and Spectrum

A pulse has a well-defined energy. An infinite sequence of pulses has infinite energy and so is ordinarily described in terms of power instead. A data signal is *cyclostationary*; its second moment (a measure of power) is time dependent. A power meter would read the *time-averaged power*, the second moment averaged over one period (one symbol interval T).

The entire infinite ensemble of all infinite data sequences for a signal $x(t)$ has a time-averaged autocorrelation function $\overline{\psi}_x(\tau)$ whose Fourier transform $\overline{\Psi}_x(f)$ is the spectral density. This ensemble spectrum can be derived analytically, but with difficulty in many cases. It depends upon the pulse shape, the data statistics, and the symbol interval.

In a simulation, only a finite segment of one realization of the signal is available. Power and spectrum are *estimated* from the finite segment. If the data sequence were altered, a different estimate would be obtained. Estimates thus exhibit *statistical scatter*, a crucial feature that demands attention in planning and interpretation of simulations.

Power is estimated by measuring the energy of a signal segment and dividing by the time span of the segment. Spectrum is estimated by a *periodogram*: the magnitude squared of a signal's Fourier transform. These estimates are pursued in later chapters.

Passband Signals

The power spectrum of the complex envelope is concentrated around zero frequency, whereas the power spectrum of the real passband signal is concentrated around $\pm f_c$, the carrier frequency.

The complex envelope $z(t)$ has double the power found in its associated real signal $s(t)$.

2.5.4 Noise

AWGN

This book concentrates on additive white Gaussian noise (AWGN). Just as with signals, noise can be baseband or passband. If passband, it can be decomposed into the same quadrature-components format as a QAM signal and can be reduced to a similar complex-envelope representation. Power in the envelope is twice the power in the real noise. Since this same factor of 2 appeared in the power of the signal envelope, the signal-to-noise ratio is the same for real signals and noise and for complex signals and noise. A periodogram of a finite segment of noise is likely to furnish a poor approximation to the ensemble spectral density of the noise.

Other Noise Types

Although AWGN is nearly omnipresent, other types of noise arise in various circumstances. Examples include:

- Impulse noise (non-Gaussian; maybe nonwhite)
- Quantum (shot) noise (non-Gaussian; embedded in the signal)
- Phase noise (nonstationary; nonwhite; multiplicative)

Methods for simulation of AWGN are well established and covered thoroughly in the literature. Other types of noise often are more difficult to simulate; research on them is still active.

3

REPRESENTATION OF ANALOG FILTERS

All communications links contain filters; a typical link simulation will include several. A major portion of the computing burden in a simulation is expended on filters. Specification and preparation of filters is one of the most demanding tasks imposed on simulationists. A good understanding of filter representations is essential for proficient application of simulation.

This chapter is devoted to the analytical representation of *analog* filters. Computer representations of analog filters are introduced in Chapter 6. *Digital* filters are assuming greater prominence in practical communications links. Introduction of digital filters is delayed to Chapter 6 and later.

3.1 CATEGORIES OF FILTER REPRESENTATION

Analog filters can be modeled mathematically by transfer functions, impulse responses, or differential equations. Conventional analysis of filters has relied heavily on transfer functions because they are algebraic operations on the signal and therefore much easier for a human being to grasp. By contrast, a computer's notions of ease of understanding differ from those of humans; computer simulations make use of all three forms of representation (or, rather, their sampled equivalents).

3.1.1 Differential Equation Representation

Let the input signal to a filter be denoted $u(t)$ and the resulting output be denoted $v(t)$. Signals can be real or complex. If the filter is composed of a finite number of lumped elements (primarily resistors, capacitors, and inductors), its operations on the signals can be represented by an ordinary,* linear differential equation of the form

$$a_m u^{(m)} + a_{(m-1)} u^{(m-1)} + \cdots + a_1 u^{(1)} + a_0 u = b_n v^{(n)} + b_{n-1} v^{(n-1)} + \cdots + b_1 v^{(1)} + b_0 v \quad (3.1)$$

where the notation $x^{(k)}$ means the kth derivative $d^k x(t)/dt^k$.

(One differential equation of nth order can be decomposed into n simultaneous first-order differential equations. Such a decomposition is the basis for a *state variable* representation of a filter. State variable representations [3.3] have a valued role in electrical engineering but have not been widely applied in simulation of communications links. They are not treated further here.)

The coefficients $\{a\}$ and $\{b\}$ do not depend upon the signals u and v in any way. A filter with signal-dependent coefficients would be nonlinear and is far more difficult to analyze than a linear filter; only linear filters are considered in this book.

For now, assume that the coefficients are all constant, so that the filter is time invariant. Many communications systems need adaptive filters (e.g., as equalizers), in which the filter coefficients are changed according to a control algorithm. Such a filter is said to be time varying. Adaptive filters can be simulated by extensions from methods to be described for time-invariant filters, but adaptive filters, as such, are not considered in this book; only time-invariant filters are included. (Typical adaptive filters for communications links are not likely to be implemented with lumped elements—but that is another matter not considered in this book.)

In familiar, physical filters for real signals, the coefficients are all real. However, when dealing with signals that are complex in the time domain (i.e., complex envelopes, I–Q baseband signals, analytic signals) it is possible to have filters with complex coefficients for their differential equations. A filter with complex coefficients can always be decomposed into several subfilters that each have only real coefficients. A decomposition will be introduced presently.

Engineers rarely calculate the filter output $v(t)$ by solving the differential equation (3.1) analytically. Much simpler, better-organized methods (e.g., Laplace transforms, as below) are far preferable for traditional analysis. But computers are different; they provide numerical rather than analytical results. One important numerical method of digital filtering used in simulations implements the difference-equation equivalent of the differential equation. Difference-equation filters are given comprehensive attention in later chapters.

3.1.2 Rational Transfer Functions

If $u(t)$ and $v(t)$ are causal (physical signals are always causal), they can be Laplace transformed. Their transforms are denoted $U(s)$ and $V(s)$, respectively, where s is the

---

*If the filter included distributed elements, partial differential equations would be required. Only ordinary differential equations are considered in this work.

transform complex variable. The differential equation can be Laplace transformed as follows. First, assume for now that all initial conditions are zero. (Initial conditions cannot be ignored in simulations; they will be reintroduced at a suitable place in the narrative.) Then use the relation

$$\mathscr{L}\left[\frac{d^k x(t)}{dt^k}\right] = s^k X(s)$$

where $\mathscr{L}[\cdot]$ means "Laplace transform of" and $X(s) = \mathscr{L}[x(t)]$. The transformed differential equation becomes

$$[a_m s^m + a_{m-1} s^{m-1} + \cdots + a_1 s + a_0]U(s)$$
$$= [b_n s^n + b_{n-1} s^{n-1} + \cdots + b_1 s + b_0]V(s) \quad (3.1a)$$

Collect terms to obtain the *transfer function*

$$H(s) \triangleq \frac{V(s)}{U(s)} = \frac{a_m s^m + a_{m-1} s^{m-1} + \cdots + a_1 s + a_0}{b_n s^n + b_{n-1} s^{n-1} + \cdots + b_1 s + b_0} \quad (3.2)$$

The transfer function depends only on the filter coefficients, not the input or output signals. In the Laplace transform domain, (3.2) can be rewritten as

$$V(s) = U(s)H(s) \quad (3.3)$$

which is far easier to solve analytically than the differential equation (3.1).

Properties of Rational Transfer Functions

If the filter is linear, time invariant, and lumped, its transfer function will be rational, consisting of the ratio of two polynomials in the Laplace variable s, as in (3.2). The degree m of the numerator cannot exceed the degree n of the denominator for a physically realizable filter.

Each polynomial in $H(s)$ can be factored into the product of its roots and rewritten as

$$H(s) = H_0 \frac{(s - z_1)(s - z_2) \cdots (s - z_m)}{(s - p_1)(s - p_2) \cdots (s - p_n)} \quad (3.4)$$

where the roots of the numerator $(z_1 \cdots z_m)$ are called the *zeros* of the transfer function and the roots of the denominator $(p_1 \cdots p_n)$ are called the *poles*. The poles and zeros, together with the multiplier H_0, define the transfer function completely and uniquely.

If the coefficients of the transfer function in (3.2) are real, the poles and zeros must either be real or must occur in complex-conjugate pairs.* No such restriction applies if coefficients are complex. A stable filter has all of its poles in the left half of the s-plane.

Partitioning of Rational Transfer Functions

Although the coefficient representation of (3.2) is perfectly correct, difficulties are encountered with numerical precision when working directly with coefficients of higher-order filters ($n > 3$ or 4). Difficulties arise in design computations and in simulations. One common expedient for evading numerical problems is to partition the larger transfer function into a cascade of filters of lower degree such that

$$H(s) = H_1(s)H_2(s)\cdots H_k(s) \qquad (3.5)$$

Requirements on precision of coefficients are much less demanding for polynomials of lower degree. For that reason, all rational transfer functions in many programs (STÆDT included) are partitioned into cascaded *biquadratic sections*, each section involving polynomials with no higher than second degree. All rational transfer functions, of any degree, can be built up as the cascade of biquadratic sections.

A biquadratic can be expressed in pole–zero format, as in (3.4), or can be stated in coefficient format, similar to (3.2). It will be convenient to specify a biquadratic in a coefficient format of

$$H(s) = H_0 \frac{a_2 s^2 + a_1 s + a_0}{b_2 s^2 + b_1 s + b_0} \qquad (3.6)$$

This format is overspecified; two of the seven coefficients can be set to unity by dividing or multiplying through appropriately. Overspecification is convenient in a computer program, since a user can enter coefficient data in any acceptable form and the computer then handles the labor of conversion to standard forms. Methods for entering filter characteristics into the computer are treated at length in Chapter 17.

Relation to Fourier Transform

Any transfer function $H(s)$ in the Laplace variable s can also be expressed equivalently as a transfer function $H(f)$ in the Fourier variable f. The relation is

$$H(f) = H(s)|_{s=j2\pi f} \qquad (3.7)$$

(Mathematicians might object to the nonrigorous notation for the substitution, but engineers will welcome the brevity.) Although it is convenient for human beings to

---

*Statements in this chapter regarding properties of rational filters are based upon [3.2]. Statements pertaining to nonrational filters are based on the theory of functions of complex variables [2.3, 2.4].

express rational transfer functions in Laplace notation, transform-domain computer operations actually employ the Fourier transform.

3.1.3 Nonrational Transfer Functions

Most physically realizable analog filters have a rational transfer function that can be defined through Laplace transforms. However, nonrealizable filters often must be simulated: for example, bandlimited filters. Response of a bandlimited filter is noncausal (response precedes application of the stimulus), so the filter is nonrealizable, and furthermore, its operation cannot be analyzed with Laplace transforms. Nonetheless, such a filter can often be analyzed with Fourier transforms and it has a perfectly good transfer function $H(f)$. For a pertinent example, refer to the bandlimited, Nyquist, cosine-rolloff filters introduced in Chapter 2.

The transfer function of a bandlimited filter will not be rational; it cannot be expressed as the ratio of two polynomials in s (or in f either, for that matter). A nonrational transfer function does not have an ordinary differential equation of the form (3.1) associated with it.

3.1.4 Impulse Response

Designate the Fourier transform of the input to a filter as $U(f)$ and the Fourier transform of the output as $V(f)$. Then the relation between input and output for a transfer function $H(f)$ is

$$V(f) = U(f)H(f) \tag{3.3a}$$

which is really the same as (3.3). The only difference here is that $H(f)$ is no longer required to be rational. Response in the time domain is found by taking the inverse Fourier transform

$$v(t) = \int_{-\infty}^{\infty} U(f)H(f)e^{j2\pi ft}\,df \tag{3.3b}$$

Let the input $u(t)$ to a filter be an impulse function $\delta(t)$, so that $U(f) = 1$. Then the Fourier transform of the output is

$$V(f) = H(f)$$

Taking the inverse Fourier transform gives the *impulse response*

$$h(t) \triangleq \int_{-\infty}^{\infty} H(f)e^{j2\pi ft}\,df \tag{3.8}$$

of the filter whose transfer function is $H(f)$. Impulse response and transfer function contain exactly the same information about the filter. They are just different ways of expressing that information: one way in the time domain and the other in the frequency domain.

A filter is said to be *causal* if its impulse response $h(t)$ is causal. A filter is said to be *real* if its impulse response is real. Single-input, single-output filters of real signals are necessarily real. Complex filters—those that have a complex $h(t)$—work with complex signals.

Convolution Integral

How can the filtering process be described entirely in the time domain? That is, given the impulse response $h(t)$ and input $u(t)$, calculate $v(t)$ without using Fourier transforms. Or, stated differently, what is the time-domain equivalent of (3.3a)?

The inverse Fourier transform of (3.3a) is

$$v(t) = \int_{-\infty}^{\infty} U(f)H(f)e^{j2\pi ft}\,df$$

$$= \int_{-\infty}^{\infty} U(f) \int_{-\infty}^{\infty} h(\tau)e^{-j2\pi f\tau}\,d\tau\, e^{j2\pi ft}\,df$$

$$= \int_{-\infty}^{\infty} h(\tau) \int_{-\infty}^{\infty} U(f)e^{j2\pi f(t-\tau)}\,df\,d\tau$$

$$= \int_{-\infty}^{\infty} h(\tau)u(t-\tau)\,d\tau \qquad (3.9)$$

where the last line is the *convolution integral*, which may also be written as

$$v(t) = \int_{-\infty}^{\infty} u(\tau)h(t-\tau)\,d\tau \qquad (3.9a)$$

Either way, convolution will be denoted by the $\otimes$ symbol:

$$v(t) = u(t) \otimes h(t) \qquad (3.10)$$

Although engineers rarely figure convolutions analytically [transform methods via (3.3) or (3.3a) are almost always simpler], convolution is sometimes the most direct method of filtering in a digital computer, particularly if the impulse response $h(t)$ is time limited. Convolution is an important tool for simulation and is explored further in later chapters.

The transfer function $H(f)$ corresponding to a strictly time-limited impulse response $h(t)$ is not rational; it cannot be expressed in the form (3.2) or (3.4), it does not arise as the transform of a differential equation of the form (3.1), nor can it be expressed as the cascade of biquadratic sections of the form of (3.6).

A time limited continuous-time impulse response $h(t)$ may be realizable as a tapped delay-line filter (transversal filter). A discrete-time impulse response can approximate a continuous-time response. Any discrete-time finite-duration impulse response can be realized as a transversal filter. Therefore, transversal filters are the primary structure for direct simulation of convolution; they will be examined again when implementation of simulation filters is considered.

(Any discrete-time impulse response can also be realized as a lattice filter [1.12, Chap. 7]. Many DSP tasks fruitfully employ lattice filters, but they have not had widespread application in communications links or their simulation. Lattice filters are omitted from further consideration in this book.)

3.1.5 Summary of Categories

Three different methods of filter representation have been identified, and three corresponding methods of computer simulation have been implied:

1 *Transfer-Function Representation.* Simulation employs the Fourier transform $H(f)$ via equation (3.3a).

2 *Impulse-Response Representation.* Simulation employs the convolution integral via equation (3.9).

3 *Differential-Equation Representation.* Simulation employs the differential equation (3.1).

All of the equations above are for time-continuous signals and filters. Simulation on a digital computer actually employs discrete-time approximations to each of them. Discrete-time representations are introduced in Chapter 6.

3.2 FILTERS FOR COMPLEX SIGNALS

Simulations make extensive use of complex signals such as analytic signals, complex envelopes, and *I–Q* baseband signals. (As will be seen later, it is not always easy—or necessary—to distinguish between these signal types.) In many circumstances, complex signals are modeled in place of real passband signals so as to avoid the unnecessary burden of computing the fine structure of a high-frequency carrier. In other circumstances, the apparatus being simulated actually generates complex signals.

A conventional filter, operating on a real passband signal, also necessarily operates on the signal's complex envelope. In many instances, one's only interest lies in the effect of the filter on the envelope; the total effect on the real signal is immaterial. All that is needed for simulation is an *envelope filter*, a process that models the effect of the real filter on the envelope of the real signal. Input to the envelope filter would be just the complex envelope of the real signal applied to a real filter, and output of the envelope filter would be the complex envelope of the real output from the real filter.

One purpose of this section is to show how envelope filters can be modeled and how their characteristics can be derived from those of the real filter. Another purpose is to expand the concept of envelope filter to encompass all filters on complex signals.

3.2.1 Bandpass Filters

Consider a bandpass filter with impulse response $h(t)$ and transfer function $H(f)$, where $h(t)$ and $H(f)$ are a Fourier pair. A filter is *bandpass* if $H(f)$ is concentrated in the vicinity of a frequency other than zero. Impulse response is real and will be assumed to be causal for purposes of this discussion.

Postulate the existence of a complex representation of the impulse response in a form resembling an analytic signal [see (2.20)] such that

$$h(t) = 2\,\mathrm{Re}[h_L(t;f_0)e^{j2\pi f_0 t}] \qquad (3.11)$$

where $h_L(t;f_0)$ is a complex function whose form is to be determined.

The analytic representation is referenced to an arbitrary frequency f_0. Any frequency whatever could be selected; a different $h_L(t;f_0)$ would arise for each different frequency. There is no implication that $H(f)$ has any symmetry about f_0 nor that f_0 must somehow be a "center" frequency.

It is usually convenient, but not necessary, to select a frequency in the region where $H(f)$ is concentrated. In that case the Fourier transform of $h_L(t;f_0)$ will be concentrated about zero frequency and the subscript L can be regarded as implying "lowpass."

In later sections, f_0 typically will be selected equal to the carrier frequency f_c of a passband signal applied to the filter. Bear in mind, though, that f_0 is a mathematical property of the filter model, while f_c is a property of the signal. The two are quite distinct and need not be equal. (*Example:* An interfering signal at an adjacent frequency does not have $f_c = f_0$.)

The factor of 2 in (3.11) is included because of foreknowledge of the ultimate outcome; it will be absorbed in the course of the following development. Decompose $h_L(t;f_0)$ into its rectangular components according to

$$h_L(t;f_0) = h_I(t;f_0) + jh_Q(t;f_0) \qquad (3.12)$$

where $h_I(t;f_0)$ and $h_Q(t;f_0)$ are both real. Henceforth, the explicit inclusion of f_0 in the arguments of the functions may be dropped, for brevity of notation. The reader should keep in mind, though, that h_L, h_I, h_Q, and every function derived from them are all dependent on the choice of f_0.

Substituting the rectangular form (3.12) into the analytic-signal format of the impulse response in (3.11) and performing some manipulations yields

$$h(t) = 2h_I(t) \cos 2\pi f_0 t - 2h_Q(t) \sin 2\pi f_0 t$$

$$= 2\sqrt{h_I^2(t) + h_Q^2(t)} \; \cos\left[2\pi f_0 t + \tan^{-1} \frac{h_Q(t)}{h_I(t)}\right] \tag{3.13}$$

for the rectangular- and polar-coordinate representations of the real impulse response $h(t)$, in terms of the envelope-filter components h_I and h_Q. Observe from the polar form that the real impulse response exhibits combined amplitude and phase modulation on a carrier at frequency f_0.

Let the Fourier transforms of $h_I(t)$ and $h_Q(t)$ be $H_I(f)$ and $H_Q(f)$, respectively. [Note that H_I and H_Q, in general, are both complex; they are not the real and imaginary parts of $H(f)$.] Take the Fourier transform of $h(t)$ in the rectangular format from (3.13) to obtain, after some algebra,

$$H(f) = H_I(f - f_0) + H_I(f + f_0) + jH_Q(f - f_0) - jH_Q(f + f_0) \tag{3.14}$$

Notice that the factor of 2 introduced in (3.11) and also appearing in (3.13) has been absorbed in passage to the frequency domain.

Equations (3.13) and (3.14) give expressions relating the real filter to the components of a complex representation. Given h_I and h_Q (or H_I and H_Q) and f_0, it is a simple matter to determine $h(t)$ or $H(f)$. To proceed in the reverse direction—to find suitable h_I and h_Q from $h(t)$ [or H_I and H_Q from $H(f)$]—is more difficult. For now, just accept the existence of the complex representation; its realization will be treated in later paragraphs.

3.2.2 Signal Transmission Through a Bandpass Filter

Let a real passband signal

$$s_i(t) = x_i(t) \cos 2\pi f_c t - y_i(t) \sin 2\pi f_c t \tag{3.15}$$

with carrier frequency f_c be applied to a real filter whose impulse response is $h(t)$ and transfer function is $H(f)$. (Subscript i indicates input; subscript o indicates output.)

The Fourier transform of the input is

$$S_i(f) = \tfrac{1}{2}[X_i(f - f_c) + X_i(f + f_c)] + j\tfrac{1}{2}[Y_i(f - f_c) - Y_i(f + f_c)] \tag{3.16}$$

where $X_i(f)$ and $Y_i(f)$ are the Fourier transforms of the baseband components $x_i(t)$ and $y_i(t)$.

Filter output can be expressed by the frequency-domain product

$$S_o(f) = S_i(f)H(f) \tag{3.17}$$

Use the complex-component representation (3.14) for $H(f)$, and set the reference frequency $f_0 = f_c$. Multiplying out (3.16) by (3.14) gives a product with 16 terms:

$$\begin{aligned} S_o(f) = \tfrac{1}{2}[&X_i(f-f_c)H_I(f-f_c) + X_i(f+f_c)H_I(f+f_c) \\ +&jY_i(f-f_c)H_I(f-f_c) - jY_i(f+f_c)H_I(f+f_c) \\ +&jX_i(f-f_c)H_Q(f-f_c) - jX_i(f+f_c)H_Q(f+f_c) \\ -&Y_i(f-f_c)H_Q(f-f_c) - Y_i(f+f_c)H_Q(f+f_c)] \\ + \tfrac{1}{2}[&X_i(f-f_c)H_I(f+f_c) + X_i(f+f_c)H_I(f-f_c) \\ +&jY_i(f-f_c)H_I(f+f_c) - jY_i(f+f_c)H_I(f-f_c) \\ -&jX_i(f-f_c)H_Q(f+f_c) + jX_i(f+f_c)H_Q(f-f_c) \\ +&Y_i(f-f_c)H_Q(f+f_c) + Y_i(f+f_c)H_Q(f-f_c)] \end{aligned} \quad (3.18)$$

Terms inside the first set of brackets [e.g., $X_i(f-f_c)H_I(f-f_c)$] are *co-spectral*. That is, the filter and signal factors in each term are concentrated in the same region of the spectrum. Terms inside the second set of brackets [e.g., $X_i(f-f_c)H_I(f+f_c)$] are *cross-spectral*; the factors in each term are concentrated at two different regions of the spectrum, separated by $2f_c$.

To simplify the model, impose a weak bandlimiting constraint:

$$\begin{aligned} &\text{For } |f| > f_c: \\ &X_i(f) = 0, \quad Y_i(f) = 0 \\ &H_I(f) = 0, \quad H_Q(f) = 0 \end{aligned} \quad (3.19)$$

[The constraint on H_I and H_Q is equivalent to $H(f) = 0$ for $|f| > 2f_c$.]

The constraint is an approximation when physical signals and filters are involved; no realizable signal or filter can be strictly bandlimited. Nonetheless, the approximation is often very good and can be applied with negligible error in the simulation results. Sometimes the carrier frequency f_c is so small compared to signal and filter bandwidths that (3.19) is not a good approximation. Then a user should reassess the situation and perhaps employ real models for the signal and filter, instead of complex models. If (3.19) is valid, all of the cross-spectral terms in (3.18) are zero for all f and only the co-spectral terms in the first set of brackets remain. Errors caused by neglecting the cross-spectral terms are analyzed in [2.1, Secs. 7-2 and 7-4].

Define filter output components by the shorthand notation

$$\begin{aligned} X_I(f) &= X_i(f)H_I(f) = F[x_I(t)] = F[x_i(t) \otimes h_I(t)] \\ Y_I(f) &= Y_i(f)H_I(f) = F[y_I(t)] = F[y_i(t) \otimes h_I(t)] \\ X_Q(f) &= X_i(f)H_Q(f) = F[x_Q(t)] = F[x_i(t) \otimes h_Q(t)] \\ Y_Q(f) &= Y_i(f)H_Q(f) = F[y_Q(t)] = F[y_i(t) \otimes h_Q(t)] \end{aligned} \quad (3.20)$$

where $F[\cdot]$ means "Fourier transform of." Taking inverse Fourier transforms of the co-spectral terms of (3.18) line by line leads to

$$s_o(t) = [x_I(t) - y_Q(t)]\cos 2\pi f_c t - [y_I(t) + x_Q(t)]\sin 2\pi f_c t \qquad (3.21)$$

Comparing this result to the definition of complex envelope in (2.19) and (2.18) reveals that the filter output $s_o(t)$ has a complex envelope

$$z_o(t) = x_o(t) + jy_o(t) = [x_I(t) - y_Q(t)] + j[y_I(t) + x_Q(t)] \qquad (3.22)$$

Expand the envelope components according to the output definitions of (3.20) to obtain the time-domain representation

$$\begin{aligned} x_o(t) &= x_i(t) \otimes h_I(t) - y_i(t) \otimes h_Q(t) \\ y_o(t) &= y_i(t) \otimes h_I(t) + x_i(t) \otimes h_Q(t) \end{aligned} \qquad (3.23a)$$

or the frequency-domain representation

$$\begin{aligned} X_o(f) &= X_i(f)H_I(f) - Y_i(f)H_Q(f) \\ Y_o(f) &= Y_i(f)H_I(f) + X_i(f)H_Q(f) \end{aligned} \qquad (3.23b)$$

Compact Notation

Equations (3.23a,b) can be written more compactly. Equation (3.12) defines $h_L(t)$ and implies that $H_L(f) = H_I(f) + jH_Q(f)$, while (2.19) defines the envelope $z(t)$ and implies that $Z(f) = X(f) + jY(f)$. With a little algebra, equations (3.23) reduce to

$$\begin{aligned} Z_o(f) &= Z_i(f)H_L(f) \\ z_o(t) &= z_i(t) \otimes h_L(t) \end{aligned} \qquad (3.24)$$

Thus the envelope of the output signal is found by operation of the envelope filter on the envelope of the input signal. There is no need to bring the high-frequency carrier into the calculations. The frequency-domain version of (3.24) can be just as simple as it appears; the complex quantity $Z_i(f)$ is multiplied by the complex quantity $H_L(f)$ at every frequency f. There is no need to separate the envelope and filter into quadrature components for those calculations. On the other hand, the time-domain version of (3.24) conceals its intricacy. Calculation of the complex output envelope in essence requires four real convolutions or differential equations (difference equations in the computer). Further information is provided in Sec. 3.3.2 and Chapter 6.

3.2.3 Filter Structure

The filtered-signal equations of (3.23) can be represented by the block diagram of Fig. 3.1. Each line in the diagram represents a real signal x or y; each box represents a real filter on a real signal. The structure contains in-phase boxes [transfer function

82 ☐ REPRESENTATION OF ANALOG FILTERS

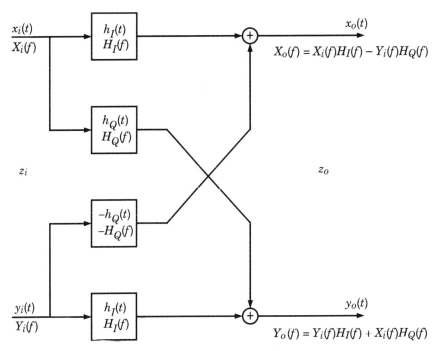

FIGURE 3.1 Structure of complex filter.

$H_I(f)$] that process x input to x output and y input to y output. It also contains quadrature boxes [transfer function $H_Q(f)$] that cross-couple x input to y output, and y input to x output. Transfer functions and impulse responses are indicated on each box.

If the model is derived from a real filter [with real impulse response $h(t)$], the structure will be symmetric; the two in-phase boxes will have the same transfer function, and the two quadrature boxes will have the same transfer function but with opposite signs. There is no inherent reason why an asymmetric structure could not be implemented; they just do not arise from modeling of real filters. No examples of asymmetric structures are included in this book.

The complex-filter structure of Fig. 3.1 affords valuable conceptual insight but is unnecessary for simulations. The most useful models of complex filters are not decomposed into the structure of Fig. 3.1; entirely different conversion methods are applied instead, as explained in the next section.

3.3 CONVERSION TO COMPLEX FILTERS

Task statement: Start with a real transfer function $H(f)$ [equivalently, $H(s)$]. Derive the complex-envelope transfer function $H_L(f,f_0) = H_I(f) + jH_Q(f)$ corresponding to $H(f)$ and the specific conversion frequency $f_0 = \omega_0/2\pi$. The objective is to convert $H(f)$ with real coefficients to $H_L(f,f_0)$ with complex coefficients. Decomposition into $H_I(f)$ and $H_Q(f)$ will be shown in the development of $H_L(f,f_0)$, but these components are not explicitly visible in the simplest models. Thus the aim is *conversion*, not *decomposition*.

Although $H(f)$ is uniquely specified once $H_I(f)$ and $H_Q(f)$ are known [see (3.14)], there is no single decomposition from $H(f)$ into unique quadrature components. Consequently, there is no unique conversion from $H(f)$ to $H_L(f,f_0)$. Many different conversions are possible—some more useful than others. Two conversion methods are described below, one for frequency domain and the other for time domain filters. Both are included in the STÆDT program library.

3.3.1 Frequency-Domain Conversion

Split the real transfer function $H(f)$ between positive and negative frequencies according to

$$H(f) = H_+(f) + H_-(f) \tag{3.25}$$

where

$$\begin{aligned} H_+(f) &= H(f); & H_-(f) &= 0, & f \geq 0 \\ H_+(f) &= 0; & H_-(f) &= H(f), & f < 0 \end{aligned} \tag{3.26}$$

Slide H_+ and H_- to the left and right, respectively, by an increment f_0 to produce the lowpass functions

$$H_{L+}(f) = H_+(f+f_0), \qquad H_{L-}(f) = H_-(f-f_0) \tag{3.27}$$

Figure 3.2 illustrates the terminology and operations.

Because the impulse response $h(t)$ is real, the transfer function $H(f)$ is conjugate symmetric; that is,

$$H(f) = H^*(-f) \tag{3.28}$$

where * indicates a complex conjugate. In consequence,

$$H_{L+}(f) = H_{L-}^*(-f) \tag{3.29}$$

84 □ REPRESENTATION OF ANALOG FILTERS

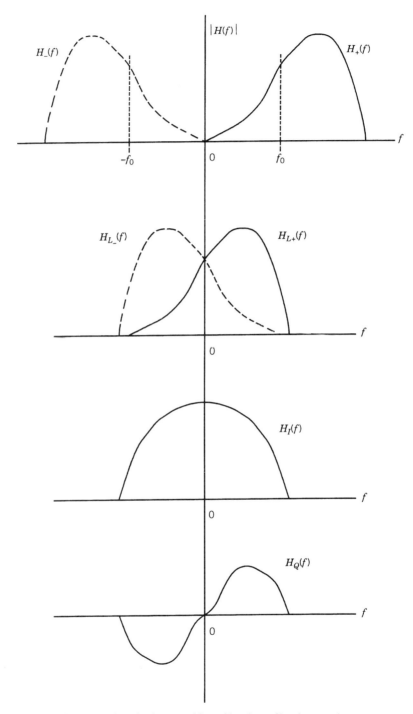

FIGURE 3.2 Frequency-domain decomposition of bandpass filter into quadrature components.

Now define the quadrature components in terms of sums and differences of H_{L+} and H_{L-}.

$$H_I(f) \triangleq \frac{H_{L+}(f) + H_{L-}(f)}{2} = \frac{H_{L+}(f) + H_{L+}^*(-f)}{2}$$

$$H_Q(f) \triangleq \frac{H_{L+}(f) - H_{L-}(f)}{2j} = \frac{H_{L+}(f) - H_{L+}^*(-f)}{2j} \quad (3.30)$$

The second equality in each line makes use of (3.28).

These specific definitions are not unique; they were chosen because they lead to a particularly simple representation and have intuitively satisfying properties. First, conjugation of (3.30) demonstrates that H_I and H_Q are conjugate symmetric; that is,

$$H_I^*(-f) = H_I(f), H_Q^*(-f) = H_Q(f) \quad (3.31)$$

Consequently, their inverse Fourier transforms $h_I(t)$ and $h_Q(t)$ are real, so the filter can be represented by the structure of Fig. 3.1.

Second, substituting the definition of $H_L(f)$ into (3.30) and applying some algebra gives

$$H_L(f) = H_I(f) + jH_Q(f) = H_{L+}(f) \quad (3.32)$$

Experimental measurements of frequency response of a filter usually provide $H_+(f)$ (most likely in polar format), from which $H_{L+}(f)$ is easily derived. Thus frequency-domain simulations can employ measured data nearly directly, with only relatively simple conditioning being needed. Data conditioning is an important topic to be examined in Chapter 17.

Notice the extreme simplicity of this procedure; $H_L(f)$ is obtained almost trivially by (1) discarding the negative-frequency portion of $H(f)$, and (2) sliding the positive-frequency portion to the left by an amount f_0. That's all there is to it. There is no need whatever to decompose $H_L(f)$ into quadrature components. Equation (3.30) shows how the quadrature components can be calculated, but they are not needed for the simulation; decomposition would only make the model more elaborate.

Third, if $H_+(f)$ is conjugate symmetric about f_0, then $H_{L+}(f)$ will be conjugate symmetric about $f = 0$. In that case, $H_{L+}(f) = H_{L+}^*(-f)$, whereupon from (3.30) one finds that $H_Q(f) = 0$ and there is no cross-coupling between x and y. Cross-coupling twists the x and y signals together in a complicated manner; only an opposite cross-coupling (complex equalization) can untwist them. In most signal formats, such cross-coupling is regarded as distortion and a designer attempts to avoid it. Consequently, filter symmetry is often specified carefully so as to hold the cross-coupling to low levels. No physical filter can have perfect symmetry about a nonzero frequency; simulation is invaluable in determining how much asymmetry can be tolerated.

3.3.2 Time-Domain Conversion

The foregoing is a simple and convenient method of representing envelope filtering in the frequency domain. Requirements also arise to perform operations in the time domain. One might suppose that a time-domain representation could be found by applying the inverse Fourier transform.

Unsatisfactory Methods

Invariably, though, the time functions $h_I(t)$, $h_Q(t)$, and $h_L(t)$ obtained by inverse-transforming (3.30) or (3.32) will be noncausal, even if $h(t)$ itself is causal. Reasons may be found in [2.1, Sec. 10-5] and [3.1]. Noncausal filters can be simulated in the frequency domain (by techniques to be shown in Chapter 6) but not in the time domain. A different approach is needed for complex decomposition in the time domain.

One expedient might be to multiply the noncausal time-domain functions by a finite-duration window and to apply a delay so that the windowed, shifted impulse responses are causal. Signals can then be filtered by convolution. Results can approximate the action of the true filter as closely as desired by selecting the window duration sufficiently long and by judicious selection of window shape.

This method suffers from two impediments: (1) a user must engage in a filter-approximation procedure each time the method is employed, and (2) the approximate transfer function will not be rational [not even if the original $H(f)$ itself is rational], which means that the differential-equation model (approximated by a difference equation in the computer) cannot be applied easily. Convolution would be the only feasible method of time-domain simulation—an awkward requirement if the window must be of long duration.

Another expedient might be to decompose $H(f)$ into differently defined $H_+(f)$ and $H_-(f)$ such that the resulting $h_I(t)$ and $h_Q(t)$ were causal. Alternative decomposition of this kind could not be found in the literature but may be worthy of further research. From [3.1] it can be concluded that a necessary condition for causal quadrature components is that $H_+(f)$ and $H_-(f)$ would have to overlap; each would have to extend over the entire infinite frequency axis. The extreme simplicity of (3.26) and (3.32) would be lost.

Rational-Function Conversion

A conversion technique inspired by [3.2] and suitable for time-domain operations is described in Appendix 3A. When the conversion is applied to causal filters with rational transfer functions, the resulting quadrature components are also causal filters with rational transfer functions and so can be modeled by differential equations (difference equations in the digital computer), not just convolutions. Once again, decomposition into distinct quadrature components would only lead to unnecessarily elaborate models. A simpler model directly gives $H_L(s, \omega_0)$ with complex coefficients. See Appendix 3A for details.

3.4 KEY POINTS

Filters are ubiquitous in communications links and in simulations thereof. This chapter deals with the representation of analog filters, placing emphasis on simulation-related issues.

3.4.1 Filter Categories

Filters can be represented by means of transfer functions, by impulse responses, or by differential equations. Electrical engineers have concentrated on the transfer-function representation (via Laplace or Fourier transforms) because analysis is much easier with the consequent algebraic formulation. A computer is numeric, however, and can deal with the other two representations just as well.

Any linear filter is fully characterized by its impulse response, denoted $h(t)$. Only linear filters are considered in this work. For any time-invariant impulse response of a linear filter, there exists a transfer function $H(f)$ that is the Fourier transform of $h(t)$ [or $H(s)$ that is the Laplace transform of $h(t)$].

A time-varying filter does not have a conventional transfer function. Time-varying filters are found in data-communication systems as interpolators (treated in Chapter 10) or as adaptive filters such as equalizers and echo cancelers (not treated in this book).

A filter constructed from lumped elements has a transfer function $H(s)$ that is a rational function of s. Impulse response of such a filter is causal. The vast majority of time-continuous analog filters for data-communications systems are implemented with lumped elements. Corresponding to every rational transfer function $H(s)$ of degree n in its denominator, there exists a linear, ordinary, constant-coefficient differential equation of degree n as an alternative representation of the same filter, characterized by the same coefficients.

Not all time-continuous filters of interest possess rational transfer functions. An important exception is the transfer function of a bandlimited filter (see Sec. 2.2.2), which might be described by piecewise transcendental functions. Impulse responses of bandlimited filters are noncausal, so they are not physically realizable. They are often employed as idealizations in analysis and can be simulated in the frequency domain (see Chapter 6) but have to be approximated by realizable causal filters in the time domain or in hardware. A nonrational transfer function does not have a corresponding ordinary differential equation.

Frequency-domain filtering is performed by multiplying the frequency-domain representation (Laplace or Fourier transform) of an input signal by the transfer function of the filter. Time-domain filtering is performed in one of two ways: (1) convolve the time-domain signal with the impulse response of the filter, or (2) drive the differential-equation representation of the filter with the signal as a forcing function. All three methods are provided in STÆDT.

It is common practice to partition rational transfer functions into a cascade of rational biquadratic sections for simulation purposes (and often in physical filters

3.4.2 Complex Filters

Real filters act on real signals; complex filters act on complex signals (i.e., complex envelopes, analytic signals, or complex baseband signals). A real filter has a real impulse response; a complex filter has a complex impulse response. One often simulates the complex envelope of a carrier signal to avoid irrelevant fine detail of the carrier wave. A complex filter representation is needed such that if the complex envelope of a real signal is applied as input, filter output will be the same as the complex envelope of the real signal output from the corresponding real filter.

Envelope properties of a real bandpass filter can be represented by a complex impulse response $h_L(t,f_0)$ or its Fourier transform $H_L(f,f_0)$, a complex transfer function [or, Laplace transform $H_L(s,f_0)$]. The complex filter is defined relative to an arbitrary frequency f_0, which is usually (but not necessarily) taken as the carrier frequency of the signal. Complex-filter characteristics depend on the choice of f_0.

The impulse response of a real filter is related to the complex-envelope impulse response by

$$h(t) = 2\,\mathrm{Re}[h_L(t,f_0)e^{j2\pi f_0 t}]$$

The complex impulse response can be decomposed into two real components in quadrature according to

$$h_L(t,f_0) = h_I(t,f_0) + jh_Q(t,f_0)$$

while the transfer function decomposes similarly as

$$H_L(s,f_0) = H_I(s,f_0) + jH_Q(s,f_0)$$

Because $h_L(t,f_0)$ is complex, the corresponding transfer function $H_L(s,f_0)$ has complex coefficients. (*Caution:* Although H_I and H_Q have real coefficients, they are themselves both complex functions of the complex variable s and are not the real and imaginary parts of H_L.)

If a real signal $s_i(t)$ with carrier frequency f_c is applied to a real filter with impulse response $h(t)$, the real output is the convolution

$$s_o(t) = s_i(t) \otimes h(t)$$

and the Fourier transforms give

$$S_o(f) = S_i(f)H(f)$$

Under mild bandlimiting assumptions, a complex envelope filter $H_L(f,f_c)$ can be associated with the real filter $H(f)$. Then the input and output complex envelopes are related in the frequency domain by

$$Z_o(f) = Z_i(f)H_L(f,f_c)$$

where $Z(\cdot)$ denotes the frequency-domain representation of the complex envelope of the real signal. This equation is the basis for simulation of frequency-domain filtering of complex envelopes.

Taking inverse Fourier transforms, the input and output complex envelopes are related in the time domain by

$$z_o(t) = z_i(t) \otimes h_L(t,f_c)$$

This equation is the basis for one method of simulation of time-domain filtering of complex envelopes.

Another time-domain method is available if the transfer function $H_L(s,f_0)$ is rational. The complex coefficients of the rational function can be applied to a differential equation with complex coefficients, which is then driven by the time-domain complex-envelope input signal. (The computer actually approximates a differential equation by a difference equation, as discussed at length in Chapter 6.)

A complex filter can be decomposed conceptually into four real filters in the cross-coupled structure shown in Fig. 3.1.

3.4.3 Conversion to Complex Filters

There is no one complex filter representation that is uniquely associated with a particular real impulse response $h(t)$ or transfer function $H(f)$, not even if the reference frequency f_0 has been fixed. There are several different ways to convert a real filter to complex; each involves certain approximations. Two methods are described in the text and both are furnished with STÆDT.

For frequency-domain simulations: Start with the frequency response $H(f)$ of the real filter, truncate and discard $H(f)$ at negative frequencies, and slide it to the left by an amount f_0 to give $H_L(f,f_0)$. This procedure is very simple, does not require decomposition into explicit components H_I and H_Q, and corresponds nicely to conventional measurements of frequency response in the laboratory.

Conversions for time-domain simulations are more complicated. Just taking the inverse Fourier transform of $H_L(f,f_0)$ does not work because the result is always noncausal, whereas time-domain simulations are always causal. Appendix 3A develops one method for real-to-complex conversion of biquadratic sections. Conversion produces another biquadratic section $h_L(t,f_0)$ that has complex coefficients. One could rationalize $h_L(t,f_0)$ and separate it into the two real impulse responses $h_I(t,f_0)$ and $h_Q(t,f_0)$, but these become quartic functions under some conditions. It is easier not to separate into two components; retain the complex coefficients instead.

3.4.4 STÆDT Processes

BQCV: converts coefficients of real rational filter (partitioned into biquadratic sections) into complex coefficients, based on user specification of conversion frequency f_0

BQFR, FRNP: convert real frequency-domain filters into complex frequency-domain filters (along with other tasks)

APPENDIX 3A: COMPLEX CONVERSION OF TIME-DOMAIN FILTERS

In this appendix we present a method of converting a real rational filter to a complex envelope filter. (A real filter has real coefficients in its transfer function and impulse response; a complex filter has complex coefficients.) Since rational transfer functions can be partitioned into biquadratic (two-pole) sections, the decomposition is developed for a biquadratic section (*biquad* for short). Any rational filter can be modeled by cascading biquads. The converted filters will also be rational, biquadratic sections, but with complex coefficients.

To begin, assume that the transfer function of the biquad is *proper*. That is, the degree of numerator is less than the degree of the denominator. Treatment of improper biquads is addressed later.

Complex Poles

Assume that the poles of the biquad are complex conjugates of one another and that the impulse response $h(t)$ of the biquad is causal and real. The transfer function $H(s)$, written in Laplace transform notation, can be expanded in partial fractions [3.2, Chap. 6] into the form

$$H(s) = \frac{r}{s + \alpha - j\beta} + \frac{r^*}{s + \alpha + j\beta} \tag{3A.1}$$

where the poles are located at $s = -\alpha \pm j\beta$, and the partial-fraction coefficients r and r^* are the residues of the poles.

The impulse response of the biquad is expressed as

$$h(t) = \begin{cases} re^{-\alpha t}e^{j\beta t} + r^*e^{-\alpha t}e^{-j\beta t}, & t \geq 0 \\ 0, & t < 0 \end{cases} \tag{3A.2}$$

which is causal. All succeeding functions to be derived from (3A.2) will also be causal; they are zero for $t < 0$. The constraint is implied henceforth and will not be repeated.

That the impulse response is real can be seen by combining the complex exponentials into sines and cosines to obtain

$$h(t) = e^{-\alpha t}[2\operatorname{Re}(r)\cos\beta t - 2\operatorname{Im}(r)\sin\beta t]$$

$$= 2e^{-\alpha t}|r|\left[\frac{\operatorname{Re}(r)}{|r|}\cos\beta t - \frac{\operatorname{Im}(r)}{|r|}\sin\beta t\right] \quad (3A.3)$$

Define the angle ξ such that

$$\cos\xi = \frac{\operatorname{Re}(r)}{|r|} \quad \text{and} \quad \sin\xi = \frac{\operatorname{Im}(r)}{|r|} \quad (3A.4)$$

to obtain

$$\begin{aligned}h(t) &= 2|r|e^{-\alpha t}(\cos\xi\cos\beta t - \sin\xi\sin\beta t)\\ &= 2|r|e^{-\alpha t}\cos(\beta t + \xi)\\ &= 2|r|e^{-\alpha t}\operatorname{Re}[e^{j(\beta t + \xi)}]\end{aligned} \quad (3A.5)$$

Now expand $h(t)$ in (3A.5) on the reference frequency $\omega_0 = 2\pi f_0$ according to

$$\begin{aligned}h(t) &= 2\operatorname{Re}[|r|e^{-\alpha t}e^{j(\beta t + \xi)}e^{-j\omega_0 t}e^{j\omega_0 t}]\\ &= 2\operatorname{Re}[h_L(t,\omega_0)e^{j\omega_0 t}]\end{aligned} \quad (3A.6)$$

whereupon the envelope-filter impulse response is found as

$$h_L(t,\omega_0) = |r|e^{-\alpha t}e^{j[(\beta - \omega_0)t + \xi]} \quad (3A.7)$$

That can be decomposed into quadrature components

$$\begin{aligned}h_I(t,\omega_0) &= |r|e^{-\alpha t}\cos[(\beta - \omega_0)t + \xi]\\ &= |r|e^{-\alpha t}\frac{e^{j[(\beta - \omega_0)t + \xi]} + e^{-j[(\beta - \omega_0)t + \xi]}}{2}\\ h_Q(t,\omega_0) &= |r|e^{-\alpha t}\sin[(\beta - \omega_0)t + \xi]\\ &= |r|e^{-\alpha t}\frac{e^{j[(\beta - \omega_0)t + \xi]} - e^{-j[(\beta - \omega_0)t + \xi]}}{2j}\end{aligned} \quad (3A.8)$$

Transfer functions are found by taking the Laplace transforms of (3A.7) and (3A.8) to obtain

92 ☐ REPRESENTATION OF ANALOG FILTERS

$$H_L(s, \omega_0) = \frac{r}{s + \alpha - j(\beta - \omega_0)} \qquad (3A.9)$$

and

$$H_I(s, \omega_0) = \frac{1}{2} \left[\frac{r}{s + \alpha - j(\beta - \omega_0)} + \frac{r^*}{s + \alpha + j(\beta - \omega_0)} \right]$$

$$= \frac{(s + \alpha)\operatorname{Re}(r) - (\beta - \omega_0)\operatorname{Im}(r)}{(s + \alpha)^2 + (\beta - \omega_0)^2}$$

$$H_Q(s, \omega_0) = \frac{1}{2j} \left[\frac{r}{s + \alpha - j(\beta - \omega_0)} - \frac{r^*}{s + \alpha + j(\beta - \omega_0)} \right]$$

$$= \frac{(s + \alpha)\operatorname{Im}(r) + (\beta - \omega_0)\operatorname{Re}(r)}{(s + \alpha)^2 + (\beta - \omega_0)^2} \qquad (3A.10)$$

Equation (3A.9) can be interpreted analogously to $H_{L+}(f)$ of (3.26), (3.27), and (3.32) from the frequency-domain decomposition. That is, (3A.9) is equivalent to discarding the negative-frequency pole of $H(s)$ in (3.23) and sliding the positive-frequency pole downward by $\omega_0 = 2\pi f_0$. Reduction in the number of poles is a simplification, at least to first appearances.

However, traditional circuit theory has been confined mostly to transfer functions with real coefficients, so there is no well-traveled road showing how to deal with the complex coefficients of (3A.9). One expedient is to rationalize (3A.9), separate it into real and imaginary parts to obtain H_I and H_Q of (3A.10), and then use H_I and H_Q in the structure of Fig. 3.1; this is a well-established approach.

Notice, though, that rationalization has caused two poles to appear in each of H_I and H_Q; the ostensible simplification that appeared in (3A.9) has been lost. Applying the complex envelope is supposed to simplify matters. Simplification should be possible for filters as well as signals. If a filter could be implemented with complex coefficients, the simpler version of (3A.9) would be more attractive than the rationalized version of (3A.10). This subject will be revisited in Chapter 6. For now just let it be said that complex-coefficient filters are easily implemented by digital means; it is not necessary or advisable to obtain an explicit separation into H_I and H_Q.

A note on terminology is appropriate. Rice [2.10] distinguishes several different kinds of envelopes: The complex envelope $z(t)$ of (2.19) is called the *natural* envelope, while the impulse response $h_L(t, \omega_0)$ of (3A.7) corresponds to his *rational spectrum* envelope.

Real Poles

A bandpass filter could conceivably include real poles, even though that is uncommon. If real poles do occur, it will be necessary to convert them into complex format.

APPENDIX 3A: COMPLEX CONVERSION OF TIME-DOMAIN FILTERS

A conversion is shown here as a formal extension of the decomposition of a complex-pole biquadratic.

Single Pole. Denote a single-pole transfer function as

$$H(s) = \frac{a}{s+a} \tag{3A.11}$$

which has the impulse response

$$h(t) = ae^{-at} \tag{3A.12}$$

Expand the impulse response on ω_0 to obtain

$$\begin{aligned} h(t) &= 2(\tfrac{1}{2} ae^{-at} e^{-j\omega_0 t}) e^{j\omega_0 t} \\ &= 2\,\mathrm{Re}[h_L(t,\omega_0) e^{j\omega_0 t}] \end{aligned} \tag{3A.13}$$

from which a definition of the envelope impulse response is

$$\begin{aligned} h_L(t,\omega_0) &= \tfrac{1}{2} ae^{-at} e^{-j\omega_0 t} \\ h_I(t,\omega_0) &= \tfrac{1}{2} ae^{-at} \cos \omega_0 t \\ h_Q(t,\omega_0) &= -\tfrac{1}{2} ae^{-at} \sin \omega_0 t \end{aligned} \tag{3A.14}$$

Taking Laplace transforms yields the transfer functions

$$H_L(s,\omega_0) = \frac{1}{2} \frac{a}{s+a+j\omega_0}$$

$$H_I(s,\omega_0) = \frac{1}{2} \frac{a(s+a)}{(s+a)^2 + \omega_0^2}$$

$$H_Q(s,\omega_0) = -\frac{1}{2} \frac{a\omega_0}{(s+a)^2 + \omega_0^2} \tag{3A.15}$$

The rationalized transfer functions for H_I and H_Q have quadratic denominators even though the original single-pole transfer function (3A.11) has a first-order denominator. The unseparated transfer function H_L has only a single pole but with complex coefficients.

Two Real Poles. Suppose that the biquadratic $H(s)$ has two distinct real poles at $s = -s_1, -s_2$ and can be expanded into partial fractions as

$$H(s) = \frac{r_1}{s+s_1} + \frac{r_2}{s+s_2} \tag{3A.16}$$

where r_1 and r_2 are the residues (both real). Impulse response is

$$\begin{aligned}h(t) &= r_1 e^{-s_1 t} + r_2 e^{-s_2 t} \\ &= 2[\tfrac{1}{2}(r_1 e^{-s_1 t} e^{-j\omega_0 t} + r_2 e^{-s_2 t} e^{-j\omega_0 t}) e^{j\omega_0 t}]\end{aligned} \tag{3A.17}$$

Following the same steps as in the preceding paragraphs, the complex-envelope impulse response is

$$h_L(t, \omega_0) = \tfrac{1}{2}(r_1 e^{-s_1 t} e^{-j\omega_0 t} + r_2 e^{-s_2 t} e^{-j\omega_0 t}) \tag{3A.18a}$$

and its rationalized quadrature components are

$$\begin{aligned}h_I(t, \omega_0) &= \tfrac{1}{2} r_1 e^{-s_1 t} \cos \omega_0 t + \tfrac{1}{2} r_2 e^{-s_2 t} \cos \omega_0 t \\ h_Q(t, \omega_0) &= -\tfrac{1}{2} r_1 e^{-s_1 t} \sin \omega_0 t - \tfrac{1}{2} r_2 e^{-s_2 t} \sin \omega_0 t\end{aligned} \tag{3A.18b}$$

Taking Laplace transforms gives the transfer functions

$$\begin{aligned}H_L(s, \omega_0) &= \frac{1}{2}\left(\frac{r_1}{s+s_1+j\omega_0} + \frac{r_2}{s+s_2+j\omega_0}\right) \\ &= \frac{1}{2}\left[\frac{s(r_1+r_2) + r_1 s_2 + r_2 s_1 + j\omega_0(r_1+r_2)}{s^2 + s(s_1+s_2+2j\omega_0) + (s_1+j\omega_0)(s_2+j\omega_0)}\right] \\ H_I(s, \omega_0) &= \frac{1}{2}\left[\frac{r_1(s+s_1)}{(s+s_1)^2+\omega_0^2} + \frac{r_2(s+s_2)}{(s+s_2)^2+\omega_0^2}\right] \\ H_Q(s, \omega_0) &= -\frac{\omega_0}{2}\left[\frac{r_1}{(s+s_1)^2+\omega_0^2} + \frac{r_2}{(s+s_2)^2+\omega_0^2}\right]\end{aligned} \tag{3A.19}$$

The expression for H_L has complex coefficients but only two poles. Rationalized expressions for H_I and H_Q are each composed of the sum of a pair of terms with quadratic denominators. If terms were combined over a common denominator, the result would be a transfer function with a quartic denominator. A quartic would have to be implemented as the cascade (or sum) of two biquadratics. Clearly, H_L with complex coefficients has a simpler format.

Suppose that poles in the real filter are coincident at $s_1 = s_2 = -a$, so that the proper original transfer function has the generic format

$$H(s) = \frac{a_1 s + a_0}{(s+a)^2} \tag{3A.20}$$

Then the complex-envelope filter is found to have a transfer function

$$H_L(s, \omega_0) = \frac{1}{2}\left[\frac{r_1 + r_2(s + a + j\omega_0)}{(s + a + j\omega_0)^2}\right] \quad (3A.21)$$

where the residues are $r_1 = a_0 - a_1 a$ and $r_2 = a_1$. Observe the quadratic denominator. By contrast, the rationalized components H_I and H_Q (not shown) inescapably have quartic denominators.

Improper Transfer Function

Let $H(s)$ be an improper biquadratic, as in (3.6) with $a_2 \neq 0$ and $b_2 \neq 0$. Divide the numerator of (3.6) by the denominator to obtain

$$H(s) = \frac{H_0 a_2}{b_2} + H_0 \frac{s(a_1 - a_2 b_1/b_2) + a_0 - a_2 b_0/b_2}{b_2 s^2 + b_1 s + b_0}$$
$$= c + H'(s) \quad (3A.22)$$

where $c = H_0 a_2/b_2$ is a real constant and $H'(s)$ is a proper rational function that can be decomposed in the same ways as outlined in the preceding paragraphs. Total transmission by $H(s)$ consists of the linear sum of transmission through the proper function $H'(s)$, plus transmission through the improper constant term.

Let's concentrate on correct treatment of complex-envelope transmission by the constant term c. Impulse response of the constant is

$$h_c(t) = c\delta(t) = 2[\tfrac{1}{2} c\delta(t) e^{-j\omega_0 t} e^{j\omega_0 t}] = 2[h_L(t, \omega_0) e^{j\omega_0 t}] \quad (3A.23)$$

whereupon the complex-envelope impulse response is

$$h_{Lc}(t, \omega_0) = \tfrac{1}{2} c\delta(t) e^{-j\omega_0 t} \quad (3A.24)$$

Taking its Laplace transform yields

$$H_{Lc}(s, \omega_0) = \frac{c}{2} \quad (3A.25)$$

If the complex-envelope transfer function of $H'(s)$ is $H'_L(s, \omega_0)$, the complex-envelope transfer function of the improper function $H(s)$ is

$$H_L(s, \omega_0) = \frac{c}{2} + H'_L(s, \omega_0) \quad (3A.26)$$

4

REPRESENTATION OF ANALOG SIGNAL PROCESSES

By the term *process* we mean an operation performed on a signal (or on noise): an operation that transforms the signal in some manner. Examples of processes are the operations performed by filters, amplifiers, modulators, detectors, and other elements that might be found in a communications link. Filters are important enough, and so intricate, that they have been given a chapter to themselves. In this chapter we examine other processes.

Computer operations that simulate the elements of a communications link will also be called processes. In addition to the link elements, simulations also include simple mathematical processes, such as addition, multiplication, and counting, as well as complicated processes, such as correlation and Fourier transforms. By extension, the program codes that implement the operations will also be called processes.

An *analog signal process* will mean a time-continuous operation performed on time-continuous, amplitude-continuous signals (although, strictly speaking, an analog signal could be sampled, or even quantized). By this definition, a *digital signal process* is performed on time-discrete signals. This chapter contains representations of analog processes that are important for simulation. Processes to be examined are modulation, demodulation, signal reception, symbol detection, and nonlinear devices. Several vital concepts are introduced, including the PAM link model, symbol strobes, matched filters, decision regions, and memoryless nonlinearities.

Computer representations of analog processes are treated in later chapters. Computer representations of digital processes are presented after the simulation of analog processes has been covered.

4.1 MODULATORS

Consider the *I–Q* baseband signals $x(t)$ and $y(t)$ that are to be modulated onto quadrature carriers at frequency f_c to form the real passband signal

$$s(t) = x(t) \cos 2\pi f_c t - y(t) \sin 2\pi f_c t \qquad (4.1)$$

A simulation could generate the cosine and sine waves, multiply them by x and y, respectively, and subtract the products to generate a simulated real passband signal. If carrier frequency f_c is low enough, this direct synthesis of the passband signal may be the best course to follow.

4.1.1 Complex Envelope

But carrier frequency is often very high compared to bandwidths of $x(t)$ and $y(t)$. If that is the case, generation of the sine and cosine involves a large amount of computing that contributes nothing useful to the ultimate results. Instead, it is more efficient to work with the complex envelope

$$z(t) = x(t) + jy(t) \qquad (4.2)$$

Most simulation programs for signal transmission employ the complex envelope (and maybe the analytic signal) extensively, even to the complete exclusion of any real passband signals. (A user should be cautious if carrier frequency is low; the complex envelope may not offer a satisfactory representation of the real signal. Chapters 2 and 3 provide guidance on the approximations involved.)

If the complex envelope is employed to represent a linearly modulated signal, no explicit modulation operation whatever need be simulated. The *I–Q* baseband signals are identical to the complex-envelope representation of the passband signal. For this reason, simulations will often lack visible modulator processes even when passband signals are being simulated.

4.1.2 Angle Modulation

Absence of a modulator is most obvious for linear amplitude modulation. If angle modulation of $\phi(t)$ is to be generated (with a constant amplitude for the modulated signal), the *I–Q* baseband signals can be represented as

$$x(t) = \cos \phi(t), \qquad y(t) = \sin \phi(t) \qquad (4.3)$$

and the complex envelope of (4.2) is still applicable.

Thus (4.3) shows one way to produce angle modulation—a way that is relatively simple in a computer. Although complex-envelope production for an amplitude-modulated signal involves no modulator process whatever, the complex envelope for angle modulation via (4.3) requires nonlinear sine and cosine processing of the real signal $\phi(t)$.

In many systems the phase $\phi(t)$ could accumulate without bound, which would become awkward numerically. To avoid unbounded phase accumulation, the phase can be reduced modulo-2π, by means exemplified in Sec. 7.4.2.

4.1.3 Analytic Signal

If the complex envelope of a signal is $z(t)$, its analytic signal is

$$s^+(t) = z(t)e^{j2\pi f_c t} \qquad (4.4)$$

In the frequency domain, entered by taking the Fourier transform of (4.4), the analytic signal is represented by

$$S^+(f) = Z(f - f_c) \qquad (4.5)$$

whereupon it becomes apparent that the analytic signal is identical to the complex envelope, except for a frequency shift. [The $^+$ superscript indicates that the baseband $Z(f)$ has been shifted in a positive direction along the frequency axis. If $Z(f) \equiv 0$ for all $|f| \geq f_c$, then $S^+(f)$ is nonzero only at positive frequencies.]

Several features can be deduced from these simple forms:

- To perform a frequency shift in the time domain, by an arbitrary frequency f_1, multiply the complex signal (complex envelope or analytic signal) by $\exp(j2\pi f_1 t)$.
- To perform a frequency shift in the frequency domain, just translate the signal spectrum by f_1 along the frequency axis.
- Because of these simple frequency-shift properties, the distinction between complex envelope and analytic signal becomes blurred in practice. It is easy to begin thinking of generic complex signals and not distinguish among complex envelopes, analytic signals, and I–Q baseband signals.

A frequency shift is equivalent to a linearly changing phase. More generally, an arbitrary phase modulation $\phi(t)$ can be applied to a complex signal by multiplying by $\exp[j\phi(t)]$. A phase modulator process is usable for frequency shifting and for phase rotating, in addition to modulation as such. Recollect that a complex exponential can be expanded into rectangular format as

$$e^{j\phi(t)} = \cos\phi(t) + j\sin\phi(t) \qquad (4.6)$$

and is likely to be simulated in that format in many applications.

4.2 DEMODULATORS

Almost any signal that has been modulated onto a carrier will require demodulation in the receiver. There are numerous ways to implement demodulation. One of those ways—I–Q frequency translation—will be described here and used as a basis for further development.

4.2.1 I–Q Demodulation

Real Signals

Introduce an arbitrary phase angle θ_i into the received signal, represented in real format as

$$s(t) = x(t)\cos(2\pi f_c t + \theta_i) - y(t)\sin(2\pi f_c t + \theta_i) \tag{4.7}$$

Multiply the signal against two quadrature local references,

$$\mu(t) = 2\cos(2\pi f_c t + \theta_o) \quad \text{and} \quad \nu(t) = -2\sin(2\pi f_c t + \theta_o) \tag{4.8}$$

Multiplier products are

$$\begin{aligned} s(t)\mu(t) &= x(t)\cos\Delta\theta - y(t)\sin\Delta\theta \\ &\quad + x(t)\cos(4\pi f_c t + \theta_i + \theta_o) - y(t)\sin(4\pi f_c t + \theta_i + \theta_o) \\ s(t)\nu(t) &= x(t)\sin\Delta\theta + y(t)\cos\Delta\theta \\ &\quad - x(t)\sin(4\pi f_c t + \theta_i + \theta_o) - y(t)\cos(4\pi f_c t + \theta_i + \theta_o) \end{aligned} \tag{4.9}$$

where $\Delta\theta = \theta_i - \theta_0$.

The mixer products of (4.9) consist of difference-frequency components with spectra concentrated near zero frequency and sum-frequency components concentrated near $\pm 2f_c$. If $X(f)$ and $Y(f)$—the Fourier transforms of $x(t)$ and $y(t)$—are bandlimited to $|f| < f_c$, the spectra of the sum- and difference-frequency products do not overlap and an ideal lowpass filter can suppress the sum-frequency terms without distorting the difference-frequency signal. Baseband signals after such filtering are

$$\begin{aligned} I(t) &= x(t)\cos\Delta\theta - y(t)\sin\Delta\theta \\ Q(t) &= x(t)\sin\Delta\theta + y(t)\cos\Delta\theta \end{aligned} \tag{4.10}$$

If $\Delta\theta = 0$, the original baseband signals $x(t)$ and $y(t)$ are recovered as $I(t)$ and $Q(t)$. Phase angle $\Delta\theta$ may be time varying or may be constant; its character has a strong influence on the behavior of the demodulator, as will be seen presently.

Complex Signals

Corresponding to the real signal $s(t)$ is an analytic signal

$$s^+(t) = [x(t) + jy(t)]e^{j(2\pi f_c t + \theta_i)} \tag{4.11}$$

and corresponding to the real references $u(t)$ and $v(t)$ there is a reference analytic signal

$$\xi(t) = \cos(2\pi f_c t + \theta_o) - j\sin(2\pi f_c t + \theta_o) = e^{-j2\pi f_c t}e^{-j\theta_o} \tag{4.12}$$

(Note the reduction of the reference amplitude by a factor of 2, needed for agreement of real-signal and complex-signal results.)

Multiply the two analytic signals to obtain a baseband complex signal

$$\begin{aligned}z(t) &= s^+(t)\xi(t) = [x(t) + jy(t)]e^{j\Delta\theta} \\ &= x(t)\cos\Delta\theta - y(t)\sin\Delta\theta + j[x(t)\sin\Delta\theta + y(t)\cos\Delta\theta]\end{aligned} \tag{4.13}$$

The real and imaginary parts of (4.13) are the same, identically, as the I and Q parts of (4.10). Furthermore, (4.13) may be recognized as the products of the complex envelopes of the signal and the quadrature references. There was no need to bring in the carrier frequency at all; the demodulation process could be described most compactly in terms of complex envelopes, ignoring the carrier frequency entirely (provided that f_c is large enough compared to the bandwidths of the baseband signals).

4.2.2 Demodulator Coherence

Demodulators can be categorized according to their *coherence:* the precision to which $\Delta\theta$ is known. There are three useful degrees of coherence:

1. *Coherent Demodulation.* Phase angle $\Delta\theta$ is known within narrow bounds.
2. *Differentially Coherent Demodulation and FM-Discriminator Demodulation.* Phase angle $\Delta\theta$ is unknown, but $d(\Delta\theta)/dt$ is small compared to the rate of change of phase of the signal.
3. *Noncoherent Demodulation.* Phase is ignored entirely; $\Delta\theta$ may be changing rapidly.

Succeeding paragraphs show how each of these categories of demodulators can be represented in terms of complex envelopes and I–Q implementation, thereby leading to useful simulation models.

Coherent Demodulators

If demodulation is coherent, the original baseband signals $x(t)$ and $y(t)$ can be recovered. The requirement for small $\Delta\theta$ usually means that the local reference must be phase-locked to the incoming signal. Locking of the reference is accomplished by the carrier synchronizer of Fig. 1.2.

Experience has shown that satisfactory coherent reception of a QPSK signal requires $\Delta\theta$ to be no more than a few degrees, and the requirement for larger constellations is even more stringent. Coherent demodulation of BPSK is substantially more tolerant of phase errors.

A carrier synchronizer is called upon to adjust the reference phase θ_o so that $\Delta\theta$ is made small. The adjustment may be modeled as a phase rotation of $-\theta_o$ on the received signal. Because of various phase shifts that may be accumulated by a simulated signal as it travels from its origin to destination, a corrective phase rotation is often needed in simulations of coherent demodulation.

Two elements are needed to perform the rotation:

1 A phase rotator, as explained above
2 Some means of determining the phase angle θ_o by which the signal is to be rotated (addressed in Chapter 11)

A simulation process to correct a constant phase error might be implemented rather simply. Nearly perfect correction could be obtained and the ensuing simulation would then estimate system performance in the absence of phase disturbances. Many simulations are run on this basis. In other circumstances, a user may wish to evaluate degradation of performance caused by phase fluctuations or may want to simulate the carrier synchronizer itself. This becomes a rather more elaborate undertaking; some insight is provided in Chapter 11.

Differentially Coherent Demodulation

If coherence is given up, the original baseband signals $x(t)$ and $y(t)$ cannot be recovered; some other characteristic of the signal must carry the information. In differentially coherent reception, the information is carried in the change of phase from one symbol to the next.

To illustrate differentially coherent demodulation, consider differentially *encoded* M-ary phase-shift keying (MPSK), introduced in Sec. 2.2.3. The mth transmitted pulse is assigned the complex amplitude c_m, selected from the unit-amplitude vectors $\{e^{\pm j\pi/M}, e^{\pm j3\pi/M}, \ldots\}$. Let the information stream be designated $\{\alpha_m\}$; $\alpha_m = \pm 1, \pm 3,$ .... Then c_m is *differentially encoded* according to

$$c_m = c_{m-1} \exp\left(j\frac{\alpha_m \pi}{M}\right) \qquad (4.14)$$

Each data symbol α_m causes the signal phase to rotate by $\alpha_m \pi/M$ from its value at the preceding symbol time.

Do not confuse differential *encoding* with differentially coherent *demodulation*, also known as *differential detection*, explained below. Differential encoding is necessary for successful operation of differential demodulation, but the two processes are entirely distinct. (Differential encoding, with subsequent decoding, is often used also in conjunction with coherent demodulation, as a means of counteracting phase ambiguity that is inherent in all symmetric constellations; differential demodulation has no place in this application of differential encoding.)

To explain the operation of differential demodulation, assume that the phase θ_i of the received signal is time varying and that the receiver-reference phase $\theta_o = 0$. After I–Q demodulation and filtering, the complex symbol strobes out of the receiver will be (neglecting noise, and normalizing signal amplitude to unity)

$$z_o(m) = c_m e^{j\theta_i(m)} \qquad (4.15)$$

Since phase $\theta_i(m)$ is completely unknown, the complex amplitude c_m cannot be retrieved from the strobe. However, now generate the product of strobes

$$\begin{aligned}\bar{z}(m) &= z_o(m) z_o^*(m-1) = c_m c_{m-1}^* e^{j[\theta_i(m) - \theta_i(m-1)]} \\ &= e^{j\alpha_m \pi/M} e^{j[\theta_i(m) - \theta_i(m-1)]}\end{aligned} \qquad (4.16)$$

where the last line was obtained by substituting (4.14) for c_m.

The angle of $\exp(j\alpha_m \pi/M)$, and α_m itself, can be retrieved from (4.16), provided that $\theta_i(m) - \theta_i(m-1)$ is small enough compared to π/M. Thus the information $\{\alpha_m\}$ can be recovered, even if the phase $\{\theta_i(m)\}$ and complex pulse amplitudes $\{c_m\}$ cannot.

Frequency Discriminators

Represent the complex envelope of a received signal in polar form as

$$z(t) = R(t) e^{j\phi(t)} e^{j\theta_i(t)} \qquad (4.17)$$

where the time variation of phase disturbance $\theta_i(t)$ is made explicit. Assume that the receiver reference phase $\theta_o = 0$. Then (4.17) is the polar representation of the complex baseband output from an I–Q demodulator.

Consider that the signal is frequency modulated; its information is carried in the rate of change of phase $\dot\phi(t)$, where the overdot indicates a derivative with respect to time. A *frequency discriminator* is a device that recovers the derivative of the phase of $z(t)$, which is $\dot\phi(t) + \dot\theta_i(t)$. Satisfactory demodulation of FM can be attained if $\dot\theta_i(t)$ is sufficiently small compared to $\dot\phi(t)$.

Frequency discrimination is closely related to differential demodulation. In each case the actual phase disturbance $\theta_i(t)$ is unknown and some feature of the message phase $\phi(t)$ is employed: phase changes for differential demodulation and rate of change of phase for frequency discrimination. In each case the short-term changes in $\theta_i(t)$ must be small compared to the short-term changes in $\phi(t)$.

Differential demodulation and frequency discrimination are often lumped into the category of noncoherent methods, since they are not fully coherent. That dichotomy is overly coarse; there are substantial differences between differentially coherent and fully noncoherent demodulation, as will be seen shortly.

Discriminator Implementation. To see how a frequency discriminator might be simulated with operations on the complex envelope, write the rectangular components of (4.17) as *I–Q* baseband signals

$$I(t) = R(t)\cos \xi(t)$$
$$Q(t) = R(t)\sin \xi(t) \quad (4.18)$$

where $\xi(t) = \phi(t) + \theta_i(t)$; the discriminator cannot separate desired information in $\phi(t)$ from the disturbance $\theta_i(t)$. The tangent of ξ obeys

$$\tan \xi(t) = \frac{\sin \xi(t)}{\cos \xi(t)} = \frac{Q(t)}{I(t)} \quad (4.19)$$

independent of magnitude $R(t)$.

Instantaneous radian frequency of the signal plus phase disturbance is

$$\dot{\xi}(t) = \frac{d}{dt}\left[\tan^{-1}\frac{Q(t)}{I(t)}\right] = \frac{I\dot{Q} - \dot{I}Q}{I^2 + Q^2} \quad (4.20)$$

Equation (4.20) shows a way to implement FM demodulation using only the complex envelope of the signal in rectangular coordinates.

Noncoherent Demodulation

No phase information whatever is available to a noncoherent receiver. All message information must be carried in the magnitude $|z(t)|$ of the complex envelope or, equivalently, in the magnitude-squared $|z(t)|^2 = R^2(t)$. Noncoherent receivers are also called "power" or "energy" receivers. Not only is the carrier phase θ_i irrecoverable, but the barrier to phase recovery extends to data characteristics. Earlier, the value of the complex *m*th symbol was represented as $c_m = a_m + jb_m$. If noncoherent reception is to be used, all useful information in the symbol must be contained in $|c_m|$. Decomposition of c_m into quadrature components is possible only in a coherent receiver, and observation of phase changes is possible only if the receiver is at least differentially coherent.

Simulation of noncoherent reception is simple enough; the *I–Q* demodulator provides

$$R^2(t) = I^2(t) + Q^2(t) \quad (4.21)$$

Examples of systems using noncoherent reception include on–off keyed optical links and many frequency-shift-keyed (FSK) radio links. A noncoherent receiver for MFSK consists of M filters, one for each signaling frequency, and a noncoherent detector on each filter.

4.3 PAM LINK MODEL; MATCHED FILTERS

A PAM link can be represented by a particularly simple model found throughout the communications literature and often applied in simulations. That model is developed here and then examined for important characteristics.

4.3.1 PAM–QAM Link Model

Figure 4.1 shows the simplest model of a PAM–QAM transmission link. Digital data are mapped into complex M-ary symbols $a_m + jb_m$, where a and b are selected from the same alphabet. (Thus the particular model to be discussed is specialized for balanced QAM.) The symbols are converted to a uniformly spaced train of complex, weighted impulses of the form

$$z_1(t) = \sum_i a_i \delta(t - iT) + j \sum_i b_i \delta(t - \epsilon T - iT) \qquad (4.22)$$

where the offset factor $\epsilon (0 \leq \epsilon < 1)$ allows for staggered QAM signal formats.

Impulses are applied to a transmitter filter with impulse response $h_T(t)$ and transfer function $H_T(f)$. A QAM signal is produced by linear amplitude modulation of a carrier by pulses from the filter. If the signal is represented by its complex envelope, no explicit attention need be devoted to the modulator or to a carrier representation. The transmitted complex envelope is

$$z_T(t) = \sum_i a_i h_T(t - iT) + j \sum_i b_i h_T(t - \epsilon T - iT) \qquad (4.23)$$

Assume that the transmission medium is transparent; it attenuates and delays the signal but does not distort it in any manner. Then the received signal will be a scaled, delayed replica of the transmitted signal. At the receiver, assume that noise with complex envelope $w(t)$ is added to the received signal so that the complex envelope of the total signal plus noise applied to the receiver is

$$r(t) = z_T(t) + w(t) \qquad (4.24)$$

neglecting attenuation and delay of the signal.

Received signal plus noise is demodulated, but that operation is omitted from

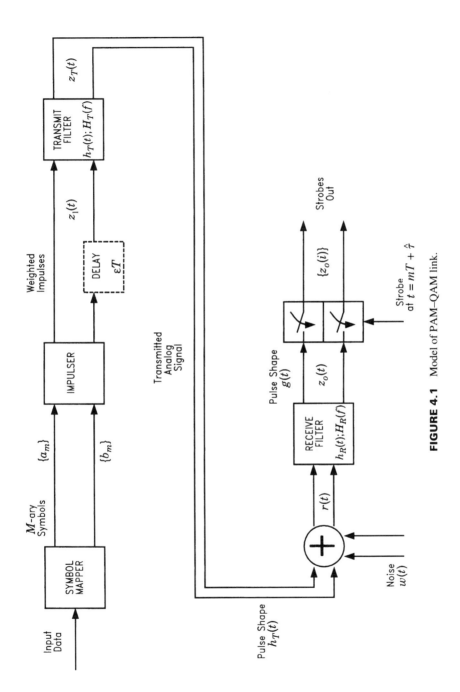

FIGURE 4.1 Model of PAM–QAM link.

the simplified model since demodulation does not affect the complex envelope. A receiver filter with impulse response $h_R(t)$ and transfer function $H_R(f)$ filters the signal plus noise and outputs a complex baseband signal

$$z_o(t) = r(t) \otimes h_R(t)$$
$$= \sum_i a_i g(t - iT) + j \sum_i b_i g(t - \epsilon T - iT) + \tilde{w}(t) \qquad (4.25)$$

where $\otimes$ denotes convolution, and

$$\tilde{w}(t) = w(t) \otimes h_R(t)$$
$$g(t) = h_T(t) \otimes h_R(t); \qquad G(f) = H_T(f)H_R(f) \qquad (4.26)$$

The function $g(t)$ is the impulse response of the overall link; the receiver filter delivers a stream of pulses of shape $g(t)$ when driven by pulses with shape $h_T(t)$.

Time-continuous output of the filter is sampled (*strobed*) once per symbol interval, at times $t = mT + \hat{\tau}$ in the I channel and $t = mT + \epsilon T + \hat{\tau}$ in the Q channel, where $\hat{\tau}$ is the receiver's estimate of the correct strobing time. Translate the time origin so that $\hat{\tau} = 0$, whereupon the complex strobes are represented by

$$z_o(m) = \sum_i a_i g(m - i) + j \sum_i b_i g(m - i) + \tilde{w}(m) \qquad (4.27)$$

Receiver equations (4.25) and (4.27) have been written with an idealized assumption of perfect phase alignment in the demodulator and perfect timing alignment of the strobes. All subsequent operations on the received signal are performed on the time-discrete strobes.

Impulse Assumption

A knowledgeable reader might object, rightly, that impulses are mathematical fictions and cannot be generated in transmitters implemented with analog circuits. Rather than attempt to approximate impulses by short pulses of finite duration, most analog-implemented transmitters employ approximately rectangular pulses, usually in NRZ format. How can this near-universal practice of the real world be adapted to the simple mathematical model described above?

A simulation could convert the M-ary symbols into weighted, rectangular, NRZ pulses instead of impulses, and then apply the rectangular pulses to a filter with transfer function $H_A(f)$. This modification is applicable when the particular $H_A(f)$ is to be investigated by simulation or when the simulated transmitter must correspond closely to hardware of interest.

As an alternative, recognize that a unit-amplitude rectangular pulse of duration T is the impulse response of a filter with transfer function

$$H_{\text{rect}}(f) = \frac{\sin \pi f T}{\pi f} \tag{4.28}$$

Furthermore, the transmitter impulse model of Fig. 4.1 can be retained, and still produce rectangular pulses, by partitioning its transmitting filter as

$$H_T(f) = H_A(f) H_{\text{rect}}(f) \tag{4.29}$$

Figure 4.2 illustrates the filter partitioning and pertinent waveforms at several locations.

Transmission literature often speaks of $x/\sin x$ compensation that is applied to the transmitting filter. That is just a reflection of (4.29), which shows that the physical transmitting filter $H_A(f)$ needed to produce a specified $H_T(f)$ is

$$H_A(f) = \frac{H_T(f)}{H_{\text{rect}}(f)} = H_T(f) \frac{\pi f}{\sin \pi f T} \tag{4.30}$$

Today, many data transmitters are being designed with digital implementation. It is not necessary, or even convenient, to employ the equivalent of analog NRZ pulses in a digital transmitter. Instead, the symbols are converted to weighted, complex, *unit samples*—one sample per symbol. A digital unit sample has strong resemblance to an analog impulse, so the impulse model of Fig. 4.1 is directly applicable to a digitally implemented transmitter, without modifications to accommodate rectangular pulses.

Other Operations

Comparison of the greatly simplified model of Fig. 4.1 to the more complete but still simplified block diagram of Fig. 1.2 reveals that numerous operations have been omitted from the model. Other elements, such as amplifiers, modulators, frequency translators, demodulators, and synchronizers are crucially important for any practical link.

Our objective in the preceding paragraphs has been to reduce the transmission model to its bare essentials, thereby facilitating study of the link. Once the fundamental characteristics of the simplified model have been established, the other elements can be added back into the model.

This pattern of extreme simplification at first and then gradual introduction of complexities is a fruitful strategy for simulations and not just for analytic modeling. Building by pieces permits each piece to be of manageable size and improves the likelihood that each piece is correct.

Figure 1.2 showed a waveform generator that accepts the mapped symbols and delivers pulse waveforms. The model of Fig. 4.1 expands that waveform generator into a weighted-impulse generator, followed by a pulse-forming filter. A physical transmission link would include filters at many locations; some are illustrated in Fig. 1.2. By contrast, Fig. 4.1 shows only two filters in the link: $H_T(f)$ in the transmitter

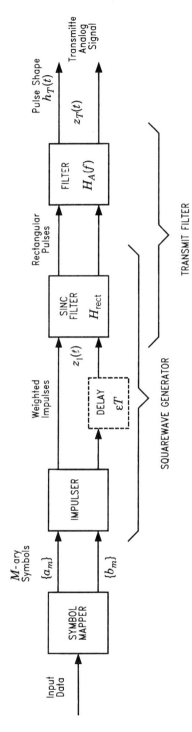

FIGURE 4.2 Partitioning of transmitting filter to produce rectangular pulses.

and $H_R(f)$ in the receiver. If the entire link is linear, all of the transmitter-side filters are combined into $H_T(f)$ and all of the receiver-side filters are combined into $H_R(f)$ for purposes of the simplified model.

If the link is nonlinear, filters must be separately located in correct relation to nonlinear elements. Nonlinear processes are introduced briefly later in this chapter and their simulation is pursued in greater detail in later chapters. Nonlinearity is an added complexity excluded from the simplified model.

4.3.2 Matched Filters

Strobes delivered by the receiver contain useful signal information but also contain noise. Best reception is obtained by maximizing the signal-to-noise ratio (SNR) of the strobes. Let the transmitted pulse shape $h_T(t)$ be fixed. What impulse response $h_R(t)$ [equivalently, transfer function $H_R(f)$] should be assigned to the receiver filter? There is an optimum transfer function, known as the *matched filter*, that maximizes the SNR as compared to all other possible filters. Its properties are described below.

White Noise Formulation

Since the simplified model of Fig. 4.1 is linear, it is permissible to consider the behavior of the system on just one isolated pulse and then superpose pulses to determine the behavior of the entire pulse train. The remainder of this section deals with isolated pulses exclusively; pulse trains are reintroduced in Sec. 4.4.

Consider a single received pulse with baseband waveshape $h_T(t)$ and energy E_p in its real signal. Energy in the complex envelope of that one pulse is $2E_p$, analogously to (2.48). Assume for now that the additive noise is white and Gaussian and that the real noise has a two-sided spectral density $N_0/2$. It follows from Sec. 2.4.1 that the spectral density of the noise complex envelope $w(t)$ is $\Psi_w(f) = 2N_0$.

The received signal and noise are processed by the receiver filter with impulse response $h_R(t)$, and the time-continuous output is strobed at an optimum time t_0 that must be established by the receiver's timing synchronizer. The strobe sample consists of a signal contribution with amplitude A and a zero-mean noise contribution with standard deviation σ_n. Signal-to-noise ratio is defined as

$$\rho^2 = \frac{A^2}{\sigma_n^2} \tag{4.31}$$

It can be shown (see any of the communications references cited in Chapter 1) that the best SNR resulting from use of the optimum $h_R(t)$ is given by

$$\text{Max } \rho^2 = \frac{2E_p}{N_0} \tag{4.32}$$

This result obtains for any transmitted pulse shape $h_T(t)$ whatever, provided that the receiver filter $h_R(t)$ is optimized for the particular h_T.

It can be further shown that the optimum receiver filter has an impulse response

$$h_R(t) = C h_T^*(t_0 - t) \qquad (4.33)$$

where C is any arbitrary complex constant; for simplicity, set $C = 1$. The optimum receiver filter then has the same impulse response as the transmit filter, except delayed by t_0, conjugated (for complex pulses), and reversed in time.

Equivalently, in the frequency domain the transfer function of the optimum receive filter is

$$H_R(f) = H_T^*(f) e^{j 2\pi f t_0} \qquad (4.34)$$

from which $|H_R(f)| = |H_T(f)|$ most clearly demonstrates the origin of the term *matched filter*. It is often said that optimum link performance is obtained by equal division of filtering between transmitter and receiver; *equal division* refers to matched filtering.

Properties

The transfer function of an overall link with a matched filter in the receiver is

$$G(f) = H_T(f) H_R(f) = H_T(f) H_T^*(f) e^{-j 2\pi f t_0}$$
$$= |H_T(f)|^2 e^{-j 2\pi f t_0} \qquad (4.35)$$

The term $|H_T(f)|^2$ is always pure real and even, irrespective of the nature of $H_T(f)$, so $G(f)$ is linear phase with delay t_0. Furthermore, the overall impulse response $g(t)$ can be shown always to be real and even-symmetric about t_0, irrespective of the shape or asymmetry of $h_T(t)$.

If $h_T(t) = \pm h_T(-t)$ (i.e., either even- or odd-symmetric) and if $t_0 = 0$, the matched filter $H_R(f)$ is exactly equal to $H_T(f)$. This condition of identical transmit and receive filters is specialized, but frequently encountered.

Although there are exceptions, a matched filter generally is nonrealizable and must be approximated, even if $H_T(f)$ is realizable. More formidably, if the pulse $h_T(t)$ has infinite duration and is causal, the matched filter $h_R(t) = h_T(-t)$ is noncausal.

A PAM receiver is particularly simple to build because it processes just one pulse shape and so requires just one matched filter. Non-PAM systems also use matched filters but may require more than one. For example, an MFSK system needs M matched filters in its receiver, one for each signaling frequency.

Matched filters have much greater significance than simply maximizing signal-to-noise ratio. The properly timed strobe sequence (correct timing is crucial) from a matched filter contains all of the useful information concerning the signal. The strobe is a *sufficient statistic* [1.4, Sec. 8.1]. Message data can be recovered optimally from properly timed matched-filter strobes without further time-continuous operations. Recovery includes the subsidiary processes of phase correction, equalization (if needed), decoding (if applicable), and symbol detection. Because they provide sufficient statistics, matched filters are widely used in communications links for all kinds of signals, not just those with PAM formats.

Examples

Several examples of input and output waveforms of matched filters are shown in Fig. 4.3. Even symmetry of the output is evident, including in particular the example for

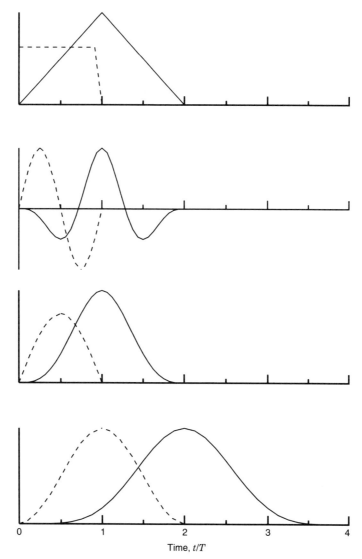

FIGURE 4.3 Examples of matched filter responses (data generated by STÆDT). *Top to bottom:* rectangular pulse, sine pulse, half-cosine pulse, raised-cosine pulse. Dashed curves, $h_T(t)$; solid curves, $g(t)$.

an odd-symmetric transmitted dipulse. All examples have finite-duration transmitted pulses, so output of the matched filter also has finite duration. It can be seen that if the transmitted pulse has duration T_1 and begins at $t = 0$, the time-continuous output of the matched filter has pulse duration $2T_1$, and its peak (where it should be strobed) occurs at $t = T_1$.

Colored Noise

If the noise spectrum $\Psi_w(f)$ is not flat, the simple formulation (4.33) or (4.34) of the matched filter is not quite correct. To obtain the matched filter in the presence of colored noise, first pass the input signal and noise through a *whitening filter* with transfer function $H_w(t)$ such that $\Psi_w(f)|H_w(f)|^2$ = constant. The Fourier transform of the signal pulse after the whitening filter will be $\Pi_1(f)H_w(f)$. Then use a receiver matched filter with the transfer function $H_R(f) = H_T^*(f)H_w^*(f)\exp(-j2\pi f t_0)$. Unless stated otherwise, any further reference to "matched filter" will imply white noise and the definition of (4.33) and (4.34).

4.4 SYMBOL DETECTION

A receiver is required to extract *estimates* $\{\hat{a}_m, \hat{b}_m\}$ (also called *decisions*) of the transmitted data symbols $\{a_m, b_m\}$ from the strobes $\{z_o(m)\}$. Provided that the receiver filter is matched to the received pulse shape and that strobe timing is correct, the strobes contain all of the information needed or available to determine the estimates.

4.4.1 Symbol-by-Symbol Detection

If (1) the *m*th strobe contains signal contributions only from the *m*th symbol, and (2) the symbols are uncorrelated with one another, the optimum estimate for the *m*th symbol is derived solely from the *m*th strobe [1.3, Chap.5]. This approach is known as *symbol-by-symbol detection*.

Sequence Estimation

If either of the conditions above is violated, symbol-by-symbol detection is not optimum. When information about one symbol is spread over multiple strobes, all of the strobes in the sequence should be examined to decide each symbol. An approach known as maximum likelihood sequence estimation (MLSE) delivers optimum decisions [4.1] when the strobes are correlated. Computationally efficient MLSE is performed using the Viterbi algorithm [4.2, 4.3].

The following are some example conditions for which information on a individual symbol appears in more than one strobe:

- In dispersive channels, unwanted intersymbol interference (ISI) causes each strobe to have contributions from more than one pulse.

- In partial response signaling, controlled ISI is introduced deliberately.
- In continuous-phase modulation (CPM), the phase at any strobe includes contributions from all symbols that have preceded the current one because of memory in the modulation scheme.
- In convolutionally encoded sequences, the encoding process introduces correlations among the transmitted symbols.

Each of these examples requires MLSE for optimum symbol estimation.

Both types of detection are often investigated by means of simulations. This work concentrates on symbol-by-symbol detection because it is much the simpler to explain and apply. A user might best become proficient in simulation principles first before attacking MLSE problems.

4.4.2 Decision Rules

A complex strobe $z_o(m)$ delivered from a coherent PAM/QAM receiver can be represented as a single point in the complex I–Q plane. In the absence of all disturbances and receiver errors, that point will coincide with the constellation data point $(a_m + jb_m)$ that was sent from the transmitter. Typical constellations were illustrated in Fig. 2.12.

Decision Regions

Symbol data must be recovered from the strobe value, but disturbances cause the strobe value to deviate from the exact constellation point. How should the strobe be interpreted to effect that recovery?

Assume that all symbols in the constellation are a priori equally likely to have been transmitted, that disturbances affect all strobes in the same statistical manner irrespective of location in the constellation, and that disturbance-caused displacement of a strobe from the constellation point is equally probable in all directions. Then it is reasonable to decide that the estimate $(\hat{a}_m + j\hat{b}_m)$ of the mth symbol should take the value of the constellation point lying closest to the strobe point. (The caret ^ indicates an estimated value.)

Each constellation point will be surrounded by an associated *decision region* in the I–Q plane. Every point inside an individual decision region lies closer to the associated constellation point than it does to any other constellation point. If a strobe falls in a given decision region, the symbol detector declares that the associated constellation point was the one that was transmitted.

Properties of Boundaries

A boundary between two decision regions is the locus of points equidistant from the nearest constellation points. Consider two adjacent constellation points. The locus of points equidistant from them is the perpendicular bisector of the line segment that joins them. Thus the boundaries for coherent QAM decisions must be composed of

114 ☐ REPRESENTATION OF ANALOG SIGNAL PROCESSES

line segments. If a strobe should fall exactly on a boundary, the decision can be assigned to either constellation point arbitrarily.

In general, a boundary of a decision region has a polygonal shape, where each side of the polygon is a line segment midway between two adjacent constellation points. Typical examples may be seen in Fig. 4.4. Polygonal boundaries can be degenerate; some polygons might have one open side, while others might be reduced to a single straight line of infinite extent.

A symbol detector is a device or algorithm that has knowledge of decision boundaries. This knowledge of the boundaries is unique to a given signal format; a different

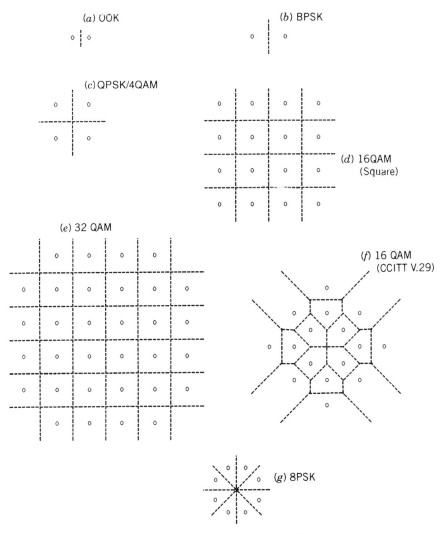

FIGURE 4.4 Decision regions of typical constellations. Dashed lines show decision boundaries.

set of decision rules is needed for each signal format to be received or simulated. The detector accepts receiver strobes, one at a time, and outputs digital numbers ($\hat{a}_m + j\hat{b}_m$) representing the decisions for each symbol. Symbol detectors are also called *slicers* or *quantizers*.

Two-dimensional constellations are associated with bandpass or complex signals. With a real baseband signal, the two-dimensional plane degenerates to a straight line. Constellation points are all located along the line; decision boundaries become threshold points lying midway between constellation points.

If the constellation points of a complex signal form a rectangular grid, the I and Q components can be decided separately, each as a one-dimensional decision along a straight line. If the modulation is offset, the I and Q strobes will be taken at separated time instants and can be decided separately also. If all points of a constellation have the same magnitude (e.g., any MPSK constellation), the decision can be made without knowledge of the signal amplitude. Binary baseband signals such as BPSK and QPSK are particularly simple; the symbol detector need only decide the polarities of the strobe's I and Q components (just one component for BPSK).

If the constellation has points of differing amplitudes, either the signal must be normalized to a standard amplitude before being applied to the symbol detector or the decision boundaries in the symbol detector must be adapted to the amplitude of the received signal. Level control is needed in physical receivers and in simulation models. Inaccuracies in level control will impair detection performance.

Decision Terminology

A symbol detector delivers so-called *hard decisions*. If data symbols are correlated (as they would be if the data had been encoded before transmission), hard decisions will destroy some information that is present in the strobes and will degrade performance of any subsequent operations (e.g., decoding) that take advantage of the correlation. Full benefit of decoders (or other devices that utilize correlation between symbols) is gained by using the strobes themselves and not the hard decisions.

Decoders are often fed with *soft decisions* instead of either unquantized strobes or hard decisions. A soft decision is a quantized digital representation of the strobe, usually to a numerical precision that is just sufficient not to unduly degrade performance of the subsequent operations. By contrast, one might think of the strobe itself as an analog quantity or as a digital number of relatively high precision.

Soft decision, and more generally quantization, is an important topic that is considered in Chapter 15, on simulation of digital processes. It really is quite distinct from the symbol detector process and the two should not be confused with one another.

Non-PAM Symbol Decisions

Many signals cannot be represented in terms of PAM–QAM constellations; MFSK is one example. Their symbol detectors are quite different from the models shown for PAM–QAM. See [1.2, Chap. 4] for further information.

4.5 NONLINEAR PROCESSES

Some processes in a link simulation are linear; others are nonlinear. Ordinary filters are a prime example of linear processes, while symbol detectors or other quantizers are inherently nonlinear. Amplitude modulation was shown to be linear and angle modulation to be nonlinear. Among simple processes, addition is linear but multiplication might be characterized either way, depending upon application. Among more complicated processes, Fourier transformation is linear but autocorrelation is nonlinear. Nonlinear processes are often encountered in simulations and frequently require special treatment not needed by linear processes. In this section we introduce nonlinear processes; in later chapters we examine techniques for their simulation.

Notation

A memoryless nonlinearity operating on a real input signal $s_i(t)$ and producing a real output signal $s_o(t)$ will be represented by

$$s_o(t) = \gamma[s_i(t)] \tag{4.36}$$

The restriction to *memoryless* (or *zero-memory*) nonlinearities is expanded on below. A memoryless envelope nonlinearity operating on a complex input signal $z_i(t)$ and producing a complex output $z_o(t)$ will be represented by

$$z_o(t) = \Gamma[z_i(t)] \tag{4.37}$$

Envelope nonlinearities are introduced later.

Superposition

Let a process be represented by $s_o(t) = f[s_i(t)]$. The process is linear if the superposition condition

$$f[s_1(t) + s_2(t)] = f[s_1(t)] + f[s_2(t)] \tag{4.38}$$

applies, and otherwise is nonlinear. As a corollary to superposition, a linear process scales linearly:

$$f[as_i(t)] = af[s_i(t)] \tag{4.39}$$

where a is any number. The scaling corollary is necessary for a process to be linear, but is not sufficient. Addition of two signals clearly meets the superposition condition and is unequivocally a linear process. Multiplication of signals is more complicated, as will be demonstrated next.

Define a multiplier by the relation $s_o = s_1 s_2$ (omitting time arguments for the sake of brevity). If $s_1 = s_{1a} + s_{1b}$, the output is $s_o = (s_{1a} + s_{1b})s_2 = s_{1a}s_2 + s_{1b}s_2$.

Since the product fulfills the superposition condition, a multiplier is linear in one input signal when that signal is considered alone. It is this property of multipliers that makes amplitude modulation a linear process.

But now apply the same signal to both ports. That is, let $s_2 = s_1$, so that $s_0 = s_1^2$, whereupon $(s_{1a} + s_{1b})^2 = s_{1a}^2 + s_{1b}^2 + 2s_{1a}s_{1b} \neq s_{1a}^2 + s_{1b}^2$. A multiplier's output does not fulfill the superposition condition, so the process is nonlinear if the same signal is applied to both ports. Similarly, it can be shown that a multiplier product does not fulfill the superposition condition if pairs of signals are added at both input ports.

Memoryless Nonlinearities

A process is said to have zero memory if output at time t depends only upon input at time t. A process is said to have memory if output at time t depends upon inputs or outputs from times earlier than t. The few example nonlinearities mentioned above are all memoryless. The notation of (4.36) and (4.37) applies properly only to memoryless nonlinearities. In general, there is no comparably straightforward way to model most nonlinearities with memory.

One expedient for nonlinearities with memory, sometimes successful, is to approximate the process as a zero-memory nonlinear device connected between linear processes (usually filters) with memory, as in Fig. 4.5. When that can be accomplished satisfactorily, the problem reduces to processes that can be modeled by manageable techniques. Another approach to nonlinearities with memory is to employ a Volterra series. The interested reader should consult [1.5, Chap. 10] and the references cited therein.

Only zero-memory nonlinearities are considered in this book.

Envelope Nonlinearities

Let's explore this topic by way of examples. Consider a real signal

$$s_i(t) = R(t) \cos\left[2\pi f_c t + \phi(t)\right] \tag{4.40}$$

and suppose that it is applied to a square-law nonlinearity $\gamma(s) = s^2$, whereupon the output signal is given by

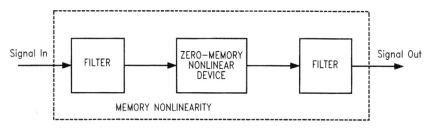

FIGURE 4.5 Approximation model of nonlinear device with memory.

$$s_o(t) = s_i^2(t) = \tfrac{1}{2}R^2(t) + \tfrac{1}{2}R^2(t)\cos[4\pi f_c t + 2\phi(t)] \qquad (4.41)$$

There are two distinct spectral components of the output: one centered at zero frequency and the other at $2f_c$—twice the input carrier frequency.

A simulationist does not want to compute the high-frequency cosine term if it can be avoided and so would like to use the complex envelope instead, if possible. The complex envelope of the input signal is $z_i(t) = R(t)e^{j\phi(t)}$. Either of two different envelope nonlinearity models might be used, depending upon which component of (4.41) constituted the useful output. In a square-law *detector*, the zero-frequency component is retained for the envelope, while in a *frequency doubler*, only the double-frequency component is of interest.

Detector model:

$$\Gamma[z(t)] = |z|^2, \qquad z_o(t) = R^2(t) \qquad (4.42)$$

Frequency-doubler model:

$$\Gamma[z(t)] = z^2, \qquad z_o(t) = R^2(t)e^{j2\phi(t)}$$

COMMENT 1: The square-law device has only two output components, so there are only two possible envelope-nonlinearity models. Higher-degree nonlinearities generate more than two components and so can have more than two models for envelope nonlinearities. A user is required to choose the applicable component and its model.

COMMENT 2: The envelope input and output for the square-law example are shown in polar coordinates, which is helpful for visualizing the nature of the output but could be computed equally well in rectangular coordinates. Other nonlinearities might be much easier to handle in one coordinate system than in the other.

For another example, consider a hard limiter with real-signal nonlinearity:

$$\gamma[s(t)] = \operatorname{sgn}[s(t)] = \begin{cases} +1, & s > 0 \\ -1, & s < 0 \\ 0, & s = 0 \end{cases} \qquad (4.43)$$

When the signal (4.40) is applied to this nonlinearity, the output

$$s_o(t) = \operatorname{sgn}[\cos(2\pi f_c t + \phi(t))] \qquad (4.44)$$

is a square wave with fundamental center frequency f_c and phase modulation $\phi(t)$. The amplitude $R(t)$ has been wiped out.

Because of its wide bandwidth, the output of the hard limiter does not have a well-defined complex envelope. Often, though, only the fundamental component of the output is passed farther down the link. The fundamental has a complex envelope of $e^{j\phi(t)}$ and implied frequency f_c. Only the phase remains; the amplitude has been wiped out.

The hard-limiter envelope nonlinearity could be modeled by $\Gamma[z(t)] = z_i(t)/R(t)$, provided that $R(t) \neq 0$. This model applies irrespective of whether z_i is in polar or rectangular coordinates. Alternatively, the model could be constructed, in polar coordinates, by replacing the varying $R(t)$ by a fixed amplitude of unity. This second method is exactly equivalent to the first [except in the rare instances when $R(t) = 0$] but is rather easier to implement. The first method has a closed-form model for $\Gamma[z(t)]$, but the second does not. Nonetheless, simulation processes can implement either method.

Further Development

Nonlinearities are a diverse group of processes with widely differing properties. From the simulation standpoint, they share the central characteristic that an output value is computed from an input signal value inserted into a function $\gamma[s]$ or $\Gamma[z]$. In this sense, to the computer all memoryless nonlinearities are the same at their core. They differ primarily in the nature of the nonlinear functions that describe them. Even among the few simple examples presented above, there appears (1) closed-form analytical formulas (square-law device), (2) piecewise switching functions (hard limiter, quantizer), and (3) replacement rules [setting $R(t) = 1$ for envelope limiter]. Often a nonlinearity is described only by experimental data in a table; function evaluation for simulation might then consist of interpolation in the table.

Characteristics of the outputs of many nonlinear processes depend upon the amplitude of their inputs; most nonlinear processes do not scale with amplitude in the sense of (4.39). A change of input amplitude could have major effects upon output waveform and spectrum. By contrast, change of input amplitude in a linear process affects nothing but the amplitude of the output. For that reason it is often necessary to control the input amplitude to a nonlinear process in a manner not required for linear processes. These matters, and others, are treated further in later chapters.

4.6 KEY POINTS

This chapter has introduced features of several diverse processes:

- Modulators
- Demodulators
- PAM link model; matched filters
- Symbol detectors
- Nonlinear processes

4.6.1 Modulators

Passband signals are often simulated with a complex-envelope model so as to avoid computing the fine, immaterial details of the sinusoidal carrier.

Amplitude Modulation. For a linear modulation, the complex envelope is identical to the I–Q complex baseband signal. Therefore, simulations often do not include an explicit amplitude modulator, even when simulating modulated signals.

Angle Modulation. The complex envelope of an angle-modulated signal is

$$z(t) = e^{j\phi(t)} = \cos \phi(t) + j \sin \phi(t) \tag{4.6}$$

where $\phi(t)$ is the desired angle modulation. The same model can be used for a phase shifter: $\phi(t) = \phi_0 =$ constant, or for a frequency shifter: $\phi(t) = 2\pi f_1 t$.

4.6.2 Demodulators

I-Q Demodulation

Real Signals. Multiply the received real signal

$$s(t) = x(t) \cos (2\pi f_c t + \theta_i) - y(t) \sin (2\pi f_c t + \theta_i) \tag{4.7}$$

against the quadrature local references

$$\mu(t) = 2 \cos (2\pi f_c t + \theta_0); \qquad \nu(t) = -2 \sin (2\pi f_c t + \theta_0) \tag{4.8}$$

to obtain the difference-frequency I–Q products,

$$\begin{aligned} I(t) &= x(t) \cos \Delta\theta - y(t) \sin \Delta\theta \\ Q(t) &= x(t) \sin \Delta\theta + y(t) \cos \Delta\theta \end{aligned} \tag{4.10}$$

where $\Delta\theta = \theta_i - \theta_0$. The original baseband components $x(t)$ and $y(t)$ are recovered if $\Delta\theta = 0$.

Equivalently, multiply the received analytic signal,

$$s^+(t) = [x(t) + jy(t)] e^{j(2\pi f_c t + \theta_i)} \tag{4.11}$$

by the reference analytic signal,

$$e^{-j2\pi f_c t} e^{-j\theta_0} \tag{4.12}$$

to obtain the complex baseband product,

$$z(t) = x(t) \cos \Delta\theta - y(t) \sin \Delta\theta + j[x(t) \sin \Delta\theta + y(t) \cos \Delta\theta] \tag{4.13}$$

The real baseband I–Q outputs are related to the complex baseband output by

$$I(t) = \text{Re}[z(t)]$$
$$Q(t) = \text{Im}[z(t)]$$

Demodulator Coherence

Three different levels of coherence may be encountered in various demodulators:

1. *Coherent Demodulation.* Phase error $\Delta\theta$ must be small. The local reference typically will be phase-locked to the incoming signal. Coherent I–Q demodulation is necessary for separate recovery of original x and y quadrature modulation components.

2. *Differentially Coherent or FM-Discriminator Demodulation.* Phase error $\Delta\theta$ is unknown. Instantaneous frequency error—the derivative $d(\Delta\theta)/dt$ of phase error—must be small compared to the rate of change of the signal phase.

    a. *Differential Demodulation.* Information is recovered from phase changes from symbol to symbol:

    $$\bar{z}(m) = z_0(m) z_0^*(m-1) \qquad (4.16)$$

    where $z_0(m)$ is the mth complex-symbol strobe from the receiver filter and $\bar{z}(m)$ is the complex-strobe output of the differential-demodulation process.

    b. *FM Discrimination.* Information can be recovered from the rate of change of phase of the baseband complex envelope according to

    $$\frac{d}{dt}\left(\tan^{-1}\frac{Q}{I}\right) = \frac{I\dot{Q} - \dot{I}Q}{I^2 + Q^2} \qquad (4.20)$$

3. *Noncoherent Demodulation.* Both phase error and its derivative are unknown. Information is carried in the magnitude

$$|z(t)|^2 = R^2(t) = I^2(t) + Q^2(t) \qquad (4.21)$$

4.6.3 PAM Link Model; Matched Filters

PAM Link Model

Data-weighted impulses drive a linear transmitter, which has a pulse-shaping filter with impulse response $h_T(t)$ and frequency response $H_T(f)$. Signal and noise are applied to a linear receiver which has a filter with impulse response $h_R(t)$ and frequency response $H_R(f)$. The overall PAM link has responses

$$g(t) = h_T(t) \otimes h_R(t)$$
$$G(f) = H_T(f)H_R(f) \qquad (4.26)$$

where $\otimes$ denotes convolution.

The receiver's filter is strobed once per symbol. Strobes have the form

$$z_0(m) = \sum_i a_i g(m-i) + j \sum_i b_i g(m-i) + \tilde{w}(m) \qquad (4.27)$$

where $a_i + jb_i$ is the transmitted ith symbol and $\tilde{w}(m)$ is the noise contribution to the mth strobe. (Timing and carrier-phase errors have been neglected.)

Matched Filters

To maximize the signal-to-noise ratio (SNR) of the strobe (4.27), use matched filters such that

$$h_R(t) = h_T^*(t_0 - t) \qquad (4.33)$$

or, equivalently,

$$H_R(f) = H_T^*(f) e^{-j2\pi f t_0} \qquad (4.34)$$

The SNR from matched filtering of an isolated pulse will be

$$\text{SNR} = \frac{A^2}{\sigma_n^2} = \frac{2E_p}{N_0} \qquad (4.31),(4.32)$$

where A is the amplitude of signal portion of the strobe, σ_n the rms noise contribution, E_p the pulse energy, and N_0 the one-sided spectral density of input white noise.

A properly timed strobe from a matched filter is a *sufficient statistic* for the associated data symbol. The strobe contains all of the information needed, or available, for recovery of the symbol value, carrier phase, decoding operations, or equalization.

4.6.4 Symbol Detection for PAM–QAM Coherent Reception

Decision Regions

Each strobe maps as a single point in the complex I–Q plane. *Decision rule:* Assign the value of the constellation point closest to the strobe point as the best estimate of the transmitted symbol. Each constellation point is surrounded by its associated decision region. Any point inside a region is closer to the associated constellation point than it is to any other point of the constellation. Any strobe falling inside a given decision region is assigned the data value of the associated constellation point.

Decision Boundaries

Boundaries between decision regions are composed of line segments, in general forming polygons. Boundary lines lie midway between the nearest constellation points.

4.6.5 Nonlinear Processes

Memoryless (or zero-memory) nonlinearities can be represented in real-signal format as

$$s_o(t) = \gamma[s_i(t)] \tag{4.36}$$

or in complex format as

$$z_o(t) = \Gamma[z_i(t)] \tag{4.37}$$

where the subscripts i and o denote input and output, respectively, and $\gamma[\cdot]$ and $\Gamma[\cdot]$ are the nonlinear functions. Several examples of zero-memory nonlinearities are provided in each format. Nonlinear processes with memory cannot be represented nearly so compactly and so are omitted from this account.

4.6.6 STÆDT Processes

DDET: differential demodulator
DIFD: differential decoder
DIFE: differential encoder
FFT: fast Fourier transform
LETS: sine and cosine for angle modulation, I^2+Q^2 for envelope detection, plus many other operations on complex signals
LMTR: various limiters, plus other nonlinear operations
MDET: symbol detector
MULT: multiplies two complex signals
RTAT: phase rotator
SHFT: signal translation (time or frequency)

5

DIGITAL SIMULATION OF ANALOG SIGNALS AND PROCESSES

In this chapter we introduce the elements of simulation of analog signals and processes on a digital computer. Models for specific analog processes are treated in later chapters. Models for digital signals and processes are addressed afterward, once the foundations for analog models have been established.

5.1 DIGITIZATION

Analog signals exist in continuous time and take on continuous values of amplitude. These physical signals often last for long time intervals and are frequently modeled analytically as having effectively infinite duration. Signal quantities are measured in units with dimensions of volts, amperes, watts, and so on. Simulation on a digital computer does away with all of these characteristics of physical analog signals:

- A simulated signal must be sampled, so it exists only at discrete times.
- The digital samples can be represented only to finite precision, so the signal amplitude is quantized to discrete numbers.
- Long simulations are costly in computer resources, so it is desirable for simulations to be short in duration. But long signal durations are needed for statistically reliable results; there is an ever-present conflict between these two requirements.
- The samples are just digital numbers and have no intrinsic physical meaning; they are dimensionless.

In short, underlying a computer simulation there is a digital mathematical model that is an approximation to an analog mathematical model that approximately describes the behavior of physical analog signals and processes. Although the multiple layers of approximations can be quite accurate in total, a simulationist must always be aware that it is abstract models that are being manipulated and must always question the applicability of the models to the specific problem being investigated.

In this chapter and in those to follow, the approximation of an analog model by a digital model will be examined repeatedly. The analog models themselves will be regarded as valid, established representations of physical reality.

5.1.1 Quantization

The amplitude-discrete character of a digital simulation can be disposed of quickly for now. Amplitude-continuous analog quantities usually are simulated by floating-point numbers, which are finely quantized in any reasonable computer or programming language. In our experience, single-precision floating-point representation (six or seven significant decimal digits) has been adequate for much communications simulation. Some exceptional situations require double precision; examples will be identified as they arise.

From a user's standpoint, floating-point representation of numbers is nearly indistinguishable from the infinite precision of continuous analog amplitudes. Computer quantization of amplitudes of analog signals will be ignored henceforth, except where quantization as such becomes an issue in the simulation model.

When dealing with digital signals, the number of bits to be used for each operation, and the scaling that must accompany limited precision, become important issues that do not arise with analog signals. Simulation is commonly employed to explore the effects of different levels of quantization and scaling. For the most part, quantization of digital signals tends to be coarser than that afforded by single-precision floating-point representation. Quantization for simulations of digital signals is deferred to Chapter 15.

5.1.2 Sampling

Consider a complex time-continuous analog signal $z_a(t)$, which has either a Fourier transform $Z_a(f)$ or a spectral density $\Psi_{za}(f)$. Subscript a indicates an analog quantity. Complex representations (either complex envelopes, complex baseband signals, or analytic signals) are employed throughout for the sake of generality and because of their importance in simulations. A real signal then is just a special case of a complex signal.

Let the analog signal be sampled by multiplying it by an infinite train of unit impulses, uniformly spaced by the *sampling interval* t_s. The resulting complex time-discrete *analog* signal is represented by

$$z_\delta(t) = z_a(t) \sum_{n=-\infty}^{\infty} \delta(t - nt_s - \tau) = \sum_{n=-\infty}^{\infty} z_a(nt_s + \tau)\delta(t - nt_s - \tau) \qquad (5.1)$$

126 □ DIGITAL SIMULATION OF ANALOG SIGNALS AND PROCESSES

Time shift τ ($0 \leq \tau < t_s$) is introduced to provide distinct time origins for $z_a(t)$ and the sampling wave. [The subscript δ is a temporary notation to indicate a sampled *analog* signal, consisting of a train of complex-weighted uniformly spaced physically nonrealizable impulses. Equation (5.1) is an intermediate representation and is not the digitized signal employed in simulations.]

If the Fourier transform $Z_a(f)$ of $z_a(t)$ exists, the Fourier transform of $z_\delta(t)$ can be obtained through formal manipulations as

$$Z_\delta(f) = \frac{1}{t_s} \sum_{k=-\infty}^{\infty} e^{-j2\pi k\tau/t_s} Z_a\left(f - \frac{k}{t_s}\right) \tag{5.2}$$

which consists of an infinite number of periodic images of $Z_a(f)$, repeated at frequency intervals $1/t_s$. Equation (5.2), written with $\tau = 0$, is found in all standard references (e.g., [1.9] to [1.12]) on sampled systems.

Signal Reconstruction

If $Z_a(f)$ is bandlimited and the sampling frequency $1/t_s$ is large enough such that

$$Z_a(f) \equiv 0, \qquad |f| \geq \frac{1}{2t_s} \tag{5.3}$$

the periodic images of $Z_a(f - k/t_s)$ do not overlap and it is possible to reconstruct $z_a(t)$ exactly from the sampled analog signal by passing $z_\delta(t)$ through an ideal rectangular filter with cutoff frequency $1/2t_s$. [If $Z_a(f)$ is a bandpass function, reduced sampling rates may be adequate. See [1.12, Sec. 6.1.5] or [5.1] for further explanation. Bandpass sampling is important in digitally implemented hardware but need not be invoked to explain the foundations of simulation.]

Since $z_a(t)$ can be recovered exactly from the samples in $z_\delta(t)$, provided that (5.3) is satisfied, all of the information contained in $z_a(t)$ must also be in the samples. Therefore, a strictly bandlimited, adequately sampled signal can be simulated digitally without any distortion whatever arising from the sampling.

5.1.3 Aliasing

No physical signal is strictly bandlimited. No signal of finite duration can be bandlimited. Although in some applications it is reasonable to idealize a signal as being almost bandlimited, in many other applications the signal is not even nearly bandlimited (e.g., the often-encountered "NRZ" rectangular pulse shape of Figs. 2.1 and 2.2).

Equation (5.3) is not satisfied for a nonbandlimited signal, so the periodic images in $Z_\delta(f)$ overlap one another and cannot be separated by any filter. The original $z_a(t)$ then cannot be reconstructed exactly from the samples of $z_\delta(t)$; *aliasing distortion* will be incurred. Spectral components from signal frequencies lying beyond the *folding* or *Nyquist* frequency $1/|2t_s|$ are translated to frequencies of less than $1/|2t_s|$. The translated frequency is an *alias* of the correct frequency.

Aliasing is fundamental; it goes to the heart of digital-computer simulation of analog signals. Some distortion due to aliasing will be present in just about every simulation; this source of error is known to exist before the simulation even begins. Yet pressure to simulate realistic (nonbandlimited) signals is unrelenting, despite the aliasing errors. What is a simulation engineer to do?

The answer is to select the sampling frequency large enough so that aliasing distortion is acceptably small. Practical communications signals have spectra that roll off at higher frequencies, so that an increase in sampling rate reduces distortion caused by aliasing. A higher sampling rate causes the periodic images to be spaced farther apart so less signal energy is aliased to the wrong frequencies. It is incumbent upon the user to assess the aliasing distortion and to select the sampling frequency large enough that the effects of aliasing are tolerably small. Guidance to aid that assessment is offered in the succeeding paragraphs.

Properties of Signal Reconstruction

A simplified model for sampling and reconstruction are shown in Fig. 5.1. The input signal $z_a(t)$ is sampled to produce the sequence of impulses $z_\delta(t)$ of (5.1). These uniformly spaced complex impulses are applied to an ideal nonrealizable lowpass filter with transfer function

$$H(f) = \begin{cases} t_s, & |f| < \dfrac{1}{2t_s} \\ 0, & |f| \geq \dfrac{1}{2t_s} \end{cases} \quad (5.4)$$

or, equivalently, impulse response

$$h(t) = \frac{\sin \pi t/t_s}{\pi t/t_s} \quad (5.5)$$

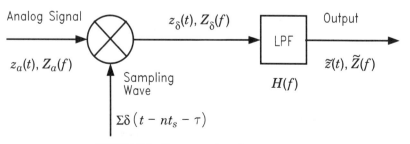

FIGURE 5.1 Reconstruction of sampled signal.

Output of the filter, denoted $\tilde{z}(t)$, is supposed to be a reconstruction of the analog input $z_a(t)$. Output in the frequency domain is given by

$$\tilde{Z}(f) = Z_\delta(f)H(f) \qquad (5.6)$$

or in the time domain by the discrete convolution

$$\tilde{z}(t) = z_\delta(t) \otimes h(t) = \sum_{n=-\infty}^{\infty} z_a(nt_s + \tau) \frac{\sin \pi \left(\frac{t-\tau}{t_s} - n\right)}{\pi \left(\frac{t-\tau}{t_s} - n\right)} \qquad (5.7)$$

Some properties of the reconstructed output $\tilde{z}(t)$ include:

- It is a *time-continuous analog* signal.
- It is *bandlimited* to $|f| < \frac{1}{2}t_s$ because of the bandlimiting filter.
- It is the *unique* bandlimited signal that interpolates the samples of the input signal. That is, $\tilde{z}(nt_s + \tau) \equiv z_a(nt_s + \tau)$.
- From the preceding property it can be seen that $\tilde{z}(t)$ depends upon τ, unless $z_a(t)$ itself is bandlimited to $|f| < \frac{1}{2}t_s$.
- There are an infinite number of different nonbandlimited input signals that could have produced the same samples $z_\delta(t)$ and therefore an infinite number of different input signals that could be reconstructed as any particular $\tilde{z}(t)$.
- The output has infinite time duration (a necessary property of any bandlimited signal), even if $z_a(t)$ is time limited.

Aliasing Error

Error of reconstruction is defined as $\tilde{z}(t) - z_a(t)$; this quantity is commonly designated *aliasing error* (but see the comment below regarding terminology). A simulation engineer needs a simple quantitative evaluation of that error to be able to determine the suitability of sampling rates for nonbandlimited signals. Methods for analysis of aliasing error are summarized in [5.2, Sec. VI.B]. The paragraphs below apply some of those methods to simulations. (Reference [5.2] is a detailed survey, with copious references, of the extensive literature on sampling that has been published over many years.)

Aliasing Energy/Power. Assume that $z_a(t)$ is stationary, so that it must be described in the frequency domain by its spectral density $\Psi_{za}(f)$. Thomas and Liu [5.3] demonstrate that the variance (i.e. "power") of the aliasing error is

$$\sigma_A^2 = \text{Avg}[|\tilde{z}(t) - z_a(t)|^2] = 2 \int_{|f| \geq 1/2t_s} \Psi_{za}(f) \, df \qquad (5.8)$$

Similarly, if $z_a(t)$ has a Fourier transform $Z_a(f)$, the aliasing energy averaged over τ is shown in Appendix 5A to be

$$\overline{E}_A = \text{Avg}_\tau \left[\int_{-\infty}^{\infty} |\tilde{z}(t) - z_a(t)|^2 \, dt \right] = 2 \int_{|f| \geq 1/2t_s} |Z(f)|^2 \, df \qquad (5.9)$$

These results have a simple physical interpretation. The integrals in (5.8) and (5.9) represent the power or energy in the signal at frequencies beyond the folding frequency $|f| = \frac{1}{2} t_s$. A reconstructed signal is distorted twice by these out-of-band components: thus the factor of 2 in each equation. One contribution arises from simple linear distortion when true components of the signal at outer frequencies are excluded from the reconstruction, and one contribution arises when outer-frequency aliases are included as interference. Appendix 5A shows the contributions to be equal, on average.

More exact terminology might denote the first contribution as *spectral truncation error* and reserve the term *aliasing error* exclusively for the folded component. The two components together constitute the reconstruction error. In principle, the aliasing component can be prevented by ideal *antialias filtering* of the signal prior to sampling. Spectral truncation cannot be avoided by any means. Moreover, although antialias filtering is possible (and usually necessary) in sampling of a physical signal, it is not possible in a computer simulation, where the signal can exist only in sampled form. Both components of reconstruction error will necessarily be incurred.

The folded component in a particular realization of $\tilde{z}(t)$ depends upon τ. Expressions (5.8) and (5.9) are statistical averages over τ; averaging is needed to obtain simple formulations independent of time shift. According to [5.4], the aliasing variance for the worst-case choice of τ does not exceed twice that shown in (5.8).

Total power or energy of the signal is given by the integral of Ψ_{za} or $|Z_a|^2$ over all frequencies. The ratio of aliasing power (energy) to total power (energy) can be interpreted as an alias-to-signal ratio and used as a compact indication of the degree of aliasing incurred with any particular choice of sampling rate.

Aliasing Magnitude. Appendix 5B and several references cited in [5.2] demonstrate that the instantaneous error of reconstruction is upper bounded by

$$|\tilde{z}(t) - z_a(t)| \leq 2 \int_{|f| \geq 1/2t_s} |Z_a(f)| \, df \qquad (5.10)$$

Although (5.10) has an attractively simple form, two features limit its applicability:

(1) the formula is an upper bound, which may greatly exceed the actual error, and (2) the integral in (5.10) does not necessarily converge. An example of an infinite bound (a highly pessimistic result) is given shortly.

Tighter Bounds. References [5.5] and [5.6] derive improved bounds on the magnitude of aliasing, but these tend to be more difficult to evaluate or are applicable only to restricted situations.

Aliasing Examples

Rectangular Pulse. Many communications links employ rectangular pulses (the so-called "NRZ" waveform), so signals based on these pulses often must be simulated. The discontinuous edges of a rectangular pulse generate a broad spectrum, requiring dense sampling if aliasing is to be constrained to tolerable bounds. Aliasing of rectangular pulses establishes an approximate worst case that will not be exceeded (or even approached) in most other circumstances.

Consider a rectangular pulse with unit amplitude and duration T:

$$z_a(t) = \begin{cases} 1, & |t| < T/2 \\ 0, & |t| > T/2 \end{cases} \tag{5.11}$$

The pulse has a Fourier transform

$$Z_a(f) = \frac{\sin \pi f T}{\pi f} \tag{5.12}$$

The pulse is to be sampled at a rate $1/t_s$ samples per second. Define the ratio

$$M_s = \frac{T}{t_s} \tag{5.13}$$

which is the average number of samples per symbol (averaged over the timing shift τ).

Notation introduced here will be employed throughout the sequel. Sampling interval will always be denoted t_s, the notation T will be used to signify the interval between symbol pulses in a data signal, and the ratio M_s is the number of samples per symbol, an essential parameter in simulations. Although M_s can take on any real positive value, it is usually much easier to generate a simulated signal if M_s is an integer. Signal generation is addressed in Chapter 7.

Aliasing energy is given by (5.9) and total energy of a unit-amplitude rectangular pulse is T. Inserting (5.12) and (5.13) into (5.9) and dividing by T gives the relative aliasing energy (alias-to-signal ratio, denoted η^2) for the rectangular pulse:

$$\eta^2 = \frac{\overline{E_A}}{T} = 4T \int_{M_s/2T}^{\infty} \frac{\sin^2 \pi f T}{(\pi f T)^2} df$$

$$= \frac{8}{\pi^2 M_s} \sin^2 \frac{\pi M_s}{2} + 2 - \frac{4}{\pi} \text{Si}(\pi M_s) \qquad (5.14)$$

where Si(·) is the sine-integral [5.7, Chap. 5]. [A one-sided integration, multiplied by 2, is correct because $z_a(t)$ is real, so $|Z_a(f)|^2$ is even-symmetric about $f = 0$.]

Aliasing ratio η^2 of a rectangular pulse is tabulated below for several values of M_s.

M_s	η^2
4	0.10
8	0.05
16	0.025
32	0.0126
64	0.0063

Relative aliasing rolls off only as $\eta^2 \propto 1/M_s$: a discouraging reward for effort expended.

When the $\sin x/x$ spectrum (5.12) of the rectangular pulse is applied to the aliasing magnitude bound of (5.10), the integral in the bound does not converge [2.1, p. 9]. Since the actual reconstruction error is always finite, this infinite bound on the error magnitude is excessively pessimistic. More to the point, the magnitude bound provides no useful information with rectangular pulses nor with any other pulse shape that has a discontinuous waveform.

A reconstruction of a rectangular pulse from four nonzero samples is shown in Fig. 5.2. Timing shift was set at $\tau = 0+$ and the sample locations are indicated by open circles. The reconstructed waveshape was calculated from (5.7).

Four samples are not sufficient for good reconstruction of a rectangular pulse. Denser sampling would provide steeper transitions and a better approximation to a flat top. The oscillations would die off more rapidly. But the peak amplitudes of the overshoots would remain unaltered; overshoot is a manifestation of the Gibbs phenomenon [2.1, 2.2] that attends the reconstruction of all discontinuous waveforms. In light of the reconstruction properties listed after (5.7), take note that the illustrated recovered waveform (1) passes through the sample values and (2) has oscillating tails before and after the main pulse.

As time shift τ increases from 0+ (as in Fig. 5.2) toward t_s, the original square pulse remains fixed while the samples and the reconstructed waveform move to the right. If M_s is an integer, the reconstructed waveform is the same for all τ; only its position changes. If M_s is not an integer, there will be either IP[M_s] or 1 + IP[M_s] samples in the pulse interval T, where IP[x] denotes the integer part of x. The recon-

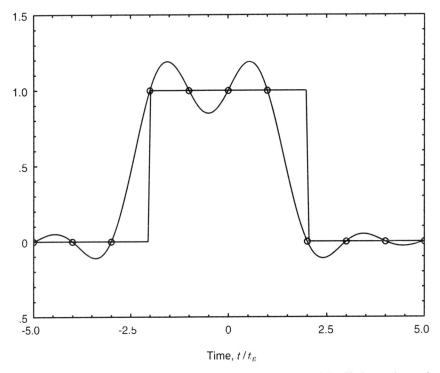

FIGURE 5.2 Square pulse reconstructed from four samples in pulse width. Circles mark samples for $\tau = 0+$.

structed waveform then depends upon τ because the number of nonzero samples changes with τ.

Figure 5.2 demonstrates that sampling imposes an implicit rise time of t_s upon an erstwhile square wave that would have zero rise time in continuous form. Therefore, the sampling interval must be chosen small enough that the reconstructed rise time is suitable to the application being simulated. Simulations of systems with nearly rectangular pulses often take rise times into account explicitly. If an analog waveform with rise time t_p is sampled at intervals t_s, the reconstructed waveform has a rise time [5.8, p. 77]

$$t_r = \sqrt{t_p^2 + t_s^2} \tag{5.15}$$

For sampling effects to be neglected, there must be at least three to five samples occurring in the rise time t_p.

Raised-Cosine Pulse. A waveshape more typical of moderately band-restricted systems is the raised-cosine pulse with duration $2T$, as shown in Table 2.1 and Figs. 2.1

and 2.4. This pulse has unit amplitude and energy $3T/4$. It is apparent from Fig. 2.4 that the spectrum of the raised-cosine pulse is much more concentrated than that of a square pulse, so aliasing will be much less. The spectrum expression from Table 2.1 was substituted into (5.9) and (5.10) to find the average energy of aliasing and the upper bound on magnitude of reconstruction error, respectively. Although the ensuing integrals can be reduced to comparatively simple looking expressions, no closed-form answer could be obtained. Results of numerical integration are tabulated below for several values of M_s.

M_s	Alias-to-Signal Ratio, η^2	Magnitude Bound
2	1.0×10^{-3}	2.8×10^{-2}
4	2.8×10^{-5}	6.5×10^{-3}
8	8.4×10^{-7}	1.6×10^{-3}

In contrast to aliasing behavior for square pulses, the aliasing from sampling of a raised cosine pulse falls off very rapidly as M_s increases. The numbers for magnitude bound suggest that the discrepancy between original signal and reconstruction would be invisible on ordinary plots of the waveforms, for $M_s \geq 4$.

Practice. In our experience, we have encountered applications using as few as four samples per symbol to as many as 32 samples per symbol for simulation of analog signals. The lower number was for routine examination of nominally bandlimited systems, while the larger number (it is not often needed) arose when exploring interference between adjacent-channel signals with rectangular pulse waveforms. Running time of a simulation is often an important consideration; to minimize running time, the sampling rate ought not be selected any larger than necessary. In that regard, a signal that was strictly bandlimited to $|f| < 1/T$ could be simulated, without error, by as few as two samples per symbol.

Signals in different sections of a communications link have different bandwidths. Accordingly, a simulation might reasonably employ different sampling rates in these different sections. Indeed, changeable sampling rate is a capability that experienced users will demand of a simulation program. Changing of sampling rate is explored further in Chapter 16.

When selecting a sampling rate, aliasing is not the only factor to consider, spectra of the signals frequently will not be known analytically, formulas (5.8) to (5.10) may not necessarily be tractable even if the spectra are known, and the formulas may not be the best criteria to use in evaluating the errors. The alias-error formulas can serve as guidelines, but a user often must select the sampling rate by experiment.

One experimental course of action might be to start with a sampling rate known to be inadequate and examine the consequences of the inadequacy with a trial short-duration simulation. Then double the sampling rate and try again. Continue to double the sampling rate until the trial results become satisfactory. A typical criterion of performance will vary with changes in sampling rate and level out to some stable value as sampling rate is increased sufficiently.

5.1.4 Other Features

Truncation Error

Although aliasing error usually receives the most attention, simulations often must truncate signals in the time domain as well as in the frequency domain. If a pulse $g(t)$ is forced to zero for $t < t_1$ and $t > t_2$ ($t_1 < t_2$), the truncated energy is

$$\int_{-\infty}^{t_1} |g(t)|^2 \, dt + \int_{t_2}^{\infty} |g(t)|^2 \, dt$$

References [5.2], [5.3], and [5.5] explore truncation and other sources of error in further detail. If $g(t)$ is bandlimited, it must be infinite in extent. Its truncated version is no longer bandlimited, so aliasing will arise even if it would have been absent on the original signal. Such aliasing should be taken into account when deciding upon suitable truncation.

Alternative Fourier Transform

For future reference, take the Fourier transform of (5.1) one term at a time, first setting $\tau = 0$:

$$Z_\delta(f) = \sum_{n=-\infty}^{\infty} z_a(nt_s) e^{-j2\pi nft_s} \qquad (5.16)$$

This same expression will occur again below, arising from a different approach. (Time shift τ was introduced to allow averages to be calculated for the aliasing-error formulas. It is dropped temporarily, until needed again later.)

5.1.5 Digital Representation

The sampled signal $z_\delta(t)$, consisting of a train of weighted impulses, is *not* the ultimate digital representation that appears in the computer; a computer has no way to represent an impulse—a singular *analog* function. A computer can only represent digital numbers. These can be contained in an ordered digital sequence $\{z_d(n)\}$, where $z_d(n)$ is the weight of the nth impulse in (5.1) and thus has the value

$$z_d(n) = z_a(nt_s) \qquad (5.17)$$

Subscript d indicates *digital*, and n will be used throughout as the designation for sample index in a time-domain sequence. Braces $\{\cdot\}$ indicate a sequence. The sequence is *ordered* in the sense that the nth sample occurs immediately prior to the

($n + 1$)th sample. Think of the digital samples as being a list of successive numbers generated by a computer or stored in computer memory.

Simulations do not operate in real time. The digital samples are not spaced by the sampling interval t_s, nor are they necessarily generated at equally spaced instants in the computer operations. Indeed, if the samples are stored in memory, there is no longer any intrinsic meaning to the idea of time spacing. Nonetheless, the sequence is still said to be a time-domain sequence.

To ascribe a uniform time spacing between digital numbers in a list is an external, user-imputed interpretation of the data that is not part of the computer operations. Furthermore, the ascribed time interval can have any value whatever and is not restricted in any way to the original sampling interval t_s. Stated differently: The same data samples can be used for simulation of signals on any arbitrary time scale or ascribed spacing. These features are a basis for great flexibility in simulations and a potential source of confusion. They will be examined repeatedly in the sequel.

5.2 DIGITIZED SIGNALS IN THE FREQUENCY DOMAIN

Just as an analog signal in the time domain can be transformed into the frequency domain by use of Fourier or Laplace transforms, a digital time-domain sequence can also be transformed into the frequency domain. The principles of applicable transforms are described in this section.

5.2.1 z-Transform

A complex sequence $\{z_d(n)\}$ has a z-transform, defined by

$$Z_d(z) = \sum_{n=-\infty}^{\infty} z_d(n) z^{-n} \tag{5.18}$$

where the complex variable z is usually defined as $z = \exp(s t_s)$ and s is the complex variable of the Laplace transform.

[An unfortunate collision of well-established nomenclatures has occurred here. The letter z is widely employed in the mathematics literature to designate complex variables generally, such as the signal $z_d(n)$, while the same letter z is used for the complex variable of the z-transform. These conflicting nomenclatures lead to dual use of z, as starkly exemplified in (5.18). We are reluctant to depart from either entrenched convention and so will use z for both quantities, while recognizing that there is a potential for mixup. The usage for signals will always include an index argument, such as $z(n)$, while the transform complex variable will always be just z. Since the z-transform is not widely employed in this book, severe collisions as in (5.18) will not appear often.]

Continuous Fourier Transform

The z-transform of the sequence $\{z_d(n)\}$ is usefully evaluated on the unit circle of the z-plane, for $z = \exp(j2\pi f t_s)$. Substituting into (5.18) gives

$$Z_d(e^{j2\pi f t_s}) = \sum_{n=-\infty}^{\infty} z_d(n) e^{-j2\pi n f t_s} \qquad (5.19)$$

which is identical to the Fourier transform (5.16) for the train of weighted impulses. Equation (5.19) is the frequency-continuous Fourier transform of the sequence $\{z_d(n)\}$. Rather than write the clumsy $Z_d(e^{j2\pi f t_s})$, the continuous Fourier transform will be denoted by the nonrigorous but simpler notation $Z_d(f t_s)$, meaning $Z_d(z)$ evaluated at $z = \exp(j2\pi f t_s)$.

Notice that the normalized, dimensionless product $f t_s$ is the only meaningful continuous-frequency variable for a sequence. Since the time interval t_s has lost its significance in the computer, the frequency f cannot have independent meaning either. Many authors define a specific symbol for the product, such as $F = f t_s$ or $\Omega = 2\pi f t_s$, but this book will retain $f t_s$ despite the lack of meaning of the constituent factors. Reasons for this departure from convention will appear in later chapters. The product $f t_s$ can be regarded as an angle, measured counterclockwise about the unit circle, taking values from -0.5 to 0.5 cycle. The continuous Fourier transform (5.19) is periodic, with period $f t_s = 1$ cycle.

5.2.2 Discrete Fourier Transform (DFT)

A computer cannot represent the continuous-frequency variable of the continuous Fourier transform, nor can it accommodate the possibly infinite extent of the sequence $\{z_d(n)\}$ in (5.19). It can only represent a finite number of frequencies for a sequence of finite length. Consider a finite sequence or *block* of N samples, designated $\{z_d(n)\}_N$. (Henceforth, the symbol N will be reserved for block length, in conformance with common practice in the digital signal-processing literature.) The discrete Fourier transform (DFT) of this sequence is defined by

$$Z_d(k) = \sum_{n=0}^{N-1} z_d(n) e^{-j2\pi nk/N} \qquad (5.20)$$

It is readily demonstrated that $Z_d(k)$ is periodic in k, with period N, so the transform is completely defined by any N adjacent values of k. There are N samples of $\{z_d(n)\}_N$ in the time domain and N samples in one period of $\{Z_d(k)\}_N$ in the frequency domain.

Index k may be regarded as a frequency-domain index. Use of k for the frequency domain and n for the time domain agrees with conventions in the literature. Most computations employ $0 \leq k \leq N - 1$, but it is often convenient to a user to deal with $-N/2 \leq k < N/2$ (for N even) or $-(N - 1)/2 \leq k \leq (N - 1)/2$ (for N odd). To distinguish between the two usages, the zero-centered index will be designated k'. Despite the two notations, k and k' are really the same index, just defined over different starting points on the same periodic frequency axis.

Compare (5.20) to (5.19). If the possibly infinite sequence $z_d(n)$ in (5.19) were nonzero only for N adjacent points, the DFT $Z_d(k)$ would constitute frequency-domain samples of $Z_d(ft_s)$ evaluated at $ft_s = k/N$. The DFT provides a finite-extent sampled frequency-domain representation, compatible with a digital computer.

Inverse Discrete Fourier Transform

Given an N-sample frequency-domain block $\{Z_d(k)\}_N$, a time-domain block $\{z_d(n)\}_N$ can be found from the inverse discrete Fourier transform (IDFT):

$$z_d(n) = \frac{1}{N} \sum_{k=0}^{N-1} Z_d(k) e^{j 2\pi k n / N} \qquad (5.21)$$

Equations (5.20) and (5.21) are the discrete Fourier transform pair; they are a vital tool in simulations and in all digital signal processing.

Because $Z_d(k)$ is periodic in k, the IDFT could be computed for any N contiguous values of k. The choice of 0 to $N-1$ is convenient; it is the choice that appears most frequently in the literature and is used in most computations, but any other choice is equally acceptable.

Examination of (5.21) shows that the IDFT $z_d(n)$ is periodic in n, with period N. That will be true even if the original time-domain block was sectioned from a longer sequence and was not considered to be periodic. As a general rule, use of the DFT imposes an implied time-domain periodicity for which a simulation engineer must be constantly alert. Failure to take account of the periodicity leads to simulation disaster. Periodicity associated with the DFT will arise repeatedly in the sequel.

Rather than thinking of sequences as periodic and infinite, it is also helpful to think of them as circular and finite. The circular concept is well illustrated by the continuous-frequency Fourier transform of (5.19), which is defined on the finite unit circle. Both aspects—periodicity and circularity—are employed interchangeably in the chapters that follow.

Computation of the DFT/IDFT

One could easily program a computer to calculate the DFT or IDFT according to the formulas (5.20) and (5.21), but more efficient methods are available. The so-called fast Fourier transform* (FFT) is a well-known algorithm that saves considerable computing effort in the transformations, particularly as N becomes large. Values

---

*Extensive descriptions of FFT implementation are found in the DSP references cited in Chapter 1.

of N commonly range from 256 to 4096, and even larger values may be appropriate in some circumstances. Any professional-quality simulation program includes FFT routines.

Computation saving and FFT program simplicity are maximized when N is an integer power of 2. For that reason the FFT routine in STÆDT only accepts block lengths N that are an integer power of 2. Restrictions arise from the constraint on block size, as will be explained in the proper places. Other block sizes are permissible for other operations in STÆDT where FFTs are not involved.

5.2.3 Time/Frequency Relations

Let a uniform time spacing t_s be ascribed to adjacent samples of $z_d(n)$; a block of N samples therefore has an imputed duration of Nt_s. In the frequency domain, one N-sample cycle of the period of $Z_d(k)$ then has an ascribed frequency span of $1/t_s$, so the frequency increment between adjacent values of k is deemed to be $1/Nt_s$.

It is often convenient to restate these relations in terms of the imputed symbol interval T and the ratio $M_s = T/t_s$—the number of samples per symbol. Thus the time-domain samples are imputed to have a spacing of T/M_s, and an N-sample block has a duration of NT/M_s. The frequency-domain representation has a periodic span of M_s/T and a frequency increment of M_s/NT.

If zero frequency is associated with $k' = 0$ (not a necessary association), the frequency span of the continuous spectrum ranges from $-M_s/2T$ to $+M_s/2T$, but the frequencies at the samples range from $-M_s/2T$ to $+M_s/2T - 1$ for N even; that is, $k' = -N/2$ to $N/2 - 1$. The sample at frequency $M_s/2T$ is in the next period of N points and is identical to the one at $-M_s/2T$. For N odd the centered sample span is $k' = -(N - 1)/2$ to $+(N - 1)/2$.

Setting zero frequency at the center of the span eases user visualization since it places positive and negative frequencies to the right and left of zero, respectively, in accordance with conventional representations. However, computations generally work with $0 \leq k \leq N - 1$ so that if $k = 0$ is associated with zero frequency, the negative-frequency portion of the spectrum is shifted to the right of the positive-frequency portion. This permutation is immaterial to the computer, but distracting to a human user. Some of these frequency-axis arrangements are illustrated in Fig. 5.3 for $N = 8$.

There is no requirement that zero frequency be associated with $k = 0$ (or that zero time be associated with $n = 0$). Since these associations are imputed by the user and do not enter into the simulation operations, any arbitrary origin whatever can be associated with a sample sequence. The STÆDT program has provision for I/O and display in user-designated time and frequency units but operates internally with dimensionless sample indexes.

Problems of Imputed Scales

Ascribing time or frequency increments to inherently dimensionless sample sequences can be ambiguous. Seemingly the most straightforward approach is to

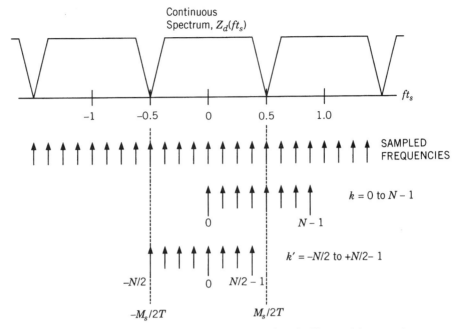

FIGURE 5.3 Arrangements of frequency-domain axis. Illustrated for $N = 8$.

ascribe a time increment t_s to the interval between time-domain samples, whereupon the frequency increment is $1/Nt_s$. But sampling density is likely to be different in different portions of a simulation, so t_s has no unique definition. Moreover, block size N need not be the same in all portions of a simulation, so the frequency increment is not uniquely defined. Time and frequency scales in the DSP literature are heavily based upon sample interval, but that is too erratic for many simulations.

Somewhat firmer ground is reached by employing the symbol interval T as the basis for time and frequency scales. In many communications links, the symbol rate does not vary from one end of the link to the other, even if the sample rate should be changed within the simulation. Furthermore, a user is more interested in symbol rate (a system parameter) than in sample rate (a simulation parameter).

The frequency span of the simulation is M_s/T, which is independent of block size N. If sample rate varies within a simulation, that variation is reflected in changes in M_s and therefore the frequency span. It is the user's responsibility to assure a large-enough frequency span (sampling rate) at all locations when planning a simulation. Signal processes within the simulation must be kept informed of the pertinent value of M_s.

A scale based upon T would be satisfactory if the definition of T were itself unambiguous. Although T is clearly defined for balanced, nonoffset PAM–QAM signals, it was shown in Sec. 2.2.3 that the definition of *symbol interval* can be ambiguous in offset QAM signals or MSK signals. Furthermore, if the QAM modulation is not balanced, there are two different symbol intervals to be considered. These ambiguities can be resolved, but with care and close attention.

In yet other circumstances, a user may prefer to normalize the frequency scale to the information bit rate. This may be the natural choice when, for example, comparing different modulation types with the same information throughput. Specification of bit rate may be less subject to ambiguity than symbol rate. However, although bit rate may be unambiguous to a recipient of simulation results, the simulation engineer still must resolve symbol-rate ambiguities for simulation planning and setup, since many fundamental simulation parameters—such as filter bandwidths and strobe-instant requirements—are symbol-interval related.

Whatever imputed scale may be adopted, problems of scale ambiguities arise solely in the interfaces between the user and the simulation. Once the simulation has been set up properly, there is no further problem internal to the computer. But planning and setup of the simulation require translation of system characteristics into simulation characteristics, which is where mistakes are commonly made. This issue will be faced again, most notably in the chapters on filters, with regard to adapting filter data from an arbitrary time/frequency scale to the dimensionless format needed by the computer.

A different problem arises if the signal is subjected to Doppler shift or other time or frequency compression. In those cases, the symbol rate of a particular signal changes within the simulation, necessitating special handling. In a related case there may be multiple signals generated in a simulation (e.g., a desired signal plus interferers), with somewhat different symbol rates for each. Although time and frequency scale ambiguities clearly arise in these situations, their dominant problem is that of rate changing. Rate changing is addressed in Chapter 16.

5.3 STRUCTURES AND NOMENCLATURE

Various common structures and formats encountered within simulation programs are introduced in this section.

5.3.1 Simulation Modules

As illustrated in the block diagram of Fig. 1.2, a communications link is modular in form and is composed of a number of connected black boxes. Each black box accepts an input signal, operates upon the signal to alter it in some manner, and then outputs the altered signal to the next box. A simulation of a communications link is modular in much the same manner. The simulation is composed of software *modules* that correspond closely to the black boxes of the hardware link. Instead of signals, the modules process digital *data*. The terms *signal* and *data* will be used interchangeably. A module accepts input data, performs an operation on the data that alters it in some manner, and delivers its output data to serve as input for the next module in line.

A hardware black box performs a particular operation with certain characteristics. For example, a filter that performs a frequency-selective operation on a signal has characteristics described by its transfer function or impulse response. Some black boxes also have *memory*; the current output depends not only upon the current input but also upon past inputs and outputs and past conditions internal to the operation itself. A filter is a prime example of a black box with memory.

A software module must closely mimic the hardware black box that it is to simulate. A module contains one or more *signal processes*, plus information on the process characteristics called the *process parameters*. Parameters describe only the process and are unrelated to the signals being processed. Parameters are specified by the user before a simulation run starts and ordinarily remain constant throughout a run. (There is an exception; parameters of adaptive systems are altered in the course of a run.) Some modules have no parameters at all, others may have exceedingly complicated parameters, and most modules lie somewhere in between.

If a process has memory, the module must also have provision for *state* information: the past inputs, outputs, and internal conditions that contribute to the current or future output. Simulation processes need not be causal; future conditions can contribute to the current output and must be accommodated in state memory when appropriate. States depend upon the signals applied and upon the nature of the process. States are dynamic in that they evolve during the course of a run, according to the signals encountered. Initial state information must be provided at the start of a run, by the user or by the program, to each module containing memory.

Some processes are memoryless, so their modules contain no state information whatever. Other processes can have extensive memory and therefore require extensive state information. In addition to these features that mimic a black box, a module must include interface provisions for data I/O, and information to establish connectivity between modules.

The chapters that follow are devoted primarily to modeling of the signal processes: first for computer simulation of analog processes and then for digital. Along the way, problems of handling parameters are examined, and state memory is treated as an outgrowth of the models of the processes. A chapter on programming issues deals with data I/O, connectivity, and other items of greater interest to programmers than to users. Necessary information on data structures is provided in the sections immediately following.

5.3.2 Data Formats

A *data point* or *element* corresponds to a *signal sample*; all of these terms will be used synonymously. The point can be in the time domain or the frequency domain. In general, a data point will be considered to be complex and can be in either rectangular or polar coordinates.

Data points can be arranged in *block* format (examples of which have been seen already in Sec. 5.2) or in *single-point* format; these terms will be explained presently. Some kinds of processes inherently can be accomplished only in block format, while others can work only with single-point. Each format has its purpose, and both are necessary in any complete simulation program.

Data format might seem to be a programming issue of little direct concern to a user. But choice of format can have a profound influence upon the progress of a simulation and upon models of the processes. Definitions and properties of formats are introduced now rather than in Chapter 19, on programming, so that the information is available when process models are described.

Format Impact on Scheduling

The modules of a simulation program can be regarded as subroutines that operate on the data that are provided as input. The program calls the first module; that module processes the data and returns. The program then calls the second module, which performs its process on the data delivered from the first process and returns. This continues until all modules have been called. Then the program starts again at the beginning with fresh input data. (The idea of a module as a subroutine is put forward for purposes of explanation. Alternative modular structures are discussed in Chapter 19; it is not necessary that the modules be arranged as true subroutines.)

The question at issue is the amount of data to be processed on each call of a module. If only one data point is processed, the format is *single point*.* If a whole block of N points is processed on each call, *block* format is being employed. A certain amount of overhead is incurred each time that a module is called. For an equal amount of total data, single-point format incurs at least N times as much calling overhead as block format. For that reason, single-point format executes more slowly than block format. Slow execution is particularly aggravated if modules must be called from disk rather than electronic memory.

Circular Versus Linear Blocks

Block format was introduced earlier via the DFT. Frequency-domain output of a DFT operation may be regarded as periodic, in which case only one period need be stored in a block since all periods are identical. Or the DFT can be regarded as *circular*, including just one period, with the highest frequency considered to be wrapped around to abut the lowest.

Because it is computed from frequency-domain samples, output from the inverse DFT must be periodic in the time domain. Therefore, any time-domain sequence obtained by inverse-transforming a sampled spectrum can also be regarded as circular. It is an N-point time-domain block where $n = 0$ is the same as $n = N$. More broadly, time index n only has meaning modulo-N. Properties of circular blocks will be described shortly. For now, recognize that they exist and that they are closely associated with DFT operations.

A time-domain block could also be *linear*; think of it as a section from a longer, nonperiodic sequence. Provided that the block was not produced from IDFT operations, there is no inherent reason why it must be circular. Operations on linear blocks can be *concatenated* to perform simulations on sequences that are longer than can be held in convenient block sizes.

---

*For reasons explained in Chapter 16, single-point format might actually handle several data points on each call.

5.3.3 Properties of the Data Formats

Circular Blocks

All frequency-domain processes operate only on circular blocks. It is not meaningful to represent sampled frequency-domain signals in any other format. All time-domain signals obtained by inverse Fourier transformation of a frequency-domain signal are produced automatically in circular-block format. The signals of a time-domain circular block must wrap around the ends of the block and not be truncated. If the time-domain circular block is produced by frequency-domain operations, the wraparound is accomplished automatically (see Chapter 6). If a user constructs a time-domain circular block in the time domain, precautions must be taken to assure correct wraparound. (Methods exist for opening a circular block into a linear block under restricted circumstances, as considered in Chapter 6.)

Because a circular block is associated with a periodic signal, any simulations with circular blocks are *steady-state*; all transients have died out. That is advantageous if the steady-state behavior alone is being investigated, since no simulation effort need be wasted on waiting for transients to disappear. If transients must be simulated, either linear-block or single-point format should be employed. A circular block cannot be used for simulation of feedback operations; single-point format is needed instead. The reason for this is given below in the section "Single-Point Format."

A run performed with a circular block is self-contained; there are no initial conditions to be provided and no state memory carried over from one block to another. Such runs tend to be the least demanding upon a user. Because of the relative ease of performing them, circular-block simulations are often favored for quick determination of qualitative behavior of a system under study.

Duration of a practical circular block is not always sufficient to permit reliable performance statistics to be gathered. More commonly, many blocks must be tried in separate runs, each run with different realizations of signals, noise, or other disturbances. Performance is then taken as an average over the multiple blocks. The simulation program should have convenient provision for multiple runs, for multiple realizations, and for averaging of results.

A circular-block simulation of a PAM-like data signal must contain a whole number of symbols in the block. Truncated pulses would require representation of a transient behavior that is not well handled in a circular block. Therefore, the number of symbols in the block, designated N_B, must divide N, the number of samples in the block. That is, N/N_B must be an integer.

There are M_s samples per symbol. The total number of samples per block is the number of samples per symbol multiplied by the number of symbols per block, or

$$N = M_s N_B \qquad (5.22)$$

If N_B divides N, then, as a rigorous constraint for circular blocks, M_s too must be an integer. In linear blocks or single-point format, M_s need not be an integer nor even rational; this topic will be discussed further in subsequent chapters. The FFT in

144 ☐ DIGITAL SIMULATION OF ANALOG SIGNALS AND PROCESSES

ST&EDT is restricted to block sizes that are integer powers of 2, which can be divided only by smaller integer powers of 2. Therefore, it is necessary that N_B and M_s be selected by the user as integer powers of 2 whenever FFT is used on the block.

Single-Point Format

Only time-domain signals can be processed in single-point format; inherently, frequency-domain signals exist solely in circular-block format. A time-domain signal that is processed in single-point format can be periodic but need not be. Usually, a point in a single-point sequence is regarded as one sample from a long, aperiodic signal. A single-point simulation will exhibit the same transient behavior as the system being simulated. If steady-state behavior is required, time must be allowed for all transients to die away. Initial conditions and state information must be provided for all memory processes in the system.

Single-point simulations can have arbitrarily long durations without making special provisions; this feature is convenient when collecting statistics of performance. Single-point computation in ST&EDT is substantially more time consuming than block operations. We recommend that single-point be employed only for feedback loops, where there is no other alternative.

Feedback Loops. Only single-point format can be used for simulating feedback loops. To see why, consider the rudimentary feedback loop shown in Fig. 5.4. It consists of two generic processes, labeled A and B, and a subtracting process that closes the loop. Signals $z_i(n)$ designate inputs and outputs of the processes.

Suppose that computation of the loop were to be attempted in block format. Represent the signal blocks in braces as $\{z_i(n)\}$. Block computation of the subtraction $\{z_1(n) - z_2(n)\}$ requires that blocks $\{z_1(n)\}$ and $\{z_2(n)\}$ be available as inputs. However, existence of $\{z_2(n)\}$ requires the prior computation of $\{z_3(n)\}$, from which $\{z_2(n)\}$ is computed via process B. But $\{z_3(n)\}$ cannot be computed until $\{z_1(n) - z_2(n)\}$ has been established. That requires $\{z_2(n)\}$, which does not yet exist. Thus nonexistence of $\{z_2(n)\}$ prevents computation of $\{z_3(n)\}$, and vice versa.

Even in single-point format it is necessary that there be a delay inside the loop, or

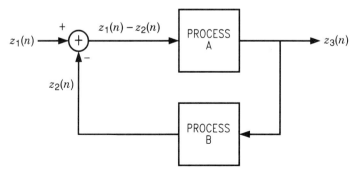

FIGURE 5.4 Rudimentary feedback loop.

else it is not computable. When computing the subtraction $z_1(n) - z_2(n)$, it is necessary that $z_2(n)$ be present at the nth computation. That is only possible if it has already been computed by the end of the $(n-1)$th computation; there must be a delay of at least one sample interval. The delay required in a feedback loop is a memory process which requires that an initial condition be supplied and that a state be updated on each computation of the process.

von Neumann Architecture. A knowledgeable reader may object that all operations in a computer with von Neumann architecture must be performed one at a time, irrespective of data format. That is certainly correct; all processes are single-point internally, even those that inherently must use blocks. From the user's standpoint, the distinction between single-point and block format lies in the manner of data handling: Should just one data point at a time be passed for computation from module to module, or should a whole block be passed? The latter, when possible, is more efficient.

Linear Blocks

For the most part, linear blocks provide a more efficient way of performing the same time-domain operations that might be accomplished in single-point format (except for feedback loops, which can only be done single-point). Linear blocks offer a means of attaining the computational benefits of block operations while avoiding the periodicity limitations of circular blocks. A linear-block simulation exhibits the same transient behavior as a single-point simulation and time must be allowed for the transient to dissipate. Initial conditions and state memory must be provided.

Simulations that are long compared to the block length can be performed by running one block through all modules, being careful to preserve all state information, and then concatenating another block to be run through the modules. This may be repeated as often as needed to obtain the statistics required. The method is an amalgam among those used for long simulations with circular blocks and with single-point formats.

A concatenated block is a section from a long time record; it contains postcursor tails of pulses that began in previous blocks and precursors of pulses whose time comes in succeeding blocks. A pulse can be split arbitrarily across block boundaries, provided that no portion of it is lost in the splitting. Concatenating linear blocks is a means of performing an extended task piece by piece.

It is not necessary that there be an integer number of symbols N_B in a linear block nor that there be an integer number of samples per symbol. It is necessary, of course, that each block contain an integer number of samples.

5.4 ENERGY AND POWER RELATIONS

Frequent need arises to determine the energy, power, or related quantities of simulated signals. This section tells how these quantities are specified for sample sequences and their relations to the energy or power of the associated time-continuous signals.

5.4.1 Energy Formulas

The energy of a time-continuous signal $z_a(t)$ is

$$E_{za} = \int_{-\infty}^{\infty} |z_a(t)|^2 \, dt \qquad (5.23)$$

in the time domain. By Parseval's rule, the energy can be determined in the frequency domain from

$$E_{za} = \int_{-\infty}^{\infty} |Z_a(f)|^2 \, df \qquad (5.24)$$

where $Z_a(f)$ and $z_a(t)$ are a Fourier pair.

Sampling of the time-continuous signal at a rate $1/t_s$ produces a sample sequence $\{z_d(n)\}$, as explained in Sec. 5.1. Energy of the sequence is defined as

$$E_{zd} = t_s \sum_{n=-\infty}^{\infty} |z_d(n)|^2 \qquad (5.25)$$

Parseval's Rule

Given a finite digital sequence $\{z_d(n)\}$ of N samples in the time domain, there will be a DFT sequence $\{Z_d(k)\}$, also with N samples, in the frequency domain. Parseval's rule for digital signals can be shown to be

$$\sum_{n=0}^{N-1} |z_d(n)|^2 = \frac{1}{N} \sum_{n=0}^{N-1} |Z_d(k)|^2 \qquad (5.26)$$

This formula relates measurement of energy (or power) in the time domain to measurement in the frequency domain.

Presence of t_s

Equation (5.25) for the energy of a digital signal contains the sampling interval t_s. That seems to contradict the argument in Sec. 5.2.3 that the computer is oblivious to time intervals; substituting arbitrary t_s into (5.25) would lead to arbitrary energy. In practice, it is usual to deal with ratios of energies (e.g., E_b/N_0) where the common factor t_s will cancel out, or with power, where t_s also cancels out (as shown below). Formula (5.25) is a point of departure for other developments; the factor t_s will be kept for now so that its later elimination will be visible.

Energy Relations Between Analog and Digital Signals

The digital energy formula (5.25) is a staircase approximation to the analog energy integral (5.23). Close agreement between the formulas ordinarily would not be expected unless t_s was quite small. In general, $E_{zd} \neq E_{za}$. However, it can be demonstrated (see Appendix 5C) that the following statements are true:

1. If $Z_a(f) = 0$ for $|f| \geq 1/2t_s$ (i.e., the analog signal is adequately bandlimited), then $E_{zd} \equiv E_{za}$.
2. If the analog signal is not adequately bandlimited, the energy in the digital sequence depends upon the timing shift τ. [See (5.1) for a definition of τ.] But the digital energy averaged over all τ is the same as the analog energy; that is, $\overline{E}_{zd} = \text{Avg}_\tau[E_{zd}(\tau)] \equiv E_{za}$.

Thus, for bandlimited signals, the energies of the analog and digital signals are equal and the simulation is exact. If the signal is not bandlimited, the energies (5.23) and (5.25) are equal on average but deviate as a function of time shift τ.

5.4.2 Energy of Analog Pulse Trains

Let the analog signal be a finite PAM train with L data pulses, represented as

$$z_a(t) = \sum_{m=1}^{L} c(m) g(t - mT) \tag{5.27}$$

See Chapter 2 for an introduction to PAM signals and their representation. Data values are designated by $c(m)$ and m will be used to denote the symbol index or, equivalently, pulse index. Note that although $z_a(t)$ is composed of a finite number of pulses, so that energy is finite, the pulse $g(t)$ might have infinite extent.

Energy of the finite PAM signal is given by

$$E_{zN}(c) = \int_{-\infty}^{\infty} \left| \sum_{m=1}^{L} c(m) g(t - mT) \right|^2 dt$$

$$= \sum_{m=1}^{L} \sum_{i=1}^{L} c(m) c^*(i) \int_{-\infty}^{\infty} g(t - mT) g^*(t - iT) \, dt \tag{5.28}$$

This energy will depend upon the data sequence $\{c(m)\}$ if either of two conditions obtain: (1) the pulses overlap or (2) $|c(m)|$ takes on multiple values. Data dependence illustrates *statistical scatter*, a feature that arises ubiquitously in signals or noise with random characteristics. The scatter is inherent to the finite analog signal and will also appear in energy measurements on simulated digital signals.

Data-Averaged Energy

Using E[·] to denote statistical expectation,* energy in (5.28) averaged over all data values is

$$\bar{E}_{zN} = \mathrm{E}[E_{zN}(c)] = \sum_{m=1}^{L} \sum_{i=1}^{L} \mathrm{E}[c(m)c^*(i)] \int_{-\infty}^{\infty} g(t-mT)g^*(t-iT)\,dt$$

$$= \sum_{m=1}^{L} \sigma_c^2 \int_{-\infty}^{\infty} |g(t-mT)|^2 \, dt = L\sigma_c^2 E_g \tag{5.29}$$

where E_g is the energy in the pulse $g(t)$ and the data have been assumed to obey the condition

$$\mathrm{E}[c(m)c^*(i)] = \begin{cases} \sigma_c^2, & i = m \\ 0, & i \ne m \end{cases} \tag{5.30}$$

Truncated Pulse Trains

The signal in (5.27) is finite in the sense that it is composed of only a finite number of pulses. However, the pulses could each have infinite duration, so the signal itself would have infinite extent. A signal of this kind can be simulated well using circular blocks. A different situation often arises with linear blocks or single-point simulations. The signal may be regarded as a section of finite extent that has been snipped out of a PAM signal of infinite extent. This finite section may be represented as

$$z_a(t) = \begin{cases} \displaystyle\sum_{m=-\infty}^{\infty} c(m)g(t-mT), & t_1 < t < t_2 \\ 0, & t < t_1 \text{ or } t > t_2 \end{cases} \tag{5.31}$$

Now the pulses within the interval (t_1, t_2) may have tails cut off, and pulses outside the interval can have tails (both trailing and leading) falling inside the interval. In addition to data dependence on the pulses inside the interval, energy measurements on the signal section will be affected by the tails of pulses outside the interval. This is another example of statistical scatter that happens in the analog signal itself and therefore also afflicts the digital simulation.

---

*Distinguish between E_z, the energy in the signal $z(t)$, and statistical expectation E[·]. This is another unfortunate collision between well-established notations; it will not be encountered often in the sequel.

5.4.3 Power

As introduced in Sec. 2.3, the power of a signal of infinite extent can be estimated from the energy of the signal contained in a finite interval, divided by the duration of the interval. For a digital sequence $\{z_d(n)\}$ of N samples, the power estimate is given by

$$P_z = \frac{E_{zN}}{Nt_s} = \frac{1}{N} \sum_{n=0}^{N-1} |z_d(n)|^2 \qquad (5.32)$$

The factor of t_s appearing in the energy of (5.25) has canceled out; power is estimated from the samples alone without concern for an externally imputed time scale. Power as defined in (5.32) is the mean square of the sequence $\{z_d(n)\}_N$. The root mean square (rms) of the sequence is simply the square root of (5.32).

Upon occasion, it is helpful to calculate the mean of a sequence

$$\frac{1}{N} \sum_{n=0}^{N-1} z_d(n) \qquad (5.33)$$

or its mean magnitude

$$\frac{1}{N} \sum_{n=0}^{N-1} |z_d(n)| \qquad (5.34)$$

5.4.4 Rationale

Why should so much attention be devoted to so mundane a task as measuring energy or power? There are several reasons, as explained next.

Importance. A communications system is evaluated almost universally on the basis of the energy or power needed to attain a specified performance. Precise evaluation requires precise determination of energy (or power), whether the experiment is performed on actual hardware or is simulated on a computer.

Scatter. Differing energy measurements on a signal can arise from different data (or noise) sequences, from shifts in the timing of the sampling wave, from different truncations of a long signal, and from overlap of pulses. These effects, and others, all lead to scatter among repeated simulations of a signal with differing realizations. A careful user will obtain an estimate of the scatter of simulation results. Statistical scatter is ever present in simulations and will be cited repeatedly in this book since it afflicts other measurements besides energy and power.

The prominence given above to energy measurements serves to alert a user to the

150 ☐ DIGITAL SIMULATION OF ANALOG SIGNALS AND PROCESSES

attention to details needed to achieve accurate measurements. It also warns an analyst that simulation results generally are not to be trusted to close absolute precision without careful scrutiny of all aspects of the models.

5.5 KEY POINTS

This chapter has covered the following topics that are fundamental to simulation of analog signals on a digital computer:

- Digitization of the analog signal
- Time-domain/frequency-domain transformations via the discrete Fourier transform (DFT)
- Simulation structures and data formats
- Energy and power relations

5.5.1 Digitization

Analog signals to be simulated must be converted to digital form. Conversion involves sampling, quantization, and truncation. Sampling introduces aliasing if the analog signal is not adequately bandlimited. Aliasing energy or power (averaged over the sampling time shift τ) is twice the signal spectral energy lying beyond frequency $|f| \geq 1/2t_s$, where $1/t_s$ is the sampling rate. Quantization can be neglected in many simulations if single-precision floating-point numbers are employed (six or seven decimal digits). Truncation typically is neglected if the simulated signal duration is long compared to system time constants.

Signals are represented in a computer as a sequence of dimensionless digital numbers. The computer ascribes no units (such as volts, watts, etc.) to the numbers. Also, the concept of time interval between samples is lost in the computer. Any arbitrary interval whatever can be imputed externally by the user, without effect upon simulation operations.

5.5.2 Discrete Fourier Transform

Given a time-domain sequence $\{z_d(n)\}$ of N samples, there exists a discrete Fourier transform (DFT) frequency-domain sequence of N samples whose values are

$$Z_d(k) = \sum_{n=0}^{N-1} z_d(n) e^{-j2\pi nk/N} \qquad (5.20)$$

Given an N-point frequency-domain sequence $\{Z_d(k)\}$ there exists an inverse DFT

(IDFT) time-domain sequence with sample values

$$z_d(n) = \frac{1}{N} \sum_{k=0}^{N-1} Z_d(k) e^{j2\pi kn/N} \qquad (5.21)$$

As an inherent consequence of sampling, the DFT $Z_d(k)$ and the time domain $z_d(n)$ recovered from the IDFT are both periodic sequences, with period N. Alternatively, they may be regarded as circular sequences. Periodicity (circularity) is imposed upon the IDFT by frequency-domain operations even if the original time-domain sequence was not regarded as periodic. A DFT typically is computed with a fast Fourier transform (FFT) algorithm to minimize the computing burden.

5.5.3 Structures and Formats

A simulation is composed of software *modules*, corresponding to the black boxes of a communications link. A module accepts an input *signal*, or *data*, executes a *process* upon the signal, and outputs the result as data to be processed by a succeeding module. Characteristics of the module's process are contained in process *parameters*. Some processes contain a memory of past (or future) signal values; this information is kept in *state* memory associated with the module.

Signal data can be arranged in *block* format or in *single-point* format. Block format is further subdivided into *circular* blocks or *linear* blocks. Frequency-domain operations inherently work with circular blocks. Feedback loops inherently require single-point format.

A circular block is a steady-state representation of the signal. There are no transients and no initial conditions. The signal waveform wraps around the ends of the block. There must be an integer number of symbols N_B in a block of N samples, so N_B must divide N. That implies that the number of samples per symbol M_s must also be an integer that divides N. Simulation of each circular block stands alone; statistics of long-duration signals are built up by separate simulations of a large number of independent circular blocks.

Simulations with linear blocks or single-point format will exhibit the same transients experienced by the system being analyzed. Initial conditions must always be specified. If steady-state behavior is sought, time must be allowed for the transients to die out. State memory is needed in all processes with memory.

Simulations of unlimited duration can be performed. With single-point format, the computer is simply allowed to run until a specified halting condition is achieved. Linear blocks of convenient size can be concatenated into runs of unlimited length. There is no wraparound of signal waveforms.

There are no requirements that a linear block contain an integer number of symbols or that there be an integer number of samples per symbol for either linear blocks or single-point format. Computation in block format is substantially more efficient than in single-point format. Block format is preferred unless feedback loops compel single-point. Different formats can be employed in different portions of the same simulation.

Time/Frequency Relations

Define the symbol interval $T = M_s t_s$, where M_s need not be an integer, except for circular blocks. Within the simulation, no explicit value inheres to t_s, nor to T. However, if the user were to ascribe a value to t_s, the following relations would apply to signals:

	Increment	Period[a]
Time domain	$\Delta t = t_s = T/M_s$	$Nt_s = NT/M_s$ (CB)
		∞ (LB, SP)
Frequency domain (CB only)	$\Delta f = 1/Nt_s = M_s/NT$	$1/t_s = M_s/T$

[a]CB, circular blocks; LB, linear blocks; SP, single point.

5.5.4 Energy and Power

Energy in an N-point time-domain sequence $\{z_d(n)\}$ is defined as

$$E_z = t_s \sum_{n=0}^{N-1} |z_d(n)|^2 \qquad (5.25)$$

The factor t_s cancels out from energy ratios, which are more important for evaluation purposes than are absolute energies.

Parseval's rule for sampled signals is

$$\sum_{n=0}^{N-1} |z_d(n)|^2 = \frac{1}{N} \sum_{n=0}^{N-1} |Z_d(k)|^2 \qquad (5.26)$$

where $\{z_d(n)\}$ and $\{Z_d(k)\}$ are a Fourier pair.

The power of a signal can be estimated from an N-point sequence as

$$P_z = \frac{E_z}{Nt_s} = \frac{1}{N} \sum_{n=0}^{N-1} |z_d(n)|^2 \qquad (5.32)$$

Energy and power estimates are subject to *statistical scatter* arising from dependence upon symbol-value realizations, timing shift τ between the analog signal and the sampling wave, signal truncation, and interference between pulse tails. All but sample-timing shift would occur in measurements on the original analog signal.

Energy or power measured with the digital formulas (5.25) or (5.32), averaged over timing shift τ, agree exactly with measurements on the underlying analog signals. If the analog signal was adequately bandlimited, the sampled measurements are independent of timing shift. Care is needed in simulated measurements of energy or power if precise quantitative results are to be attained.

5.5.5 STÆDT Processes

DFT: discrete Fourier transform

FFT: fast Fourier transform

POWR: measures summed square or mean summed square of sample sequence

SP .. SPND: delimit single-point mode within a simulation (default is block mode)

APPENDIX 5A: ALIASING ENERGY

Refer to Fig. 5.1 and equations (5.1) to (5.4). Reconstruction error is defined by $\tilde{z}(t) - z_a(t)$, and the energy in the error is

$$E_A(\tau) = \int_{-\infty}^{\infty} |\tilde{z}(t) - z_a(t)|^2 \, dt \tag{5A.1}$$

where dependence on timing shift τ is made explicit.

Spectra

Input signal $z_a(t)$ is specified to have a Fourier transform $Z_a(f)$. The sampling wave can be written in a Fourier series expansion as

$$\sum_{n=-\infty}^{\infty} \delta(t - nt_s - \tau) = \frac{1}{t_s} \sum_{k=-\infty}^{\infty} e^{-j2\pi k\tau/t_s} e^{j2\pi kt/t_s} \tag{5A.2}$$

whereupon the spectrum of the sampled signal is found to be

$$Z_\delta(f) = \int_{-\infty}^{\infty} z_\delta(t)e^{-j2\pi ft}\,dt$$

$$= \frac{1}{t_s}\int_{-\infty}^{\infty} z_a(t)\sum_{k=-\infty}^{\infty} e^{-j2\pi k\tau/t_s}e^{j2\pi kt/t_s}e^{-j2\pi ft}\,dt$$

$$= \frac{1}{t_s}\sum_{k=-\infty}^{\infty} e^{-j2\pi k\tau/t_s}\int_{-\infty}^{\infty} z_a(t)e^{-j2\pi ft}e^{j2\pi kt/t_s}\,dt$$

$$= \frac{1}{t_s}\sum_{k=-\infty}^{\infty} e^{-j2\pi k\tau/t_s} Z_a\left(f - \frac{k}{t_s}\right) \tag{5.2}$$

The ideal reconstruction filter has a rectangular-shaped transfer function $H(f)$ given by (5.4). The reconstructed signal $\tilde{z}(t)$ has a Fourier transform

$$\tilde{Z}(f) = Z_\delta(f)H(f) = \begin{cases} \sum_{k=-\infty}^{\infty} e^{-j2\pi k\tau/t_s} Z_a\left(f - \frac{k}{t_s}\right), & |f| < \frac{1}{2t_s} \\ 0, & |f| \geq \frac{1}{2t_s} \end{cases} \tag{5A.3}$$

Error-Energy Expression

Equation (5A.1) may be expanded as

$$E_A(\tau) = \int_{-\infty}^{\infty} [\tilde{z}(t)\tilde{z}^*(t)\,dt + z_a(t)z_a^*(t) - z_a(t)\tilde{z}^*(t) - z_a^*(t)\tilde{z}(t)]\,dt \tag{5A.4}$$

Each term may be transformed to the frequency domain according to the following:

$$\int_{-\infty}^{\infty} \tilde{z}(t)\tilde{z}^*(t)\,dt = \int_{-\infty}^{\infty} \tilde{z}(t)\int_{-1/2t_s}^{1/2t_s} \tilde{Z}^*(f)e^{-j2\pi ft}\,df\,dt$$

$$= \int_{-1/2t_s}^{1/2t_s} \tilde{Z}^*(f)\int_{-\infty}^{\infty} \tilde{z}(t)e^{-j2\pi ft}\,dt\,df$$

$$= \int_{-1/2t_s}^{1/2t_s} \tilde{Z}^*(f)\tilde{Z}(f)\,df \tag{5A.5a}$$

$$\int_{-\infty}^{\infty} z_a(t)z_a^*(t)\,dt = \int_{-\infty}^{\infty} Z_a(f)Z_a^*(f)\,df$$

$$= \int_{-1/2t_s}^{1/2t_s} Z_a(f)Z_a^*(f)\,df + \int_{|f|\geq 1/2t_s} Z_a(f)Z_a^*(f)\,df$$

(5A.5b)

$$\int_{-\infty}^{\infty} z_a(t)\tilde{z}^*(t)\,dt = \int_{-\infty}^{\infty} \tilde{z}^*(t) \int_{-\infty}^{\infty} Z_a(f)e^{j2\pi ft}\,df\,dt$$

$$= \int_{-\infty}^{\infty} Z_a(f) \int_{-\infty}^{\infty} \tilde{z}^*(t)e^{j2\pi ft}\,dt\,df$$

$$= \int_{-\infty}^{\infty} Z_a(f)\tilde{Z}^*(f)\,df = \int_{-1/2t_s}^{1/2t_s} Z_a(f)\tilde{Z}^*(f)\,dt \quad (5A.5c)$$

$$\int_{-\infty}^{\infty} z_a^*(t)\tilde{z}(t)\,dt = \int_{-1/2t_s}^{1/2t_s} Z_a(f)Z_a^*(f)\,df \quad (5A.5d)$$

The expressions above are simply applications of Parseval's rule. The integrals in (5A.5c,d) are reduced to the interval $-1/2t_s$ to $1/2t_s$ because $\tilde{Z}(f) \equiv 0$ outside those limits.

Substitute (5A.5) into (5A.4) and clear terms to obtain

$$E_A(\tau) = \int_{|f|\geq 1/2t_s} Z_a(f)Z_a^*(f)\,df$$

$$+ \int_{-1/2t_s}^{1/2t_s} [Z_a(f)Z_a^*(f) + \tilde{Z}(f)\tilde{Z}^*(f) - Z_a(f)\tilde{Z}^*(f) - Z_a^*(f)\tilde{Z}(f)]\,df$$

$$= \int_{|f|\geq 1/2t_s} Z_a(f)Z_a^*(f)\,df$$

$$+ \int_{-1/2t_s}^{1/2t_s} [\tilde{Z}(f) - Z_a(f)][\tilde{Z}^*(f) - Z_a^*(f)]\,df \quad (5A.6)$$

The first integral in (5A.6) is the signal energy cut off by the ideal lowpass filter and the second integral is the energy of the alias components folded over into the passband of the filter.

Substitute (5A.3) for $\tilde{Z}(f)$ in (5A.6) to get

$$E_A(\tau) = \int_{|f| \geq 1/2t_s} Z_a(f) Z_a^*(f) \, df$$

$$+ \int_{-1/2t_s}^{1/2t_s} \left[\sum_{k=-\infty}^{\infty} e^{-j2\pi k\tau/t_s} Z_a\left(f - \frac{k}{t_s}\right) - Z_a(f) \right]$$

$$\cdot \left[\sum_{l=-\infty}^{\infty} e^{j2\pi l\tau/t_s} Z_a^*\left(f - \frac{l}{t_s}\right) - Z_a^*(f) \right] df \qquad (5A.7)$$

Average Energy

Energy averaged over τ is defined as

$$\overline{E}_A = \text{Avg}_\tau[E_A(\tau)] = \frac{1}{t_s} \int_0^{t_s} E_a(\tau) \, d\tau \qquad (5A.8)$$

where τ is assumed to be random and uniformly distributed over the interval $[0, t_s)$. Equation (5A.7) can be expanded into the sum of five separate integrals; average energy is the sum of the averages of each individual term. Integrations over τ and f may be interchanged, so the averaging can be performed on the integrands. Results of those averages are

$$\frac{1}{t_s} \int_0^{t_s} Z_a(f) Z_a^*(f) \, d\tau = Z_a(f) Z_a^*(f) \qquad (5A.9a)$$

$$\frac{1}{t_s} \int_0^{t_s} \sum_{k=-\infty}^{\infty} e^{-j2\pi k\tau/t_s} Z_a\left(f - \frac{k}{t_s}\right) Z_a^*(f) \, d\tau$$

$$= \frac{1}{t_s} \sum_{k=-\infty}^{\infty} Z_a\left(f - \frac{k}{t_s}\right) Z_a^*(f) \int_0^{t_s} e^{-j2\pi k\tau/t_s} \, d\tau$$

$$= Z_a(f) Z_a^*(f) \qquad (5A.9b)$$

The integral of the exponential in (5A.9b) is zero for $k \neq 0$ and t_s for $k = 0$.

$$\frac{1}{t_s} \int_0^{t_s} \sum_{l=-\infty}^{\infty} e^{j2\pi l\tau/t_s} Z_a^*\left(f - \frac{l}{t_s}\right) Z_a(f) d\tau = Z_a(f)Z_a^*(f) \quad (5A.9c)$$

$$\cdot \frac{1}{t_s} \int_0^{t_s} \sum_{k=-\infty}^{\infty} \sum_{l=-\infty}^{\infty} e^{-j2\pi(k-l)\tau/t_s} Z_a\left(f - \frac{k}{t_s}\right) Z_a^*\left(f - \frac{l}{t_s}\right) d\tau$$

$$= \frac{1}{t_s} \sum_k \sum_l Z_a\left(f - \frac{k}{t_s}\right) Z_a^*\left(f - \frac{l}{t_s}\right) \int_0^{t_s} e^{-j2\pi(k-l)\tau/t_s} d\tau$$

$$= \sum_{k=-\infty}^{\infty} Z_a\left(f - \frac{k}{t_s}\right) Z_a^*\left(f - \frac{k}{t_s}\right) \quad (5A.9d)$$

The simplification in (5A.9d) comes about because the last integral over τ is nonzero only if $l = k$.

Upon inserting the results of (5A.9) into the expansion of (5A.7) and combining like terms, the average alias energy is found to be

$$\overline{E}_A = \int_{|f| \ge 1/2t_s} Z_a(f)Z_a^*(f) df$$

$$+ \int_{-1/2t_s}^{1/2t_s} \left[\sum_{k=-\infty}^{\infty} Z_a\left(f - \frac{1}{t_s}\right) Z_a^*\left(f - \frac{k}{t_s}\right) - Z_a(f)Z_a^*(f) \right] df$$

$$= \int_{|f| \ge 1/2t_s} Z_a(f)Z_a^*(f) df + \int_{-\infty}^{\infty} Z_a(f)Z_a^*(f) df - \int_{-1/2t_s}^{1/2t_s} Z_a(f)Z_a^*(f) df$$

$$= 2 \int_{|f| \ge 1/2t_s} Z_a(f)Z_a^*(f) df \quad (5.9)$$

The simplification in developing (5.9) arises because the infinite sum of adjacent finite-limit integrals of the same integrand is just one integral over infinite limits.

APPENDIX 5B: UPPER BOUND ON MAGNITUDE OF ALIASING ERROR

The demonstration starts with the definition of the magnitude of aliasing error and progressively imposes increasing (looser) bounds until a simple format is attained. From the definitions of aliasing error earlier in the chapter and of $\tilde{Z}(f)$ in (5A.3), the magnitude of aliasing error can be written as

158 □ DIGITAL SIMULATION OF ANALOG SIGNALS AND PROCESSES

$$|\tilde{z}(t) - z_a(t)| = \left| \int_{-1/2t_s}^{1/2t_s} \sum_k e^{-j2\pi k\tau/t_s} Z_a\left(f - \frac{k}{t_s}\right) e^{j2\pi ft}\, df - \int_{-\infty}^{\infty} Z_a(f) e^{j2\pi ft}\, df \right|$$

$$= \left| \int_{-1/2t_s}^{1/2t_s} \left[\sum_k e^{-j2\pi k\tau/t_s} Z_a\left(f - \frac{k}{t_s}\right) - Z_a(f) \right] e^{j2\pi ft}\, df \right.$$

$$\left. - \int_{|f| \geq 1/2t_s} Z_a(f) e^{j2\pi ft}\, df \right|$$

$$\leq \left| \int_{-1/2t_s}^{1/2t_s} \left[\sum_k e^{-j2\pi k\tau/t_s} Z_a\left(f - \frac{k}{t_s}\right) - Z_a(f) \right] e^{j2\pi ft}\, df \right|$$

$$+ \left| \int_{|f| \geq 1/2t_s} Z_a(f) e^{j2\pi ft}\, df \right|$$

$$\leq \int_{-1/2t_s}^{1/2t_s} \left| \left[\sum_k e^{-j2\pi k\tau/t_s} Z_a\left(f - \frac{k}{t_s}\right) - Z_a(f) \right] e^{j2\pi ft} \right| df$$

$$+ \int_{|f| \geq 1/2t_s} |Z_a(f) e^{j2\pi ft}|\, df$$

$$= \int_{-1/2t_s}^{1/2t_s} \left| \left[\sum_{k=-\infty}^{\infty} e^{-j2\pi k\tau/t_s} Z_a\left(f - \frac{k}{t_s}\right) - Z_a(f) \right] \right| df$$

$$+ \int_{|f| \geq 1/2t_s} |Z_a(f)|\, df$$

$$= \int_{-1/2t_s}^{1/2t_s} \left| \sum_{k=1}^{\infty} \left[e^{-j2\pi k\tau/t_s} Z_a\left(f - \frac{k}{t_s}\right) + e^{j2\pi k\tau/t_s} Z_a\left(f + \frac{k}{t_s}\right) \right] \right| df$$

$$+ \int_{|f| \geq 1/2t_s} |Z_a(f)|\, df$$

$$\leq \int_{-1/2t_s}^{1/2t_s} \left| \sum_{k=1}^{\infty} e^{-j2\pi k\tau/t_s} Z_a\left(f - \frac{k}{t_s}\right) \right| df$$

$$+ \int_{-1/2t_s}^{1/2t_s} \left| \sum_{k=1}^{\infty} e^{j2\pi k\tau/t_s} Z_a\left(f + \frac{k}{t_s}\right) \right| df + \int_{|f| \geq 1/2t_s} |Z_a(f)|\, df$$

$$\leq \int_{-1/2t_s}^{1/2t_s} \left| \sum_{k=1}^{\infty} Z_a\left(f - \frac{k}{t_s}\right) \right| df + \int_{-1/2t_s}^{1/2t_s} \left| \sum_{k=1}^{\infty} Z_a\left(f + \frac{k}{t_s}\right) \right| df$$

$$+ \int_{|f| \geq 1/2t_s} |Z_a(f)| \, df$$

$$= \sum_{k=1}^{\infty} \int_{-1/2t_s}^{1/2t_s} \left| Z_a\left(f - \frac{k}{t_s}\right) \right| df + \sum_{k=1}^{\infty} \int_{-1/2t_s}^{1/2t_s} \left| Z_a\left(f + \frac{k}{t_s}\right) \right| df$$

$$+ \int_{|f| \geq 1/2t_s} |Z_a(f)| \, df$$

$$= \sum_{k=1}^{\infty} \int_{(-1-2k)/2t_s}^{(1-2k)/2t_s} |Z_a(f)| \, df + \sum_{k=1}^{\infty} \int_{(-1+2k)/2t_s}^{(1+2k)/2t_s} |Z_a(f)| \, df$$

$$+ \int_{|f| \geq 1/2t_s} Z_a(f)| \, df$$

$$= \int_{-\infty}^{-1/2t_s} |Z_a(f)| \, df + \int_{1/2t_s}^{\infty} |Z_a(f)| \, df + \int_{|f| \geq 1/2t_s} |Z_a(f)| \, df$$

$$= 2 \int_{|f| \geq 1/2t_s} |Z_a(f)| \, df \qquad (5.10)$$

APPENDIX 5C: ANALOG/DIGITAL ENERGY RELATIONS

Digital Energy Formula (5.25)*

The reconstructed signal delivered by the ideal lowpass filter of Fig. 5.1 is given by

$$\tilde{z}(t) = \sum_{n=-\infty}^{\infty} z_d(n) \frac{\sin \pi(t/t_s - n)}{\pi(t/t_s - n)} \qquad (5C.1)$$

---

*This development of (5.25) is due to Umberto Mengali; it is much shorter and more direct than our original version.

which has the Fourier transform

$$\tilde{Z}(f) = \sum_{n=-\infty}^{\infty} z_d(n) \int_{-\infty}^{\infty} \frac{\sin \pi(t/t_s - n)}{\pi(T/t_s - n)} e^{-j2\pi ft} dt$$

$$= t_s \sum_{n=-\infty}^{\infty} z_d(n) e^{-j2\pi n f t_s}, \qquad |f| \leq 1/2t_s$$

$$= 0, \qquad |f| > 1/2t_s \tag{5C.2}$$

Energy in the reconstructed signal is

$$E_{\tilde{z}} = \int_{-\infty}^{\infty} |\tilde{z}(t)|^2 \, dt = \int_{-\infty}^{\infty} |\tilde{Z}(f)|^2 \, df$$

$$= \int_{-1/2t_s}^{1/2t_s} t_s^2 \sum_{n=-\infty}^{\infty} \sum_{m=-\infty}^{\infty} z_d(n) z_d^*(m) e^{j2\pi f t_s (m-n)} \, df$$

$$= t_s^2 \sum_{n=-\infty}^{\infty} \sum_{m=-\infty}^{\infty} z_d(n) z_d^*(m) \int_{-1/2t_s}^{1/2t_s} e^{j2\pi f t_s (m-n)} \, df$$

$$= t_s^2 \sum_{n=-\infty}^{\infty} \sum_{m=-\infty}^{\infty} z_d(n) z_d^*(m) \frac{\sin \pi(m-n)}{\pi t_s(m-n)}$$

$$= t_s \sum_{n=-\infty}^{\infty} \sum_{m=-\infty}^{\infty} z_d(n) z_d^*(m) \delta(m-n)$$

$$= t_s \sum_{n=\infty}^{\infty} |z_d(n)|^2 \tag{5.25}$$

That is, the energy of the reconstructed signal $\tilde{z}(t)$, computed from (5.23), is always identical to the energy computed from the samples using (5.25). If the original signal $z_a(t)$ is bandlimited to $|f| < 1/2t_s$, then $\tilde{z}(t) \equiv z_a(t)$, so in that case the energy of the original analog signal $z_a(t)$ is exactly the energy computed for the sampled digital signal.

Average over τ

Make use of Parseval's rule and (5A.3) to obtain the τ-dependent energy of the reconstructed signal $\tilde{z}(t)$ as

$$E_{\tilde{z}}(\tau) = \int_{-1/2t_s}^{1/2t_s} \tilde{Z}(f)\tilde{Z}^*(f)\,df$$

$$= \int_{-1/2t_s}^{1/2t_s} \sum_{l=-\infty}^{\infty} \sum_{k=-\infty}^{\infty} e^{j2\pi\tau(l-k)/t_s} Z_a\left(f - \frac{k}{t_s}\right) Z_a^*\left(f - \frac{l}{t_s}\right) df \quad (5C.3)$$

Now take the average of (5C.3) over τ in the range 0 to t_s:

$$\overline{E}_{\tilde{z}} = \frac{1}{t_s} \int_0^{t_s} E_{\tilde{z}}(\tau)\,d\tau$$

$$= \int_{-1/2t_s}^{1/2t_s} \left[\sum_{l=-\infty}^{\infty} \sum_{k=-\infty}^{\infty} Z_a\left(f - \frac{k}{t_s}\right) Z_a^*\left(f - \frac{l}{t_s}\right) \frac{1}{t_s} \int_0^{t_s} e^{j2\pi\tau(l-k)/t_s}\,d\tau \right] df$$

$$= 0, \quad l \neq k$$

$$= \int_{-1/2t_s}^{1/2t_s} \sum_{k=-\infty}^{\infty} Z_a\left(f - \frac{k}{t_s}\right) Z_a^*\left(f - \frac{k}{t_s}\right) df, \quad l = k \quad (5C.4)$$

The last line in (5C.4) is the infinite sum of adjacent, finite integrals with identical integrands and so reduces to a simple infinite integral

$$\overline{E}_{\tilde{z}} = \int_{-\infty}^{\infty} Z_a(f) Z_a^*(f)\,df = \int_{-\infty}^{\infty} |Z_a(f)|^2\,df = E_{za}$$

6

COMPUTER MODELING OF ANALOG FILTERS

The next few chapters treat computer modeling of signal processes. Logical development of the subject could have been based on a block diagram of a communications system, such as in Figs. 1.1 and 1.2, starting with the message source and going through the transmitter, the medium, and the receiver, in that order. We soon discovered that filters did not fit well into the logical order, for several reasons: Filters occur at numerous locations within a system, but they need be explained just once. Requirements arising from filter modeling prove to have substantial impact upon modeling of other processes. Simulation of filters accounts for the greatest single amount of computer burden and user effort of any process. Filter models are more complicated than those for most other processes. For all of these reasons, we lead off the chapters on computer modeling with an explanation of filters.

In Chapter 3, analog filters were represented in three different ways:

1. As a transfer function $H_a(s)$ or frequency response $H_a(f)$ in the frequency domain. (Laplace representation uses the complex variable s; the equivalent Fourier representation uses the frequency variable f.)
2. As an impulse response $h_a(t)$ in the time domain.
3. As a differential equation in the time domain.

Subscript a indicates an analog quantity. Only linear, time-invariant filters are considered here.

Corresponding to each analog representation, there is also a digital representation:

1. As a transfer function $H_d(z)$ or sampled frequency response $H_d(k)$ in the frequency domain. [The continuous complex variable z is associated with the z-transform; the discrete frequency index k applies to the discrete Fourier transform (DFT). The computer can deal only with discrete representations, but both forms will be used in the text.]
2. As a response $h_d(n)$ to a unit-sample stimulus, where n is the time-domain index.
3. As a difference equation in the time domain.

Subscript d indicates a digital quantity.

Several questions arise in using these models:

- What algorithm is employed in a process for each kind of model?
- How are the characteristics of the digital model derived from the analog filter that is to be simulated?
- What facilities does a user need to translate analog filter data into digital representations?
- What restrictions are imposed on the structure of a simulation by use of these filter models?
- What precautions must a user observe when selecting filter models?

These matters, and others, are examined in this and succeeding chapters.

6.1 FILTER ALGORITHMS

Process algorithms for each of the filter models are examined in this section, in the order introduced above.

6.1.1 Frequency-Domain Filters

Subscripts d and a will be dropped temporarily, and reintroduced when needed. Consider a digital filter with frequency-sampled transfer function $H(k)$. If a frequency-domain signal sequence $\{Z_{in}(k)\}$ is applied as input to the filter, the resulting output sequence in the frequency domain is

$$\{Z_{out}(k)\} = \{Z_{in}(k)H(k)\} \tag{6.1}$$

The kth element of the output sequence consists of the kth element of the input sequence multiplied by the kth element of the transfer function. Braces $\{\cdot\}$ indicate block data structures. Indexes of the elements must line up correctly. A transfer function $H(k)$ is represented in the computer by a finite number N of points.

Therefore, the signal sequence also must contain exactly N points for computation of (6.1).

Computation of Filter Output

Suppose that the input signal to a frequency-domain filter is in the time domain and output is also required to be in the time domain. How is a frequency-domain computation to be arranged?

1. Designate the time-domain input signal sequence as $\{z_{\text{in}}(n)\}$. The sequence is an N-point block.
2. Using the DFT, transform $\{z_{\text{in}}(n)\}$ to the frequency domain $\{Z_{\text{in}}(k)\}$.
3. Multiply $Z_{\text{in}}(k)$, one element at a time, by the transfer function $H(k)$ to obtain $\{Z_{\text{out}}(k)\}$, as in (6.1).
4. Apply the inverse DFT to find the time-domain output $\{z_{\text{out}}(n)\}$.

Circularity

The sequence $\{z_{\text{out}}(n)\}$ is a circular block; it wraps around from the $(N-1)$th element to the zeroth element and beyond, even if $\{z_{\text{in}}(n)\}$ is a linear block and does not wrap around. This time-domain circularity is inherent to performing operations in the frequency domain (see Secs. 5.2 and 5.3). Circularity must be taken into account whenever frequency-domain operations are applied. Methods exist, applicable in special circumstances, whereby $\{z_{\text{out}}(n)\}$ can be opened out into a linear block, even after frequency-domain filtering. These methods are outlined in Sec. 6.1.5, after the other kinds of filter algorithms have been introduced.

Data Handling

It is reasonable to suppose that the blocks $\{Z_{\text{in}}(k)\}$ and $\{H(k)\}$ are held in arrays in computer memory and that the algorithm of (6.1) operates on the array elements. Arrays must be two-dimensional, since the elements of both Z and H are complex.

Once the output array $\{Z_{\text{out}}(k)\}$ has been computed, there usually will be no more need for the input array $\{Z_{\text{in}}(k)\}$. Rather than consume the memory needed for a third N-element array, it is more economical of resources to store the output in the same array used for the input; the output may be computed in place. Similarly, the FFT routine used for the DFT or IDFT almost invariably will be computed in place.

Computation in place destroys the data in the input array. If those data are to be used elsewhere, they must be saved in other storage prior to computation and held until needed. Most simulation facilities (STÆDT included) make provision for necessary storage while placing minimal burden upon the user.

Coordinate System

Computations may be carried out in rectangular coordinates or polar; which is best? In rectangular coordinates, each point of (6.1) is a complex multiplication which involves four real multiplications and two additions. That may be seen from the following:

$$Z_{in}(k) = X_{in}(k) + jY_{in}(k), \qquad H(k) = H_{re}(k) + jH_{im}(k) \qquad (6.2a)$$

(Subscripts *re* and *im* refer to real and imaginary parts, respectively.) Rectangular components of the filtered product are

$$X_{out}(k) = X_{in}(k)H_{re}(k) - Y_{in}(k)H_{im}(k)$$
$$Y_{out}(k) = X_{in}(k)H_{im}(k) + Y_{in}(k)H_{re}(k) \qquad (6.2b)$$

where $Z_{out}(k) = X_{out}(k) + jY_{out}(k)$.

In polar coordinates, the signal and filter are represented by

$$Z_{in}(k) = |Z_{in}(k)|e^{j\phi_i(k)}, \qquad H(k) = |H(k)|e^{j\phi_h(k)} \qquad (6.3a)$$

so the product is

$$Z_{out}(k) = |Z_{in}(k)|\,|H(k)|e^{j[\phi_i(k) + \phi_h(k)]} \qquad (6.3b)$$

where $\phi(k)$ is the phase of the signal or filter at index k. Magnitude would be stored in one dimension of the two-dimensional arrays and phase in the other. There is only one real multiplication and one real addition in polar coordinates. These features seem to favor the use of polar coordinates for frequency-domain filtering.

Influence of DFT. But before drawing hasty conclusions, consider where the input data come from and how the output data are to be used. The frequency-domain input probably will be computed via the DFT (FFT) from a time-domain input. Furthermore, the frequency-domain output likely will be transformed back to the time domain via the IDFT (IFFT). Each elemental operation of the DFT or FFT consists of a *sum* of products; see (5.20) and (5.21) for the DFT, or [1.9] to [1.12] for the FFT. Each operation requires one complex multiplication, whose product is then added to the results of other operations. Multiplication could be performed in polar or rectangular coordinates (it is much simpler in polar), but addition can only be performed in rectangular. If the multiplication were performed in polar coordinates, a polar-to-rectangular conversion would be needed for the addition. Coordinate conversion demands much more computation than complex multiplication in rectangular coordinates, so an FFT or DFT almost surely will be designed for rectangular-coordinate inputs and outputs.

Coordinate Conclusion. As shown in the next two sections, operations in time-domain filters also consist of sums of products, so they too are computed most efficiently in rectangular coordinates. Because of the central importance of filters, most signal processes in a simulation are designed to work in rectangular coordinates. Coordinate conversions are applied for the few processes that might require polar representations.

6.1.2 Time-Domain Convolution with Unit-Sample Response

Corresponding to the impulse response $h_a(t)$ of an analog filter, a digital filter has a unit-sample response designated $h_d(n)$; or, again dropping the subscript d, just $h(n)$.

Unit Sample

The *unit-sample sequence* is defined as

$$\delta(n) = \begin{cases} 1, & n = 0 \\ 0, & n \neq 0 \end{cases} \quad (6.4)$$

It is tempting to see the unit sample as the digital representation of the unit impulse $\delta(t)$ that is so useful in analysis of analog systems. Indeed, $h_d(n)$ is often called* the impulse response of the digital filter.

Despite the similar notation and the resemblances in appearance, the two concepts are quite different mathematically. The unit sample is a well-behaved sequence of discrete digital numbers in which only the one at $n = 0$ has a nonzero value. A unit impulse $\delta(t)$ is a singular function in continuous time that cannot be represented in sampled form. Conversely, the unit sample is *not* a sampled version of the unit impulse.

Convolution

Apply a time-domain input sequence $\{z_{in}(n)\}$ to a digital filter with unit-sample response $h(n)$. The nth output point is given by the convolution

$$z_{out}(n) = \sum_{i=0}^{N_h - 1} z_{in}(n - i) h(i) \quad (6.5)$$

where N_h is the number of samples in the filter sequence $\{h(n)\}$ and argument i is the *filter index*. For (6.5) to be computable, N_h must be a finite number—the filter must have finite "impulse" response (FIR). Not all filters of interest are FIR; many have infinite impulse response (IIR) and these are treated in the next section. The convolution formula is also valid analytically for IIR filters but is not computable because of the infinite limits. Coefficients $\{h(n)\}$ will be complex for filters of complex signals. Experienced readers will recognize that (6.5) describes a *transversal filter*—a powerful tool that is used extensively in DSP hardware.

---

*Incorrectly, but sometimes we do it here, too.

Computation

Texts on DSP use block diagrams and signal-flow graphs to show different configurations of transversal filters. Simulationists are more interested in methods of computation. Accordingly, consider this simplistic fragment of pseudocode for computing (6.5):

FIR Filter

```
FOR n = n₁ TO n₂
    SUM = 0
    FOR i = 0 TO Nₕ- 1
        SUM = zin(n - i) × h(i) + SUM
    NEXT i
    zout(n) = SUM
NEXT n                                                          (6.6)
```

The exact form of the code, and its interpretation, depend on the data format: single-point, linear block, or circular block. (Data formats were introduced in Sec. 5.3.2.)

Single-Point Format. A single-point simulation continues for an arbitrary number of points, so there should be no prespecified limits on sample index n; the outer FOR-NEXT loop would be omitted. In fact, the index n itself would be omitted since z_{in} and z_{out} each would be handled as individual complex variables as far as the signal process was concerned, and not as array elements.

Computation of SUM requires that N_h complex filter coefficients $\{h(i)\}$ be in filter-parameter memory and N_h complex signal values $\{z_{\text{in}}(-i)\}$ be in state memory. At the beginning of a simulation, the content of state memory is undefined. A user must provide $N_h - 1$ complex initial conditions to state memory to get the run started. For filters in the main signal path it is commonplace to set all initial conditions to zero (as is done in STÆDT) and simply to wait out the transient until the filter relaxes to steady state. Other applications might call for nonzero initial conditions; the user is then responsible for providing the information and the program must be arranged to accept the user-specified initial conditions.

Notice that the filter index goes from 0 to $N_h - 1$. It could have been taken from 1 to N_h instead and the same output signal would have been produced—except delayed by one sample interval. With the chosen limits on the sum, the nth input contributes to the nth output. With a delay, the effect of the nth input does not appear in the output before the $(n + 1)$th computation. Excess delay ought not be incorporated in the filter unless it is wanted.

Linear Blocks. A linear signal block has N complex points $\{z_{\text{in}}(n)\}$ stored in an array. The outer FOR-NEXT loop of (6.6) would have limits of $n_1 = 0$ and $n_2 = N - 1$. Treatment of state memory and initial conditions is much the same as for single-point operations. Except for the I/O management of z_{in} and z_{out}, the filter process cannot distinguish between single-point and linear-block formats. The output block could be computed in place (overwrite the input block) with the aid of the temporary complex variable SUM.

Circular Blocks. A circular block also has N complex points in its input array and employs the outer FOR-NEXT loop with limits from 0 to $N-1$. But a circular block has no state variables and no initial conditions. The signal values $z_{in}(n-i)$ are taken from the input array itself. To effect wraparound on the block, the index $(n-i)$ is interpreted modulo-N. That is, if $(n-i)$ is negative, it is replaced by $(N-n+i)$ for computation of SUM.

An input point must not be overwritten while it is still needed, yet output points are generated immediately, beginning with $n = 0$. Therefore, it is necessary either to provide separate input and output arrays or to provide N_h elements of auxiliary storage until there is room to begin overwriting the input array. This is just the storage used for state information with single-point and linear-block formats.

A circular convolution is closely related to the frequency-domain filtering of (6.1). If $\{H(k)\}$ is the DFT of $\{h(n)\}$ and $\{Z_{in}(k)\}$ is the DFT of $\{z_{in}(n)\}$, then $\{Z_{out}(k)\}$ of (6.1) is the DFT of the circular block $\{z_{out}(n)\} = \{z_{in}(n) \otimes h(n)\}$.

Adaptive Filters

Coefficients of $h(n)$ tacitly have been assumed to be constant in the discussion above. In many applications requiring adaptive equalizers or other adaptive filters, characteristics of the filter are not fixed but evolve according to some adaptation algorithm. A transversal configuration is the most widely used arrangement for hardware adaptive filters.

Linear convolution, as above, is a starting point for development of simulation of adaptive filters. Although simulation is heavily employed in the study of adaptive filters, they are not treated in this book.

6.1.3 Time-Domain Recursive Difference Equations

Representation of filters by means of differential equations was described in Chapter 3, equations (3.1) to (3.3). Taking the Laplace transform of the differential equation led directly to a transfer function representation. A difference equation for a digital filter corresponding to (3.1) has the form

$$z_{out}(n) + B_1 z_{out}(n-1) + B_2 z_{out}(n-1) + \cdots + B_{n_b} z_{out}(n-n_b) \\ = A_0 z_{in}(n) + A_1 z_{in}(n-1) + A_2 z_{in}(n-2) + \cdots + A_{n_a} z_{in}(n-n_a) \quad (6.7)$$

which can be rearranged and compacted to give a computable form

$$z_{\text{out}}(n) = \sum_{i=0}^{n_a} A_i z_{\text{in}}(n-i) - \sum_{i=1}^{n_b} B_i z_{\text{out}}(n-i) \qquad (6.8)$$

Observations

1. The coefficients $\{A_i, B_i\}$ of the difference equation (6.7) are not the same as the coefficients $\{a_i, b_i\}$ of the differential equation (3.1). Relations between coefficients are developed subsequently.

2. The first summation in (6.8) is the same as the convolution of (6.5), except for notation. Therefore, simulation by convolving a signal with the filter's impulse response also involves a difference equation, albeit simpler than (6.8).

3. Because differential equations are cumbersome to handle analytically, engineers rarely employ them directly to analyze a filter's time response. However, a computer is well suited to performing the operations in the difference equation (6.8).

4. When past output samples are employed in (6.8) to determine the current output, the filter is said to be *recursive*. By contrast, the convolution of (6.5) employs only input samples and is said to be *nonrecursive*.

5. The response of (6.8) to a unit-sample input never dies out completely for nondegenerate coefficient sets. The filter has infinite impulse response (IIR).

6. The coefficients $\{a_i, b_i\}$ of differential equations like (3.1) are real for everyday analog filters, but the coefficients $\{A_i, B_i\}$ in (6.8) will be complex for filters of complex signals.

Relation to Transfer Function

Apply the z-transform to each term of (6.7) and collect terms to obtain

$$Z_{\text{out}}(z) = \frac{\sum_{i=0}^{n_a} A_i z^{-i}}{1 + \sum_{i=1}^{n_b} B_i z^{-i}} Z_{\text{in}}(z) \qquad (6.9)$$

from which the transfer function is defined as

$$H(z) = \frac{Z_{\text{out}}(z)}{Z_{\text{in}}(z)} = \frac{\sum_{i=0}^{n_a} A_i z^{-i}}{1 + \sum_{i=1}^{n_b} B_i z^{-i}} \qquad (6.10)$$

Although computations are performed with (6.8) or an equivalent, it is convenient to use the transfer function (6.10) for explanation and for purposes of filter setup. Both forms occur in the following text.

Cascaded Biquadratic Sections

Numerical difficulties with coefficients become severe if the number of terms n_a or n_b grows large. It is common practice to partition the overall transfer function into smaller sections and either cascade the sections or connect them in parallel. This book deals only with cascaded sections, although simulations of paralleled sections would be equally feasible. Section representations have the same denominators but different numerator coefficients, depending on whether they are to be cascaded or paralleled.

All larger filters can be realized as a cascade of *biquadratic* sections; that is, $n_a = n_b = 2$. All further discussion of recursive difference-equation filters will concentrate on biquadratic sections. The transfer function for a biquadratic section is

$$H(z) = \frac{A_0 + A_1 z^{-1} + A_2 z^{-2}}{1 + B_1 z^{-1} + B_2 z^{-2}} \qquad (6.11)$$

which has two zeros and two poles. Any individual coefficients might be zero, so the biquadratic form includes first-order forms as special cases.

Computation

Formula (6.8) with $n_a = n_b = 2$ is the difference equation for a biquadratic filter, corresponding to (6.11). Because (6.8) is a sum of products, computation in rectangular coordinates avoids coordinate transformations, so is more efficient than polar coordinates. The difference equation would be computed in a manner similar to the pseudocode of (6.6), but expanded to account for two summations. Four complex state variables (two past inputs and two past outputs) plus the current input contribute to the current output. Straightforward computation of (6.6) is equivalent to the configuration that the DSP literature calls "direct form 1," illustrated in Fig. 6.1 for complex coefficients. (See Section 6.1.4 for a discussion of complex coefficients.)

Data Format. Multiple cascaded sections in a filter are computed by applying the signal output of one section as the input of the succeeding section. Signals could be transferred between sections in block format or in single-point format. That is, compute an entire block of samples in each section before going to the next section, or take one point at a time through the entire cascade before starting the next point.

In a library of simulation processes, it is preferable to have a single IIR filter process that can handle external inputs and outputs in either format. If that one process is to work with external single-point simulations, it must be single-point internally. Thus data transfer between cascaded IIR filter sections needs to be in single-point format even if external I/O signals are in blocks.

6.1.4 Complex Coefficients

The difference equations (6.5), (6.6), and (6.8) for time-domain filtering, either FIR or IIR, may appear rather innocuous. However, the signals and the coefficients may

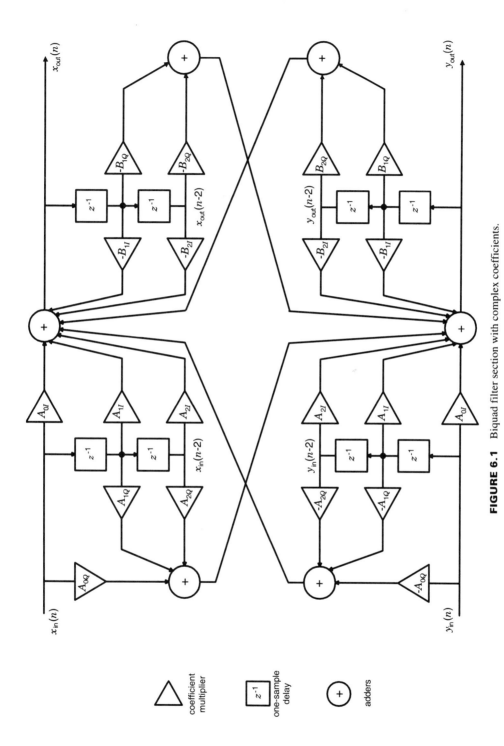

FIGURE 6.1 Biquad filter section with complex coefficients.

be complex, whereupon the formulas and code fragment shown above have been deceivingly simplified. Actual procedures must take complex quantities into account.

As an example, suppose that a biquadratic IIR filter section has complex coefficients $A_i = A_{iI} + jA_{iQ}$ and $B_i = B_{iI} + jB_{iQ}$ and complex signals $z(n) = x(n) + jy(n)$, where $A_{iI}, A_{iQ}, B_{iI}, B_{iQ}$, and x and y are all real. Then the real computations to be performed would be

$$x_{\text{out}}(n) = A_{0I}x_{\text{in}}(n) - A_{0Q}y_{\text{in}}(n) + A_{1I}x_{\text{in}}(n-1) - A_{1Q}y_{\text{in}}(n-1)$$
$$+ A_{2I}x_{\text{in}}(n-2) - A_{2Q}y_{\text{in}}(n-2) - B_{1I}x_{\text{out}}(n-1) + B_{1Q}y_{\text{out}}(n-1)$$
$$- B_{2I}x_{\text{out}}(n-2) + B_{2Q}y_{\text{out}}(n-2)$$
$$y_{\text{out}}(n) = A_{0Q}x_{\text{in}}(n) + A_{0I}y_{\text{in}}(n) + A_{1Q}x_{\text{in}}(n-1) + A_{1I}y_{\text{in}}(n-1)$$
$$+ A_{2Q}x_{\text{in}}(n-2) + A_{2I}y_{\text{in}}(n-2) - B_{1Q}x_{\text{out}}(n-1) - B_{1I}y_{\text{out}}(n-1)$$
$$- B_{2Q}x_{\text{out}}(n-2) - B_{2I}y_{\text{out}}(n-2) \tag{6.12}$$

Similar division into real and imaginary parts will arise for FIR time-domain filters. Filter-builder routines (see Chapter 17) generate these complex coefficients as appropriate. If the coefficients are entered directly by the user (a commonly offered option, but quite laborious), the user is responsible for supplying the correct real and imaginary parts. Some guidance on the coefficients is provided in subsequent pages.

Filter Configurations

Figure 6.1 is a pictorial illustration of the structure of (6.12). This structure is a variation on that of Fig. 3.1, which was derived as the complex-envelope filter corresponding to a general real bandpass filter. These complex structures are *balanced* when derived from ordinary analog filters. That is, the straight-through paths (I to I and Q to Q) are identical, while the cross paths (I to Q and Q to I) differ only in sign. Balanced structures are provided in the time-domain filter processes of most simulation programs, including STÆDT.

But why must a filter model be balanced? What might be the meaning or purpose of an unbalanced model? How can a user be provided with an unbalanced model when needed? These questions are pursued below.

An unbalanced model implies that the real filter underlying the model has more than one real input and/or more than one real output. A familiar example is found in the 90° phase-splitting networks employed in some single-sideband transmitters (e.g., [1.4, p. 17]), which accept a single input signal and deliver two output signals that have 90° phase difference. In a sense, that filter can be regarded as converting a real physical signal into a complex physical signal. Filters of this kind (often called *Hilbert transformers*) are used frequently in DSP applications.

To alter the structure of Fig. 6.1 to an unbalanced model, just assign a distinct value to each multiplier coefficient, so that I-to-I transmission is no longer equal to Q-to-Q, and Q-to-I transmission is no longer the negative of I-to-Q. Capabilities for unbalanced filters could be built into the FIR and IIR filter processes in a simulation library. The principles of coding for unbalanced filter structures should be plain enough from the earlier discussions. Yet the capability imposes an unnecessary burden on the user when it it not required: Should it be imposed upon all simulations with filters for the comparatively few instances in which it might be needed? That is a question that program authors need to decide.

Another approach is to provide ready capability for the user to construct more complicated processes out of simpler processes. Then all complex filters in the library could be balanced and the user would build unbalanced models from interconnections of balanced filters. That is the approach taken in STÆDT.

6.1.5 Linear Blocks from Frequency-Domain Filters

Frequency-domain multiplication via (6.1) is a powerful tool for filtering a signal. Furthermore, the FFT is an efficient method for transforming signals between time and frequency domains. Unfortunately, invoking a discrete frequency-domain representation of a signal imposes a periodicity upon the associated time-domain representation; only circular signal blocks are delivered. That constraint is in conflict with the frequent need to simulate signals of great length, much longer than feasible for practical block sizes.

Linear Convolution

Under restricted circumstances it is possible to avoid the circular-block constraint and to apply frequency-domain filtering to concatenated linear blocks. When feasible, that approach retains the convenience of transfer-function specification of the filter and the efficiency of FFT computations. Consider a filter with finite unit-sample response $h(n)$ of duration N_h. Let a long signal sequence $\{z_{in}(n)\}$ be divided into shorter linear blocks of duration N_s. Designate the rth short block as $\{z_{in}(n)_r\}$. Convolution of a short block with $h(n)$ gives an output sequence of duration $N_h + N_s - 1$ points. This linear convolution could be performed according to the transversal-filter methods of Sec. 6.1.2, but it can also be performed by frequency-domain filter methods, as will now be explained. Resorting to frequency-domain operations is computationally efficient if $h(n)$ has a duration greater than 25 to 30 points.

(Recognize that linear convolution via frequency-domain filtering is possible only if the filter has finite impulse response. In practice, it is often feasible to truncate the time response of an IIR filter to obtain an FIR approximation. Truncation introduces distortion; evaluation of distortion and choice of appropriate truncation are responsibilities of the user.)

As a first step, append zero-valued points to both sequences to pad them out to N points each, where $N \geq N_h + N_s - 1$. Padded-length N should be selected to be efficient for FFT operations (most efficient but not essential: an integer power of 2). Take the DFT of each padded sequence (via the FFT) to obtain the filter transfer function $H(k)$, and the signal transform sequence $\{Z_{in}(k)_r\}$. Multiply the two frequency-domain sequences together to obtain

$$\{Z_{out}(k)_r\} = \{H(k)Z_{in}(k)_r\} \qquad (6.13)$$

A time-domain block $\{z_{out}(n)_r\}$ results from inverse-transforming $\{Z_{out}(k)_r\}$. This is the contribution to the time-domain output from the rth input block.

Observe that $H(k)$ need be computed from $h(n)$ just once; it is the same for all subsequent blocks. Indeed, the filter might even be specified originally in terms of $H(k)$, so that the impulse response need be known only to verify that $h(n)$ is FIR and to ascertain its duration.

Circularity Circumvented

The time-domain output block is necessarily periodic and circular. Its tails wrap around and overlap the main block. This property is inescapable in any time-domain block computed from the DFT. However, because $h(n)$ and $\{z_{in}(n)_r\}$ are both sufficiently short (before padding to N points), the overlapping tail samples all have zero values and so have no effect when added to the N-point main block. Thus, by subterfuge, a linear convolution can be performed with the inherently circular frequency-domain filtering.

Recombination of Short Blocks

Duration of the output segment $\{z_{out}(n)_r\}$ exceeds that of the input segment $\{z_{in}(n)_r\}$ from which it was computed; it will overlap one or more following output segments. Two methods for combining output blocks are described in the DSP literature [1.9, Chap. 3; 1.11, Chap. 2]:

- *Overlap and add.* Employ nonoverlapping short blocks $\{z_{in}(n)_r\}$ at the input. Add the overlapping samples of output blocks.
- *Overlap and save.* Employ overlapping short blocks at the input. Discard any samples from one output block that overlap the next.

Of the two, overlap and add imposes somewhat less computing burden. See the references for details.

6.2 ANALOG-TO-DIGITAL FILTER TRANSFORMATION

So far in this chapter it has been assumed tacitly that the filter characteristics were already available in digital format; only the computation algorithms have been described. However, the analog representation must be transformed to digital before the algorithms can be applied. That transformation is the subject of this section.

Three distinct approaches to transformation are described, corresponding to the three filtering algorithms presented above:

- *Frequency-Domain Transformations.* Frequency reponse $H_a(f) \to H_d(k)$.
- *Time-Domain IIR Transformations.* Differential equation $\to$ difference equa-

tion. [These equations are most conveniently represented by their associated transfer functions $H_a(s)$ and $H_d(z)$; see Sec. 6.1.3.]
- *Time-Domain FIR Transformations.* Impulse response $h_a(t) \to h_d(n)$.

Subscripts *a* and *d* refer to *analog* and *digital*, respectively.

Transformations are performed by *builder routines* upon data previously entered into the computer with the aid of *entry routines*. Builder and entry routines are treated in Chapter 17; only the analytical aspects of transformations are examined in this subchapter. The boundary between filter building and entry on one hand, and filter design on the other, is an indistinct line that is not always easy to pick out. Filter design is a synthesis problem detached from simulation; it is not reasonable to expect a simulation program to perform filter design, too. Some experienced users contend that entry and building are a part of filter design and ought not be considered part of a simulation program. Our opinion is that they are indispensable adjuncts to a comprehensive simulation suite.

6.2.1 Frequency-Domain Transformations

At its simplest, the transformation of the frequency response $H_a(f)$ of an analog filter merely involves taking samples of $H_a(f)$ at N uniformly spaced frequencies and assigning these samples as the computer representation $H_d(k)$. But frequency-domain sampling is nearly the last step in the transformation; other preliminary steps also may be required before the sampling takes place. Any or all of the following actions might be needed:

- Select the appropriate filter *characteristics*. (This is a design problem, outside the scope of a simulation program.)
- Perform *conditioning* upon raw data of the selected filter.
- *Transform* a lowpass filter to bandpass.
- *Convert* a real filter to complex representation.
- *Rescale* the frequency axis of $H_a(f)$.
- *Compute* and store the samples $\{H_d(k)\}$. (This is the actual A/D transformation. Everything beforehand constitutes preparatory actions on analog filters.)
- *Check* the response characteristics of the transformed filter.
- *Cascade* the partitioned sections to produce a composite filter.

These actions and others are discussed in the paragraphs immediately following and in later portions of the book. Not all actions are needed for every preparation of a filter. Many of these actions are also required for setup of filters to be simulated in the time domain. They are introduced at this point in the narrative because the issues tend to be somewhat easier to explain in the frequency domain.

Forms of Filter Characteristics

The raw data on filter characteristics might be entered in any of several different forms. The main divisions are (1) tabulated data (typically, the results of experimental measurements), and (2) analytical formulas. Tabulated data are usually presented as magnitude and phase versus frequency or as magnitude and delay versus frequency. Delay must be *integrated* to phase before it can be used in the simulation. It is often necessary to *fill in* missing data points, such as dc response of a lowpass filter. *Interpolation* is applied to tabular data to produce a continuous approximation of the raw characteristics. These *conditioning* operations are covered in Chapter 17.

Analytical formulas might be described by poles and zeros, by partitioned biquadratic sections, by coefficients of rational functions (but be wary of numerical difficulties), or by transcendental expressions (such as for bandlimited Nyquist filters). The first three forms might be standard filters (Butterworth, Chebyshev, Bessel, elliptic), or they might be any arbitrary analog filters with lumped elements.

STÆDT accepts tabular data and biquadratic coefficients. It also includes in its library formulas for particular standard all-pole and transcendental filters. To incorporate additional formulas into STÆDT requires that new routines be written. Other simulation programs might take pole–zero locations, permit entering of new formulas without overt programming, or accept higher-order rational functions without prior factoring.

Lowpass-to-Bandpass Transformations

Bandpass filters are often designed as transformations of lowpass prototype filters. This lowpass-to-bandpass (LP–BP) transformation is clearly a filter-design matter and not a necessary part of a simulation program. Nonetheless, the need for a LP–BP transformation arises so persistently that it is often included among the filter builder routines. Also, it is very easy to apply in the construction of a frequency-domain filter.

The LP–BP transformation changes an analog lowpass filter, with frequency response $H_l(f)$ and specified bandwidth B_l, into an analog bandpass filter with frequency response $H_p(f')$, specified bandwidth B_p, and "center frequency" f_p. The ' notation distinguishes frequency f' in the bandpass expression from frequency f in the lowpass expression.

If the lowpass prototype is realizable, the ensuing bandpass filter is also realizable. Details may be found in the literature, such as [1.11, Chap. 4]. Elements of the transformation are depicted schematically in Fig. 6.2. Analytically, the transformation is given by

$$H_p(f') = H_l\left[\frac{B_l f_p}{B_p}\left(\frac{f'}{f_p} - \frac{f_p}{f'}\right)\right] = H_l\left[r_p\left(f' - \frac{f_p^2}{f'}\right)\right] \qquad (6.14)$$

where $r_p = B_l/B_p$. The single frequency $f = B_l$ transforms to the two frequencies

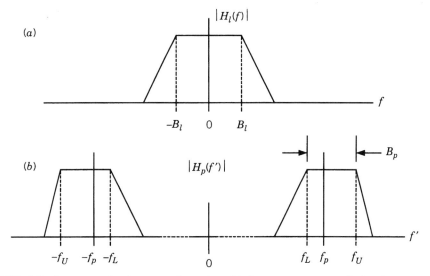

FIGURE 6.2 Lowpass-to-bandpass transformation: (a) lowpass prototype; (b) bandpass transformation.

$$f' = -f_L = +\frac{B_p}{2} - \sqrt{f_p^2 + \left(\frac{B_p}{2}\right)^2} \quad \text{and} \quad f' = f_U = +\frac{B_p}{2} + \sqrt{f_p^2 + \left(\frac{B_p}{2}\right)^2} \tag{6.15}$$

while $f = -B_l$ transforms to $f' = f_L$ and to $f' = -f_U$. The "center frequency" is

$$f_p = \sqrt{f_U f_L} \tag{6.16}$$

and the bandpass bandwidth is

$$B_p = f_U - f_L \tag{6.17}$$

"Center frequency" is enclosed in quotation marks because f_p lies at the geometric mean of the two bandedge frequencies f_U and f_L and not midway between at the arithmetic mean $\frac{1}{2}(f_U + f_L)$. Furthermore, there is no reason why carrier frequency of a signal need be set to coincide with either definition of center frequency.

The LP–BP transformation preserves the amplitude and phase of the lowpass prototype but distorts the frequency scale of the bandpass result. Figure 6.2 illustrates distortion in stylized fashion. A flat top in the passband is nicely preserved, but linear phase and arithmetic symmetry are both lost. The latter two are important in most communications links, so the LP–BP transformation may not be satisfactory unless the relative bandwidth ratio B_p/f_p is small. Other methods of design for bandpass

178 ☐ COMPUTER MODELING OF ANALOG FILTERS

filters are often needed. The decision to apply the transformation is a matter of filter design and lies outside the purview of a simulation program.

Definitions of Bandwidth

Bandwidth of the lowpass prototype was labeled as B_l, and bandwidth of the bandpass transformation was labeled as B_p. Yet the meaning of "bandwidth" was not specified. Many different definitions of bandwidth might be needed under different circumstances. Some common examples are −3 dB bandwidth; −6 dB bandwidth; ripple bandwidth; noise bandwidth; frequency of first null; full-width, half-maximum (FWHM); and others that are less common.

The definition of bandwidth is imposed by a user; the simulation is oblivious to the definition. Definition of bandwidth is primarily a filter-design issue that must be resolved outside the simulation program. This indeterminancy of definition applies to all filters and is not unique to the LP–BP transformation. Whatever bandwidth definition is selected, the LP–BP transformation applies the same definition to both B_l and B_p.

Real-to-Complex Conversion

Raw descriptions of filter characteristics typically will be stated in terms of real filters, whereas simulation of carrier signals most often is performed using complex envelopes and complex filters. (*Reminder:* A *real* filter has a real impulse response and real coefficients in its transfer function; a *complex* filter has a complex impulse response and complex coefficients in its transfer function. Transfer functions of most filters evaluate as complex, no matter whether its coefficients are real or complex.)

If complex signals are to be simulated, the real description of the raw filter characteristic must be converted to a complex description. Conversion for filters described in the frequency domain is the subject of this section. Conversion of time-domain descriptions is covered in a later section. Frequency-domain conversion is much simpler, by far.

The main elements of frequency-domain conversion were explained in Sec. 3.3.1. Refer to Fig. 6.3 for an illustration of conversion steps. The conversion proceeds as follows:

1. Designate the frequency response of the real filter as $H(f)$.
2. Split the response of the real filter into postive- and negative-frequency halves:

$$H_+(f) = H(f), \quad f \geq 0$$
$$H_-(f) = H(f), \quad f < 0 \qquad (6.18)$$

3. Discard $H_-(f)$.
4. Select a conversion frequency $f_c > 0$. (Selection of f_c is examined in greater detail below.)

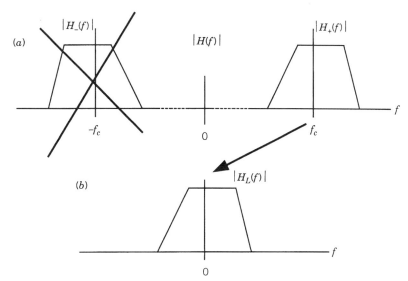

FIGURE 6.3 Real-to-complex conversion in frequency domain: (*a*) response of real filter; (*b*) response of complex filter.

5 Translate $H_+(f)$ to the left by an amount f_c. The resulting complex-signal filter has a frequency response

$$H_L(f) = H_+(f + f_c) \tag{6.19}$$

Conversion Properties. The response that $H_+(f)$ exhibits at $f = f_c$ is translated to $f = 0$ in $H_L(f)$. Response $H_L(f)$ of the complex filter retains the exact shape of the original $H_+(f)$ without distortion. If $H_+(f)$ is asymmetric about f_c (as is true in any realizable filter), $H_L(f)$ will be identically asymmetric about $f = 0$.

Experimental Data. Only positive frequencies are employed in laboratory measurements of frequency response. Thus experimental data include only $H_+(f)$. The negative-frequency portion $H_-(f)$ of the filter has already been discarded.

When performing measurements upon a bandpass filter, it is common, although not invariable practice to record the measurement frequency f as the deviation from some reference frequency f_r, particularly if the filter bandwidth is small compared to the center frequency. In that case the measured data are already in the form of a complex frequency response $H_L(f)$, and no additional real-to-complex conversion is needed if $f_r = f_c$.

Pseudobandpass Filters

If the bandwidth of a bandpass filter is very much smaller than its center frequency, it is possible that the filter response can be nearly arithmetically conjugate symmetric

180 □ COMPUTER MODELING OF ANALOG FILTERS

about a center frequency f_p. In that case, a simulationist often employs a real lowpass filter to approximate the complex-envelope response of the narrowband bandpass filter. We have dubbed the lowpass filter response used in this manner a *pseudobandpass* filter.

Let the lowpass filter have frequency response $H_1(f)$ and lowpass bandwidth B, as sketched in Fig. 6.4a. The first step (never actually carried out) in forming the pseudobandpass filter is to translate $H_1(f)$ to the right by an amount f_p, as shown in Fig. 6.4b to produce a bandpass complex filter with frequency response $H_2(f) = H_1(f - f_p)$. The bandwidth of the bandpass filter is $B_p = 2B$.

The next step (Fig. 6.4c) is to translate the complex bandpass filter to the left by an amount f_1 to obtain a complex-envelope filter $H_3(f) = H_2(f+f_1) = H_1(f-f_p+f_1)$. If $f_1 = f_p$ (as is usually the case), then $H_3(f) \equiv H_1(f)$. Thus no actual operations are needed to convert a lowpass filter into a pseudobandpass filter for complex envelopes.

Despite the fact that no changes may have been performed on the filter characteristic, the definition of bandwidth has changed. The lowpass filter has a bandwidth

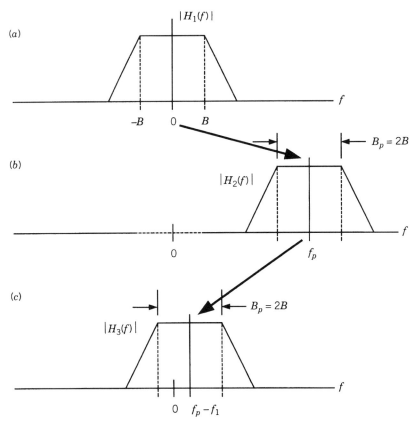

FIGURE 6.4. Pseudobandpass filter.

B defined on the one-sided lowpass spectrum, but a bandpass filter has a bandwidth $B_p = 2B$ that is defined to both sides about the carrier frequency. This ambiguity of bandwidth definition is a potential source of confusion and must be handled with care. Ambiguity arises only in constructing a pseudobandpass filter; it is a consequence of tricky use of a lowpass filter characteristic to masquerade as something that it is not. No similar ambiguity arises with an actual bandpass filter or with the LP–BP transformation.

Frequency Rescaling

Raw data for a particular filter are described in terms of its own frequency scale: a scale that may be completely unrelated to the simulation problem at hand. One essential task for a builder routine is to rescale the raw filter data to conform to the dimensionless filter characteristic needed for the on-line filter process. Although rescaling might superficially seem to be a trivial matter, it turns out to be unexpectedly intricate and can be a source of confusion to the unwary. Rescaling is explored more fully in Chapter 17; only the bare facts are stated here.

The fundamental frequency-rescaling rule is

$$H_u(f) = H_a(\gamma f) \tag{6.20}$$

where γ is the frequency-rescaling factor. Some properties of rescaling are:

- The entire frequency axis is rescaled (see Fig. 6.5a and b).
- Rescaling is linear, so the relative shape of the filter's frequency response is unchanged.
- Amplitude and phase are preserved, unchanged, at the scaled frequency. Similarly, the rectangular components of frequency response are unchanged by rescaling.
- Group delay τ, which is the derivative of phase, is given by

$$\tau_u(f) = \frac{1}{\gamma} \tau_a(\gamma f)$$

   Thus rescaling does not introduce delay distortion that was not already present in the raw data.
- Rescaling preserves relative bandwidth (i.e., bandwidth divided by center frequency) of a bandpass filter. Equivalently: Rescaling alters bandwidth and center frequency by the same ratio.
- If the response $H_a(f)$ of the original filter is physically realizable, the response $H_u(f) = H_a(\gamma f)$ is also physically realizable. It is only necessary to scale component values of the original filter to obtain the rescaled filter.

Sometimes a user may want to rescale the bandwidth of a bandpass filter but

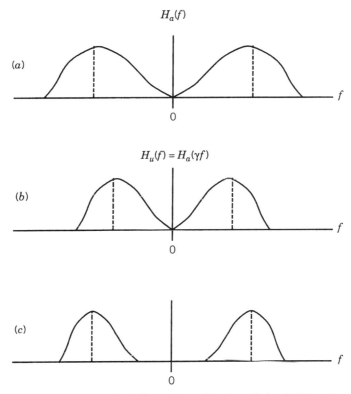

FIGURE 6.5 Frequency rescaling of real filter: (*a*) raw filter data; (*b*) bandwidth and center frequency rescaled identically (physically realizable); (*c*) bandwidth rescaled, center frequency held constant (not physically realizable).

hold the center frequency unchanged, as illustrated in Fig. 6.5*c*. That kind of rescaling cannot be accomplished on a realizable bandpass filter; a change of fractional bandwidth necessitates redesign of the entire filter.

Nonetheless, bandwidth can be scaled at will if the center frequency lies at $f = 0$. Where modulated signals are involved, that frequency assignment can arise only if $H_a(f)$ represents a complex-envelope filter (or a real lowpass filter), not a real bandpass filter. The rescaled complex-envelope filter $H_u(f)$ will be physically realizable as a real filter only upon reconversion to a center frequency that has been rescaled by the same factor as the bandwidth. This rescaled center frequency cannot be the same as that associated with the real filter that produced the complex filter $H_a(f)$.

A related issue arises in conjunction with communications links in which the signal undergoes frequency translations. Radios typically include frequency convertors and one or more intermediate frequencies with bandpass filters centered at each of those frequencies. For simulation purposes it is helpful to refer all such filters to a single frequency. That common reference is easily attained by converting all bandpass filters in the system to complex-envelope representation, with zero center frequency. Each filter then can be rescaled freely for simulation purposes, although the rescaled versions may not be realizable in real bandpass filters centered at the actual frequencies of the system.

Sometimes it is necessary to shift the center frequency of a filter while holding the bandwidth unchanged. Shifting is not the same as rescaling. Shifting cannot be performed simply in a physically realizable manner on the response of a real bandpass filter. Center frequency of a complex-envelope filter can be shifted by any desired amount, but a model of the real equivalent of the shifted filter generally requires inclusion of a heterodyne operation to be physically realizable. Pitfalls of frequency shifting are taken up in Chapter 10.

Analog-to-Digital Transformation

All of the foregoing operations are preliminary to the actual A/D transformation, which is accomplished merely by sampling the frequency response of the analog filter. That is,

$$H_d(k') = H_a[k' \Delta f] \qquad (6.21)$$

Index k' runs from $IP[-(N-1)/2]$ to $IP[(N-1)/2]$, where $IP[x]$ means the integer part of x. Refer to Sec. 5.2.3 and Fig. 5.3 for a discussion of frequency indexing. Frequency increment is defined as $\Delta f = 1/Nt_s = M_s/NT$, where N is the number of elements in the filter block $\{H_d(k')\}$, $1/t_s$ is the sampling rate, $1/T = 1/M_s t_s$ is the symbol rate, and M_s is the number of samples per symbol (must be integer for circular-block frequency-domain operations; see Sec. 5.3.3).

The user specifies N and either $1/t_s$ or M_s and $1/T$ (STÆDT requires M_s and $1/T$). Simulated time/frequency scales often are normalized to the symbol rate, which implies that $1/T = 1$. Other choices are possible, though, as discussed briefly in the next section and in greater detail in Chapter 17.

Combined Operations

Various operations have been presented above: LP–BP transformation, real-to-complex conversion, frequency rescaling, and the analog-to-digital transformation itself. Although these operations have been described separately, they can all be combined together within the filter builder routine with just one computation per frequency-domain point. The formula that accomplishes all that is

$$H_d(k') = H_a(f_k) \qquad (6.22)$$

and the sampled frequency f_k is defined by

$$f_k = r_p \left(\frac{\gamma k' M_s}{NT} + f_0 - \frac{f_p^2}{f_0 + \gamma k' M_s / NT} \right) \qquad (6.23)$$

where r_p and f_p are parameters of LP–BP transformation, f_0 is the reference frequency for complex conversion and γ is the coefficient for frequency rescaling. To omit LP–BP transformation, set $r_p = 1$ and $f_p = 0$; to omit complex conversion, set $f_0 = 0$; to omit frequency rescaling, set $\gamma = 1$.

Frequency f_0 is chosen such that $f_0 - f_p$ is transformed to the origin $k' = 0$ of $\{H_d(k')\}$. If $\{H_d(k')\}$ represents a complex-envelope filter, f_0 ordinarily would coincide with the carrier frequency of the signal. If an analytic signal is to be simulated, f_0 has an application-dependent offset from the carrier frequency. If a baseband signal is to be simulated, typically $f_0 = 0$. The frequency scale for f_0 and f_p is determined by $H_a(f)$ and the LP–BP scaling factor r_p.

From (6.23) it is evident that parameters $f_0, \gamma, r_p, f_p, N, T$, and M_s must be provided. If $H_d(k')$ is built before the simulation is constructed, all simulation parameters must be entered into the transformation by the user. Care must be taken to assure agreement between these parameters in the filter builder and in the simulation. Alternatively, if $H_d(k')$ is built within the simulation, some of the parameters may be extractable from data already in the simulation's storage, thereby reducing potential for erroneous conflicts.

Other Frequency-Domain Matters

Cascading Sections. The raw filter data might be in the form of partitioned biquadratic sections; each section most likely would be transformed individually. Rather than compute the on-line filtering process (6.1) one section at a time, it is much more efficient to have the builder routine cascade the sections and provide just one on-line filter block $\{H_d(k')\}$. If the analog-filter sections are designated $H_{a1}(f)$, $H_{a2}(f), \ldots$, and the transformed sections are designated $\{H_{d1}(k')\}, \{H_{d2}(k')\}, \ldots$, the cascaded filter block is

$$\{H_d(k')\} = \{H_{d1}(k') \times H_{d2}(k') \times \cdots\} \qquad (6.24)$$

Frequency Window. A simulation provides a frequency *window* with finite span $1/t_s$. Unless strictly bandlimited within the window, part of the spectrum of the signal or frequency response of the analog filter will fall outside the window and be truncated. Truncation causes distortion. It is incumbent upon the user to verify that the distortion is not excessive. If filter response is substantially attenuated at the window edges (a typical situation with lowpass or bandpass filters), distortion is likely to be small. If filter transmission is not small at the window edges (e.g., highpass filters), the signal's spectrum might require close scrutiny in evaluating truncation-caused distortion.

6.2.2 Time-Domain Transformations of IIR Filters

An analog IIR filter is described by a differential equation (Sec. 3.3.1) or, exactly equivalently, by a transfer function $H_a(s)$, as explained in Sec. 3.3.2. A digital IIR filter is described by a difference equation or, equivalently, a transfer function $H_d(z)$. This section employs transfer functions as convenient surrogates for the differential and difference equations. The problem to be examined is stated as:

> *Given:* an analytical specification of a rational transfer function $H_a(s)$ of an analog filter
>
> *Find:* coefficients of a digital filter $H_d(z)$ that will closely simulate the behavior of the analog filter

Transfer function $H_d(z)$ is an A/D transformation of $H_a(s)$. Time-domain filtering in the computer is performed by cascaded biquadratic difference equations of the form shown in (6.8); coefficients of the difference equation are those of $H_d(z)$, related to (6.8) by the form of (6.11).

Various actions associated with frequency-domain transformations were described in the last section; many of the same actions are also needed for time-domain transformations. However, the methods employed for performing the actions are much different for the two domains. Explanation of the time-domain actions parallels the order employed for the frequency domain.

Filter Characteristics

Raw data are assumed to be provided only in analytical form; tabulated experimental data cannot be applied easily to produce time-domain IIR filters. The transfer function $H_a(s)$ is a rational function in Laplace transform notation. Its impulse response is causal. Transfer functions of this kind apply to linear time-invariant lumped-element physically realizable filters. Many standard filter characteristics, such as Butterworth, Bessel, Chebychev, and elliptic, are described by rational functions. Transfer functions composed of transcendental expressions [e.g., cosine-rolloff Nyquist functions as in (2.8)] cannot be treated by the methods of this section.

Assume that raw data for $H_a(s)$ have been entered into the computer in a form either of a listing of its poles and zeros and a gain coefficient, or a listing of the coefficients of its partitioned biquadratic sections. Pole–zero format is readily converted to biquad format, so most of the following discussion will deal with biquads. Complicated transfer functions can be partitioned into biquads in many different arrangements, with different zeros associated with various poles and with different orderings of cascaded sections. These are important filter-design questions but are decided outside the simulation program.

A rational function could also be entered in expanded format, as the ratio of two polynomials. Expanded formats are not considered here; $H_a(s)$ is assumed to have been decomposed into factors before being entered into the computer.

Lowpass-to-Bandpass Transformation

Reasons for employing the LP–BP transformation were proffered for frequency-domain filters; the same reasons apply to time-domain filters. Given the transfer function $H_l(s)$ of a lowpass prototype filter, its bandpass transformation is given by

$$H_p(s') = H_l\left[\omega_p \frac{B_l}{B_p}\left(\frac{s'}{\omega_p} + \frac{\omega_p}{s'}\right)\right] \qquad (6.25)$$

where $\omega_p = 2\pi f_p$, and the other notation is the same as for (6.14).

The frequency-domain LP–BP transformation (6.14) is readily combined with the frequency-domain rescaling and A/D transformation as in (6.22) and (6.23), and so can be accomplished with but few extra resources in the builder routine. No similar happy combination occurs when building time-domain filters; appreciable extra resources are needed, as outlined briefly below.

Suppose that $H_l(s)$ is a biquad with real coefficients. Then $H_p(s')$ is a bi*quartic* function, also with real coefficients. To accommodate a biquartic, it would be necessary to perform one of two actions:

1. Provide an on-line difference-equation structure of fourth degree.
2. Factor the biquartic into the cascade of two biquadratics.

Because of its demand for extra resources, we have elected to omit time-domain LP–BP transformation from STÆDT.

Real-to-Complex Conversion

Section 3.3.2 and Appendix 3A have already provided the foundations for conversion from real to complex representations for time-domain filters. There it was shown that time-domain complex conversion is substantially more elaborate than frequency-domain complex conversion. The conversion formulas from Appendix 3A have been incorporated into a STÆDT builder routine and need no further attention here.

Observe that the complex-converted filter is still an analog filter; transformation to digital remains to be accomplished. That is contrary to the building of frequency-domain filters, where the operations are all combined together.

Time-Domain A/D Transformations

In the frequency domain, A/D transformation consisted merely of sampling a windowed, scaled version of the frequency response of the analog filter. The transformation itself is so simple as to nearly disappear into the welter of other details that accompany it. Because of its simplicity, it is hard to imagine alternative frequency-domain transformations.

The situation is quite different for time-domain A/D transformation. There is a multitude of candidate transformations to be considered. Each candidate has significant defects; none yields distortion-free simulation. In this section we describe four candidates for the time-domain A/D transformation and compare their properties.

Analog-Filter Representation. In the discussion to follow, two different representations of the analog-filter transfer function $H_a(s)$ will be needed. One is the partial-fraction expansion

$$H_a(s) = r_0 + \sum_i \frac{r_i}{s - p_i} \qquad (6.26)$$

where the p_i are the poles of the filter and the coefficients r_i are the residues of the poles. [The representation of (6.26) is valid if all poles are distinct. Format modification is needed to accommodate repeated poles.] Coefficient r_0 is nonzero only if the number of zeros is the same as the number of poles—only if $H_a(s)$ is improper. If $H_a(s)$ is a proper rational function, then $r_0 = 0$. A biquadratic section has no more than two poles.

The other representation of the same transfer function is in factored pole–zero format:

$$H_a(s) = H_0 \frac{(s - z_1)(s - z_2) \cdots}{(s - p_1)(s - p_2) \cdots} \qquad (6.27)$$

where H_0 is a gain multiplier and the z_i are the zeros of the filter. A biquadratic section has no more than two zeros.

In a real filter, if a pole p_1 is complex, another pole, $p_2 = p_1^*$ (* denotes complex conjugate) must also be present. If a filter has been converted from real to complex, no such restriction applies. For example, a first-order filter can have a single complex pole, or a second-order filter can have two complex poles that are not complex conjugates of one another. The same comments also apply to zeros.

Candidate Transformations. Four plausible A/D transformations have received varying degrees of attention in the literature:

1. Impulse-invariant transformation [1.9, Chap. 5; 1.11, Chap. 4; 6.1, Chap. 7]
2. Step-invariant transformation [1.9, 6.2]
3. Matched-z transformation [1.11, 6.1]
4. Bilinear transformation [1.9, 1.11, 6.1]

Each candidate will be defined individually and their properties compared.

A strong case is often argued for the *impulse invariant* transformation as the most natural choice. If the impulse response of the analog filter is designated $h_a(t)$ and the unit-sample response of the digital filter is designated $h_d(n)$, impulse invariance is defined by

$$h_d(n) = h_a(nt_s) \qquad (6.28)$$

Impulse response waveforms are preserved with this definition—a desirable property when waveforms are being simulated.

The impulse-invariant definition can apply only if $h_a(t)$ has no impulsive component in its impulse response. Any proper transfer function will have an impulse-free response to an impulse excitation. But any improper transfer function has an impulse component $r_0 \delta(t)$ in its response to unit impulse input. Since a sampled system cannot represent an analog impulse, the impulse-invariant transformation can be applied only to analog filters with proper transfer functions.

Highpass, allpass, and band-rejection filters all are improper and cannot be represented by the impulse-invariant transformation. Moreover, if filters are partitioned into biquadratic sections, some of the sections are likely to be improper and thus not amenable to impulse-invariant treatment, even if the overall analog transfer function itself is proper. For these reasons, an impulse-invariant transformation cannot be employed as the sole A/D time-domain transformation in a simulation program.

Subject to $r_0 = 0$, the digital transfer function of an impulse-invariant transformation can be shown to be

$$H_d(z) = \sum_i \frac{r_i}{1 - z^{-1} e^{p_i t_s}} \qquad (6.29)$$

The poles p_i and residues r_i are those of the analog filter $H_a(s)$, as in (6.26). Poles in the s-plane at $s = p_i$ are transformed to the z-plane at $z = e^{p_i t_s}$. Sampling interval t_s is discussed below. Transformation of zeros depends upon the residues, the pole locations, and the sampling interval. No simple rule can be stated for the transformed locations of the zeros.

Antoniou [6.1] describes a *modified* impulse-invariant transformation that circumvents the problem with improper transfer functions but at the cost of increasing the order of the digital filter. A biquadratic analog filter will not transform into a biquadratic digital filter if you use the modified transformation. Moreover, transformation procedures taken to assure a stable digital filter are likely to have an adverse impact upon group delay; the modified procedure may not preserve waveforms very well.

The *step-invariant* transformation is a different waveform-preserving method that avoids the failure of an impulse-invariant transformation on improper transfer functions. The response of a physically realizable analog filter to a step function contains no impulse components. Therefore, the response can be sampled without encountering unrepresentable singularities.

Let an analog unit-step function $u(t)$ be applied to an analog filter, and designate the response of the filter as $s_a(t)$. Let a unit-step sequence $u(n)$ be applied to a digital filter, and designate the response of the filter as $s_d(n)$. The unit steps are defined by

$$u(t) = 0, \quad t < 0; \quad u(t) = 1, \quad t > 0$$
$$u(n) = 0, \quad n < 0; \quad u(n) = 1, \quad n \geq 0 \tag{6.30}$$

Then if the digital filter is the step-invariant transformation of the analog filter, the step responses will be related by

$$s_d(n) = s_a(nt_s) \tag{6.31}$$

A step-invariant transformation preserves waveforms resulting from rectangular pulses applied to the filter, since a rectangular pulse is just the sum of two steps. That property is attractive when a system using rectangular pulses is to be simulated.

Using the partial-fraction representation of (6.26), the step-invariant transformation can be shown [6.2] to yield a digital-filter transfer function

$$H_d(z) = r_0 + \sum_i \frac{r_i}{-p_i} \frac{z^{-1}(1 - e^{p_i t_s})}{1 - z^{-1} e^{p_i t_s}} \tag{6.32}$$

This expression has the same pole locations as the impulse-invariant transfer function of (6.29), but the zeros lie elsewhere since the residues are different.

Notice the z^{-1} in the numerator of (6.32). Because of it, response to the nth input appears no sooner than in the $(n + 1)$th output (provided that $r_0 = 0$). This is precisely the behavior to be expected from a proper filter function in response to any nonsingular input. Notice also that an impulse-invariant digital filter responds to the nth input on the nth output, which is rather too soon.

An example of this behavior is illustrated in Fig. 6.6 for step response of a single-pole filter. As expected, the step-invariant samples match the analog step response exactly. Output of the impulse-invariant digital filter is already nonzero at $n = 0$, whereas $s_a(0) = 0$. Gain of the impulse-invariant response has been normalized by t_s; its gain discrepancy in steady state is explained several pages hence.

Impulse and unit-sample responses of the same single-pole filter are shown in Fig. 6.7. The ordinate has been scaled so that the analog impulse response is unity at $t = 0$. Impulse-invariant samples fall exactly on the continuous analog response, so those samples are not plotted. Samples of the response of a step-invariant digital filter to a unit sample input are marked by triangles. That response corresponds to a delayed analog waveform. The delay is a fractional interval, so the samples miss the sharp peak of the continuous waveform. (A single-pole filter has a discontinuity in its response to an impulse, and discontinuities are represented poorly by discrete samples. Responses from filters with more poles would look better because they have no discontinuities.)

These examples demonstrate that although digital filters derived from waveform-invariant transformations are exact for the waveform selected, they are not exact when stimulated by other waveforms.

Step invariance eliminates some of the deficiencies of an impulse-invariant trans-

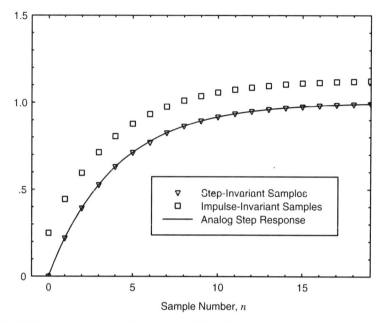

FIGURE 6.6 Step response of analog and digital filters. $H_a(s) = a/(s+a); at_s = 0.25$.

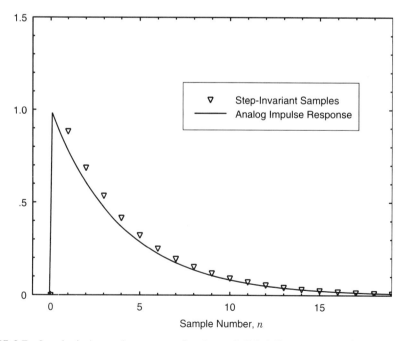

FIGURE 6.7 Impulse/unit-sample response of analog and digital filters. $H_a(s) = a/(s+a); at_s = 0.25$. Ordinate is normalized so that analog impulse response is unity at $t = 0$.

formation but retains others (to be discussed later). Some of the remaining deficiencies can be removed by using a *matched-z* transformation, wherein a pole or zero at $s = s_i$ in $H_a(s)$ is transformed to a pole or zero, respectively, at $z = e^{s_i t_s}$ in $H_d(z)$. The resulting digital transfer function is given by

$$H_d(z) = D_0 \frac{(1 - z^{-1}e^{z_1 t_s})(1 - z^{-1}e^{z_2 t_s}) \cdots}{(1 - z^{-1}e^{p_1 t_s})(1 - z^{-1}e^{p_2 t_s}) \cdots} \quad (6.33)$$

where D_0 is a gain coefficient to be considered further below. (Another unfortunate collision has occurred here between two established nomenclatures: z is the complex variable of the z-transform, and z_i denotes a zero location in the s-plane. There are not enough letters in the alphabet to avoid such collisions.)

The last candidate, the *bilinear* transformation, is seductively attractive because of its mathematical elegance. The transformation is defined by

$$H_d(z) = H_a\left(\frac{2}{t_s} \frac{1 - z^{-1}}{1 + z^{-1}}\right) \quad (6.34)$$

It accomplishes a one-to-one mapping of the left half of the s-plane into the interior of the unit circle in the z-plane and of the right half of the s-plane into the exterior of the z-plane unit circle. Other attractions of the bilinear transformation are examined below, along with its deficiencies.

Stability. All four candidates transform a stable analog filter into a stable digital filter. Specifically, a pole in the left half of the s-plane is always transformed to the interior of the unit circle in the z-plane.

Aliasing. Three of the candidate transformations—the exception is the bilinear transformation—incur aliasing in the frequency domain. Aliasing was described for sampled signals in Sec. 5.1.3. Qualitatively similar statements can be made for aliasing of frequency response of sampled filters. Frequency response of the analog filter $H_a(f)$ is not bandlimited; its spectrum has infinite extent. When the filter is transformed, frequency response of the digital filter [i.e., $H_d(z)$ evaluated for $z = e^{j 2\pi f t_s}$] is periodic with period $1/t_s$. Any portion of the frequency response $H_a(f)$ that falls outside one period, or *window*, of extent $1/t_s$ is truncated and also aliased into the window, in the manner explained for signals in Sec. 5.1.3. Sampling rate $1/t_s$ must be large enough so that truncation and aliasing are not excessive. [Frequency-domain A/D transformation (see Sec. 6.2.1) also suffers truncation but not frequency aliasing. That is because the analog frequency response is truncated before sampling, and the sampling is performed directly in the frequency domain.]

Aliasing and truncation are minimized if frequency response of the analog filter is strongly attenuated beyond the folding frequency $1/2t_s$. Accordingly, aliasing suffered with transformation of a many-pole filter would be less than that suffered from individual transformations of partitioned biquadratic sections. Thus even though we advocate biquadratic partitioning to avoid numerical difficulties and to simplify the on-line IIR filter processes, there is an aliasing penalty incurred from partitioning.

There is another form of aliasing that arises with complex poles. Consider an s-plane complex pole at $s = -\alpha \pm j\beta$ that transforms to a z-plane pole at $z = e^{-\alpha t_s} e^{\pm j\beta t_s}$. The pole itself will be aliased into the next frequency window (shifted around the end of the circular window) unless

$$|\beta t_s| < \pi \tag{6.35}$$

The same restriction must also be applied to zeros for the matched-z transformation.

No aliasing or truncation occurs in the frequency response of the bilinear-transformed filter because the transformation squeezes the entire infinite-extent imaginary axis of the s-plane onto the finite unit circle. The frequency response of this digital filter is still periodic, but there are no overlapping contributions within any one period. Instead of aliasing, the bilinear transformation suffers from pronounced frequency warping, which will be examined shortly.

Gain of Frequency Response. The step-invariant and bilinear transformations both preserve dc gain, in the sense that $H_d(1) = H_a(0)$. For the matched-z transformation, the gain coefficient D_0 can be adjusted to establish any gain that might be wanted. However, dc gain is not preserved by the impulse-invariant transformation. That can be seen from (6.26), from which dc gain of the analog filter is $H_a(0) = \Sigma_i r_i/(-p_i)$, assuming that $r_0 = 0$. From (6.29), dc gain of the impulse-invariant digital filter is $H_d(1) = \Sigma_i r_i/(1 - e^{p_i t_s})$, which is not the same as $H_a(0)$.

For t_s sufficiently small, dc gain of the impulse-invariant digital filter is approximated by $H_d(1) \simeq \Sigma_i r_i/(-p_i t_s)$, which is $1/t_s$ times the dc gain of the analog filter. If t_s is small enough, the dc gain of a lowpass filter can become large.

It is customary to modify the impulse invariant transformation by multiplying it by t_s, thereby preventing large dc gains and nearly preserving the dc gain, for large sampling rates. The definitions become

$$h_d(n) = t_s h_a(n t_s) \tag{6.28a}$$

and

$$H_d(z) = t_s \sum_i \frac{r_i}{1 - z^{-1} e^{p_i t_s}} \tag{6.29a}$$

This modified transformation still does not preserve dc gain exactly; the discrepancy is most noticeable at low sampling rates. A discrepancy is particularly unsettling if $H_a(0) = 0$ and the signal to be filtered contains a dc component that is to be blocked.

Frequency Warping. Because the bilinear transformation squeezes the entire imaginary axis of the s-plane onto the finite unit circle in the z-plane, without overlapping, the frequency scale is subjected to a highly nonlinear compression as a result of the transformation. Therefore, filter characteristics that fall at one frequency in the analog filter will be shifted to a different frequency in the digital filter. As an especially blatant example, the squeezing shifts the passband of a real bandpass filter from its rightful center frequency. See [6.1, Fig. 7.10] for a warning.

Analysis shows that bilinear transformation preserves a transmission null even though it shifts the frequency of the null. Also, bilinear transformation of an analog allpass filter produces a digital allpass filter but with a warped delay versus frequency characteristic. Bilinear transformation will not preserve linear phase.

The transfer function of the analog filter can be prewarped such that any selected nonzero frequency on the imaginary axis of the s-plane maps to any desired angle (not 0 or π) on the unit circle in the z-plane [6.1, Secs. 7.6 and 8.1]. Prewarping can preserve the center frequency of a bandpass filter or the frequency of one null.

Neither the step-invariant nor the impulse-invariant transformation will preserve a null occurring in the frequency response of the analog filter. The impulse-invariant transformation fails entirely for a biquadratic null network (because the transfer function is improper), while the step-invariant transformation moves the pure-imaginary zeros that cause a null in the s-plane to z-plane locations off the unit circle [6.2]. Similarly, the impulse invariant transformation fails for a biquadratic allpass network (also improper), while the step-invariant transformation of an analog allpass filter does not produce a digital allpass filter. Both invariant transformations preserve pole locations.

The matched-z transformation preserves locations of both poles and zeros. Analog null and allpass filters transform to digital null and allpass filters. Frequencies of nulls are not shifted; this property may be crucial if a filter has been designed with a null to attenuate a specific frequency.

Choice of Method. How is a transformation method to be selected? Clearly, no one method is outstanding. As far as we are aware, no study has been published that compares the candidates quantitatively and identifies the best (if there even is a method that is best for all applications). In the discussions above, we have found the fewest faults in the matched-z transformation, but are reluctant to rely upon it alone because of its poor repute among the cited references (mainly because of aliasing). In the past we have had favorable experience with the step-invariant transformation but only with all-pole filters and under limited circumstances. Other simulationists insist that impulse invariance is the only "natural" method and should be the one chosen even if additional poles have to be inserted to counteract improper biquads.

Some simulation programs provide only the bilinear transformation without offering much justification or any alternatives; it seems adequate for many purposes and raises no mathematical obstacles if frequency warping is not excessive. STÆDT permits a user to select any of the four transformations.

Frequency Rescaling. All of the transformations include the sampling rate $1/t_s$ directly in their definitions. To perform the transformation, a builder routine needs a user-provided number for $1/t_s$. Since $1/t_s = M_s/T$, STÆDT builders call for specification of M_s and $1/T$ instead. In addition, a frequency scaling factor γ is provided; see Chapter 17 for an extended discussion of frequency scaling.

6.2.3 Time-Domain Transformations of FIR Filters

A few analog filters exhibit finite impulse response. One prominent example arises in filters based on surface acoustic wave (SAW) filters [6.3]; these may be regarded as analog transversal filters. It is probable, though, that they would be simulated through their frequency response in many instances.

Another, less-prevalent example is found in charge-transfer filters. These are *sampled* analog transversal filters and are well simulated by the time-domain convolution of (6.5). Otherwise, the FIR filter process is not widely used for time-domain simulations of analog filters. The most extensive use of the FIR process is for simulation of digital FIR filters, which are important in DSP applications.

Nevertheless, the FIR process is employed in some specialized areas of analog-system simulation. Two examples are:

1 As waveshaping processes for generating finite-duration pulses. (covered in Chapter 7)
2 As a truncated (or windowed) approximation of an IIR filter response

Truncated Approximations of IIR Responses

Suppose that a filter is known only from experimental measurements of frequency response but the user wishes to perform simulations in the time domain. Finding a rational transfer function (implementable by a recursive difference equation as in the preceding section) from the frequency-domain data could be difficult. By contrast, finding the impulse response is a simple matter of applying the inverse DFT to the frequency-domain data. If the impulse response falls off sufficiently rapidly, it can be truncated (either abruptly or with a tapered window) and its samples used as coefficients in the FIR filter process. This is a variation on the impulse-invariant transformation but now applied to filters whose transfer functions are not necessarily rational.

As before, impulse invariance cannot be applied if the analog impulse response has an impulsive component. More stringently, [6.1] states that the analog impulse response must satisfy $h(0+) = 0$ if a satisfactory approximation is to be attained without excessive aliasing. Equivalently, the frequency response must fall off adequately steeply toward the edges of the frequency-domain window. Occasionally, a filter might be known directly from its impulse response. That response can be sampled and the samples used as the coefficients in the FIR filter without further conditioning.

Digital FIR Filters

Digitally implemented communications systems predominantly use digital FIR filters, to the near exclusion of digital IIR filters. One reason is that FIR filters can be constrained easily to have exactly linear phase, an impossible condition for an IIR filter. Design methods for digital FIR filters are well developed, as described at length in the DSP references cited in Chapter 1.

6.3 KEY POINTS

This has been a long chapter, containing numerous small details that are not readily distinguished from one another, particularly on first reading. As a matter of fact, only two main topics have been included: (1) on-line processes for simulation of analog filters, and (2) methods for preparation of filter characteristics for use by on-line processes.

6.3.1 On-Line Filter Processes

Three different filter processes have been presented. Their equations are listed below.

Frequency Domain

$$\{Z_{out}(k)\} = \{Z_{in}(k)H(k)\} \tag{6.1}$$

where $\{Z_{in}\}$ and $\{Z_{out}\}$ are the input and output complex signal blocks and $\{H(k)\}$ is the transfer-function block.

Time Domain
FIR convolution:

$$z_{out}(n) = \sum_{i=0}^{N_h - 1} z_{in}(n - i)h(i) \tag{6.5}$$

where the N_h complex coefficients $\{h(i)\}$ represent the time-domain response of the filter to a unit-sample sequence.
IIR recursive difference equation:

$$z_{out}(n) = A_0 z_{in}(n) + A_1 z_{in}(n-1) + A_2 z_{in}(n-2) - B_1 z_{out}(n-1) - B_2 z_{out}(n-2) \tag{6.8}$$

It is the difference equation that is executed in the simulation, but the filter can be represented more compactly by its associated transfer function

$$H_d(z) = \frac{A_0 + A_1 z^{-1} + A_2 z^{-2}}{1 + B_1 z^{-1} + B_2 z^{-2}} \tag{6.11}$$

The difference equation and transfer function have been specialized to biquadratic sections, which may be cascaded to form a rational transfer function of any size. All coefficients in these equations can be complex, in general.

Process Features

Coordinate System. All three processes listed above involve sums of products. Because of the need for additions, the computations on complex signals are substantially more efficient in rectangular coordinates than in polar. To minimize the necessity for conversions between coordinate systems, most other processes will also be computed in rectangular coordinates instead of polar.

Block Structure. Frequency-domain filtering typically enforces circular data blocks, which may be opened out into linear blocks by the overlap-and-save or overlap-and-add techniques. Time-domain filters can accept either kind of data-block structure, or single-point operation. Even if input to an IIR filter is in circular blocks, its output consists of linear blocks.

6.3.2 Filter Preparation

Characteristics (e.g., frequency response, impulse response, coefficients of difference equation) of a filter are supplied as parameters to the on-line filter process. The parameters themselves are prepared separately in builder routines. This chapter has concentrated on the analytical models underlying the preparation of characteristics. Details of builder routines are given in Chapter 17.

Characteristics Formats

STÆDT has provisions to accept filter data in the following formats, which we have emphasized in the text:

- Numerical (measured) samples of frequency response
- Numerical (measured) samples of impulse response
- Coefficients of partitioned biquadratic sections

Partitioned biquads are restricted to filters with rational transfer functions. Numerical data can belong to filters that are not described by rational functions.

Other simulation programs might accept additional formats, such as pole–zero locations, unpartitioned higher-order rational transfer functions, or explicit, nonrational formulas for frequency or impulse response. STÆDT is distributed with a small number of nonrational formulas in its library; others may be added by programming new routines. The filter library in STÆDT also contains normalized biquad information on several standard kinds of filters with rational transfer functions.

Filter Preparation Processes

Domain Selection. Many filters can be implemented in either the time or frequency domain, as selected by the user.

Lowpass–Bandpass Transformation. Some programs include the LP–BP transformation as an option. (STÆDT offers it only for frequency-domain filters.) The user specifies bandpass center frequency and the ratio of the bandwidth of the bandpass filter to that of the lowpass prototype. Refer to (6.14) for the analytical form of the frequency-domain transformation.

Real–Complex Conversion. If signals are to be simulated in complex format, real filters must be converted to complex format. Conversion is trivially simple for frequency-domain filters [see (6.19) and Fig. 6.3] but more intricate for time-domain filters (see Sec. 3.3.2 and Appendix 3A). The user need only provide the conversion frequency, in the frequency scale of the raw data; the builder routine handles all calculations.

Frequency Rescaling. Filter data need to be rescaled to conform to the dimensionless on-line filter process. The scaling parameters needed are the sample rate, or the symbol rate and number of samples per symbol. If a frequency-domain filter is being built, block size must also be provided. Parameters specified to the builder, such as N and M_s, must be consistent with parameters used in the simulation. These features were treated at length earlier and are treated again in Chapter 17. Scaling must be treated with meticulous care; its specification is vulnerable to user error.

A/D Transformation. Frequency-domain A/D transformation consists merely of sampling the analog-filter characteristic within the simulation's frequency window. Time-domain A/D transformation is more complicated; it might be performed in any of a number of ways, four of which have been described and are offered by STÆDT.

Concatenation. If a frequency-domain filter is built in partitioned sections, it is usually more efficient to concatenate those sections into a single filter characteristic rather than deal with multiple, cascaded filters.

6.3.3 STÆDT Processes

Filter Processes

DFT: discrete Fourier transform

FFT: fast Fourier transform

FIR: finite impulse response convolution, time-domain filter

IIR: infinite impulse response difference equation, time-domain filter

LNCNLV: linear convolution of time-domain blocks via frequency-domain multiplication of their Fourier transforms and the overlap-and-add algorithm

MULT: multiplies signal block $\{Z(k)\}$ by filter block $\{H(k)\}$ to accomplish frequency-domain filtering

Builder Routines

BQAD: A/D transformation of biquadratic coefficients for time-domain IIR filters

BQCV: real-to-complex conversion of analog biquadratic filters

BQFR: prepares frequency-domain filter block $\{H(k)\}$ from biquadratic coefficients

BQST: prepares biquadratic coefficients for standard filters: Butterworth, Chebychev, or Bessel

FRNP: prepares frequency-response block $\{H_d(k)\}$ from table of conditioned frequency-response data

FRNQ: prepares frequency-response block $\{H_d(k)\}$ for cosine-rolloff root Nyquist filter

FRPC: preconditions measured frequency-response data into format required by STÆDT processes

LNCFLTR: prepares filter characteristics for LNCNLV

7

GENERATION OF DATA SIGNALS

Now that a basis for filters has been established in Chapter 6, the remaining processes in a communications link can be pursued in logical order, beginning at the transmitter and progressively moving downstream. This chapter is devoted to transmitter processes, which generate signals for use in simulations. One's own transmission is the desired signal; unwanted transmissions by others are designated as interference or jamming. All transmissions, wanted or unwanted, are generated by similar techniques.

Objective

The objective in simulating a transmitter is to create a sample sequence $\{z(n)\} = \{x(n) + jy(n)\}$ representing the complex envelope of a data signal. Attention will be concentrated on linear-modulation (QAM) signals in which the rectangular components $x(n) = \Sigma a_m g_T(n - mM_s)$ and $y(n) = \Sigma b_m g_T(n - mM_s)$ are PAM pulse streams. These signals were introduced in Chapter 2.

Although the objective is simple to state, you could easily lose sight of it in the welter of implementation details treated in this chapter. Much of what follows will appear to be only distantly related to the stated objective. Nonetheless, this material is all meaningful information to be taken into account when planning a simulation.

Notation is as follows: $c_m = a_m + jb_m$ is the mth complex symbol to be transmitted, m is the symbol index, n is the sample index, M_s is the number of samples per symbol, and $g_T(n)$ is the sampled waveshape of the transmitted pulse.

To simulate a real baseband signal, just set $y(n) \equiv 0$. To simulate a nonlinear modulation (examples will be shown), $z(n)$ is created as a nonlinear function of the PAM sequences. Real modulated signals (either linear or nonlinear modulations) are readily obtained from their complex envelopes, as will be demonstrated.

In addition to data signals, most programs also provide facilities for generating

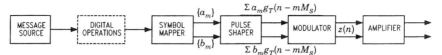

FIGURE 7.1 Transmitter processes.

auxiliary nondata signals, such as impulses, steps, square pulses, ramps, sines and cosines, and so on. Auxiliary signals are not considered here (but are provided in STÆDT).

Transmitter Arrangement

Figure 7.1 shows the main processes to be modeled in a transmitter. Markings on the diagram indicate points of origination for various signal components identified above. Observe that a substantial number of preliminary operations take place before the actual signal components emerge. Transmitter processes include:

- *Message Source:* generates a bit stream.
- *Digital Operations:* encoding, interleaving, scrambling, and so on. Digital operations are touched upon only lightly here, and only in connection with other processes in the transmitter.
- *Symbol Mapper:* generates complex symbols $c_m = a_m + jb_m$.
- *Waveform Generator:* generates PAM pulse streams $\Sigma c_m g_T(n - mM_s)$.
- *Modulator:* absent for baseband signals or for complex envelopes of linearly modulated carrier signals; required to be present for nonlinear modulations.
- *Amplifier:* trivial to simulate if linear. Material in this chapter concentrates upon nonlinear models.

These processes are shown in an order that facilitates discussion, but the reader should be aware that processes can interchange their orders and even intermix in any particular system.

7.1 MESSAGE SOURCE

A simulation starts with a data message that is to be conveyed through the system model. Initially, regard the message as consisting of a stream of bits, denoted equivalently by the Boolean sequence $\{B(n)\}$, where $B(n) = 0$ or 1, or by the real-number sequence $\{d(n)\}$, where $d(n) = +1$ or -1. To convert from $B(n)$ to $d(n)$, or vice versa, associate a Boolean 0 with +1 and a Boolean 1 with -1. Simulations typically consume substantial numbers of message bits; a simulation program needs a process that will generate them efficiently.

Any practical application will have its own data format, including preambles, frame markers, idle patterns, and so on. Therefore, a need exists to be able to generate (and repeat) arbitrary finite data sequences. From the user's point of view, this problem reduces to a matter of external data entry, which can be tedious and error-prone for long sequences but conceptually simple. External sequences are not considered further.

More often, analysis deals with random message streams. The remainder of this section shows how "random" streams are generated on the computer, and various subtle points that arise in the employment of those streams.

7.1.1 Binary Shift-Register Sequences

Binary shift registers of m stages can be arranged in linear feedback* configurations to provide periodic sequences whose period is as large as

$$N_p = 2^m - 1 \tag{7.1}$$

Sequences of this length are known as m-sequences. The sequences so generated are completely deterministic, but have properties closely resembling those of the ensemble of all random sequences; the shift registers are said to be pseudorandom noise (PRN) generators. Pseudorandom properties are outlined below.

A vast literature has accumulated on shift-register sequences. Original accounts are found in [7.1] and [7.2]. Reference [7.3] is an often-cited, readable, brief tutorial with an extensive bibliography. Reference [7.4] provides a much longer summary, also with an extensive bibliography. Tables of connections for useful binary PRN generators are included in each of [7.1] to [7.5] and in other publications cited in their bibliographies.

For the most part it is sufficient that a simulationist understand the communications-related properties of the PRN sequences and select the desired period N_p. Shift-register connections can be looked up in the tables to construct a functioning generator.

Pseudorandom Properties

Various interesting properties of binary m-sequences are listed below, abstracted from the cited references. These properties resemble those of truly random binary sequences, as may be produced by flipping a fair coin. Along with the listing of each specific PRN property, there is a comparison to the corresponding behavior of a random sequence.

Length. An m-sequence has a period N_p, given by (7.1). If more than N_p bits are taken from the generator, the sequence repeats, exactly. A true random sequence is not periodic and has no characteristic length.

---

*Feedback is *linear* because it is conducted through exclusive-OR gates rather than AND or OR gates. Exclusive-OR is a linear operation on Boolean variables.

Balance. An m-sequence has exactly one more 1 than 0. Random coin flips, *on average*, have the same number of 1's as 0's (or heads and tails). A particular random sequence can have any proportion of 1's and 0's, including the complete exclusion of one or the other. Frequency of occurrence of 1's (0's) is binomially distributed in random sequences.

Subsequences. Every m-bit subsequence, except $000 \cdots 000$, appears exactly once over the period N_p of an m-sequence (regarded as circular, wrapped around its ends). Since there are $2^m - 1$ positions for m-bit subsequences in an N_p-bit circular sequence, each subsequence appears with frequency $1/(2^m - 1)$. The probability of any specific m-bit subsequence (including all 0's) occurring in m independent coin flips is $1/2^m$.

Autocorrelation. Define the periodic autocorrelation function of a sequence $\{d(n)\}$ of length N_p as

$$\psi(i) = \sum_{n=0}^{N_p - 1} d((n+i))d(n) \tag{7.2}$$

where the double parentheses $((n+i))$ indicate that $(n+i)$ is to be taken modulo-N_p. A binary m-sequence has an autocorrelation of

$$\psi(i) = \begin{cases} N_p, & i = 0 \bmod\text{-}N_p \\ -1, & i \neq 0 \bmod\text{-}N_p \end{cases} \quad \text{(PRN sequence)} \tag{7.3}$$

The *average* of the periodic autocorrelations of all possible N_p-long random sequences of ± 1's is

$$\psi(i) = \begin{cases} N_p, & i = 0 \bmod\text{-}N_p \\ 0, & i \neq 0 \bmod\text{-}N_p \end{cases} \quad \text{(average of random sequences)} \tag{7.4}$$

If N_p is large, these two autocorrelations are nearly the same. Be mindful that the autocorrelation of any particular random sequence is itself random and that large deviations from the average of (7.4) will be encountered.

Spectral Density. The discrete Fourier transform of the periodic autocorrelation is the spectral density. For the m-sequence, that becomes

$$\Psi(k) = \begin{cases} 1, & k = 0 \bmod\text{-}N_p \\ N_p + 1, & k \neq 0 \bmod\text{-}N_p \end{cases} \quad \text{(PRN sequence)} \tag{7.5}$$

whereas the DFT of (7.4) is

$$\Psi(k) = N_p \quad \text{for all } k \quad \text{(average of random sequences)} \quad (7.6)$$

These are nearly the same except at $k = 0$. Just as with the autocorrelation, the spectral densities of N_p-long random sequences (or squared magnitudes of the DFTs of these sequences) are themselves random, and most depart greatly from the average given in (7.6).

Relation to Continuous Descriptions. In many accounts of PRN generators, the autocorrelation is shown as a triangular continuous periodic function. In the continuous-frequency domain, the spectrum then consists of delta functions with frequency spacing $1/N_p T$ and envelope proportional to $(\sin x/x)^2$. This altered representation applies to a sequence of rectangular NRZ pulses, each of duration T, taking on values ±1 according to the successive outputs of an m-sequence feedback shift register. The alterations in forms are due to pulse shaping; still different forms would be obtained from other pulse shapings. In this chapter, message generation and pulse shaping are kept separated from one another by employing the sequence correlation (7.3) and sequence spectrum (7.5).

Completed Sequences. An m-sequence can be completed to 2^m bits by inserting just one more bit into the sequence. Inserting a 0 gives exact balance between 1's and 0's. Inserting the extra 0 into the subsequence consisting of $m - 1$ 0's provides a balance of subsequences and is also readily programmed in software.

Strong reasons to complete the sequence often arise. For example, FFTs may be used when simulations are performed in both time and frequency domains. Maximal FFT efficiency is obtained when block size is an integer power of 2, whereas efficiency suffers for a block size of $2^m - 1$. However, completing the sequence destroys the regular PRN autocorrelation and spectral density properties of (7.3) and (7.5). An extra bit of either polarity, inserted at any location in the sequence, induces wild variations in the correlation and spectrum. See [7.6] for examples and analysis. Destruction of regularity introduces strange-looking artifacts into signal spectra, artifacts that tend to distract attention. Spectra are covered further in Chapter 18.

Comment. A single PRN sequence has properties that closely approximate the *average* properties of random sequences. Few random sequences of finite duration have correlations or spectra that are nearly so regular as that of a PRN sequence of the same length. In that sense a PRN sequence is even better than most individual random sequences; the narrative will revert to this theme shortly.

7.1.2 Representative Sequences

Now that PRN sequences have been introduced, let's examine the message needs of typical simulations. Binary signals are considered first, followed by the additional needs imposed by M-ary signals.

Binary Transmissions

Intersymbol Interference (ISI) causes substantial impairment in many band-restricted systems. Experience with ISI-contaminated simulations has shown that the greatest impact on performance arises at the few symbols in a sequence that are exposed to the worst ISI. Thus it is important that the worst ISI be included in the simulation.

Let signal pulses have a duration of L symbol intervals, so that any L consecutive pulses overlap and interfere with one another. The parameter L is called the *memory* of the system. One way to probe ISI is to simulate exhaustively all possible interference patterns of up to L symbols in extent.

In reality, memory is always infinite since no physical signal is strictly time limited. Practically speaking, though, pulses sufficiently far away from the one of immediate interest will have small-amplitude tails and thus contribute negligible ISI. Memory in a signal path might range from very few symbols up to 10 to 20 symbols in duration, or even more.

There are 2^L possible binary patterns in L bits. All (or almost all) of them can be included in a simulation simply by employing an m-sequence generated by a feedback shift register of L stages. That is an efficient method of exhaustively providing all L-bit patterns without much thought or effort. The user can consider the circumstances in deciding whether to complete the sequence or to omit the L-zeros pattern from the simulation. This application of maximal-length sequences appears in [7.7], where the name *representative sequence* is proposed.

Memory extent is often uncertain; any value for L is likely to be nebulous. That uncertainty can be absorbed by using more than L stages in the shift register. However, adding just one stage doubles the number of bits in the sequence and could double the required simulation time. A memory extent of $L \leq 5$ places trivial demands on computing time; $L \approx 10$ bits is quite reasonable, but if L approaches 20 bits the computing time tends to become burdensome.

Suppose that L were to grow as large as 100 bits, or even more: What then? This situation is not likely to develop in the main signal path, but it is commonplace in receiver synchronizers whose bandwidths may be but a small fraction of the symbol rate. Synchronizers are explored in Chapter 11.

If the simulated message sequence does not include all significant ISI patterns, changing the sequence will yield a different realization of ISI and therefore a different estimate of system performance. Discrepancies so incurred are a form of *statistical scatter* brought on by nonexhaustive coverage by the data patterns. Statistical scatter arises from many causes; data insufficiency is just one of them. Scatter sources will be identified repeatedly as this account progresses. Notice that the correlation and spectral properties of m-sequences do not arise in this recital of simulation requirements on sequence length.

M-ary Transmissions

Now suppose that each symbol is taken from an alphabet of M characters. In most communications links, M will be an integer power of 2, but other values are also seen. The preceding arguments for exhaustive coverage of ISI in a binary simulation apply with equal force to *M*-ary systems. Consequently, it is desirable to have a message source capable of generating maximal-length *M*-ary sequences. As a matter of fact, *m*-sequences with length $M^m - 1$ do exist and can be produced according to methods explained in [7.1] to [7.4]. References to tables for *M*-ary sequences are cited in [7.3, Sec. IV-A], and heuristic methods for adapting binary generators are suggested in [7.8] and [7.9]. The latter shows a specific realization of an 8-ary generator.

In an *M*-ary system, there are M^L possible data patterns of *L*-symbol duration. If either *M* or *L* is not small, the number of patterns becomes too large for exhaustive simulation. For example, $L = 10$ would be reasonable for binary simulations but needs $8^{10} \simeq 10^9$ different patterns for exhaustive coverage of 8-ary simulation. Or, if $M = 64$ (a constellation in use for terrestrial digital radios and proposed for cable digital TV), a system memory of just five symbols is equally burdensome.

If exhaustive coverage of data patterns is too demanding on computer time, a simulationist is forced to use shorter message sequences and to accept the resulting statistical scatter. For this reason, many simulation programs provide only binary PRN generators, whose output streams are easily grouped as described further below.

Pattern-caused scatter cannot be avoided entirely if the message-generator period is too short, but the simulationist need not surrender completely. By using the longest *M*-ary sequence consistent with acceptable simulation duration, at least the closest adjoining symbols are covered exhaustively; the closest neighbors tend to contribute the largest ISI. Furthermore, employing a maximal-length complete *M*-ary sequence assures that all *M* symbols will be exercised—an assurance that is lacking otherwise. If *M* is an integer power of 2—the predominant condition—*M*-ary sequences can be derived from binary generators, as in [7.8].

7.2 SYMBOL MAPPING

Classification of transmitter processes into digital operations, symbol mapping, pulse shaping, and modulation is an artificial categorization, undertaken primarily to ease understanding and explanations. In physical hardware, the processes might be separated or they may be intermixed. Separation was the norm in earlier years, but intermixing has become more prevalent, especially as engineers have come to realize that system performance can be enhanced if coding and modulation are combined rather than separated.

Authors of a simulation program are called upon to devise a comparatively small number of library processes that can be combined readily by a user to simulate the great majority of unforeseen applications that will be encountered. Those simulations need to be workable for a separated configuration, where all processes are distinct, or for an intermixed configuration, where the processes might overlap.

In this section we identify processes associated with the mapping of data into transmission symbols. The processes identified will be seen to encroach upon digital operations and message generation at the input side and upon pulse shaping and modulation at the output side. Discussion begins with the simplest PAM/linear-modulation signal formats and moves on to more complicated formats.

7.2.1 Mapping for Linear Modulations

If the system employs linear amplitude modulation (or is a baseband system entirely lacking carrier modulation), the complex envelope of the passband signal is indistinguishable from the *I–Q* baseband signal. The transmitter simulation might be satisfied with the *I–Q* baseband signal alone, whence no visible modulator need be employed at all. That is the configuration explored initially.

Constellations

The notion of a constellation was introduced in Chapter 2 and examples are shown in Fig. 2.12. An *M*-ary constellation is, most simply, a collection of *M* points in two dimensions; each point is described by its Cartesian coordinates or, equivalently, by a complex number. The complex numbers $c_m = a_m + jb_m$ associated with the points of the constellation are the data symbols to be transmitted. (Index *m* is now the symbol index, not the length of a PRN shift register.) Constellations can reduce to just one dimension, as in Fig. 2.12*a* and *b* or as in any real baseband signal.

For purposes of this discussion it is helpful to distinguish symbol mapping proper from preliminary operations such as *bit grouping*, *differential encoding*, or any other encoding. Consider that characters in an *M*-ary alphabet are assigned real integer labels $i(m) = 0$ to $M - 1$. These integers might be equal to the binary number represented by the data bits, but that is not necessary. The symbol mapper accepts an integer number $i(m)$ and outputs a complex number $c_m = a_m + jb_m$ corresponding to a constellation point.

A symbol mapper that accepts *M*-ary inputs works equally well with any digital operation that delivers *M*-ary numbers, such as an encoder, a bit grouper, or an *M*-ary message source. This modular separation of mapping, grouping, encoding, and message generation is highly advantageous in simulation, but keep in mind that the processes are likely to be arranged and combined quite differently in a physical implementation.

Bit Grouping

The size *M* of a constellation most often is an integer power of 2. Since data streams almost inevitably are binary, there usually will be an integer number of data bits per symbol. A bit grouper assembles groups of $\log_2 M$ bits into *M*-ary integers equal to the binary value of the assembled bits. Grouping may be the simplest way of generating *M*-ary numbers from a binary sequence.

Assume that $M = 2^J$, so that *J* bits are transmitted in each symbol. Input to a bit grouper may be regarded as a stream of bits $B(n) = 0$ or 1. The bit grouper takes the input in groups of *J* bits $\{B(mJ), B(mJ - 1), \cdots, B(mJ - J + 1)\}$ and delivers output as an integer $i(m)$ whose binary representation is

$$i(m) = B(mJ) \cdot 2^{J-1} + B(mJ-1) \cdot 2^{J-2} + \cdots + B(mJ-J+1) \qquad (7.7)$$

This integer will be called the "natural" grouping in the sequel. Software representation of $i(m)$ may be in any number system convenient in the programming language used for implementing the grouping process; it need not be binary.

Combinatorial Symbol Mapping

Symbol mapping is defined as the assignment of unique coordinates of a point in an M-ary constellation to each element from an M-ary alphabet. With M points in the constellation and M elements 0 to $M-1$ to be assigned, there are $M!$ different ways to map to the points. Not all ways are necessarily distinct: For example, if the constellation is P-ambiguous rotationally, each assignment exists in P indistinguishable rotations.

There are many possible assignments for each constellation, there are different constellations for each value of M, and there are many different values of M that will be of interest to different workers. It is not feasible for program authors to anticipate all constellations and assignments that will be needed and provide distinct maps (assignment rules) for each. Instead, provisions are needed to allow a user to specify mappings for particular applications. That flexibility is afforded by placing the desired mapping rules into lookup tables constructed by the user. The symbol mapper then is a process that just looks up the constellation coordinates in a table designated by the user. For each input $i(m)$, the table returns an output $c(m) = a(m) + jb(m)$—the coordinates of the mapped constellation point.

A user-friendly simulation program will have provisions for easy production of new tables and easy accessing of a specified table by the mapper process. Symbol mapping tables are used in the transmitter for generating signals, but also in the receiver so that the symbol detector can know the decision boundaries to apply. Demapping processes are treated in Chapter 12.

Mapping Rules. Numerous different mappings are possible, especially for large constellations. Some empirical guidelines are provided here for generating mapping tables, but ultimately the rules pertinent to the application at hand will prevail.

Tables 7.1 and 7.2 exemplify nonunique tables devised for data streams that are not differentially encoded. The guidelines employed in their construction are listed below. Other guidelines, leading to different tables, would be equally valid.

1. *Sign/Quadrant Consistency.* For real constellations, the most significant bit determines the sign of the mapped point. A binary 0 maps to + and binary 1 maps to −. For complex constellations, the two most-significant bits of the integer $i(m)$ (written as a binary number) determine the quadrant of the mapped point. (Least-significant bits could have been used instead.)

2. *Gray Encoding.* Mapping has been assigned so that the binary representation of $i(m)$ for any given point in the constellation differs in just one bit

TABLE 7.1: Symbol mapping tables; various signal formats

Signal Format	Binary Data	$i(l)$	I	Q
Unipolar–binary	0	0	0	0
OOK	1	1	1	0
Bipolar–binary,	0	0	+1	0
BPSK	1	1	−1	0
Bipolar–quaternary	00	0	+3	0
	01	1	+1	0
	10	2	−3	0
	11	3	−1	0
QPSK	00	0	+1	+1
	01	1	+1	−1
	10	2	−1	+1
	11	3	−1	−1
8PSK	000	0	$\cos\frac{3\pi}{8}$	$\sin\frac{3\pi}{8}$
	001	1	$\cos\frac{\pi}{8}$	$\sin\frac{\pi}{8}$
	010	2	$\cos\frac{-3\pi}{8}$	$\sin\frac{-3\pi}{8}$
	011	3	$\cos\frac{-\pi}{8}$	$\sin\frac{-\pi}{8}$
	100	4	$\cos\frac{5\pi}{8}$	$\sin\frac{5\pi}{8}$
	101	5	$\cos\frac{7\pi}{8}$	$\sin\frac{7\pi}{8}$
	110	6	$\cos\frac{-5\pi}{8}$	$\sin\frac{-5\pi}{8}$
	111	7	$\cos\frac{-7\pi}{8}$	$\sin\frac{-7\pi}{8}$

from those of any of the closest points nearby. Therefore, the most probable transmission errors cause just one bit error.

3. *Scaling.* Constellations built on a square grid are scaled so that closest points are distance 2 apart. This scaling assigns integer coordinates to those points and harmonizes with nomenclature customary in the communications literature. Circular constellations for MPSK ($M > 4$) are scaled for unit amplitude, which is also customary.

Special Handling for Square Constellations

If an M-ary constellation is square, the I and Q data streams can be independent. Under those circumstances, one sometimes applies two $\sqrt{M}$-ary data streams inde-

TABLE 7.2: Symbol mapping table, 16QAM signal

Signal Format	Binary Data	$i(l)$	I	Q
16-QAM	0000	0	+3	+3
(square	0001	1	+1	+3
constellation)	0010	2	+3	+1
	0011	3	+1	+1
	0100	4	+3	−3
	0101	5	+1	−3
	0110	6	+3	−1
	0111	7	+1	−1
	1000	8	−3	+3
	1001	9	−1	+3
	1010	10	−3	+1
	1011	11	−1	+1
	1100	12	−3	−3
	1101	13	−1	−3
	1110	14	−3	−1
	1111	15	−1	−1

pendently to I and Q directly. Durations of maximum-length sequences on $\sqrt{M}$ are much shorter than on M, so it may be feasible to reduce simulation time substantially. This expedient is valid only when there is negligible cross-channel interference between I and Q. If cross interference is significant, using $\sqrt{M}$-ary components does not exhaustively exercise all cross patterns.

Independent sequences are particularly simple to apply in QPSK simulations, requiring only two binary message generators: two invocations of the binary PRN generator present in almost all simulation programs. The two channels should be uncorrelated as a precaution against peculiar results that might accompany correlated data. In general, two arbitrarily chosen, different, PRN sequences of the same or different periods do not exhibit good cross-correlation properties [7.10]. The best possible (least) cross-correlation is between two copies of the same PRN sequence that are shifted relative to one another; the magnitude of their cross-correlation over the entire PRN period is guaranteed to be small. A shift somewhat in excess of the system memory duration is typical. (But remember, if the sequence has been completed, the PRN correlation properties have been destroyed and large peaks of cross correlation may be found.)

7.2.2 Differential Encoding

Many constellations exhibit rotational symmetry; see Fig. 2.12 for a few examples. The constellation is *P-ambiguous* if rotation by $2\pi/P$ radians yields a constellation that is indistinguishable from the original. Square constellations are 4-ambiguous and MPSK constellations are M-ambiguous. Bipolar baseband signals are 2-ambiguous if transmission polarity can be inverted, such as by reversal of wire-pair connections.

Something has to be done to resolve the ambiguity; otherwise, the message data cannot be retrieved. Ambiguity cannot be resolved from observations on the signal itself because phase rotation is a relative quantity that cannot be measured absolutely. The receiver cannnot extract knowledge of the transmitter's absolute reference phase from measurements on the received signal alone.

In simulations, the received data could be compared to the transmitted data, which are always available. The ambiguous rotation of the constellation can be determined from the comparisons (see Appendix 11A for a method) and the received data counterrotated to correct for the ambiguity. But transmitted data sequences are not often known to receivers in physical links, so other antiambiguity measures have to be pursued. In practice, ambiguity is counteracted by including special information (e.g., preambles, training sequences, embedded symbols) in the data stream. The special information is recognized by the receiver and used to resolve the ambiguity.

Another common antiambiguity method is *differential encoding*, in which the message data are encoded into *changes* between the transmitted symbols instead of into the individual symbols themselves. The receiver then retrieves the message data by *decoding* the received symbols in pairs, or by differentially *detecting* the changes that occur from one symbol to the next. (In Chapter 12 we explain the distinction between differential decoding and differential detection.) With this expedient, ambiguity in the individual symbols becomes immaterial; the changes between symbols are not ambiguous.

Differential encoding is treated here because of its immense practical importance and because of the frequent need to incorporate its effect into evaluations of simulated communications links. This treatment departs from our regular policy of not dealing with coding in this book.

Differential Encoding Rules

Differential encoding over a pair of adjacent symbols is accomplished readily when the data have been transformed into a sequence of M-ary integers. Denote the uncoded M-ary data sequence as $\{i_D(m)\}$. Differential encoding follows the simple rule

$$i_C(m) = [i_D(m) + i_C(m-1)] \bmod\text{-}M \tag{7.8}$$

where $\{i_C(m)\}$ is the sequence of differentially encoded M-ary integers to be transmitted. Differential encoding is a *sequential* mapping process since the current output depends upon memory of past outputs. By contrast, the symbol mapping described earlier is a *combinatorial* process, without memory.

The differentially encoded integers $\{i_C(m)\}$ are to be mapped onto a constellation by a symbol mapper.

COMMENT 1: This scheme, so easily simulated on the computer for any M, is rarely implemented in such a manner in practical hardware. Instead, differential encoding in digital hardware is more likely to be performed in Boolean-logic circuits on numbers encoded in binary representations.

COMMENT 2: In the special case of binary data [$M = 2; i_D(m) = 0$ or 1], the Boolean and the integer procedures coincide. Addition of two binary digits modulo-2 is the same as the Boolean exclusive-OR.

COMMENT 3: The algorithm presented in (7.8) performs differential encoding on pairs of adjacent transmitted symbols. Less commonly, differential encoding might span more than two symbols [1.7, Chaps. 7 and 8]; that situation is not treated here.

COMMENT 4: Differential encoding does not generally preserve PRN properties of a message sequence. The fact that $\{i_D(m)\}$ has a desirable autocorrelation function does not mean that $\{i_C(m)\}$ also has a desirable autocorrelation.

COMMENT 5: Complications arise, so that the simple differential encoder of (7.8) is often insufficient by itself. Complications are explored further below and in Appendix 7A.

Encoding State. Since the message resides in the changes between symbols, an arbitrary starting value of $i_C(0)$ has to be provided so that the first data symbol $i_D(1)$ can induce a change to generate $i_C(1)$. That is, i_C is a state variable and $i_C(0)$ is its initial value.

Coded elements in the sequence $\{i_C(m)\}$, including $i_C(0)$, are presented to a symbol mapper for transmission. If N_D data elements of $\{i_D(m)\}$ are contained in a group for contiguous transmission, there would be N_D+1 coded elements of $\{i_C(m)\}$. Block-mode operations have to make allowance for this potential irregularity in block size.

Let N_s data elements be contained in each linear block applied to the differential encoder, and let N_s symbols be contained in each linear block delivered by the differential encoder to the symbol mapper. Then the last element in each data block is not consumed with the other elements of its block but is held over to be the first element of the next block. The held-over element is the state variable. The last element of the last data block is never used; it will be discarded.

In circular data blocks, the differentially encoded sequence $\{i_C(m)\}$ has to close properly; that is, $i_C(0) = i_C(N_s)$ is required. Closure obtains if

$$\sum_{m=1}^{N_s} i_D(m) \equiv 0 \bmod\text{-}M \qquad (7.9)$$

This requirement may prove to be a nuisance to impose; it does not occur naturally in arbitrary data sequences. A simulationist may prefer not to employ circular blocks in conjunction with differential encoding.

Gray Encoding

Just as it is desirable to impose Gray encoding on absolute mapping of symbols, it is also desirable to impose Gray encoding on differential encoding. If two J-bit groups differ in just one bit, their M-ary data integers $i_D(m)$ also should differ by 1 mod-M. With that Gray-encoded constraint, their differentially encoded integers $i_C(m)$ will also differ by an amount 1 mod-M, provided that $i_C(m-1)$ is the same for both groups.

As an example, consider Gray-encoded differential encoding for QPSK, as shown in Table 7.3. The same principles can be extended to other MPSK signals. With four phases to be mapped, each bit group is composed of $J = 2$ bits (one dibit). The natural grouping [labeled $i(m)$ in the table] is simply the integer represented by the binary bit pattern.

But the natural bit pattern does not lead to Gray encoding when mapped directly to phase increments $\Delta\phi$. (A phase increment is the phase change caused by the current symbol.) Gray encoding requires that patterns that differ in just one input bit cause phase increments that differ by 90°.

There is no unique Gray-encoding assignment; Table 7.3 shows two that have been selected for different CCITT standard formats [1.8, Chap. 4]. Each standard employs a different reflection of the natural order of bit grouping. (Gray encoding is also known as reflected encoding.) How can this rearrangement be accomplished readily in simulations?

The rearrangements are simply combinatorial mappings from the natural grouping to the Gray-encoded patterns. *Data mapping* could be performed by lookup in a table just like the symbol mapping tables described earlier, with $i(m)$ as the input and $i_D(m)$ as the output. Data mapping tables differ from symbol mapping tables in that the former always will have real outputs, never complex; otherwise, they are the same in principle. A simulation program might have a generic combinatorial mapping process capable of performing either task.

The mapped data sequence $\{i_D(m)\}$ is then applied to the differential-encoding algorithm (7.8), whose output is applied in turn to a symbol mapper to obtain the correct constellation points. Figure 7.2 illustrates the sequence of processes.

TABLE 7.3: Gray-encoded differential encoding for QPSK

Dibits	Natural		CCITT V.26		CCITT V.22	
	$i(m)$	$\Delta\phi$	$i_D(m)$	$\Delta\phi$	$i_D(m)$	$\Delta\phi$
00	0	0°	0	0°	1	90°
01	1	90°	1	90°	0	0°
10	2	180°	3	270°	2	180°
11	3	270°	2	180°	3	270°

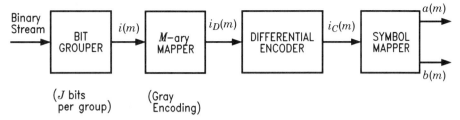

FIGURE 7.2 Process for Gray-encoded differential encoding.

7.2.3 Mapping for Nonlinear Modulations

Symbol mapping is most clearly associated with linear modulations, where recognizable constellations can be identified. Signals generated by nonlinear modulations do not necessarily have constellations at all and may not have any distinct mapping process either. Some examples are instructive. In each of the following, assume that a bit grouper or other digital operation delivers an M-ary integer $i(l)$.

- *Pulse Waveform Transmission (PWT)*. Each integer datum selects one of M distinctive waveforms.
- *PPM*. The mapping is from $i(l)$ to a time slot in an M-slot frame. But the selection of the slot is a modulator function; modulator input logically should be $i(l)$ itself. Thus there is no explicit symbol mapper needed. (Alternatively, symbol mapping could be regarded as a part of modulation; this is a broadly applicable premise, not restricted to PPM.)
- *MFSK*. The M-ary datum is mapped to one of M signaling frequencies, which again is a modulator function.
- *CPM*. Prior to modulation, a CPM signal can be represented as a baseband PAM signal; see (2.30) and associated text. It might be necessary to convert the always-positive $i(l)$ to balanced symbols: ± 1 for binary, $\pm 1, \pm 3$ for quaternary, and so on. That can be accomplished by a symbol mapping table of the kind discussed previously.

Any further consideration of symbol mapping for nonlinear modulations will be deferred until modulators are scrutinized below.

7.3 PULSE SHAPING

A pulse shaper produces a sequence of uniformly spaced PAM pulses, all of identical shape and differing only in their complex amplitudes. Operations to be performed are (1) assigning amplitudes c_m, (2) establishing a time base, and (3) actually shaping the individual, possibly overlapping pulses $g_T(n)$. Pulse trains constitute PAM complex baseband signals $\Sigma c_m g_T(n - mM_s)$.

The basic method employed, with some variations to be expounded upon, is the following: Generate a sequence of complex-weighted impulses, uniformly spaced by the same symbol interval, and send those impulses through filters to produce the desired pulse shapes. A model was shown in Fig. 4.1. Alternative approaches could be followed, but this one is extremely flexible and makes good use of processes that will also be employed in other parts of a simulation.

7.3.1 Generating Impulse Trains

Prior to generating the impulse train, the data might best be regarded as an ordered list of information to be transmitted. Assigning the data as amplitudes of an impulse train produces a signal. Operations prior to the impulse train could be deemed to be data handling; beginning with the impulse train, subsequent processes are signal handling.

Time Base

Designate the complex amplitudes of the symbols delivered by the symbol mapper as $c_m = a_m + jb_m$, where m is the symbol index. The mth PAM pulse to be generated begins as a complex unit sample* that is weighted by c_m. A new pulse is to be started after every M_s samples. An impulse train consists of the mth impulse with amplitude c_m, followed by $M_s - 1$ samples of zero amplitude, followed by the $(m+1)$th impulse with amplitude c_{m+1}, and so on (see Fig. 7.3a). Samples have index n, which increases by M_s over each symbol interval. The signal is said to have M_s samples per symbol.

Generating the impulse train is tantamount to laying out the data on a discrete time base, where each nonzero sample carries the information belonging to one data symbol. For the mth symbol, the impulse in the real channel is assigned amplitude a_m and the impulse in the imaginary channel is assigned the amplitude b_m.

In the computer, the time intervals between samples are dimensionless. A time increment t_s could be imputed to the intervals (thereby also establishing a symbol interval $T = M_s t_s$), but the on-line processes do not use that imputed interval in any way. See Chapter 6 for the principles of time and frequency scales. In STÆDT there is no provision for an on-line process to accept a value for imputed sample interval.

The impulse method of pulse generation requires that M_s be an integer. Many, but not all, simulations can be carried out satisfactorily with integers for M_s throughout. If a noninteger M_s is required (either larger or smaller than the original value), it is obtained by interpolating the PAM signal (not the impulse train) and, in effect, establishing a new sampling rate. Signal interpolation is covered in Chapter 16.

---

*Henceforth called an *impulse*, to agree with popular nomenclature, but true impulses cannot be represented in discrete time.

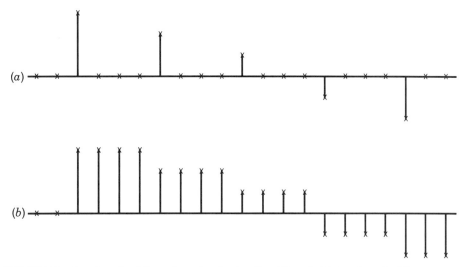

FIGURE 7.3 Sample trains delivered to pulse-shaping filter ($M_s = 4$): (*a*) impulse train; (*b*) square-wave train.

Square Pulses

Because square pulses need to be simulated so frequently, and because they are readily generated, most simulation programs (including STÆDT) also provide for trains of square pulses from the time-base process, as an optional choice in place of impulses. Rather than filling in $M_s - 1$ samples of zero amplitude after the *m*th impulse, a square-pulse generator produces M_s samples of amplitude c_m for the *m*th pulse (see Fig. 7.3*b*).

7.3.2 Pulse-Shaping Filters

A shaped pulse train is obtained by applying an impulse train or square-pulse train to a pulse-shaping filter. Principles of filters have been explained at length in Chapters 3 and 6; only a few additional points will be raised at this juncture. These points generally apply to all uses of filters, not just to pulse shaping. Any of the several different kinds of filters described earlier are acceptable for pulse shaping. Select whichever kind best serves your purpose.

FIR Time-Domain Filters

An FIR time-domain filter (see Sec. 6.1.2) is convenient for generating time-limited pulses. Just assign the desired pulse shape to the filter coefficients $h(n)$ and drive the filter with an impulse train. Output is a PAM pulse train, with the pulse waveform determined by the filter and the pulse amplitudes determined by the symbol values delivered from the symbol mapper.

216 ☐ GENERATION OF DATA SIGNALS

Simple examples generated by STÆDT are shown in Fig. 7.4. The top line of the figure shows a binary-weighted impulse train, with $M_s = 8$. This same input is applied to each of two filters: one whose impulse response is a half cosine of one-symbol duration, and the other whose impulse response is a raised cosine of two-symbol duration. Both pulses are time limited—the first to just one symbol interval so the pulses do not overlap (as on line b). The longer-duration pulse has overlap between adjacent symbols (as on line c).

When pulses overlap, each point in the filter output consists of the sum of contributions from multiple pulses. The filter structure takes care of all bookkeeping arising from overlap, even if many pulses are involved simultaneously; the user is not concerned with those details.

An FIR filter of N_h taps computes the current output point from the current input and $N_h - 1$ past inputs, per (6.5). Past inputs are state variables that are preserved in

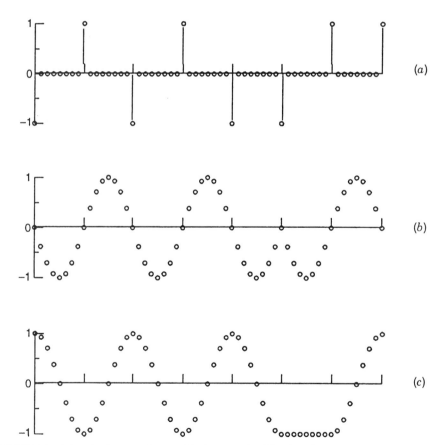

FIGURE 7.4 Pulse-shaping waveforms with time-domain FIR filter (STÆDT-generated, $M_s = 8$): (a) impulse train (binary weighted); (b) filtered pulse train, half-cosine pulses; (c) filtered pulse train, raised-cosine (2RC) pulses. Time axes (abscissas) marked in symbol intervals T.

memory until consumed. Programs might require the user to help manage the state memory. Three situations can arise:

1. For either linear blocks or single-point operations, the state memory requires initial values to start the process. It is commonplace to assign all-zero initial values to signal processes on the signal main line (as is done in STÆDT), on the assumption that transients die out quickly and that the simulation can wait out the transient without much run-time penalty. The simulation needs to know that a new run is starting and that the state memory should be initialized to zero; user intervention (supplying a parameter) may be required for initialization.

2. Once the initial transient has disappeared, the filter just takes samples from state memory, as needed. If concatenated linear blocks are used, the simulation needs to know that the blocks are indeed being concatenated and that a new block does not mean that a new run is starting. User intervention complementary to that in situation 1 is needed.

3. If a circular block is to be filtered, the state variables are the circular wraparound of the input signal block itself. The filter must be notified that operation is circular so that it knows where to look for its state variables. The examples of Fig. 7.4 were computed with a circular block, although only part of the block is plotted. No starting transient occurs with circular blocks.

Many other processes incorporate state memory and thus have similar requirements for user inputs. Typically, the user supplies the needed information as parameters when the simulation is being prepared.

IIR Time-Domain Filters

Many physical systems employ rectangular pulses and lumped-constant analog filters in their transmitters. These can be simulated by generating square-pulse trains as above and shaping them with time-domain IIR filters with rational transfer functions (described in Sec. 6.1.3). In that way there is direct correspondence between waveforms in the hardware and those in the simulation. An example of waveforms generated in this manner by STÆDT is shown in Fig. 7.5. A linear block format was used in this simulation; the starting transient can be seen at the left end of the filtered waveform. Identical waveforms would result from a single-point format.

An IIR filter with rational transfer function cannot deliver a circular block as its output, because of the need to wrap around the end of the block to obtain state variables for computing the present output. State variables for a rational filter consist of both past inputs and past outputs. Circular operation would require that computations at the beginning of the block use outputs from the end of the block, which have not yet been computed. A circular block can be applied as input to a rational filter (repetitively, if needed in a long simulation run). When so driven, the filter process simply interprets the input as a linear block (or as concatenated identical linear blocks) and ignores the circularity.

218 ☐ GENERATION OF DATA SIGNALS

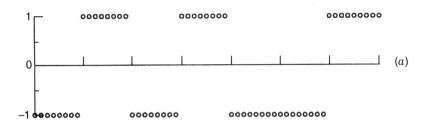

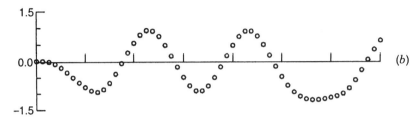

FIGURE 7.5 Pulse-shaping waveforms with time-domain IIR filter (STÆDT-generated): (*a*) square-pulse train; (*b*) filtered pulse train, three-pole Butterworth filter. Time axes (abscissas) marked in symbol intervals T.

Frequency-Domain Filters

Pulse shaping with frequency-domain filters (Sec. 6.1.1) combines the following separate processes:

- Transform the impulse train, or square pulse train, $\{z_{in}(n)\}$ to the frequency-domain representation $\{Z_{in}(k)\}$ using the DFT (FFT for efficient computation).
- Multiply the transformed impulse/square-pulse train by the filter transfer function $H(k)$ to obtain a frequency-domain representation $\{Z_{out}(k)\} = \{Z_{in}(k)H(k)\}$ of the shaped-pulse train.
- Transform back to the time domain using the inverse DFT (FFT) to obtain the time-domain representation $\{z_{out}(n)\}$ of the shaped-pulse train.

Because of the use of the discrete Fourier transform, the output $\{z_{out}(n)\}$ is a circular block, even if input $\{z_{in}(n)\}$ was not intended to be circular. Frequency-domain filtering automatically produces circular blocks without any intervention being required from the user. In some situations the circular blocks can be opened out into linear blocks; see Sec. 6.1.5 for further information.

FIR Filters. Circular blocks can be produced by frequency-domain filtering or by FIR time-domain filtering. Results are identical if the same filter is used in both approaches. Why might one approach be selected in preference to the other?

- If the filter characteristics are already specified in the frequency (time) domain, user effort may be reduced by performing the simulation in the frequency (time) domain.

- If succeeding operations are performed in frequency (time) domain, doing the filtering in the same domain may save on Fourier transformations.

- One method might compute faster than the other. An FIR time-domain filter is quicker if the filter impulse response is short, while the FFT/frequency-domain filter is quicker if the impulse response is long. The dividing line of nearly equal effort lies at 20 to 30 taps in the time-domain filter, depending on signal-block length. (If block length cannot be an integer power of 2, so that an efficient FFT can be employed, the dividing line will rise to favor somewhat longer time-domain filters.)

Rational Filters. A lumped-element analog filter has a rational transfer function with infinite extent in the frequency domain. To simulate such a filter in the frequency domain requires that its frequency response be truncated to fit the simulation's frequency window. An example of frequency-domain pulse shaping with a rational filter is illustrated in Fig. 7.6, using a square-pulse sequence and the filter transfer function employed for Fig. 7.4c. Since frequency-domain filtering produces circular blocks, the transient seen at the start of the shaped-pulse train in Fig. 7.5 is absent from Fig. 7.6.

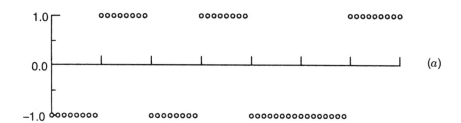

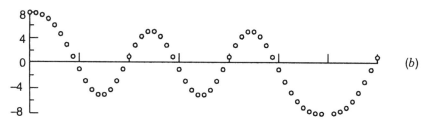

FIGURE 7.6 Pulse-shaping waveforms with frequency-domain filter (STÆDT-generated): (*a*) square-pulse train; (*b*) filtered-pulse train, raised-cosine (2RC) filter. Time axes (abscissas) marked in symbol intervals T.

Relation to DSP Interpolation

The processes of (1) interspersing $M_s - 1$ zero-valued samples between adjacent samples of an input sequence, and (2) filtering to smooth the signal together are known as *interpolation* in the DSP literature [1.12, Sec. 8.5.2; 7.15]. This is a well-known method for upsampling a discrete-time signal by a factor of M_s. The same signal-generator processes used for producing the impulse train and shaping the pulses can also be used for upsampling. Sample-rate changing is explored in Chapter 16, where the name *interpolation* is given a meaning distinct from *upsampling*.

Amplitude Peculiarities

Consider a lowpass discrete-time pulse shaping filter with transfer function $H(k)$ whose frequency response at zero frequency is $H(0)$. When driven with unit-amplitude square pulses, its output waveshapes will exhibit an amplitude of approximately $H(0)$. [More precisely: The steady-state time-domain response of a lowpass filter with transfer function $H(k)$ to a step function of amplitude A is $AH(0)$.]

The unit-sample response of the same filter will tend to have a peak amplitude of approximately $H(0)/M_s$, where M_s is the number of samples per symbol for which the filter has been designed. Thus output amplitude from an impulse-driven filter might be disconcertingly small, particularly if M_s is large. Scale the signal by M_s to establish a more-comfortable amplitude. Absolute amplitudes are irrelevant in simulations; only ratios matter (e.g., signal-to-saturation level ratio in a nonlinear device, or signal-to-noise ratio).

7.4 MODULATORS

Linear modulations are considered briefly, followed by examples for nonlinear modulations.

7.4.1 Linear Modulation

To repeat a previous statement: No modulator process is employed in simulation of the complex envelope of a linearly modulated signal. The complex envelope is identical to the *I–Q* baseband signal. All significant effects of an ideal linear modulator are imposed by the symbol mapper and are contained in the complex envelope. Upon occasion, a user wishes to simulate an analytic signal—a complex envelope that has been frequency shifted. Frequency can be shifted by multiplying the signal by a complex exponential $\exp(j2\pi n f_c t_s)$. Shifting is covered in greater detail in Chapter 10.

A need sometimes arises to simulate the imperfections of a linear modulator. Imperfections include nonlinearities, carrier leakage, quadrature error between *I* and *Q* channels, unbalance between channels, and others. These effects are best simulated by applying various additional processes to the ideal complex envelope. Suitable processes are included in many transmission-simulation programs.

Real Modulated Signals

Although most passband signals are adequately simulated by their complex envelopes, there will be systems in which a real signal must be simulated instead. A real signal is particularly appropriate if the carrier frequency is comparable to, or smaller than, the bandwidth of the baseband signal. In this section we examine modeling of real carrier signals.

Represent the complex envelope (or I–Q baseband signal) as

$$z(n) = x(n) + jy(n) \tag{7.10}$$

The corresponding real signal is

$$\begin{aligned} s(n) &= x(n)\cos(2\pi n f_c t_s + \theta_0) - y(n)\sin(2\pi n f_c t_s + \theta_0) \\ &= \text{Re}[z(n)e^{j(2\pi n f_c t_s + \theta_0)}] \end{aligned} \tag{7.11}$$

where f_c is the carrier frequency and θ_0 is an arbitrary phase angle. The real-signal representation of (7.11) applies to any carrier modulation, either linear or nonlinear. Specific nonlinear modulations are explored in later pages. Phase rotators (multiplication by $e^{j\theta}$) are described in Chapter 10.

Sampling Rate. Let $x(n)$ and $y(n)$ have a one-sided (baseband) bandwidth of B. To avoid aliasing, the sampling frequency $1/t_s$ for the modulated real signal of (7.11) must exceed $2(B+f_c)$. From this requirement it is clear why modulated signals with high carrier frequencies are rarely simulated in real-signal format. Fortunately, a need for simulation of real modulated signals arises mostly for low carrier frequencies.

Computer operations are dimensionless; only the product $f_c t_s$ appears in (7.11) and in the on-line processes. For simulation of data transmission, a user is more likely to be involved with symbol rate $1/T$ than with sample rate, so would prefer to specify the product $f_c T$ instead of $f_c t_s$. Then $f_c t_s = f_c T/M_s$, where M_s is the number of samples per symbol.

Phase Accumulation. Define the accumulated carrier phase as

$$\Theta(n) = 2\pi n f_c t_s + \theta_0 \tag{7.12}$$

This phase is the argument of the sine, cosine, and exponential functions in (7.11). Accumulated phase incorporates the sample index n explicitly. That is not a good practice, particularly in a long simulation where $\Theta(n)$ could grow without bound. Instead, it is better to define phase recursively, as

$$\begin{aligned} \theta(n) &= \Theta(n)\,\text{mod-}2\pi = [\theta(n-1) + 2\pi f_c t_s]\,\text{mod-}2\pi \\ \theta(0) &= \theta_0 \end{aligned} \tag{7.13}$$

where $\theta(n-1)$ acts as a state variable.

Phase could be defined in cycles, instead of radians. Let $p(n)$ be phase in cycles. Then the carrier phase is written as

$$\theta(n) = 2\pi p(n)$$
$$p(n) = [p(n-1) + f_c t_s] \bmod\text{-}1$$
$$p(0) = p_0 = \theta_0/2\pi \tag{7.14}$$

Representation modulo-1 is especially advantageous when $f_c t_s$ is a fraction that is exactly representable in the computer's number system. Multiple cycles of carrier then fall on exactly the same phase values for every cycle. Exact repetition is not possible when using (7.13) since π is irrational and is rounded at each iteration.

Equation (7.14) is the difference equation of a *number-controlled oscillator* (NCO), a device that comes up repeatedly in this book and throughout DSP applications.

7.4.2 Nonlinear Modulation

Communications engineers encounter many different types of nonlinear modulations; no simulation program could make provision for every one of them. A simulationist has to piece together several elementary processes to accommodate modulation types not expressly included in the program library. Three examples that illustrate the underlying principles are examined below. These models generate the complex envelope of the modulated signal. If a real modulated signal is required, just apply (7.11) to the complex envelope.

Pulse Position Modulation

Refer to Sec. 2.1.3 for a description of PPM and a continuous-time model for it. "Modulation" in a PPM system consists of generating a standard pulse at the correct time instant. Therefore, the modulator is closely intertwined with the time-base process that follows symbol mapping.

A PPM symbol interval of duration T is divided into M equal *slots* numbered 0 to $M-1$. At the mth symbol, the standard pulse $g_T(n)$ is generated in the a_mth slot, in accordance with the value of the mth data symbol $a_m \in \{0, \ldots, M-1\}$. Let the simulation employ M_s samples per slot so that there are $M_s M$ samples per symbol in PPM. (Bandwidth expansion of PPM is clearly revealed by the increased number of samples compared to PAM of the same symbol rate.) The discrete-time version of (2.10) is

$$x(n) = \sum_m g_T[n - a_m M_s - m M_s M] \tag{7.15}$$

As described here and as typically implemented, the complex envelope $x(n)$ of a PPM signal is actually real.

Conceptually, a simple way to generate the pulse train (7.15) is to establish the time base of samples with $M_s M$ samples per symbol interval, assign unit amplitude to the one sample in each symbol interval where $n = a_m M_s + m M_s M$, and assign zero amplitude to all $M_s M - 1$ other samples in the symbol interval. This is an M-ary PPM impulse train. The desired pulse train is produced by applying the impulses to a shaping filter with impulse response $g_T(n)$.

Alternatively, a square-pulse M-ary PPM train can be generated by assigning unit amplitude to M_s consecutive samples, beginning where $n = a_m M_s + m M_s M$. The desired shaped pulse train is produced by applying the square-pulse train to a shaping filter with suitable impulse response. The filter can be any of the three kinds described in Sec. 7.6.3. Just because (7.15) is written in terms of filter impulse response does not restrict the filter to be FIR.

Neither the sample index n nor the symbol index m should appear explicitly in a well-crafted PPM modulator process. Instead, a sample counter with slot subperiod M_s and symbol period $M_s M$ would be employed to identify slots and symbols. The M_s-counter keeps track of samples within each slot and the $M_s M$-counter keeps track of slots within a symbol interval. In the mth symbol interval, a single unit-amplitude sample (or M_s consecutive unit-amplitude samples for a square pulse) would be generated when the $M_s M$ count reached a_m. Otherwise, all other samples from the PPM modulator have zero amplitude.

Frequency-Shift Keying

Refer to Secs. 2.1.4 and 2.2.3. The real-signal time-continuous representation of (2.13) can be rewritten to obtain the complex-envelope time-discrete representation of an MFSK signal as

$$z(n) = \sum_m \rho(n - m M_s) \exp[j 2\pi(n - m M_s)(f_0 t_s + a_m \Delta f t_s) + j \theta_m] \qquad (7.16)$$

where M_s is the number of samples per symbol interval T, M the number of signaling frequencies uniformly spaced by Δf, and f_0 an offset frequency to be specified by the user. A complex-envelope format permits negative frequencies to be employed and can reduce the required sampling rate by half compared to a real representation.

For purposes of this section, the envelope $\rho(n)$ will be assumed to be a rectangular pulse of unit amplitude and duration of exactly one symbol. Signal envelope then is constant—the classical FSK signal format. With this simplification only one value of m contributes to each $z(n)$. Complexity increases precipitously if pulses overlap and must be summed.

Sampling Rate. Select the bias frequency as $f_0 = \Delta f(1 - M)/2$, so that the signaling frequencies are disposed equally to either side of zero frequency. This choice provides the most compact signal spectrum and thus minimizes the required sampling rate.

A frequency-modulated signal can never be bandlimited, even if the baseband signal itself should happen to be bandlimited (which it certainly is not for simple FSK). Therefore, no matter how high the sampling rate, some aliasing will still be incurred. To gain an estimate of the required sampling rate, it is helpful to invoke a modification of Carson's rule [7.11]:

$$\frac{1}{t_s} > (M-1)\Delta f + \frac{\kappa}{T} \tag{7.17}$$

where κ is an empirical coefficient (typically greater than 1 and less than about 50), selected by the user to account for the bandwidth of the baseband pulse stream.

Phase Accumulation. The argument of the exponential function in (7.16) is an accumulating phase, which should be computed recursively modulo-1 cycle, as explained in connection with (7.14). Another NCO is needed for this task.

Since the computer deals with dimensionless quantities, the times and frequencies appear only as the dimensionless products $f_0 t_s$ and $\Delta f t_s$. These are related to the symbol rate $1/T$ by $t_s = T/M_s$, so the user-specified quantities are most likely to be $f_0 t_s = f_0 T/M_s$ and $\Delta f t_s = \Delta f T/M_s = h/M_s$, where h is the modulation index.

Phase θ_m is associated with the mth symbol interval. For continuous-phase FSK $\theta_m = \theta_0$ is a constant; it is the value applied if the NCO is to be allowed to accumulate phase without disruptions. However, some FSK systems emit a discontinuous-phase signal; the phase takes a random jump at the start of each new symbol. Each symbol then has a new value for θ_m.

Phase-continuous phase accumulation obeys the difference equation

$$\theta(n) = 2\pi p(n) = 2\pi \left[p(n-1) + \frac{f_0 T}{M_s} + \frac{a_m h}{M_s} \right] \text{mod-1} \tag{7.18}$$

To simulate phase-discontinuous FSK, add a random, uniformly distributed* phase jump of $\theta_m/2\pi$ cycles to the NCO content at the start of the mth symbol interval (unless $a_m = a_{m-1}$). As the last step, the unit-amplitude complex envelope is formed by computing

$$z(n) = e^{j\theta(n)} = \cos\theta(n) + j\sin\theta(n) \tag{7.19}$$

Continuous-Phase Modulation

A continuous-time representation for CPM has been given in (2.30) to (2.32). Assume that $h_m = h$ is a constant and let $\{a_m\}$ be a symbol sequence from the alphabet $\{\pm 1, \pm 3, \ldots, \pm M/2\}$. The symbol sequence $\{a_m\}$ is mapped from grouped, differentially encoded message bits, exactly as explained already in Secs. 7.1 and 7.2. The only novelty is that the symbol mapping is from real integers $i(m)$ to real ampli-

---

*Methods for generating uniformly distributed random numbers are presented in Chapter 9.

tudes a_m; the imaginary-part amplitude b_m encountered previously is set to zero in the mapping table.

The complex-envelope discrete-time representation for a CPM signal can be derived from (2.30) as

$$z(n) = e^{j\theta(n)} = \exp\left[j\theta_0 + j\pi h \sum_{i=0}^{n} \sum_{m} a_m q(i - mM_1)\right] \quad (7.20)$$

where M_1 is the number of samples in the CPM symbol interval T_1 and $\theta_0 = \theta(0)$. The phase argument for (7.20) was obtained by discretizing the frequency pulse $q(t)$ to $q(n)$ and replacing the integral of (2.31) by a summation. The time-discrete frequency pulse must obey the constraint

$$\sum_{n=-\infty}^{\infty} q(n) = 1 \quad (7.21)$$

which is equivalent to

$$Q(1) = 1 \quad (7.22)$$

where $Q(z)$ is the z-transform of $q(n)$.

In (7.20), the inner summation is simply a PAM pulse stream, generated by the methods described in Sec. 7.3, with $g_T(n)$ replaced by $q(n)$. The outer summation is realized by a phase-accumulating NCO. These matters are expanded upon below.

Pulse Shaping. To generate a pulse stream $\Sigma a_m q(n - mM_1)$, the symbol sequence $\{a_m\}$ weights an impulse train in a time base, and the impulse train is applied to a filter with impulse response $q(n)$. Any of the three types of filters of Sec. 7.3.2 is acceptable. These are identical to the operations described earlier for generating a PAM pulse stream, with only a minor change of notation.

A square-pulse input to the filter can be used instead of an impulse train, just as shown earlier for a PAM pulse train. If that is done, the filter should be regarded as the cascade of two subfilters according to $Q(z) = Q_1(z)Q_2(z)$, where $q_1(n)$, the impulse response of the first filter, is a square pulse of M_1 samples duration (see Sec. 4.3.1). The constraint $Q(1) = 1$ is imposed on the product $Q_1(z)Q_2(z)$.

Slight modifications are needed if a frequency-domain filter is employed. Parallel to (7.22), the sampled frequency response $\{Q(k)\}$ must conform to $Q(0) = 1$. That is equivalent to requiring the sum of $q(n)$ over an N-sample period to be unity.

If the simulation is run with circular blocks, the frequency-domain filter assures that the baseband pulse train is circularly continuous around the end of the block. That is not sufficient to assure circular continuity of the phase; the user must also select a symbol sequence that maintains circular continuity of phase. A user who does not want to be burdened with this requirement can avoid it by filtering with linear-block or single-point format in the time domain.

Continuous-phase MFSK is a special case of CPM, produced with a rectangular NRZ pulse shape $q(n)$ of duration M_1 samples and amplitude $1/M_1$. Minimum-shift keying (MSK) is a further specialization of continuous-phase MFSK with $M = 2$ and modulation index $h = 0.5$.

Phase Accumulation. The outer summation of (7.20) is a phase accumulation; the sample values of the PAM baseband pulse train are simply added together and scaled by πh to produce the phase at the nth sample. Phase for CPM should be summed recursively in an NCO according to

$$\theta(n) = 2\pi p(n) = 2\pi \left[p(n-1) + \frac{h}{2} \sum_m a_m q(n - mM_1) \right] \text{mod-1} \qquad (7.23)$$

where $p(n)$ is the state variable of the NCO and may be thought of as phase in cycles. The $h/2$ factor inside the brackets cancels the 2 in the 2π outside, to cause the PAM increments to be scaled by πh, as required. Once $\theta(n)$ for CPM has been computed, its complex envelope is generated from (7.19).

7.5 AMPLIFIERS

Amplifiers can be linear or nonlinear. Simulation of fixed-gain linear amplifiers is trivial and simulation of adaptive-gain linear amplifiers is relatively straightforward. Most of this section is devoted to the greater complexity inherent to nonlinear amplifiers.

7.5.1 Linear Amplifiers and Attenuators

Amplification or *scaling* is modeled by multiplying every point of a signal by a real coefficient K. If $|K| > 1$, the device being simulated is an amplifier; if $|K| < 1$, the simulated device is an attenuator. If K is negative, the amplifier is inverting. The equation for simulation of an amplifier/attenuator is simply

$$z_{\text{out}}(n) = K z_{\text{in}}(n) \quad \text{or} \quad Z_{\text{out}}(k) = K Z_{\text{in}}(k) \qquad (7.24)$$

Scaling is performed identically in either the time or frequency domain.

Scaling could be complex, but only real scaling is considered here. To simulate a frequency-dependent linear amplifier, use a filter model. To simulate a complex value for gain, either use a simple filter model or connect an amplifier having real gain in cascade with a phase shifter. Phase shifting is treated in Chapter 10.

If gain is to be fixed throughout a simulation run, K is entered as a parameter by the user. As a convenience to the user, the interface should provide options to enter K itself, or $20 \log K$—the gain in dB.

Adaptive Scaling

A user can specify K or any other parameter only if its value is known prior to a simulation run and if it is to remain constant throughout the run. In many instances the parameter is to be altered adaptively within the simulation run; a prominent example would be a variable-gain amplifier used in an automatic gain control circuit. Some means must be provided in the simulation for communicating a variable-gain coefficient to the amplifier.

Control Signals. A parameter within one signal process can be specified from another process within the simulation; the means for transferring the specification will be called a *control signal*. The control signal for a variable-gain amplifier is the gain value K. Numerous applications employ control signals; they will be found repeatedly in processes yet to be described and explained in depth in Chapter 14.

7.5.2 Nonlinear Amplifiers

All amplifiers are nonlinear if driven hard enough; they all exhibit saturation effects, whereby output does not increase with increasing input for large enough input. Power amplifiers typically are operated close to saturation, or even saturated, so as to maximize power efficiency. Saturation is an example of a nonlinear effect, one that can have serious repercussions on the signal to be amplified. This section delves into modeling of saturating amplifiers—a topic important in its own right and one from which lessons can be drawn for other nonlinear processes.

Nonlinear Model

If a carrier signal is applied to a nonlinear amplifier, the nonlinearity generates distortion products at harmonics of the carrier frequency and also distorts the envelope of the fundamental frequency component. Assume that the modulation bandwidth is sufficiently narrow compared to the carrier frequency so that spectra of the harmonic bands do not overlap the spectrum of the fundamental component. Assume further that only the fundamental component constitutes useful output signal—that harmonics are nuisances to be suppressed. Under those conditions, the harmonics can be ignored in the transmission simulation and attention concentrated on distortion of the complex envelope of the fundamental output component.

Nonlinear processes were introduced in Sec. 4.5, where a restriction to memoryless nonlinearities was imposed. That restriction is retained in the modeling of saturating amplifiers, even though many physical amplifiers have decided memory effects. Limited guidance to modeling of memory-containing nonlinearities may be found in Sec. 4.5 and [7.12].

Amplifier Characterization. Power amplifiers typically are characterized by power output and phase shift, expressed as functions of power input. Figure 7.7 (adapted from measurements) shows an example for a traveling-wave tube amplifier (TWTA). The power output curve is conventionally known as the *AM–AM conversion* and the phase curve is known as the *AM–PM conversion* (AM = amplitude modulation; PM = phase modulation). These curves show the nonlinear effect on the complex envelope in polar coordinates.

Comment: It is also possible to define the nonlinear functions in rectangular coordinates. Polar coordinates provide a better match to the usual representation of a nonlinearity (AM–AM and AM–PM), but the computer will work with either model. Only polar coordinates are considered here.

Implied in this presentation is an assumption that the nonlinearity depends solely upon amplitude of the input—that the output amplitude and phase shift can be represented as single-variable functions of the input amplitude alone.

Nonlinear Process. Represent the input complex envelope in polar coordinates as

$$z_{in}(n) = R_{in}(n) e^{j\phi_{in}(n)} \tag{7.25}$$

and the output complex envelope similarly as

$$z_{out}(n) = R_{out}(n) e^{j\theta_{out}(n)} \tag{7.26}$$

Then the output can be represented in terms of the input according to

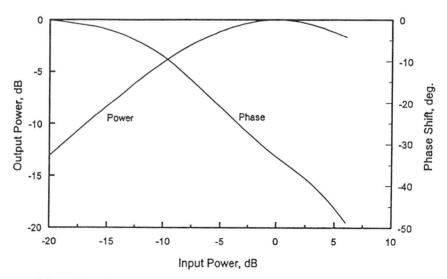

FIGURE 7.7 Typical amplitude and phase characteristics of a saturating amplifier.

$$R_{\text{out}}(n) = \gamma_R[R_{\text{in}}(n)]$$
$$\phi_{\text{out}}(n) = \gamma_\phi[R_{\text{in}}(n)] + \phi_{\text{in}}(n) \tag{7.27}$$

where $\gamma_R[R_{\text{in}}(n)]$ and $\gamma_\phi[R_{\text{in}}(n)]$ are nonlinear functions of $R_{\text{in}}(n)$. The operations can be performed meaningfully only in the time domain.

A nonlinear simulation process takes $z(n)$ as its signal input and requires specification of the two functions γ_R and γ_ϕ. If, as is likely, input is in rectangular coordinates, a rectangular-to-polar conversion must be applied to the signal. Then output amplitude and phase are computed from (7.27) and the resulting polar-coordinate envelope is converted back to rectangular coordinates for output handling.

Various different nonlinear functions will be needed for simulating different devices, but the same fundamental procedure is used for all. A plausible architecture, as in Fig. 7.8, incorporates the signal input–output and coordinate conversions (plus amplitude normalization, as described below) into a skeleton process, which in turn is combined with nonlinear functions according to user specification. Types of functions that might be used are discussed next.

Nonlinear Functions

Idealized nonlinearities might be representable by closed-form expressions. Several examples, such as hard limiters, were discussed in Sec. 4.5. Furthermore, soft limiters can be approximated by hyperbolic tangents—an excellent fit to the behavior of many transistor amplifiers. But these amplitude expressions do not account for AM–PM conversion; a separate phase function is still needed and it is unlikely that a closed form will be known a priori for phase.

More commonly, amplifier nonlinearity will be known only empirically from measurements, such as those used for the plot in Fig. 7.7. Raw data are entered into the computer as a table of discrete points on the curves. Only rarely will an amplitude $R(n)$, corresponding to a signal point, coincide with a datum point in the table. Some means must be provided to determine $\gamma[R]$ for values of R that do not appear in the table. Two methods commonly applied are: (1) interpolation and (2) curve fitting.

Interpolation. Polynomials classically have been used to interpolate in tables. Linear interpolation suffices if the points are spaced closely enough; otherwise, cubic polynomials are better. Techniques for interpolation are explained at length in countless texts on numerical methods.

Interpolation with a polynomial of degree higher than linear is problematical at the two end intervals of a table. One simple expedient is to perform only linear interpolation in those two intervals. At the small-amplitude end of the table, linear interpolation agrees very well with the AM–AM characteristic of many practical amplifiers.

Another problem is that the simulated amplitude can exceed the largest entry in the table, so that extrapolation is needed. The function must return a number $\gamma[R]$ for any argument $R \geq 0$ that arises in the simulation. Failure to return $\gamma[R]$ may cause the program to crash. [Notice especially that it is crucial to include $\gamma(0)$ in the table—a commonly overlooked point.]

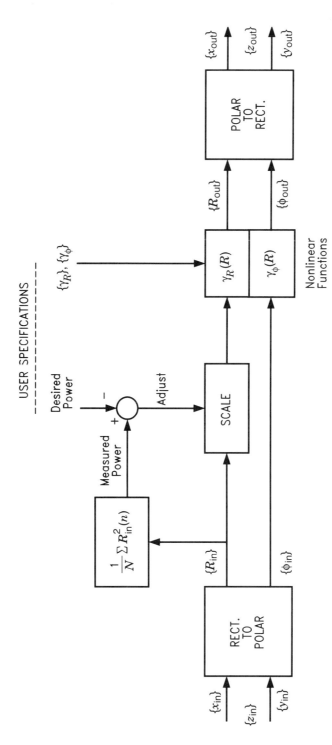

FIGURE 7.8 Processes for saturating nonlinearity.

Polynomials are an extremely poor model for extrapolation of a saturating function. For sufficiently large argument, the magnitude of any polynomial becomes very large, while a saturating amplifier holds constant amplitude or even falls off toward zero for large inputs. A nonpolynomial rule is needed in place of polynomial extrapolation for function values outside the table. A reasonable candidate might be constant amplitude beyond the table end, or even zero amplitude. Extreme R values are likely to occur only rarely, so that little difference in statistical performance of a simulated system will be discernible among widely differing saturation-style rules.

Tables of empirical data include measurement noise, which an interpolator accepts as part of the true characteristic of the simulated device. More subtly, even in the absence of measurement noise, a polynomial interpolator contributes small noise to the output signal because the interpolation slopes are not continuous from one table interval to the next. Noise contributions of this kind tend to be invisible in time-domain waveforms but appear in the frequency domain as elevated noise floors that ought not be present.

Some of the discontinuous-slope noise can be reduced by using cubic splines instead of classical interpolating polynomials. A cubic spline matches the first and second derivatives of adjacent tabular intervals so that slope noise arises only from discontinuities in higher derivatives. Splines do nothing to reduce measurement noise.

Despite its shortcomings, interpolation is a practical method (and sometimes the only practical method) for evaluating a function between tabular entries. Interpolation underlies several routines in the STÆDT library.

Curve Fitting. Polynomial interpolation provides a local approximation to a tabulated function but a poor global approximation. A rational function provides a much better global approximation to saturating nonlinearities. Rational interpolation is discussed in the textbooks, but a better approach in many instances is to obtain a least-squares fit of a rational function to the tabulated data. That has several advantages:

- Only the relatively few coefficients in the function need be stored for on-line processes, not the entire table of measurements.
- A rational function is likely to be simpler to evaluate than an interpolation at each signal point.
- Problems with end intervals and wild extrapolation are avoided.
- A rational function is smooth, thereby doing away entirely with the slope-discontinuity noise of interpolation.
- Least-squares curve fitting smooths measurement noise.

In general, finding the least-squares coefficients involves numerical solutions to simultaneous nonlinear equations, a task well outside the scope of a simulation program. See texts on numerical methods (e.g., [7.13]) for information on the established techniques.

Saleh [7.12] has devised a remarkably simple rational model that fits traveling-wave tube amplifiers rather well. It has the form

$$\gamma_R[R] = \frac{a_1 R}{1 + b_2 R^2}$$

$$\gamma_\phi[R] = \frac{c_2 R^2}{1 + d_2 R^2} \tag{7.28}$$

Moreover, he provides closed-form expressions for the values of the coefficients determined from least-squares fitting to the tabulated points. The Saleh model is often cited in papers subsequent to [7.12].

Better fit to tabulated TWTA data can be obtained by using more terms in the rational functions; for an example, refer to [7.14]. Also, (7.28) is specific to TWTAs; other kinds of saturating amplifiers, such as transistors, have substantially different characteristics and require entirely different functions for accurate models. Despite Saleh's felicitous results, closed-form expressions for the coefficients ordinarily cannot be derived for higher-order or nonrational functions—numerical solutions are needed instead.

Rational functions are not the only closed-form expressions that might be employed. A simple irrational function for modeling of monotonic saturating nonlinearities is given by

$$\gamma_R[R] = \frac{AR}{\sqrt{1 + BR^2}} \tag{7.29}$$

Coefficient A is the small-signal gain of the amplifier; $A/\sqrt{B}$ is the asymptotic saturation level. These characteristics can often be determined by inspection of measured data.

This irrational function is appropriate for a real signal as well as for a complex envelope; just substitute $x_{in}(n)$ for $R_{in}(n)$ in (7.29). Equation (7.29) is applicable as an AM–AM nonlinearity; a different function would be needed for the AM–PM nonlinearity.

Data Conditioning

Tabulated experimental data rarely are provided in a form that is directly usable by a simulation program. Assorted *conditioning* actions are required to convert the raw data to useful format. Some actions need to be performed by a user, who can exercise judgment, while others are perfunctory numerical calculations that are best delegated to the computer.

Raw data for a nonlinear amplifier is likely to be in terms of unnormalized logarithmic power measurements—dBm or dBW—whereas simulation is in terms of normalized amplitudes. A user is required to specify the reference (saturation) levels for both input and output, in terms of the units of the raw data. The computer then normalizes data to the specified references and recalculates the table as relative (voltage) amplitudes instead of logarithmic power.

To ease the burden on subsequent applications, it is common practice to normalize the amplitudes such that the normalized reference amplitudes are $R_{in} = 1$ and

$R_\text{out} = 1$—that is, 0 dB. A reference level of unit amplitude also corresponds to unit power. Definition of a reference level is a subjective matter, decided outside the simulation. A TWTA characteristic such as that in Fig. 7.7 probably would associate the reference level with the peak of the output power, whereas the reference level for a transistor amplifier might be selected as the 1-dB compression point, or similar distortion measure.

Since raw data usually will have phase in degrees, and since programming-language functions employ radians, degrees must be converted to radians. The user need not be bothered with the calculations. The computer can fill in $R_\text{out} = 0$ corresponding to $R_\text{in} = 0$ (a point necessarily missing from a dB input–output table), but the user needs to specify the phase shift corresponding to $R_\text{in} = 0$. If the on-line nonlinear processes interpolate in the converted table, there must be a user-specified extrapolation rule for input amplitudes that exceed the largest table entry. (Perhaps select among a choice of several rules offered by the simulation program.)

These actions are performed by and with the aid of builder routines similar to those already introduced for filters. Raw data are entered into the computer with the aid of entry routines. Builder and entry routines are considered further in Chapter 17.

Signal-Level Adjustment

The effects of saturating nonlinearities on a signal's envelope depend strongly on the amplitude of the signal. If the peak amplitude is small compared to the saturation level, the effect is slight. If the amplitude is large, the effect might be devastating. Simulation of the effect of saturation therefore requires that the signal be adjusted to a user-specified level relative to the saturation reference level. The signal-level setting is variously known as *input loading* or *input backoff*.

Amplitude Measurement. Signal-generation methods, as described in the earlier portions of this chapter, do not ordinarily lead to signal amplitudes that are accurately or easily predictable by the user. Thus to be able to adjust the amplitude wisely, it must first be measured. Average power is the quantity most commonly measured. Unless a signal has a constant envelope, different estimates of power will be obtained with different data patterns; the measurements will exhibit statistical scatter. If the signal constellation has multiple amplitudes (e.g., 16 QAM), the data pattern used for measurement should include outermost constellation points only.

Calibration. Amplitude measurement points up the need for *calibration* of the simulation. Here it is the signal amplitude (or power) that is being calibrated and adjusted, but other elements lying elsewhere in a simulation also need to be calibrated. Calibration will be encountered again in later chapters. Calibration often is performed with the aid of special signals (e.g., outermost points in the constellation instead of random symbols) that would not be employed in the main body of a simulation run. Adjustments from calibrations typically are held constant in the subsequent simulation run. A calibration is likely to be performed on a signal of duration much shorter than used in a simulation run. For these and other reasons, calibrations tend to be

234 □ GENERATION OF DATA SIGNALS

different from the simulation runs that follow them. Well-organized simulation programs include calibration facilities to ease the user's burden.

Spectral Spreading

Nonlinearities cause the spectrum of a signal to spread out. Figure 7.9 shows an example produced by STÆDT. A sampling rate sufficient for the input signal may

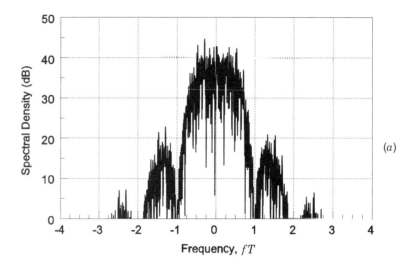

(a)

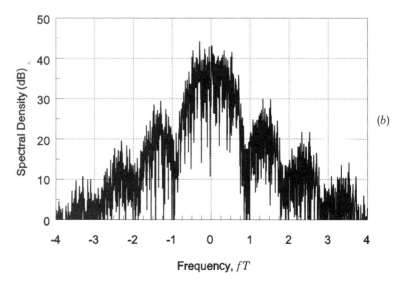

(b)

FIGURE 7.9 Spectrum spreading due to saturating amplifier (STÆDT generated).

not be enough for the output signal. It is the user's responsibility to establish sampling density adequate to hold output aliasing to tolerable levels. High-density sampling has to be applied at the input to the nonlinearity, before the spectrum has been spread. It is too late to increase the sampling rate after the nonlinearity has introduced aliasing. Furthermore, magnitude and phase conversions are nonlinear processes themselves. Even if the input signal is strictly bandlimited, amplitude and phase will not be bandlimited; a signal in polar coordinates cannot be upsampled to a higher sampling rate as effectively as can the same signal in rectangular coordinates. Do any upsampling in rectangular coordinates before conversion to polar.

Sampling rates can be chosen experimentally. Run an abbreviated simulation of the nonlinearity at several different sampling rates and examine the output spectrum at each rate. If a sampling rate is adequate, there will be little change in the spectrum of the output when the rate is increased further.

7.6 KEY POINTS

This chapter has considered the generation of signals for simulations. Greatest attention has been devoted to generating complex envelopes for linear-modulation PAM–QAM signals $z(n) = x(n) + jy(n)$, where $x(n) = \Sigma a_m g_T(n - mM_s)$ and $y(n) = \Sigma b_m g_T(n - mM_s)$. Real signals and nonlinear modulations have been presented as offshoots from complex envelopes and PAM pulse streams.

7.6.1 Message Source

Shift Register Sequences

Binary streams of 0's and 1's are efficiently generated by linear feedback shift registers. With the proper feedback connections, an m-stage shift register delivers an m-sequence with period $2^m - 1$. The sequence can be completed to maximal length by inserting an additional 0. These m-sequences are nearly balanced (just one more 1 than 0), they include all m-length subsequences (except all-0's, which can also be included if the sequence is completed), and they have a nearly impulsive autocorrelation function and nearly constant spectrum (but only if not completed). They serve as a source of pseudorandom message bits, with better apparent randomness properties than might be obtained from truly random finite sequences of the same length. Most programs for simulation of data signals will include a pseudorandom bit generator.

Representative Sequences

Designate the duration of a signaling pulse as LT, where T is the pulse interval. The system is said to have memory of L pulse intervals; there are L pulses overlapping one another. In a binary system, the overlaps cause intersymbol interference (ISI) with any of 2^L different data patterns. To be certain that the worst ISI is taken into account when evaluating system performance, it is necessary to simulate all 2^L patterns. That can be assured by using a *representative sequence*—a maximal-length sequence with period 2^L.

In an M-ary system, there are M^L possible patterns in L overlapping pulses. An M-ary representative sequence is needed to assure that all patterns are exercised in a simulation. If M is an integer power of 2, an M-ary maximal-length sequence can be generated from a binary maximal-length generator. If M^L is a large number, as it easily might be, considerations of simulation running time compel a shortened sequence. A smaller value for L should be used if possible, but retain a maximal-length sequence, to assure that all ISI patterns from close-by pulses (presumably the strongest ISI) are taken into account. Simply truncating a longer sequence almost certainly will miss some close-in ISI patterns.

7.6.2 Symbol Mapping

Mapping for Linear Modulations

No explicit modulator process is needed to generate the complex envelopes of linearly modulated signals. A *symbol mapper* assigns M-ary symbols to points $c_m = a_m + jb_m$ in the signal's constellation. Messages are almost invariably generated as binary bit streams; the bits must be grouped to form M-ary integers $i(m)$ before the symbols can be mapped into the constellation.

Bit grouping might be accomplished by generating a maximal-length M-ary sequence from binary (as in the preceding paragraph), by an encoding operation that generates M-ary symbols from binary, or simply by grouping $\log_2 M$ bits into arbitrary M-ary symbols. The last two methods give no assurance of producing representative sequences that will exercise all ISI patterns or even of generating all M points of a constellation.

Symbol mapping itself is simply a table lookup; for each M-ary number used as an address, the table stores coordinates of a constellation point. Different constellations are defined by different tables.

Differential Encoding

Many signals are ambiguous; they have to carry known patterns in their data to resolve the ambiguity. *Differential encoding* is a common method for ambiguity resolution, whereby message data are encoded into changes between transmitted symbols instead of into the symbols themselves.

Differential encoding is readily simulated by the rule

$$i_C(m) = [i_D(m) + i_C(m-1)] \,\text{mod-}M \qquad (7.8)$$

where $\{i_D(m)\}$ is the M-ary sequence to be encoded and $\{i_C(m)\}$ is the resulting differentially encoded M-ary sequence. The latter will be mapped onto the signal constellation by a symbol mapper.

Often, one wants to combine Gray encoding with differential encoding. That can be accomplished in two steps: (1) perform Gray coding on M-ary integers using a combinatorial table lookup similar to symbol mapping, and (2) perform differential encoding on the Gray-encoded M-ary integers according to (7.8).

Nonlinear Modulations

A carrier signal produced by nonlinear modulation generally does not have a constellation. Typically, M-ary numbers are delivered to the modulators without any mapping process intervening.

7.6.3 Pulse Shaping

To produce a train of signaling pulses, perform the following actions:

1. Generate complex-weighted M-ary impulses. Weights are the coordinates of the constellation points corresponding to the symbol values $c_m = a_m + jb_m$.
2. Establish a sampled time base for an impulse train. An impulse train consists of a complex-weighted sample for one symbol, followed by $M_s - 1$ samples with zero amplitude, followed by a complex-weighted sample for the next symbol, and so on, where M_s is the number of samples per symbol.
3. Send the impulse train through filters to produce a train of shaped pulses.

Square pulses are an important, special case which can easily be generated without a tangible filter. Just apply the same complex weighting to all M_s samples in each symbol interval instead of filling $M_s - 1$ zeros.

Filters can be simulated in the time domain or frequency domain; FIR or IIR; single point, linear block, or circular block. Each kind has its favorable attributes and its limitations. A user selects the filter model according to the needs of the system to be simulated.

Selective filters with rational transfer functions typically have maximum gain of approximately unity. Unit-sample response of such a filter inherently has a low peak amplitude of about $1/M_s$. If the small amplitude is disconcerting, scale the pulses to any desired larger amplitude.

7.6.4 Modulators

Real Carrier-Modulated Signals

Starting from the complex envelope (or I–Q baseband signal) $z(n) = x(n) + jy(n)$, a real modulated signal is generated by multiplying by sine and cosine carriers to form

$$\begin{aligned} s(n) &= x(n)\cos(2\pi n f_c t_s) - y(n)\sin(2\pi n f_c t_s) \\ &= \text{Re}[z(n)e^{j 2\pi n f_c t_s}] \end{aligned} \qquad (7.11)$$

where f_c is the carrier frequency, n the sample index, and t_s the sample interval.

This operation is the final step in generating a real modulated signal, whether the modulation is linear or nonlinear. More often, the real modulation will be omitted and only the complex envelope $z(n)$ will be simulated.

Sampling Rate. If $x(n)$ and $y(n)$ have one-sided bandwidth B, the sampling rate $1/t_s$ must exceed $2(B+f_c)$, to avoid aliasing. On-line computer processes are dimensionless; they deal only with normalized times and frequencies. Only the product $f_c t_s = f_c T/M_s$ appears in the process.

Phase Accumulation. Carrier phase is defined as

$$\Theta(n) = 2\pi n f_c t_s + \theta_0 \tag{7.12}$$

which grows without bound as n increases. It is better to generate the phase recursively, modulo-2π radians according to

$$\begin{aligned}\theta(n) &= \Theta(n) \bmod 2\pi = 2\pi p(n) \\ p(n) &= [p(n-1) + f_c t_s] \bmod 1 \\ p(0) &= \theta_0/2\pi \end{aligned} \tag{7.14}$$

Equation (7.14) is the difference equation for a *number-controlled oscillator* (NCO), an important tool for simulations and for digital signal processing.

Nonlinear Modulations

Many different nonlinear modulations are found in different systems. Simulation programs accommodate the variety by piecing together elementary processes to form particular modulation schemes. Some examples are noted below. The examples all produce complex envelopes, which can be turned into real modulated signals via (7.11).

PPM. Use counters to establish symbol and slot boundaries among the samples. In one slot per symbol, selected according to the M-ary symbol value, generate a standard signaling pulse.

MFSK. Frequency-modulate an NCO with an M-level NRZ pulse train to recursively generate a phase

$$\theta(n) = 2\pi p(n) = 2\pi \left[p(n-1) + \frac{f_0 T}{M_s} + \frac{a_m h}{M_s} \right] \bmod 1 \tag{7.18}$$

If the MFSK signal is supposed to be phase-discontinuous, introduce random phase jumps θ_m at data transitions; otherwise, employ continuous phase across symbol boundaries. Form the complex envelope as $z(n) = e^{j\theta(n)}$.

CPM. Phase of a CPM signal can be accumulated recursively according to

$$\theta(n) = 2\pi p(n) = 2\pi \left[p(n-1) + \frac{h}{2} \sum_m a_m q(n - mM_1) \right] \text{mod-1} \quad (7.23)$$

The summation is just a PAM pulse stream with frequency pulse $q(n)$, M-ary symbol stream $\{a_m\}$, and M_1 samples per CPM symbol interval T_1. The frequency pulse is constrained by the requirement $\sum_{-\infty}^{\infty} q(n) = 1$, and h is the modulation index. Recursion in (7.23) can be regarded as phase accumulation in an NCO, and p is the NCO phase state (in cycles) with a range of 0 to 1. A complex envelope is formed by $z(n) = e^{j\theta(n)}$.

7.6.5 Amplifiers

Linear Amplifiers and Attenuators

To simulate a linear amplifier with gain $K(> 1)$ or an attenuator with a loss ($K < 1$), simply multiply every signal sample by K. Linear amplification or attenuation is just an amplitude scaling process. Scaling can be performed equally well in the time or frequency domains.

Adaptive Scaling. Instead of a fixed user-specified scaling factor, it is often necessary to simulate a varying scaling factor whose value is established in some other portion of the simulated system. An automatic gain control loop is a familiar example of an application of adaptive scaling.

Nonlinear Amplifiers

Practical amplifiers saturate on large-enough input signals. Nonlinear effects on carrier signals are often presented in terms of the complex envelope. Curves of output power versus input power (AM–AM distortion) and phase shift versus input power (AM–PM conversion) are commonplace. Such presentations imply that (1) the nonlinearities are memoryless and (2) the nonlinear effects depend solely on the instantaneous amplitude of the complex envelope. Nonlinearities of this class can be simulated by on-line evaluation of mathematical functions that approximate the measured behavior. These functions have instantaneous input amplitude as their independent variable. Two kinds of functions have been employed: interpolation of the measured data, and curves fitted to the data. Interpolation is simpler to include in a program, but curve fitting provides superior performance.

Level Adjustment. The effect of saturation upon a signal depends on the amplitude of the signal relative to the saturation level. Signal amplitude must be adjusted to achieve a specified *loading* or *backoff*. Adjustment is performed in a calibration procedure prior to the simulation run. In a calibration, input-signal amplitude is measured and then scaled to the user-specified relative level.

Spectral Spreading. Nonlinearities cause spreading of the spectrum of any signal whose envelope amplitude is not constant. Simulations permit the spreading to be displayed for analytical study. Sampling at the input to the nonlinearity must be dense enough that excessive aliasing is not incurred in the spread spectrum at the output.

7.6.6 STÆDT Processes

Signal Generation Processes

DIFE: differential encoder

GRP: bit grouper

MAP: symbol mapper

MAPxxx.PRM: parameter files for MAP

PRBS: pseudorandom bit sequence

PRBS##.PRM: parameter files for PRBS (## indicates number of stages in shift register)

TMBS: time-base symbol spreader (generates train of impulses or square pulses)

Filter Processes

FFT: fast Fourier transform

FIR: finite impulse response, time-domain filter

IIR: infinite impulse response, time-domain filter

MULT: used for frequency-domain filters

Nonlinear Processes

LETS: various processes, some nonlinear

LMTR: limiters

NLA: nonlinear amplifier

NLCD: conditioner for data of nonlinear device

NLFN: nonlinear functions (interpolation; spline; rational-function evaluation)

SALH: fits rational function to TWTA data

SPLN: generates spline table

Other Processes

GAIN: amplifier or attenuator

NCO: number-controlled oscillator

POWR: measures power of sample sequence

APPENDIX 7A: DIFFERENTIAL ENCODING EXAMPLES

The differential encoding algorithm of (7.8) may be overly simplistic; real-world applications tend to be more complicated. Examples of greater complication are presented below. Additional processes needed to accomplish the encodings emerge from the examples; it is the additional processes that are the items of main interest in this presentation, not the examples themselves.

Forced Changes

Some differentially encoded MPSK signals employ $2M$ phases to transmit M-ary data. Denote the class as M-2MPSK. A good practical example is 4-8PSK, also known as $\pi/4$-QPSK. In this class of modulations, the phase is forced to advance by an odd multiple of π/M at each symbol interval. Advantages of the scheme include: (1) there is never a phase change of π radians from one symbol to the next, so the envelope never goes through a null; (2) a pulse transition occurs at each symbol time, which is highly favorable for clock recovery; and (3) differential detection in the receiver is simplified in some degree. Although $2M$ phases are transmitted, the method has the same data throughput and yields the same BER performance as achieved by standard differential MPSK.

Differential encoding and the forced phase advance can be performed together by the differential-encoding algorithm

$$i_C(m) = [2i_D(m) + i_C(m-1) + 1] \bmod \text{-}2M \qquad (7A.1)$$

where $i_D(m)$ is M-ary and $i_C(m)$ is $2M$-ary. The elements $i_C(m)$ are to be mapped onto a 2MPSK constellation.

Square Constellations

A square constellation is ambiguous as to quadrant but not within a quadrant. Therefore, common practice [1.8, Chap. 4] for square constellations is to apply differential encoding to just two bits of a group of J bits to resolve the quadrant ambiguity and to omit differential encoding within a quadrant. Both the differential encoding and the mapping within a quadrant typically are Gray encoded.

Reference [1.8] shows how to accomplish these functions with logic operations in hardware. For simulations, it may be simpler to work on the bit-grouped M-ary numbers instead of the individual bits. One way to operate on the M-ary numbers is outlined below.

- Designate the M-ary data element delivered by the bit grouper (or M-ary data source) as $i(m)$.
- Let the two least-significant bits in each group be devoted to specifying the quadrant changes. These two bits are to be differentially encoded. They can

be retrieved as a 4-ary number by the operation

$$i_q(m) = [i(m)] \bmod\text{-}4 \qquad (7A.2)$$

which discards all but the two least-significant bits.

- The other $J-2$ bits will be used to specify a point within the quadrant. Mapping rules are the same within all quadrants except for rotation by multiples of $90°$. Mapping within a quadrant is not subjected to differential encoding. These $J-2$ bits are retrieved as an M-ary number by the operation

$$i_a(m) = i(m) - [i(m)] \bmod\text{-}4 \qquad (7A.3)$$

which discards the two least-significant bits. Observe that $i_a(m)$ is always a multiple of 4.

- To accomplish Gray encoding on top of the differential encoding, map the 4-ary quadrant specifier $i_q(m)$ into $i_g(m)$ according to any desired 4-ary mapping rule. The two rules of Table 7.3 are examples that could be used. The mapped result $i_g(m)$ is the number of quadrants (0-3) to be rotated in going from the $(m-1)$th to the mth symbol.

- Apply differential encoding to the mapped quadrant specifier according to

$$i_e(m) = [i_e(m-1) + i_g(m)] \bmod\text{-}4 \qquad (7A.4)$$

This function will be recognized as identical to (7.8), with $M = 4$ and a minor change of notation.

- The transmission symbol to be mapped onto the square M-ary constellation is

$$i_C(m) = i_a(m) + i_e(m) \qquad (7A.5)$$

An example of mapping for 16-QAM is shown in Fig. 7A.1. The arrangement of these processes is depicted in Fig. 7A.2. The only new operations introduced (compared to Fig. 7.2) are the modulo-4 reduction of (7A.2) and trivial additions and subtractions. If Gray coding is wanted inside the individual quadrants, it is accomplished by assignment rules in the symbol mapper. The standard mapping rules illustrated in Fig. 7A.1 impose Gray encoding within each quadrant but not between adjacent points in adjoining quadrants.

APPENDIX 7A: DIFFERENTIAL ENCODING EXAMPLES □ **243**

```
       11(01)      01(01)    |   10(00)       11(00)
         •            •      |     •             •
         13           5      |     8             12


       10(01)      00(01)    |   00(00)       01(00)
         •            •      |     •             •
         9            1      |     0             4
    ───────────────────────────────────────────────────
       01(10)      00(10)    |   00(11)       10(11)
         •            •      |     •             •
         6            2      |     3             11


       11(10)      10(10)    |   01(11)       11(11)
         •            •      |     •             •
         14           10     |     7             15
```

FIGURE 7A.1 Symbol mapping for CCITT V.22 bis (per [1.8, Chap. 4]). Labels show assignments of $i_C(m)$ for each point, in binary and decimal formats. Bits in parentheses have been differentially encoded from the original data groups.

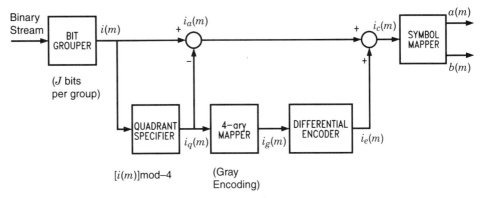

FIGURE 7A.2 Mapping for square constellations. Differential encoding between quadrants and Gray encoding in quadrants.

8

OVERVIEW OF RECEPTION PROCESSES

Chapter 7 dealt with generation of signals—with processes that are related primarily to transmitters. In subsequent chapters we treat individual processes for receiver operations. In this chapter we furnish an overview of receiver processes and illuminate the interrelations among them, interrelations that tend to be obscured when concentrating on the individual processes. Various aspects of the transmission (or storage) medium are examined briefly, to establish requirements for receiver processes and for models of the medium. Also, several instances of simulation practices are introduced.

8.1 SIMPLE CHANNELS

In the simplest conditions ordinarily encountered, the medium has only the following effects on the signal:

- Power is attenuated.
- The signal is delayed.
- Phase is shifted (in a carrier signal).
- Noise is added to the receiver input.

Statistics of the additive noise are stationary, and attenuation, delay, and phase shift are time invariant for the simplest cases. The medium is *transparent* if its attenuation and delay are independent of frequency and the medium is linear.

8.1.1 Model of the Medium

Noise statistics and the scaling of signal and noise are the only medium-related items to be modeled for the simplest channels.

Noise Model

Additive noise is most commonly modeled as white and Gaussian and denoted by the acronym AWGN. Methods for generating AWGN samples in a computer are described in Chapter 9, and evaluation of receiver performance in the presence of AWGN is investigated in Chapter 13. Evaluation of performance is one major reason for computer simulation of a problem; performance in AWGN is the dominant model for many realistic systems.

Amplitude Scaling

In a linear medium, absolute level of the signal is immaterial; the user can set any convenient level whatever. All that matters is the signal-to-noise ratio, the setting of which is considered at length in later chapters on noise.

8.1.2 Processes Required in Receiver

Numerous operations take place in a receiver simulation, even for the simplest transmission channels (refer to Fig. 1.2). Receiver functions needed with a simple channel are also employed with more-complicated channels. Some of the operations performed on the mainline signal are:

- Front-end functions: frequency translation (carrier signals only) and amplification
- Demodulation (carrier signals only)
- Filtering
- Symbol strobing
- Symbol detection
- Symbol demapping

In addition, several essential auxiliary functions may be required, off the signal's mainline:

- Amplitude setting: automatic gain control (AGC)
- Frequency adjustment: automatic frequency control (AFC)
- Phase adjustment
- Timing adjustment

Frequency and phase adjustment are meaningful only with carrier signals. Frequency adjustment is needed if a frequency error has to be corrected; phase adjustment is required for coherent demodulation. Adjustments for frequency, phase, and timing constitute *synchronization* of the receiver. These mainline and auxiliary functions are discussed briefly in the paragraphs to follow. Many of the functions are treated in depth in later chapters.

Amplification

Linear amplification (or attenuation) is modeled by scaling a signal. Absolute signal level ordinarily is immaterial within a simulation, so the numerous amplifiers found in typical receivers are not usually included in a signal-transmission model. Signal scaling needed with nonlinearities most often will be encapsulated inside the module for the nonlinearity (see Sec. 7.5.2).

Frequency Translations

For various good reasons, radio receivers commonly are designed with one or more frequency translations in the signal path. If the carrier frequency (at the input and at intermediate frequencies) is large compared to bandwidth of the signal, the signal can be represented by its complex envelope. Modeling with the complex envelope entirely ignores the actual frequencies and their translations. Thus much of the radio portion of a receiver is likely to be omitted from transmission simulations.

If the carrier frequency is not large compared to the signal's bandwidth, frequency translations may introduce spectrum folding, which is not modeled by a complex envelope. Real signals must be simulated when spectrum folding is to be taken into account. Frequency translation of real signals is examined in Chapter 10. Imperfections of frequency translators, such as nonlinearity in the mixers (see Sec. 7.5.2) or phase noise in the conversion oscillators (see Secs. 2.4.2 and 9.2.4), sometimes require simulation. These are best handled in separate processes that model the nonlinearity or the phase noise rather than lumping them into an inclusive mixer process.

Demodulation

A demodulator recovers the original modulation that was impressed on the carrier at the transmitter. Many receivers employ *I–Q* demodulators (see Sec. 4.2.1) which recover a complex baseband signal from a passband signal. Additional demodulation steps may be required, depending on the nature of the signal and the degree of coherence attained in the receiver (see Sec. 4.2.2). A physical *I–Q* demodulator involves a final frequency translation from the last intermediate frequency down to baseband. If a signal is represented by its complex envelope, the frequency translation is only implied; it is not simulated. The complex envelope is indistinguishable from the complex baseband signal, so the *I–Q* demodulator might be omitted entirely if not needed for other purposes. This same omission of an essential component was observed in Chapter 7 with linear modulators.

In addition to a final frequency translation (not modeled when signals are represented by their complex envelopes), the *I–Q* demodulator may also be called upon to perform a frequency correction, a relatively small frequency translation to compensate for frequency errors in oscillators in the system. If the signal has been simulated with a frequency error, frequency correction is necessary even if the passband signal has been represented as a complex analytic signal. This necessity is in contrast to the omission of most heterodyne conversions from simulations made with complex signals.

If coherent reception is to be simulated, the phase of the signal must be adjusted properly (see Sec. 4.2.2). The demodulator is the place where phase corrections are performed. Phase shifting is closely related to frequency shifting; both have been introduced in Sec. 7.4 and are treated further in Chapter 10.

Differentially coherent and noncoherent reception schemes require further demodulation steps following the *I–Q* demodulator. These additional demodulation steps often can be carried out on symbol strobes, one per symbol interval, instead of on the more densely sampled signal prior to strobing. Analog models have been given in Sec. 4.2.2; simulation techniques for detection are introduced in Chapter 12.

Filtering

Receivers incorporate filters for excluding noise and interference and for shaping the received pulses. In a simple channel, the receiver filter is usually matched to the received pulse shape, and the overall impulse response of the channel will be Nyquist. Filters have been examined at length in Chapters 3, 6, and 7. Receiver filters might be concentrated in the passband prior to the *I–Q* demodulator, in the baseband following the demodulator, or distributed between the two locations in any desired proportions.

Symbol Strobing

A receiver delivers one strobe output in each symbol interval. Transmitted symbols are estimated from the strobes. Optimum strobing is performed at a specific time instant within the symbol interval; any departure from the optimum strobe time will degrade performance of the receiver.

Optimum strobe times do not necessarily coincide with any of the discrete-time samples of a digitally simulated signal, unless special measures are taken to force coincidence. Usually, a simulation must interpolate among the nearby samples to extract the correct strobe value. Several methods for signal interpolation are examined in Chapter 10.

Strobing in a time-continuous modem typically is adjusted by shifting the timing of a receiver clock, but that method of adjustment is impossible once a signal has been sampled. Simulation of timing adjustment by interpolation therefore necessarily differs considerably from the actual analog-hardware implementation. Digital-hardware modems—digital matters are covered in Chapter 15—must also interpolate to find the correct strobe value and so are more closely related to the techniques needed for simulations.

Symbol Detection

Symbol estimates—also called hard decisions—are produced by quantizing signal space into M different decision regions (for M-ary symbols) and deciding which of the regions each strobe falls into. For example, binary strobes might be quantized into two intervals $(-\infty, 0]$ and $(0, \infty)$; a strobe value falling in $(-\infty, 0]$ is decided as -1, while a strobe value falling in $(0, \infty)$ is decided as $+1$.

More generally, if the modulation is two-dimensional, the decision regions are likely to be two-dimensional also. For a linear-modulation QAM signal, the decision regions each contain one point of the signal constellation. Decision regions are important in the evaluation of performance of symbol-by-symbol detection; evaluation is covered exhaustively in Chapter 13. Symbol detection takes place after all demodulation operations have been performed, including nonlinear operations associated with differentially coherent or noncoherent reception.

A symbol detector, or hard-decision device, may be absent if the receiver includes a decoder. Decoded symbols are not decided individually but as part of a group or sequence of symbols. More signal information is available and better decoding decisions are possible if unquantized strobes are used instead of M-ary quantized strobes. More commonly, the strobes are subjected before decoding to a "soft decision" process in which they are quantized into many more than M regions. A sufficient number of levels is used for soft decisions so that the quantizing causes negligible loss in decoder performance. Soft decisions are simply quantizations imposed to permit analog strobes to be processed by digital decoders and are not really decisions at all. Quantization is an important topic for simulation of digitally implemented modems and is treated in Chapter 15.

Symbol Demapping

Once symbol detection has been performed, each symbol decision is translated back into a sequence of data bits—the inverse of the symbol mapping process explained in Sec. 7.2, working on virtually the same principles. Both mapping and demapping routines and tables would be used in simulating all but the simplest of signal formats. Symbol demapping is essential in M-ary ($M > 2$) communications systems whose underlying messages are in binary streams; it is also required when evaluating bit-error, as opposed to symbol-error, probabilities.

Gain Control

Receivers habitually are built with AGC loops to compensate for variations of signal amplitude or for variations of amplifier gain. An operating AGC loop continually monitors the signal amplitude and adjusts it to a designated level. In simulations in which signal level remains invariant, it is more common to set the signal level just once in a calibration procedure and to hold it unchanged throughout the ensuing simulation run. A system-level simulation is not likely to include AGC unless performance of the AGC itself was being studied explicitly. A versatile simulation program allows a user to put together lower-level modules to produce an AGC loop or other control functions. Pertinent lower-level modules are described in several later chapters.

Synchronization

Frequency, phase, and timing of an incoming signal may have to be adjusted to achieve proper reception. In addition to the adjusting processes (frequency shifters, phase shifters, and interpolators), other processes are needed to determine the amount of adjustment to be applied and to communicate that information to the adjustment processes. Adjusting and amount-determining processes taken together constitute synchronizers.

Most receiver simulations require synchronization in some form. For many investigations, performance of synchronizers as such is not of interest; in these cases the correct timing, phase, and frequency can be adjusted to ideal values in a calibration procedure and then held constant throughout a simulation run. In other studies, the influences of realistic synchronizers on system performance or the behaviors of the synchronizers themselves are to be investigated. These cases require that entire synchronizers be modeled. Simulation of synchronizers is covered in Chapter 11.

Feedback Loops

Numerous feedback loops are employed within receivers; an AGC loop is one example. Other examples arise in many synchronizers arranged in a feedback configuration and in adaptive equalizers adjusted via feedback. Simulation of a feedback loop is subject to restrictions not imposed on feedforward configurations. It has been pointed out (Sec. 5.3.3) that a feedback loop can only be simulated in single-point format in the time domain; loop simulation is not possible in block format or the frequency domain. In Chapter 14 it is shown that a sampled feedback loop must contain a delay to be computable and that special measures are needed for computation of the first pass through the loop.

Most feedback loops of interest in communications systems include control functions (Sec. 7.5.1) that are off the signal mainline. Feedback of data signals themselves is rare. Specialized processes dedicated to control functions find a useful place in simulation programs. Features of control processes are explored in Chapter 14.

Digital Implementations

At the time of writing of this book, digital implementation of modems was being explored vigorously by many investigators. Digital implementation is becoming widespread as the cost of digital hardware continues to drop, as speed and versatility increase, and as communications engineers become better acquainted with digital methods. Simulation is an effective way to explore the behavior of digital modems. To minimize cost, a digital modem is likely to be implemented with fixed-point processors. That raises word-length issues—quantizing, overflow, and underflow—to be addressed in simulations. Also to minimize cost, digital hardware typically is run at the lowest acceptable sampling rate for each process. In consequence, sampling rate is altered at various points throughout a modem, raising decimation and interpolation issues. These issues, and others, are examined in Chapters 15 and 16.

Observation

All of the foregoing receiver processes (and even more: e.g., decoders), along with the transmitter processes of Chapter 7, are employed in conjunction with simulation of the very simplest channels. Moreover, the processes needed for performance evaluation (see below) or for displays (Chapter 18) have not yet been introduced but constitute a substantial addition to the essential receiver processes. Evidently, a simulation can be highly intricate, even if the channel is extremely simple.

Suppose that the channel becomes more complicated. Does the complexity of the simulation grow beyond practical bounds? That can happen, depending on the resources built into the simulation program, the size and speed of the computer, and the user's patience in waiting for results. A more-complicated channel requires more processes to model it (see below for some examples). Additionally, a more complicated channel may impose more complication in the receiver (e.g., adaptive equalizers or echo cancelers) to cope with the ill effects associated with complicated channels.

In later sections of the chapter we examine the modeling of some complicated channels and indicate their demands on receiver complexity. Despite concerns over excessive complexity, it will become apparent that the great bulk of simulations can be performed using those processes mentioned above (including equalizers and decoders), plus a few more to be introduced. The foundations of simulation can be learned through study of the processes used with the simplest channels.

8.1.3 Performance Evaluation

Evaluation of the performance of a proposed communications system is one of the main reasons for undertaking simulations. The most common measure of performance is the bit-error ratio (BER) used as an estimate of the probability of receiving a message bit in error. Other measures are sometimes used, such as symbol-error probability, word-error probability, packet-error probability, mean number of symbols between errors, and others. This book concentrates on techniques for BER and symbol-error evaluation, but the other measures can be simulated by kindred techniques.

Configuration

A complete communications system, from message source to message sink and all appropriate operations in between, as in Fig. 1.2, is set up in a simulation. A known message is produced by the message source and applied to the system, which generates and processes signals. Representative disturbances are applied at suitable points in the signal path. Detected messages or strobes are taken as outputs from the simulated receiver.

Evaluations are performed by counting or estimating the proportion of decision errors occurring in the receiver output. Error ratio is evaluated as a function of signal-to-noise ratio and for different values of parameters specific to the system under investigation. Methods of performance evaluation are outlined briefly below and are examined at length in Chapter 13.

Brute-Force Error Counting

One important approach is to simulate fully the entire system plus all disturbances that impinge upon it, including additive noise. Detected message data at the receiver output are compared, bit by bit or symbol by symbol, against the message data applied to the transmitter. All errors are counted and the BER is taken as the ratio of bit errors to the total number of bits simulated. This technique is known as the *error counting* or *Monte Carlo* method of performance evaluation. Under many circumstances it is the only method that is applicable.

A consumer of evaluation information (such as a systems engineer) needs to know how much trust to place in the observed BER; its statistics, such as standard deviation and confidence intervals, are needed. A simulationist needs to know how many bits to run in a simulation to attain desired confidence intervals. Clearly, smaller BERs require longer simulations to achieve acceptable confidence levels. For this reason, Monte Carlo evaluation is exceedingly time consuming for small BERs.

Quasianalytic Estimation

Another evaluation technique, customarily known as *quasianalytic* or *semianalytic estimation*, can provide BER estimates with far shorter simulations, but only under restricted conditions. For reasons that will become apparent, we have called it the *phantom noise* method in our own work.

If noise contributes strictly additively to the receiver strobes, and if the probability density function (pdf) of the noise in the strobe is known analytically, it may not be necessary to simulate the noise at all or to count bit errors. Instead, it may be sufficient to simulate only the signal (plus, perhaps, other disturbances in the system) and produce a set of noise-free strobes. Given each noise-free strobe value and the noise pdf, it is conceptually possible to calculate the probability that additive noise would cause a noisy strobe to cross over a decision boundary and cause a decision error.

Quasianalytic estimation determines the effects of additive noise as an ensemble statistic, and not by sampling the noise space as in the Monte Carlo approach. As a result, the BER estimate does not suffer statistical scatter from the noise but only from the signal realizations or from other disturbances that may be simulated.

For the most part (with extensions noted in a later chapter), this method has been applied for linear receivers exposed to additive white Gaussian noise as the main disturbance, and for regular constellations whose decision regions permit easy calculations of error probability. Despite the restrictions, the conditions under which it is applicable are of great practical importance, it permits huge reductions in computing time, and so the method sees widespread use.

8.2 MORE-COMPLICATED CHANNELS

Departure from the simplest channels opens up an enormous diversity of channel characteristics. In this section we provide a few examples to illustrate the variety of simulation questions that can arise. For further depth the reader is referred to the communications literature (copious for channel descriptions; comparatively sparse for simulation of channels). An excellent reprint collection [8.1] provides a helpful introduction to many different channel models and their statistical behaviors.

Channel simulation is a very broad topic, deserving at least a chapter (if not an entire book) of its own. Nothing more than a superficial glimpse of the subject is offered here. Ultimately, workers in specialized areas are compelled to devise their own specialized simulation techniques. Fortunately, techniques developed for simple channels can be carried over to more-complicated models, as pointed out in the sequel.

8.2.1 Nonwhite, Non-Gaussian, and Nonadditive Disturbances

Although incidence of AWGN predominates over any other disturbance, many other kinds of disturbance arise and demand simulation. Simulation aspects of several are outlined here.

Colored Gaussian Noise

If noise does not have a constant spectral density, it is said to be *colored noise*. Colored Gaussian noise is readily generated by passing white Gaussian noise through a filter. Appropriate selection of filter characteristics provides just about any noise spectrum that might be desired. Gaussian noise has a special property that makes it easier to simulate than other noise distributions: If Gaussian noise is applied to a linear process (e.g., a linear filter), the output of the process is also Gaussian. Other distributions are not preserved upon passage through linear processes. A filter alters the spectrum (equivalently, autocorrelation) and variance of a Gaussian input, but these effects can be calibrated. Thus quasianalytic error estimation can be applied with colored Gaussian noise as well as AWGN. Other noise distributions tend to force use of Monte Carlo evaluation methods.

Interference and Jamming

In many environments, interference (caused unintentionally) or jamming (caused deliberately) are the worst disturbances to communication. The term *interference* will be used generically here to mean both interference and jamming. Interference can be simulated by exactly the same techniques used for generating a desired signal, as explained in Chapter 7. Interference is either *co-channel* (at the same center frequency as the desired signal) or *adjacent channel* (centered at a nearby, different frequency). After its generation, interference is added to the desired signal at a suitable position in the simulation model.

When simulating carrier systems, the desired signal may be represented by its complex envelope. In such cases, adjacent-channel interference is most conveniently generated also as a complex envelope (i.e., as co-channel interference) and then subjected to a frequency shift before being added to the desired signal. Peculiarities of frequency shifting are examined in Chapter 10.

In some environments (e.g., crosstalk between wire pairs in a cable) the interference-coupling mechanism is frequency selective. Such coupling can be simulated by filtering the interference before adding it to the desired signal.

Communications systems employing frequency-division multiplexing are commonplace. In these systems the adjacent channels are occupied by interferers that are statistically the same as the desired signal, which is likely to have been generated with a well-planned pseudorandom message sequence. Interference should also have a good, representative sequence, for the reasons detailed in Sec. 7.1. One way to make the interference look equally random is to use the same message sequence for both the desired signal and for the interference. Just impose sufficient relative delay between the message sequences so that the signal and interferences are not correlated.

If a signal and an interferer have been generated identically, their symbol timings will be synchronized with one another; the same relative timing phase will occur throughout the simulation. Adverse effects of interference depend upon the relative timing phase; an adequate evaluation will investigate performance at all timing phases. One method to obtain various timings is to repeat a simulation run with different relative timings but otherwise identical conditions. Timing between two signals is adjusted by delaying one with respect to the other. Delay processes are examined in Chapter 10.

Another approach resamples the interferer at a different sampling rate, in effect changing the data rate slightly. The interferer timing relative to the desired signal then slowly changes during the course of the simulation run, exposing the desired signal to interference at all timing phases. Rate changing is more complicated than simple delay adjustments, requiring interpolation of new sample points. Interpolation and rate changing are treated in Chapter 16.

Interference is non-Gaussian; its instantaneous amplitude is bounded for bounded interferer power. Although the sum of multiple interferers might produce a fair approximation to Gaussian statistics (e.g., crosstalk from many wire pairs in the same cable [8.2]), the approximation is unduly pessimistic for just one or two interferers [8.3]. Numerous reprints of interference-related papers have been collected in [8.4]; several of the papers address simulations directly.

Impulse Noise

Some channels are afflicted by *impulse noise:* large-amplitude, short-lived events that occur at generally well-separated instants. The impulse model might consist of a single sample per event or the response of a filter to a single sample. Use filtering to model a disturbance that takes place over a time longer than the sampling interval in the simulation or to represent frequency selectivity in the mechanism that couples the disturbance into the communications channel. Impulses might recur periodically (radar pulse trains, fluorescent-lamp radiation, vehicle-ignition interference) or they may occur randomly (lightning flashes, relay actuations). Impulse amplitudes might be uniform (radar pulse trains) or they may be random (lightning flashes). The amplitude statistics are not Gaussian. Some flavor of the complexity of random impulse-noise statistics can be gained from [2.24] and [8.5]. Techniques for modeling impulse noise and noise suppressors are found in [8.6].

Shot Noise

If individual events in impulse noise occur randomly with Poisson arrival statistics and if the waveforms largely overlap, the result is known as *shot noise*. Lightwave communications is an area of great interest where shot noise (also called *quantum noise*) plays an important role. Each shot event marks the emission of a photoelectron in a photodector. Reference [2.25] provides an excellent tutorial introduction and an extensive bibliography on lightwave reception.

Shot noise is not Gaussian, although it approaches Gaussian if each received signal pulse contains a large number of photons. Its non-Gaussian distribution cannot be neglected in efficient direct-detection receivers of weak signals where only a small number of photons may be contained in each received pulse.

Signal-related shot noise in an optical system is not additive; it is contained in the received light itself. Regard optical shot noise to be a fluctuation of the signal amplitude. Increasing the light intensity increases the shot noise. Because the light intensity varies in accordance with the waveform of the signal pulse, the shot noise also varies with the instantaneous pulse waveform; its variance is not stationary.

Different optical configurations demand different simulation expedients. Coherent lightwave receivers include an optical-frequency local oscillator whose shot noise overwhelms all other noise sources. Because of the very large number of photoelectrons produced in response to the local oscillator illumination, the dominant shot noise is well approximated as AWGN [2.25], the quantum character of the received light signal can be neglected, and the evaluation techniques of Sec. 8.1.3 can be applied. In particular, quasianalytic estimation can be used if the receiver is linear.

Local oscillators are absent in noncoherent direct-detection receivers. If an ordinary nonamplifying photodiode is employed, the thermal noise (which is additive and Gaussian) of the following preamplifier generally predominates over the shot noise, so the evaluation techniques of Sec. 8.1.3 are again adequate.

Other receivers employ an avalanche photodetector (APD), which has internal gain sufficient that the amplified shot noise exceeds thermal noise of the preamplifier. Avalanche multiplication is a random process, so the photocurrent delivered by the APD is accompanied by two non-Gaussian, nonadditive sources of fluctuation: the quantum nature of the incident light and the random avalanche gain. The simple methods of Sec. 8.1.3 are not applicable to noise of this character. Fiber-guided lightwave systems typically work with BER of 10^{-9} or less, so that straightforward Monte Carlo evaluation requires simulation of enormous numbers of bits. To simulate each individual detected photon would consume unthinkable hours of computer time. Practical methods of simulation all incorporate schemes for reduction of computing effort. Examples are found in [8.7] to [8.13] and other papers cited in these references.

Magnetic Media Noise

Media for magnetic recording typically consist of tiny magnetic particles dispersed on a nonmagnetic substrate. During recording, a remanent magnetization pattern is imposed upon the particles. Upon playback, the motion of the magnetized medium past a playback head induces an electrical voltage in the coils of the head. The number and size of particles under a head vary with position on the medium so that magnetization varies similarly. As a result, the playback signal includes *media noise*—a fluctuation of the amplitude of the recovered signal due to the particulate nature of the magnetization.

Media noise can be measured by comparing the noise from an erased medium (nominally zero remanent magnetic field) with that from a dc-recorded medium (nominally constant remanent field at all points). Substantially larger noise is observed from the dc-recorded medium [8.14, Chap. 12]. Media noise is nonadditive, in the same manner as shot noise (which it resembles). Analyses of media noise can be found in [8.15] to [8.17], leading to expressions for noise power spectra. Other statistical properties (such as pdf) do not appear to have been investigated. For simulation purposes, the media noise might be modeled as stationary, additive, Gaussian, and colored; the stationary and additive properties are definitely approximations and the Gaussian assumption probably is also an approximation.

8.2.2 Frequency-Selective Channels

Many channels exhibit frequency selectivity (*dispersion*, in optical parlance). Examples include:

- Metallic transmission lines, where attenuation in decibels typically increases in proportion to square root of frequency.
- Magnetic recording systems, where the induced playback voltage is proportional to the time derivative of flux in the playback head. Also, finite gap width in the head causes severe rolloff of response at upper frequencies.
- Radio systems with narrowband antennas.
- Portions of the radio spectrum where the atmosphere/ionosphere are not transparent.
- Optical fibers, where attenuation and transmission velocity depend on wavelength of the light.
- Multipath channels, where constructive and destructive addition of multiple, relatively delayed versions of the signal interfere with one another.

For linear time-invariant channels the selectivity can be modeled by linear time-invariant filters, which have been treated in Chapters 3, 6, and 7.

Time-varying and multipath channels are discussed in the next section. Some of the channels—transmission lines, atmospheric propagation in radio links, optical fibers—exhibit distributed selectivity. Longer path lengths cause more pronounced selectivity. If the selectivity is uniform in each unit path length, overall channel response is the product of the frequency response in a unit length multiplied by the length of the path to be simulated. Provision for building frequency-domain filters from unit-length responses and user-specified path length is readily incorporated into filter-builder routines.

Optimum reception in frequency-selective channels is not obtained from the simple matched-filter/symbol-by-symbol detection approach put forward in Chapter 4. The best linear receiver consists of a filter matched to the transmitted pulse, followed by a transversal equalizer. The best nonlinear receiver employs maximum-likelihood sequence estimation [4.1]. Equalizers [8.19] play a major role in reception on selective channels, either by themselves or in conjunction with a sequence estimator.

8.2.3 Time-Varying Channels

Characteristics of some channels vary over time. Several examples are examined here, along with the implications for their simulation.

Doppler Shift

If a transmitter and receiver are in relative motion, the signal frequency at the receiver is different from that at the transmitter. The received signal has undergone a *Doppler shift*. Doppler shift on a carrier is readily simulated by the frequency-shift methods of Chapter 10. If the transmitted carrier frequency is f_c and the relative velocity along the path length is $v(t)$, the received carrier frequency is $f_c + f_d$, where the carrier Doppler shift is

$$f_d \simeq -\frac{f_c}{c} v(t) \qquad (8.1)$$

and c is the speed of light. The approximation is valid if $|v| \ll c$.

Modulation also undergoes a Doppler shift; substitute $1/T$ for f_c in (8.1) to determine the Doppler shift on the symbol rate. Simulation of a shifted modulation rate may require interpolation of new samples from the unshifted sampled-signal representation. Sample-rate conversion usually increases or decreases the number of samples to be processed; there must be provision for added or deleted samples. These matters are examined in Chapter 16. Doppler shift is usually a small fraction of the symbol rate (e.g., a commercial airliner travels at less than 1 part per million of the speed of light), so the shift on modulation is often (but not always) ignored in simulations.

Tape Flutter

Symbol rate of the signal recovered from a magnetic recording depends on the velocity between the magnetic medium and the heads, both in playback and in recording. In tape recorders, that velocity fluctuates significantly from periodic disturbances caused by irregularities in rotating mechanical elements and by resonant vibrations in the tape. Tracking of the flutter is a vital task for the timing synchronizer.

The original signal (prior to recording) is represented by equally spaced samples. After the flutter of recording and playback, the signal points at the original sample times are no longer equally spaced; resampling with uniformly spaced new samples requires interpolation among the old—a tedious chore. Flutter is normally small enough that it has little influence upon data-signal recovery, other than through its effects upon synchronization. Since synchronizers can often be treated analytically, simulation of flutter is avoided more often than not.

Multipath Channels

In many different systems, a signal travels from transmitter to receiver over multiple parallel paths. Some examples are:

- High-frequency (HF) radio transmission via ionospheric reflection [8.20]
- Echos in wire-line transmission
- Reflections from structures in mobile radio service [8.21]
- Reflections from Earth's surface (or other astronomical body) in communication with an aircraft or spacecraft [8.22–8.24]
- Simultaneous ground-wave and skywave reception of low-frequency radio signals
- Multimode propagation in optical fibers
- Refraction in different tropospheric layers for line-of-sight terrestrial microwave radio [8.25, 8.26]
- Tropospheric scatter for beyond-line-of-sight transmissions

Each of these examples, and other multipath topics, has its own extensive literature; the references cited above provide only a meager sampling. Since this book deals with simulation, only the generic multipath model is considered here. References [8.27] and [1.2, Chap. 7] contain thoroughgoing accounts of the model and its characterizations.

A multipath transmission medium is modeled as a tapped transversal filter with finite impulse response. If the signal representation is complex (as it usually will be for carrier signals), the transversal filter is also complex. Tap spacing is required to be fine enough to provide adequate resolution of the multipath structure. The total span of the transversal filter is the same as the delay spread of the multipath propagation.

Complex tap-multiplier coefficients are selected according to the impulse response of the multipath channel. Some channels are well modeled with just a few taps (e.g., terrestrial microwave with two or three), while others may need many more taps (e.g., HF ionospheric propagation).

Multipath channels typically exhibit *fading* [8.27, 1.2, Chap. 7] because channel characteristics are time varying. Fading is simulated by varying the tap coefficients during a simulation run. If the tap coefficients change slowly compared to the symbol rate, the channel exhibits *slow fading*. If the coefficients change rapidly, the channel undergoes *fast fading*.

If delay spread is small compared to symbol interval T (may imply just a single tap in the multipath model), the frequency response is nearly constant over the signal bandwidth; the channel will exhibit *flat fading*. If delay spread is significant compared to symbol interval, the channel will have a frequency response with substantial variations over the signal bandwidth and the received signal will exhibit *selective fading*.

In many fading channels, the tap-weight variation is well modeled as a complex Gaussian time-varying function. The real and imaginary components of the tap weight are each represented as independent stationary colored Gaussian variables, with the same statistical properties in both components. Generation of Gaussian variates is discussed in Chapter 9. Signal amplitude out of such a tap exhibits Rayleigh statistics and signal phase shift is uniformly distributed.

If all taps in the model are driven by Gaussian time-varying weights, the channel is said to exhibit *Rayleigh fading*, since the magnitude of a complex-Gaussian process has a Rayleigh distribution. Tropospheric scatter and terrestrial-mobile services provide examples of Rayleigh fading. In some situations, there may be one path—a direct ray—that is nonfading, along with one or more Rayleigh fading paths. This mixture of a direct path and Rayleigh paths gives rise to a *Ricean fading* channel. Surface reflections added to a direct path on a satellite- or aircraft-to-ground link will cause Ricean fading.

Examples of simulation of multipath channels may be found in [8.28] to [8.34]. The multipath model, in all of its many variations, is readily constructed from elementary processes—delays, multipliers, adders, noise generators—that are included in most transmission-simulation programs.

Receivers for multipath-propagated signals often employ *diversity reception* to combat ill effects of multipath. A diversity system contains two or more receivers, each processing an independently obtained version of the multipath signal. Receiver outputs are combined to produce a composite version of the signal that is better than any of the individual versions.

Each separate receiver is conventional; only the diversity combiner needs additional types of simulation processes. Combiners typically are relatively simple devices and can be simulated by compositions of elementary processes that will be included anyhow in a simulation suite. Computer resources may be strained though by the need to simulate multiple independent propagation paths and multiple receivers prior to the combiner. Reprints of papers on diversity can be found in [8.35]. Also, [1.2, Sec.7.4] explains diversity concisely.

8.3 SPREAD-SPECTRUM PROCESSES

In some systems the spectrum of the transmitted signal is deliberately spread to a bandwidth much wider than the symbol rate. In a spread-spectrum receiver, the signal is despread by an operation complementary to the spreading process in the transmitter. The desired signal is then recovered in its original narrow bandwidth, while narrowband interferers, or other uncorrelated signals, are spread much wider than the desired signal. Spreading, along with subsequent despreading, is effective as a jamming countermeasure or as a means of multiple access [code-division multiple access (CDMA)] for numerous users to share the same spectrum.

Two kinds of spreading are common: frequency hopping (in which the frequency of transmission is changed often over a wide span) and "direct sequence" spreading (in which the narrowband signal is multiplied—that is, modulated—by a wideband pseudorandom sequence).

Spread-spectrum researchers have produced a huge literature. A useful overview is available in [8.36] and [8.37], while greater detail may be found in [8.38]. Citations in these references point to an enormous number of additional articles and books on the subject.

Spreaders and despreaders are modulators and demodulators, respectively. Despreaders have to be synchronized to the incoming spread signal. These processes work in the same fundamental manner as ordinary (nonspread) modulators, demodulators, and synchronizers. The most distinctive feature of spread-spectrum simulations is that they require a very large number of samples for each symbol of the narrowband signal. There is a strong incentive to employ a low sampling rate in the narrowband portions of the model and to upsample to a high rate only in the wideband portions. Chapter 16 examines sample-rate changing.

Although spread-spectrum methods are not treated further in this book, the STÆDT program has been employed successfully to simulate spread-spectrum systems.

8.4 AFTERWORD

Simulations of phase noise, of fading channels, and of spread-spectrum systems have one important feature in common: They all contain processes that proceed at vastly different time scales. As examples:

- Phase-noise fluctuations are typically very slow compared to the data-symbol rate.
- Fading is typically slow compared to the data-symbol rate.
- Spectrum-spreading bandwidth is typically large compared to the data-symbol rate, and bandwidths of despreading synchronizers typically are small compared to the symbol rate.

Sampling rate has to be large enough to accommodate the wideband signals, while duration of the simulation run has to be long enough compared to the slowest process to assure that adequate statistics are gathered. Extremely long simulation runs may be required. Careful thought and planning need to be given to the requisite duration before one embarks upon lengthy simulations.

This incomplete overview of reception has identified various elementary processes that are required for simulation of the simplest channels and signals and for more complicated channels and signals as well. Upon examining more complicated channels and signals, it was found that their simulations employ a greater number of processes, but to a large extent these can be the same kinds of processes needed for the simplest applications. Only a few instances of different kinds of processes (e.g., special noise generators for impulse or shot noise) arise for special situations.

From these observations we have concluded that the foundations of simulation can be learned by concentrating upon the elementary processes employed in all models; that is the course followed in the remainder of this book. Specialized processes that obey the same underlying principles, but differ in particulars, are left to more advanced study.

9

GENERATION OF NOISE

Additive white Gaussian noise (AWGN) occurs extensively as a disturbance to reception of data signals. Any consequential simulation program includes a process to generate AWGN samples.

Gaussian samples are obtained from transformations on pairs of uniformly distributed samples. Thus a fundamental problem of noise generation is to produce uniformly distributed samples with acceptable characteristics. In this section we first discuss how uniformly distributed samples are generated and then how to derive Gaussian samples with the desired variance. Upon occasion, the uniformly distributed samples themselves might be used for other purposes and so should be accessible to the simulationist. One example of such use appeared in Chapter 7 in connection with discontinuous-phase FSK.

9.1 UNIFORMLY DISTRIBUTED SAMPLES

Desired Properties

Just as random data sequences are approximated by deterministic pseudorandom periodic shift-register sequences, so are random numbers generated by deterministic pseudorandom periodic recursive numerical operations. What properties are wanted from such sequences?

- The samples in the sequence and in most subsequences of adequate length should indeed be very nearly uniformly distributed.
- The period of the generator should be extremely long so that the sequence would not repeat inadvertently in a simulation of any likely duration.

- Granularity of amplitude quantization should be so fine as to appear virtually continuous to the systems to be simulated. Granularity is closely related to length of period in practical generator algorithms.
- It should be possible, by simple means, to produce identically the same sequence in simulation after simulation.
- The sequence should appear random.

Knuth [9.1] devotes a long chapter to random numbers, with emphasis on methods for deterministic generation of random-appearing, uniformly distributed numbers, and the many pitfalls associated therewith. He also describes various tests that probe the quality of randomness of a sequence. Read that chapter to understand the importance of knowing the pedigree of a random-number generator.

Uniform Number Generators

The most common algorithm for deterministic generation of uniformly distributed pseudorandom numbers is the *linear congruential method*, a simple recursion of the form

$$\xi(n) = [a\xi(n-1) + c] \bmod m \tag{9.1}$$

where n is the sample index, $\xi(n)$ the pseudorandom variate, a a fixed multiplier, c an additive constant, and m the modulus of the generator. All quantities are integers.

With proper choice of m, a, and c, the discrete integers $\{\xi(n)\}$ are very nearly equally distributed between 0 and m. To provide numbers $\lambda(n)$ distributed between 0 and 1, the output of the generator is taken as a floating-point fraction

$$\lambda(n) = \xi(n)/m \tag{9.2}$$

To obtain a long period and fine quantization, it is clear that m should be large. By choosing the multiplier a correctly [9.1, Sec. 3.2.1.2], it is possible to obtain sequences with period $m-1$ and to set $c = 0$. In personal computers, (short) signed integers are stored in words of 16 bits while long signed integers have 32 bits. Using only the positive integers, these word lengths afford maximum sequence lengths of $2^{15} - 1$ and $2^{31} - 1$, respectively, from a simple linear recursion. The first is not long enough; it will repeat during long simulation runs and its granularity is not adequately fine.

A sequence length of $2^{31} - 1$ would be attractive for many practical simulations on a small computer. However, for STÆDT we have followed [9.2], which describes composite generators that deliver much longer sequences. The STÆDT number generator UNIF contains three linear congruential recursions, each arranged to produce a subsequence of period slightly less than 2^{15}. Outputs of the three are combined to form a sequence of length about 6.9×10^{12}. Few simulations will run long enough to consume so many samples. Reference [9.2] applies some of Knuth's randomness tests to the composite generator and concludes that its performance is excellent.

A noise generator requires a starting value—a *seed*—for $\xi(n)$. UNIF provides default seeds for all three composite generators; the same sequence will be generated for every simulation run of a given length if the default seeds are employed. To generate different sequences, provision is made for user specification of one of the three seeds.

9.2 GAUSSIAN NOISE GENERATION

Transformation of uniform variates to Gaussian should be accomplished by the computer, invisibly to the user. Appropriate scaling of the resulting Gaussian samples requires user inputs. Sections 9.2.1 and 9.2.2 are concerned with the generation of "white" Gaussian noise: that is, sequences of independent, identically distributed complex samples with Gaussian statistics. In Secs. 9.2.3 and 9.2.4 we briefly treat "colored" Gaussian noise (white noise passed through a filter) and the related problems of simulating phase noise.

9.2.1 Transformation Algorithm

Knuth [9.1, Sec. 3.4] explains various methods for producing sequences with nonuniform distributions, employing samples from a uniform generator. Several Gaussian algorithms are included. STÆDT employs the polar method of [9.1, p. 117] for uniform-to-Gaussian transformation. Each call to AWGN generates two standard Gaussian samples ν_1 and ν_2, each with zero mean and unit variance. The samples can be used sequentially if real noise is being simulated, or as a pair if complex noise is needed. Noise samples appear independent, so the spectrum of the noise ensemble is white. The transformation proceeds as follows:

TRANSFORMATION, UNIFORM-TO-GAUSSIAN

- Input: two independent random samples λ_1 and λ_2, uniformly distributed on $(0, 1)$.
- Rescale: $V_i = 2\lambda_i - 1; i = 1, 2$. V_i is uniformly distributed on $(-1, 1)$.
- Compute magnitude: $S = V_1^2 + V_2^2$.
- Test magnitude: If $S \geq 1$, discard current sample pair and start over with new λ_1, λ_2 input.
- Compute Gaussian samples:

$$\nu_1 = V_1 \sqrt{\frac{-2 \ln S}{S}}, \quad \nu_2 = V_2 \sqrt{\frac{-2 \ln S}{S}}$$

9.2.2 Scaling

Unit variance (power) of the standard Gaussian samples can easily be scaled to any desired variance σ_ν^2, simply by multiplying each sample by σ_ν. The noise process AWGN has an optional provision for accepting a user-specified scaling parameter. Unfortunately, information readily available to a user is often insufficient for specifying a scaling factor, unless one performs arcane, error-prone calculations. Instead, it is much more convenient for the user to specify a required signal-to-noise ratio (SNR) and then let the computer calculate the necessary noise scaling based upon formulas and measurements described below.

Since computer quantities are all dimensionless, the SNR specified should also be dimensionless. For data signals in white noise, the applicable measures are the energy ratios E_b/N_0 or E_s/N_0, where E_b is the average energy per bit, E_s the average energy per symbol, and N_0 the one-sided spectral density of the white noise. The user decides whether E_b, E_s, or some other signal energy is to appear in the SNR definition.

Let's work out the pertinent equations first for a real signal and noise and then establish the equations for complex signals and noise. Power and energy were treated in Secs. 2.3 and 5.4 and Appendix 5C. Material from those sections is applied here to establish the procedures for scaling noise.

Real Signals and Noise

Consider a block of N samples $\{x(n)\}$ of a real signal. Energy in that block is given by

$$E_N = t_s \sum_{n=1}^{N} x^2(n) \tag{9.3}$$

where t_s is the time interval between samples. Remember that t_s is not known to the computer; nevertheless, carry it along for now.

Block energy E_N depends on the data pattern in the block. Change the data pattern and E_N might change, too. Measurements of E_N exhibit statistical scatter whose extent is a function of the signal format; energy measurements are independent of data pattern in a constant-envelope signal but are strongly data dependent in a multi-amplitude signal. It is the user's responsibility to supply an adequately representative signal so that a measurement of E_N lies close to the average energy of the signal ensemble. See Sec. 7.1.2 for a discussion of representative data sequences.

Assume that there are M_s samples per symbol so that the block contains N/M_s symbols. If E_s is the ensemble average energy per symbol, an *estimate* of E_s can be taken from the block energy measurement as

$$\hat{E}_s = \frac{E_N}{N/M_s} = t_s \frac{M_s}{N} \sum_N x^2(n) \tag{9.4}$$

Clearly, $\hat{E}_s$ exhibits statistical scatter.

Noise variance σ_ν^2 is an ensemble-average statistic; measurements on any finite sequence $\{\nu(n)\}$ of noise samples would have statistical scatter about σ_ν^2. An exact ensemble scaling is possible on the noise because it is generated with an exactly known (unity) ensemble variance. Signals are seldom generated this precisely, so signal energy must be measured in a finite block to get an estimate of its ensemble value.

Frequency span of the simulation is $1/t_s$ whence the constant, two-sided noise spectral density over that span is

$$\Psi_\nu(f) = \frac{N_0}{2} = \frac{\sigma_\nu^2}{1/t_s} = \sigma_\nu^2 t_s \qquad (9.5)$$

A constant spectral density applies only to the noise ensemble. Wild fluctuations would be observed in the estimated spectral density (periodogram) computed for any finite sequence delivered by the noise generator (see, e.g., Fig. 2.15).

With constituents from (9.4) and (9.5), the estimated symbol energy-to-noise density ratio is

$$\frac{\hat{E}_s}{N_0} = \frac{t_s M_s/N}{2\sigma_\nu^2 t_s} \sum_N x^2(n) = \frac{M_s}{2N\sigma_\nu^2} \sum_N x^2(n) \qquad (9.6)$$

Sampling interval t_s has canceled out from the energy ratio. Observe that

$$\frac{1}{N} \sum_N x^2(n) = \hat{P}_N \qquad (9.7a)$$

is the signal power estimated over the N-length time-domain block. Alternatively, by Parseval's theorem, the power in a frequency-domain block is

$$\hat{P}_N = \frac{1}{N^2} \sum_N |X(k)|^2 \qquad (9.7b)$$

Suppose that estimated signal power has been measured and M_s is known. The task at hand is to set σ_ν^2 to establish a user-specified energy ratio (E_s/N_0). That SNR is a simulation parameter, not a measured estimate. The needed scaling factor σ_ν for the noise generator is obtained from (9.6) as

$$\sigma_\nu^2 = \frac{M_s}{2N(E_s/N_0)} \sum_N x^2(n) = \frac{M_s}{2(E_s/N_0)} \hat{P}_N \qquad (9.8)$$

The user enters the desired ratio (E_s/N_0); the computer measures $\hat{P}_N$ and knows M_s.

The computer calculates σ_v^2 and applies the desired scaling to the noise generator without the user ever knowing, or caring about, the specific value of σ_v. Notice that the empirically determined noise scaling factor σ_v exhibits statistical scatter because of the necessity to measure the data-dependent quantity $\hat{P}_N$.

To set the noise in terms of the bit energy E_b, let there be b bits per symbol. Then $E_b = E_s/b$ and the noise scaling factor needed for a specified ratio (E_b/N_0) is found from

$$\sigma_v^2 = \frac{M_s}{2bN(E_b/N_0)} \sum_N x^2(n) = \frac{M_s}{2b(E_b/N_0)} \hat{P}_N \qquad (9.9)$$

If the symbols are taken from an M-ary alphabet and the message has been encoded with code rate R_c, then

$$b = R_c \log_2 M \qquad (9.10)$$

With this definition, E_b is the energy per information bit, not signaling bit.

Complex Signals and Noise

Several factors of 2 arise in the treatment of complex signals and noise. Their origins are not readily distinguished from one another nor are they easily described in words; they are a potential source of long-lasting confusion. To avoid difficulties in trying to remember different factors of 2, the expressions for E_s/N_0, E_b/N_0, and the associated noise-generator scaling factors for complex signals are derived in Appendix 9A.

The results are

$$\sigma_u^2 = \sigma_v^2 = \frac{M_s}{2N(E_s/N_0)} \sum_N |z(n)|^2 = \frac{M_s}{2(E_s/N_0)} \hat{P}_N$$

$$= \frac{M_s}{2bN(E_b/N_0)} \sum_N |z(n)|^2 = \frac{M_s}{2b(E_b/N_0)} \hat{P}_N \qquad (9.11)$$

which formally has the same pattern as (9.8) and (9.9); the factors of 2 all cancel out in the expressions for the individual noise variances σ_u^2 and σ_v^2.

In (9.11), $z(n)$ represents the samples of the signal's complex envelope, and the complex additive noise $w(n) = [u(n) + jv(n)]$ consists of the scaled pairs of noise samples returned from each call to the AWGN generator. Sample pairs are generated internally with zero mean and unit variance for each component, and then the same standard deviation is applied as a scale factor to each component.

9.2.3 Colored Noise

Gaussian noise has the unique property that it remains Gaussian after being passed through any linear process, such as a filter. Thus, to generate Gaussian noise with any specified nonconstant spectrum, just pass white Gaussian noise through a filter with the desired frequency response. But this approach might be overly simplistic in some applications. Reference [9.3] contends that for correct simulation, the autocorrelation function of the discrete-time colored noise sequence should sample the continuous-time autocorrelation function of the continuous-time colored noise that is being simulated. That equality of autocorrelations does not come about from matching of frequency responses, because of aliasing. If your application demands that the correct autocorrelation function be simulated, refer to [9.3] before setting out to generate colored noise.

9.2.4 Phase Noise

Spectrum of phase noise is often represented as a finite summation of components of the form

$$\Psi_\phi(f) = \sum_\nu \frac{h_\nu}{f^\nu} \qquad (9.12)$$

where the h_ν are constants that define the noise spectrum of a particular device. It is commonplace to assign integer values from 0 to 4 to ν, but (with exceptions) there is no reason to suppose that integer values have any better validity than fractional values.

Noise with a spectrum shape of h_ν/f^ν is variously known as fractional noise, flicker noise, or pink noise. The very meaning of *spectrum* is called into question for this class of noise, since the noise is nonstationary and has no autocorrelation function for $\nu \geq 1$. Reference [9.4] offers an interpretation of spectrum as the noise variance observed through a narrow bandpass filter.

Generate phase-noise components with slope $\nu < 2$ directly and use them to modulate the phase of a signal in a phase rotator. See Sec. 10.1.1 for phase rotators. To generate phase noise components with slope $\nu \geq 2$, generate fractional noise with slope $0 \leq \nu < 2$ and apply that as part of the frequency-control word of a number-controlled oscillator (NCO; see Secs. 7.4 and 10.2.1). The NCO integrates its input and so increases the spectral slope by 2. Output phase of the NCO is applied to a phase rotator that modulates the phase of a signal. Approximations to fractional-spectrum Gaussian noise with slopes $0 < \nu < 2$ can be generated with the aid of filters described in [9.3].

9.3 NON-GAUSSIAN VARIATES

Sometimes random numbers with a distribution other than uniform or Gaussian are needed. See Chapter 8 for discussion of non-Gaussian impulse and shot noise and references for their generation. Knuth [9.1] describes methods for generating numbers with exponential, gamma, beta, chi-square, F, t, geometric, binomial, and Poisson distributions. They all work by transforming uniform deviates according to particular algorithms. Random numbers with these distributions are rarely needed for a signal-transmission simulation.

Impulse noise has to be simulated for some applications. Because impulse noise is non-Gaussian, its modeling and analysis are rather more intricate than the well-established methods for Gaussian noise. The tails of the distribution of impulse noise are much heavier than the tails of Gaussian noise. Reference [9.5] cites numerous publications on the subject.

In that reference, a class of distributions known as *symmetric α-stable* (SαS) is proposed for simulation of impulse noise. Depending on the value of the α parameter of the distribution, the class includes Gaussian ($\alpha = 2$) and Cauchy ($\alpha = 1$) distributions as special cases. By relatively simple means outlined in [9.5], it is possible to produce SαS variates from uniformly distributed variates. If you need to simulate impulse noise, refer to [9.5] for methods and representative results.

9.4 KEY POINTS

9.4.1 Noise Generation

Complex white Gaussian noise samples typically are generated in two steps:

1 Generate a pair of independent variates, uniformly distributed on (0,1).
2 Transform the uniform pair to a zero-mean Gaussian pair, with unit variance.

Pseudorandom uniform variates are generated by linear-congruential algorithms. Unit-variance Gaussian samples have to be scaled to the variance required in the simulation. More often than not, a user wants a particular signal-to-noise ratio and does not directly know the noise variance required. Auxiliary processes are used to measure the signal and to calculate the correct noise variance from the user's specfication of desired signal-to-noise ratio. Colored Gaussian noise is generated by passing white Gaussian noise through a linear filter. Many kinds of non-Gaussian noise can be generated by transformations on uniform variates.

9.4.2 STÆDT Processes

AWGN: additive Gaussian noise

LETS: various processes on data signals, some of which might be useful for transforming uniform variates to other distributions

POWR: measures power of sample sequence

SNR: calculates scaling for AWGN from measured signal power and user specification of signal format and desired signal-to-noise ratio

UNIF: uniformly distributed numbers

APPENDIX 9A: COMPLEX ENVELOPE OF BANDPASS GAUSSIAN NOISE

The noise spectra shown in Fig. 9A.1 will help clarify the several factors of 2 that are encountered.

Real, Lowpass Noise

Start with the spectrum of real, Gaussian, lowpass, time-continuous noise $\eta(t)$ as in line a of Fig. 9A.1. A real sequence $\{\eta(n)\}$ from the Gaussian noise generator is a sampled version of this noise. Spectral density $\Psi_\eta(f)$ is constant at $N_0/2$ for $|f| < B$ and zero for $|f| > B$. (Division by 2 appears because N_0 is the customary notation for one-sided spectral density, whereas the two-sided spectrum is plotted here.) Noise

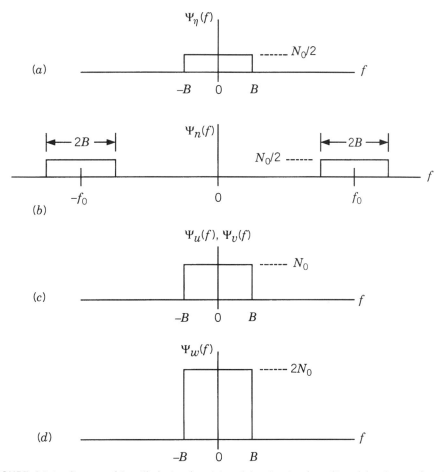

FIGURE 9A.1 Spectra of bandlimited noise: (a) real baseband noise; (b) real bandpass noise; (c) quadrature components of bandpass noise; (d) complex envelope of bandpass noise.

variance is the integral (area) of the spectrum

$$\sigma_\eta^2 = 2B \times N_0/2 = BN_0 \tag{9A.1}$$

Alias-free sampling can be accomplished with a sampling rate

$$1/t_s = 2B \tag{9A.2}$$

With this choice of sampling rate, spectral images in the periodic spectrum of the sampled noise will abut one another without overlaps or gaps. Although the time-continuous noise is bandlimited, the spectrum of the sampled noise is constant at all frequencies; the noise samples are white.

Real, Bandpass Noise

Bandpass Gaussian noise was introduced in Chapter 2, equations (2.71) to (2.73). The noise can be represented in real format as

$$n(t) = u(t) \cos 2\pi f_0 t - v(t) \sin 2\pi f_0 t \tag{9A.3}$$

where the nomenclature $u(t)$ and $v(t)$ has been substituted for the quadrature components $n_c(t)$ and $n_s(t)$ of (2.71).

A spectral representation of bandpass noise is shown in line b of Fig. 9A.1. The spectral density $\Psi_n(f) = N_0/2$ (the same as for the real, lowpass noise) for $||f| - f_0| < B$ and is zero otherwise. Noise variance is

$$\sigma_n^2 = 2 \times (2B \times N_0/2) = 2BN_0 \tag{9A.4}$$

which is double the variance of the lowpass noise in line a. This factor of 2 arises because the bandwidth of the bandpass noise is defined to be double that of the lowpass noise. A reason for the bandwidth definition will emerge shortly.

To perform alias-free sampling of the bandpass noise in line b requires a sampling rate $1/t_s = 2(f_0 + B)$. Sampled noise with this spectrum cannot be generated by simple noise generators alone; additional measures, such as spectral-shaping filters, must also be applied. When real signals are simulated, the real lowpass noise in line a [of the same spectral density $N_0/2$ but increased bandwidth $2(f_0 + B)$] is much more practical to generate. The real bandpass model of (9A.3) and line b is essential for analysis and a vital link in formulation of the complex envelope, but it is not quite so easy to simulate.

Baseband Components

The two quadrature components of the noise, $u(t)$ and $v(t)$, have identical spectral densities $\Psi_u(f)$ and $\Psi_v(f)$, as defined in (2.74). [The cross spectrum $\Psi_{uv}(f) \equiv 0$

APPENDIX 9A: COMPLEX ENVELOPE OF BANDPASS GAUSSIAN NOISE   271

if $u(t)$ and $v(t)$ are referenced to f_0, because of the symmetry of $\Psi_n(f)$ about f_0.] These component spectra are sketched in Fig. 9A.1c. Each component has spectral density N_0 for $|f| < B$ and is zero for $|f| > B$. Their variances are

$$\sigma_u^2 = \sigma_v^2 = 2BN_0 = \sigma_n^2 = 2\sigma_\eta^2 \tag{9A.5}$$

The spectra of line c can be sampled, without aliasing, at a sample rate $1/t_s = 2B$—the same sampling rate as that needed for the real noise of line a. The bandwidth of bandpass noise in line b was defined as shown, so that these sampling rates would coincide. The baseband components can be produced as sampled sequences $\{u(n)\}$ and $\{v(n)\}$ by the same noise generator used for the real noise samples $\{\eta(n)\}$.

Complex Envelope

The complex envelope of the bandpass noise is defined as

$$w(t) = u(t) + jv(t) \tag{9A.6}$$

and its spectrum $\Psi_w(f)$ is sketched in Fig. 9A.1d. That spectrum has density $2N_0$ for $|f| < B$ and zero for $|f| > B$. The variance of the complex envelope is

$$\sigma_w^2 = 2N_0 \times 2B = 4N_0B = 2\sigma_n^2 \tag{9A.7}$$

Here is another factor of 2; it is the same power/energy doubling that always appears in complex envelopes and that was introduced in Chapter 2.

SNR: Power Ratio

Consider a real time-continuous bandpass signal

$$s(t) = x(t)\cos 2\pi f_0 t - y(t)\sin 2\pi f_0 t \tag{9A.8}$$

which has a complex envelope

$$z(t) = x(t) + jy(t) \tag{9A.9}$$

These were introduced in Chapter 2.
  Variances (average powers) of the signal and its envelope are

$$\sigma_s^2 = \tfrac{1}{2}\sigma_x^2 + \tfrac{1}{2}\sigma_y^2 \tag{9A.10}$$
$$\sigma_z^2 = \sigma_x^2 + \sigma_y^2 \tag{9A.11}$$

The ratios of these signal powers to the corresponding noise powers from (9A.4) and (9A.7) are

$$\frac{\sigma_s^2}{\sigma_n^2} = \frac{\frac{1}{2}(\sigma_x^2 + \sigma_y^2)}{2BN_0}$$

$$\frac{\sigma_{\tilde{s}}^2}{\sigma_w^2} = \frac{\sigma_x^2 + \sigma_y^2}{4BN_0} \tag{9A.12}$$

which demonstrates that the power SNR is the same for the real signal and noise as it is for their complex envelopes. The factors of 2—those that appear when translating between a real signal and its complex envelope—cancel out from the ratios.

SNR: Energy Ratio

The signal $s(t)$ consists of symbols generated at an average rate of $1/T$, so the average energy per symbol is

$$E_s = \frac{\sigma_s^2}{1/T} = \sigma_s^2 T \tag{9A.13}$$

An estimate for σ_s^2 can be obtained from the average of its squared samples and (9A.10) as

$$\hat{\sigma}_s^2 = \frac{1}{2}\frac{1}{N}\sum_N [x^2(n) + y^2(n)] \tag{9A.14}$$

from which the energy per symbol can be estimated by

$$\hat{E}_s = \hat{\sigma}_s^2 T = \frac{T}{2N}\sum_N [x^2(n) + y^2(n)]$$

$$= \frac{T}{2N}\sum_N |z(n)|^2 \tag{9A.15}$$

From (9A.5), variance of the noise components is $\sigma_u^2 = \sigma_v^2 = 2BN_0$, and from (9A.2), $B = 1/2t_s$. Combining these relations gives

$$N_0 = \sigma_u^2 t_s \tag{9A.16}$$

so that the estimated energy ratio is

$$\frac{\hat{E}_s}{N_0} = \frac{(T/2N) \sum_N |z(n)|^2}{\sigma_u^2 t_s} \qquad (9A.17)$$

Using $T/t_s = M_s$, the noise variance scaling $\sigma_u^2 = \sigma_v^2$ needed in the AWGN generator to establish an energy ratio (E_s/N_0) is

$$\sigma_u^2 = \frac{M_s}{2N(E_s/N_0)} \sum_N |z(n)|^2 = \frac{M_s}{2(E_s/N_0)} \hat{P}_N \qquad (9A.18)$$

This is the same form as in (9.8) for the variance of real noise; the various factors of 2 have all canceled out from the formal expression.

10

SHIFTERS AND INTERPOLATORS

In this chapter we treat phase shifters, frequency shifters, time shifters, and interpolators. These functions are closely related to one another in that:

- They all operate on data signals.
- They are regulated by control signals.
- They are often associated with modulators, demodulators, and receiver synchronization.
- Interesting dual formats exist such that similar computations are applicable to different functions.

10.1 PHASE SHIFTERS

Shifting the phase of a complex signal is conceptually a simple operation, explained in detail below. Shifting the phase of a real signal is not as simple. Only the difficulties are explained in this section; a method for phase shifting a real signal is deferred to Sec. 10.2. Phase shifting, and frequency shifting as well, are meaningful only with carrier-modulated signals (real or complex) or with complex baseband signals.

10.1.1 Phase Shifting of Complex Signals

An input complex signal $z_{\text{in}}(n)$ is shifted in phase by multiplying it by a unit-amplitude complex exponential $\exp[j\theta(n)]$ to generate a phase-shifted output

$$z_{\text{out}}(n) = z_{\text{in}}(n)e^{j\theta(n)} \qquad (10.1)$$

The phase $\theta(n)$ can take on a different value for each n; $\theta(n)$ is a *control signal* generated by some other process in the simulation. For many simulations, though, $\theta(n)$ will be a fixed parameter, established at the start of a run and held constant thereafter. Examples of both conditions will be seen in this and later chapters.

Phase shifting is a zero-memory operation; $z_{out}(n)$ depends only on the nth samples of z_{in} and θ and not on any earlier or later samples. Phase shifting causes little or no bandwidth expansion, provided that $\theta(n)$ is slowly varying compared to $z_{in}(n)$. Thus a sampling rate sufficient for z_{in} is usually sufficient for the process (10.1) and z_{out}. If $\theta(n)$ were rapidly varying, the sampling rate would have to be large enough to accommodate the sum of the bandwidths of the signals $\exp[\theta(n)]$ and $z_{in}(n)$.

Computation

Phase shifting is extremely simple to compute if z_{in} and z_{out} are in polar coordinates; $\theta(n)$ is simply added to the phase of z_{in} to produce z_{out}. However, for reasons explained in Sec. 6.1.1, complex signals are much more likely to be handled in rectangular coordinates in the form $z(n) = x(n) + jy(n)$. Actual computations then become

$$x_{out}(n) = x_{in}(n)\cos\theta(n) - y_{in}(n)\sin\theta(n)$$
$$y_{out}(n) = x_{in}(n)\sin\theta(n) + y_{in}(n)\cos\theta(n) \qquad (10.2)$$

In many applications, the desired operation is $z_{out} = z_{in}\,e^{-j\theta}$; the direction of rotation is to be reversed from (10.1). That can be accomplished either by conjugating $e^{j\theta}$ prior to performing the multiplications of (10.2), or by negating θ before calculating $\sin\theta$ and $\cos\theta$.

Phase Shifting in the Frequency Domain

Equation (10.1) is a time-domain operation. A dual process in the frequency domain is

$$Z_{out}(k) = Z_{in}(k)e^{j\phi(k)} \qquad (10.3)$$

which formally is identical to (10.1) and can be accomplished by the same process. However, $\phi(k)$ is a *frequency-varying* phase—an allpass filter function—and is not time varying. Most emphatically, $\phi(k)$ is not the Fourier transform of $\theta(n)$ from (10.1). Thus although (10.3) is formally the same as (10.1), these two formulas describe completely different operations on the signal.

For the special case of a fixed phase, independent of time index n or frequency index k, the phase shift can be applied equally well in the time or frequency domains—that is, either by (10.1) or (10.3). Both processes then accomplish identically the same task of rotating the complex signal by a fixed angle θ or ϕ.

10.1.2 Phase Shifting of Real Signals

A real passband signal is represented as

$$s_{\text{in}}(t) = x(t) \cos(2\pi f_c t + \theta_c) - y(t) \sin(2\pi f_c t + \theta_c) \tag{10.4}$$

where $x(t)$ and $y(t)$ are the real and imaginary components of the complex envelope, f_c the carrier frequency, and θ_c an arbitrary fixed phase. Continuous-time notation is used because it is slightly more compact, but it should be understood that the computer representation $\{s_{\text{in}}(n)\}$ is time discrete.

Phase shifting by a phase $\theta(t)$ gives a shifted signal representation

$$s_{\text{out}}(t) = x(t) \cos[2\pi f_c t + \theta_c + \theta(t)] - y(t) \sin[2\pi f_c t + \theta_c + \theta(t)] \tag{10.5}$$

Equation (10.5) is all very well for analytical purposes, but the computer only has $s_{\text{in}}(t)$ and $\theta(t)$ [actually, $\{s_{\text{in}}(n)\}$ and $\{\theta(n)\}$] to work with if the signal representation is real. It does not know $x(t), y(t), f_c$, or θ_c, so cannot add $\theta(t)$ to the arguments of the sine and cosine as required by (10.5). Equation (10.5) is not obtained by multiplying $s_{\text{in}}(t)$ by $e^{j\theta(t)}$.

Analog-hardware phase shifters are implemented by various means including (1) the Armstrong method [10.1, 10.2] whereby a suppressed-carrier double-sideband amplitude-modulation signal is converted to phase modulation by quadature reinjection of a carrier and subsequent limiting, (2) voltage-variable delay or reactance networks, or (3) phase-locked loops [10.3]. These all have limited ranges of phase deviation compared to an unbounded range for (10.1) and are subject to distortion. Any of them might be simulated in special situations if they were an integral part of a system under investigation, but none is versatile or simple enough to provide a model for general-purpose simulation with real-signal representations. Phase shifting of real signals will be considered again below in conjunction with frequency shifting of real signals.

10.2 FREQUENCY SHIFTERS

Frequency shifting a complex signal in the time domain proves to be an extension of phase shifting. A frequency shifter consists of a phase shifter driven by a number-controlled oscillator (NCO). Frequency shifting in the frequency domain is quite different and requires interpolation to achieve fine resolution. Frequency shifting with the discrete Fourier transform (DFT) for circular-block frequency-domain signals is introduced and shown to be hazardous. Frequency shifting of real signals is examined, particularly in regard to simulating spectral foldover. One means of imposing a phase shift on a real signal is found as an incidental benefit.

10.2.1 Frequency Shifting of Time-Domain Complex Signals

Shift Operations

A frequency shift f_ϵ is imposed upon a complex signal $z_{\text{in}}(n)$ by the process

$$z_{\text{out}}(n) = z_{\text{in}}(n) e^{j 2\pi n f_\epsilon t_s} \tag{10.6}$$

Whereas phase shifting with a constant $\theta(n)$ *rotates* the entire complex signal about the time axis, frequency shifting with a constant f_ϵ uniformly *twists* the complex signal in a spiral around the time axis.

The argument of the exponential in (10.6) is just a phase that grows with time index n; frequency shifting can be performed in the time domain by a phase shifter, provided that the accumulating phase angle can be produced by other processes. Since (10.6) and (10.1) employ the same phase-shift operation, phase shifting and frequency shifting can be combined into a single rotation just by summing the phase angles $\theta(n)$ and $2\pi n f_\epsilon t_s$, sample by sample. Computer quantities are all normalized; neither f_ϵ nor t_s are known separately within the shift process. Only their product $f_\epsilon t_s$ is available.

Rotation speed (amount of frequency shift) and even direction will be aliased if $|f_\epsilon t_s|$ is too large. Avoidance of aliasing requires that $|f_\epsilon t_s| < 0.5$. This limit makes no allowance for nonzero bandwidth of the signal being shifted; a lesser bound on $|f_\epsilon t_s|$ must be observed if spectral wraparound is to be avoided.

Phase Accumulation: NCOs

In (10.6) the offset frequency f_ϵ itself might be time varying—that is, a function of index n. In that situation, $f_\epsilon(n)$ is incorporated into the phase $2\pi p(n)$, as defined below. Phase accumulation was encountered in Sec. 7.4.1, where it was recommended that the accumulation be performed recursively in a number-controlled oscillator (NCO) according to the difference equations

$$p(n+1) = [p(n) + W(n)] \bmod\text{-}1$$
$$\theta(n) = 2\pi p(n)$$
$$p(0) = \theta(0)/2\pi$$
$$W(n) = f_\epsilon(n) t_s \tag{10.7}$$

where $\theta(n)$ is the time-varying phase delivered to the phase rotator that performs the frequency shift.

The accumulated-phase state variable $p(n)$ is measured in cycles and reduced modulo 1 cycle. Reduction modulo 1 cycle is permissible since the sine and cosine are periodic in one cycle. Reduction is necessary so that $p(n)$ not accumulate so large as to overflow the computer's word size in long simulations. Measurement of $p(n)$ in cycles, instead of radians, is convenient when thinking of frequency in hertz instead of radians per second. Multiplication by 2π is necessary because most programming languages require the angle arguments of sine and cosine to be in radians.

Frequency control of the NCO is effected by $W(n)$, which in general can vary from sample to sample. A steady frequency is generated by setting $W(n)$ to a constant $f_\epsilon t_s$. The quantities $\theta(n)$, $p(n)$, and $W(n)$ are all real control signals, whose treatment is considered in Chapter 14.

10.2.2 Frequency Shifting of Frequency-Domain Signals

Corresponding to a complex time-domain signal $z_\text{in}(n)$ there is a frequency-domain signal $Z_\text{in}(k)$, related by the discrete Fourier transform (DFT). Both $z_\text{in}(n)$ and $Z_\text{in}(k)$ exist in N-point blocks, which may be regarded interchangeably as periodic or circular (see Secs. 5.2.2 and 5.3.2).

Shift Operation

To find the frequency-domain equivalent of (10.6), take the DFT to obtain

$$Z_\text{out}(k) = \sum_{n=0}^{N-1} z_\text{out}(n) e^{-j2\pi nk/N}$$

$$= \sum_{n=0}^{N-1} z_\text{in}(n) e^{j2\pi n f_\epsilon t_s} e^{-j2\pi nk/N}$$

$$= \sum_{n=0}^{N-1} z_\text{in}(n) e^{-j2\pi n(k - Nf_\epsilon t_s)/N}$$

$$= Z_\text{in}(k - Nf_\epsilon t_s) \tag{10.8}$$

Time-domain multiplication by $\exp(j2\pi f_\epsilon t_s)$ is equivalent to a frequency-domain translation by $Nf_\epsilon t_s$.

STÆDT includes a process named SHFT that performs a circular shift of any integer number of points on any block, either frequency domain or time domain. SHFT can also be used to perform a frequency translation on frequency-domain filters and not just on signals.

If $Nf_\epsilon t_s$ is not an integer and if a shift is to be performed entirely by frequency-domain operations, the samples of $Z_\text{in}(k)$ have to be interpolated to establish the samples of $Z_\text{out}(k)$. Interpolation is treated below after some other matters have been considered first.

The signal block is circular; if frequency shift is excessive, the signal spectrum disappears off one end of the block and wraps around to the other end of the block. In some instances, wraparound may be an indication that the shift is too large for the frequency span—that the sampling rate is too low. In other circumstances, wraparound has no particular significance; it is simply a manifestation of the circularity of the block.

Fine Resolution

Rather than interpolate $Z_{in}(k)$ to determine $Z_{out}(k)$ for the shift $Nf_\epsilon t_s \neq$ integer, it is possible to accomplish a continously variable shift through transforms, as follows:

1. Take IDFT of $Z_{in}(k)$ to obtain $z_{in}(n)$ as a time-domain circular block.
2. Apply (10.6) to perform a time-domain frequency shift and obtain $z_{out}(n)$.
3. Take DFT of $z_{out}(n)$ to obtain a frequency-domain frequency-shifted $Z_{out}(k)$.

This transform scheme is a way to interpolate frequency with the aid of the DFT; a dual approach for performing time-domain interpolation will be encountered afterward.

At first sight the transform frequency-shifting method seems promising; it would avoid the labor of explicit interpolation on each signal point. To test its behavior, Figs. 10.1 to 10.4 show the time- and frequency-domain plots of a simple exponential signal $\exp(j2\pi n f_\epsilon t_s) = \cos(2\pi n f_\epsilon t_s) + j \sin(2\pi n f_\epsilon t_s)$ with $Nf_\epsilon t_s = 4$ and 4.5.

Each time-domain waveform (Figs. 10.1 and 10.3) shows unexceptional sines and cosines, with 4 cycles in the block in Fig. 10.1 and 4.5 cycles in the block in Fig. 10.3. In a three-dimensional depiction, it would be seen that each complex exponential traces out a spiral: one with 4 turns in its block and the other with 4.5.

The spectrum for $Nf_\epsilon t_s = 4$ appears in Fig. 10.2; it is a single line at $k = 4$, as would be expected from a complex exponential. But when the frequency is shifted

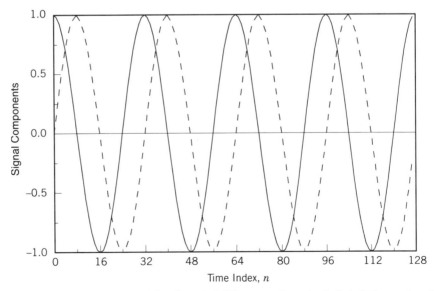

FIGURE 10.1 Complex exponential; $Nf_\epsilon t_s = 4$. Solid curve, real part (cosine); dashed curve, imaginary part (sine).

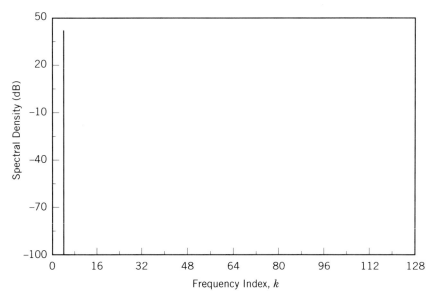

FIGURE 10.2 Spectrum of complex exponential; $Nf_\epsilon t_s = 4$.

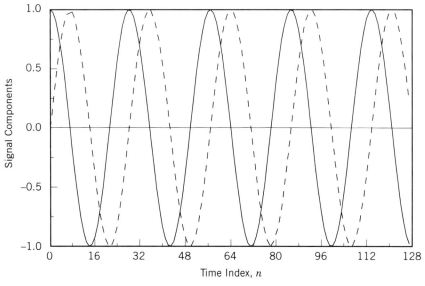

FIGURE 10.3 Complex exponential; $Nf_\epsilon t_s = 4.5$. Solid curve, real part (cosine); dashed curve, imaginary part (sine).

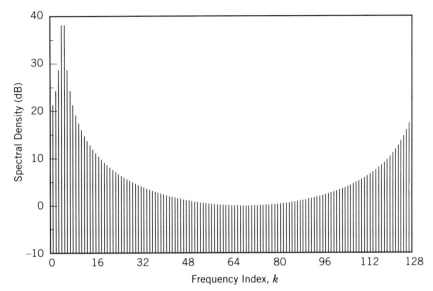

FIGURE 10.4 Spectrum of complex exponential; $Nf_\epsilon t_s = 4.5$.

to $Nf_\epsilon t_s = 4.5$, the outcome is devastating. The shifted spectrum (Fig. 10.4) now has enormous widespread sidebands that were not present originally.

What causes the sidebands to appear? To understand their origin, consider the waveforms in Figs. 10.1 and 10.3. Visualize each as one period of a periodic waveform of infinite extent. In Fig. 10.1, with an integer number of cycles per block, the right edge of each waveshape in one period is continuous with the left edge in the next. Infinite repetition of each block would produce seamless sine and cosine waves.

But the waveforms of Fig. 10.3 are not continuous around the edges. Repetition of the block produces large discontinuities at the seams. Discontinuities have high harmonic content; these are aliased into the broad spectrum that is so startling in Fig. 10.4. As an equivalent viewpoint: The complex three-dimensional spiral of a circular block closes around on itself for an integer number of twists but does not close properly for a noninteger number of twists. Failure to close properly leaves harmonic-rich discontinuities in the waveshapes.

There are many applications in which these sidebands would be unacceptable. The transform method of frequency shifting is not permissible in these situations, except for integer frequency shifts. Frequency-domain frequency shifting should be performed by frequency-domain interpolation to accomplish noninteger shifts. Simulate only integer-frequency shifts by the transform method.

Linear blocks or single-point simulations require time-domain frequency shifting via (10.6). Spectrum measurements on signals shifted in this manner will exhibit spurious sidebands if the analysis block length does not lead to an integer frequency shift. Spectral spreading is familiar from DSP spectral estimation (where the spreading is known as *spectral leakage*). When trying to measure a spectrum, it is common DSP practice to weight signals by time-domain windows [10.4] to suppress the regions of discontinuity. Windowing may not be an acceptable expedient for most simulations because of its severe distortion of the waveform. Windowing for spectrum measurement is treated further in Chapter 18.

10.2.3 Frequency Shifting of Real Signals

The preceding material on frequency shifting applies to complex time-domain signals. Situations arise in which a frequency-shifted real signal has to be simulated, requiring techniques that are modified from those shown so far. In this section we first show why real passband signals might have to be simulated and then show how the simulation can be modeled compactly.

Translation of Real Signals

Consider a real passband signal with carrier frequency f_{in} and phase $\theta_{in}(t)$ to be the input to a receiver. Translation to a lower frequency is achieved by mixing the input against a real local-reference signal with frequency f_r and phase $\theta_r(t)$ to form sum and difference mixer products at frequencies $f_{in}+f_r$ and $f_{in}-f_r = f_d$. Assuming that only the difference frequency is wanted, a filter suppresses the sum frequency components and passes the difference frequency components, delivering the real signal

$$s_{out}(t) = x(t)\cos[2\pi f_d t + \theta_{in}(t) - \theta_r(t)] - y(t)\sin[2\pi f_d t + \theta_{in}(t) - \theta_r(t)] \quad (10.9)$$

Sketches of the pertinent spectra are shown in Figs. 10.5 and 10.6. In Fig. 10.5 the difference carrier frequency f_d is adequately large compared to the modulation

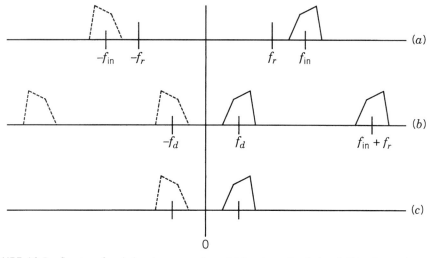

FIGURE 10.5 Spectra of real signals, no overlaps: (*a*) input passband signal; (*b*) mixer output, sum and difference components; (*c*) filter output, difference component only.

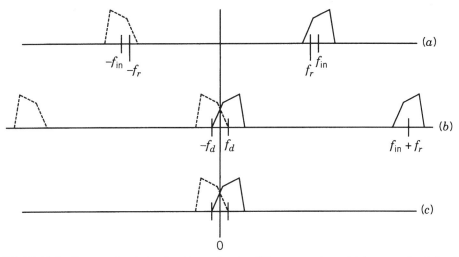

FIGURE 10.6 Spectra of real signals with overlapping difference products: (a) input passband signal; (b) mixer output; (c) filter output.

bandwidth, so the positive and negative portions of the translated-signal spectrum do not overlap. If considered from a one-sided spectral viewpoint, the difference-frequency spectrum does not fold over at zero frequency. This condition—carrier frequency large compared to signal bandwidth—permits the signal to be represented by its complex envelope without regard for the actual carrier frequency.

In Fig. 10.6 the difference frequency is not large enough with respect to the signal bandwidth; the positive and negative portions of the translated spectrum do overlap. Unthinking use of the complex envelope would wrongly neglect the overlap and possibly lead to incorrect evaluation of system performance. The complex envelope representation preserves only one overlapping component of the two shown in the bottom line of Fig. 10.6. The real signal, the signal that actually appears in physical hardware, contains both components. How can the real signal be simulated efficiently?

Representation of Real Signal

If the input frequency f_{in} and reference frequency f_r are large compared to their difference f_d, it is clear that the mixing operation should not be simulated directly; the sampling rate needed to represent both sum- and difference-frequency components would be greatly excessive. All that is wanted anyhow is the difference-frequency signal; no useful purpose is served by simulating a sum-frequency component that is promptly discarded.

The objective, then, is to simulate the real signal of (10.9) without simulating the mixing operation. One approach is just to compute (10.9) directly—a task well within the capabilities of any good simulation program. That is an efficient approach, provided that no further frequency or phase shifts must be performed. (As noted in

Sec. 10.1.2, the phase of an already existing real signal is not easily shifted since there is no direct access to the carrier phase angle. Similarly, to perform frequency shifting on a real signal requires mixing and filtering, as above; these computationally burdensome actions should be avoided when possible.)

Another approach is more versatile, although it requires more effort to generate. Consider the analytic signal corresponding to (10.9), given by

$$s_{\text{out}}^+(t) = [x(t) + jy(t)]e^{j2\pi f_d t}e^{j[\theta_{\text{in}}(t) - \theta_r(t)]} \quad (10.10)$$

This corresponds to the positive-frequency translate of Fig. 10.6c, that portion of the spectrum drawn with solid lines. The real signal is just the real part of the analytic signal:

$$s_{\text{out}}(t) = \text{Re}[s_{\text{out}}^+(t)] \quad (10.11)$$

Frequency and phase shifts are easily applied to the analytic signal $s^+(t)$, in the manner indicated in (10.10). Those shifts then appear on the real signal (10.11) extracted as the real part of $s^+(t)$. This use of the analytic signal is expedient whenever carrier frequency is not large compared to signal bandwidth; it is not restricted to frequency-translation applications.

10.3 TIME SHIFTERS

Numerous applications require that a signal be shifted in time. Some examples include modeling nonstationary channels, generating staggered I and Q modulated signals, synchronizing the timing of a received signal, constructing digital filters, providing signal memory, and changing the sampling rate. Time can be shifted over a whole number of sample intervals or over a fractional interval. In the latter case, time shifting is accomplished by interpolation.

10.3.1 Discrete Time Shifts: Delays

Circular blocks can be shifted by an integer number of samples, in either direction around the circle, using the same SHFT process as that introduced above for circular frequency-domain shifts. Concatenated linear block or extended single-point operations are causal; generally, they can shift only toward later times, not earlier.* To perform a delay of n_d sample intervals, a signal sample $z(n)$ is stored in state memory at sample time n and retrieved at $n + n_d$. The difference equation of the discrete

---

Exception: Finite-duration linear-block signals can be shifted in either direction, provided that the shifted signal does not lap around the end of the block. Methods for circular operations on linear blocks were outlined in Sec. 6.1.5.

delay process is

$$z_{\text{out}}(n) = z_{\text{in}}(n - n_d) \qquad (10.12)$$

10.3.2 Continuous-Time Shifts via Transforms

Shift Process

Perform the following steps [10.5] to impose a circular, arbitrary time shift τ on an N-sample signal block $\{z_{\text{in}}(n)\}$:

1. Take the DFT of $\{z_{\text{in}}(n)\}$ to obtain $\{Z_{\text{in}}(k')\}$, where $-N/2 \le k' < N/2$ if N is even, or $-(N-1)/2 \le k' \le (N-1)/2$ if N is odd.

2. Compute

$$Z_{\text{out}}(k') = \begin{cases} Z_{\text{in}}(k') e^{-j2\pi k' \tau / N t_s} = Z_{\text{in}}(k') e^{-j2\pi k' M_s \tau / NT}, & k' \ne -N/2 \\ Z_{\text{in}}(-N/2) \cos(\pi \tau / t_s), & k' = -N/2 \end{cases} \qquad (10.13)$$

(Notice that $k' = -N/2$ only if N is even; refer to Sec. 5.2.3 for notation and indexing.)

3. Take the inverse DFT of $\{Z_{\text{out}}(k')\}$ to obtain $\{z_{\text{out}}(n)\}$.

This procedure is a discrete-time approximation to an ideal delay line with transfer function $H(f) = \exp(-j2\pi f \tau)$. For the discrete approximation, $H(f)$ is sampled in the frequency domain at N equispaced frequencies $f = k'/Nt_s$. [Block ends of $\{Z_{\text{in}}(k')\}$ may require special treatment, as noted below.]

On-line processes are dimensionless; only the ratio τ/t_s (or $\tau/T = \tau/M_s t_s$) is known by the process. The ratio τ/t_s need not be an integer, nor need it be positive; any real number can be used.

Dual of Frequency Shift

Examination of the transform-method time shifter reveals that it is a dual of the transform-method frequency shifter of Sec. 10.2.2. The latter was found to generate intolerable spurious spectral lobes for noninteger frequency shifts, as shown in Fig. 10.4. Does the time-shift method introduce unacceptable spurious temporal lobes?

The answer is: No; the time shifter generates negligible or nonexistent spurious temporal lobes. In the frequency-shifter application, the time-domain signal generally has full amplitude at the block ends. Twisting of the time-domain block induced by a noninteger frequency shift causes a large discontinuity of the signal around the end of the block, leading directly to the spurious spectral lobes.

By contrast, the spectra of most signals to be simulated will be nearly zero at the block ends. The time shifter induces spectrum twisting, which may be discontinuous around the ends of the block, but the vanishing magnitude of the spectrum at the block ends causes the discontinuity to be very small. Thus the spurious time lobes are also vanishingly small, so the transform-method time shifter usually is practicable where its dual, the transform-method frequency shifter, is not.

Treatment of Block Ends

In (10.14) the endpoint at $k' = -N/2$ (occurs if and only if N is even) is treated differently from all other points in the block. The shifter has the allpass linear-phase transfer function $H(k') = \exp(-j2\pi k'\tau/Nt_s)$, from which $H(-N/2) = \exp(j\pi\tau/t_s)$ and $H(N/2) = \exp(-j\pi\tau/t_s)$ arise as the tentative values for the endpoints. But $H(k')$ is a circular array—$H(-N/2)$ is really the same point as $H(N/2)$, which means that both have to have the same value. The tentative values are not equal except for special choices of τ/t_s; there is an essential discontinuity around the ends of the transfer-function block.

At a discontinuity, the value of a function is commonly defined as the average of its upper and lower limits at the jump. For the delay function, that defines $H(-N/2) = H(N/2) = \frac{1}{2}[\exp(j\pi\tau/t_s) + \exp(-j\pi\tau/t_s)] = \cos(\pi\tau/t_s)$, as in (10.14).

10.4 INTERPOLATION

To compute $\{z(n-\tau/t_s)\}$ from $\{z(n)\}$ when τ/t_s is not an integer involves interpolation of new samples from the old. In that sense the transform-method time shifter performs an interpolation. However, the transform method is applicable only when $\{z(n)\}$ can be transformed readily to the frequency domain; transformation may be awkward to apply to concatenated linear blocks and is impossible in a feedback loop with single-point signal format. Other interpolation techniques are needed that can be used strictly in the time domain.

Besides implementing delays, interpolation is needed for computing the signal strobe at a receiver's output (Chapter 11) and for altering the sampling rate of a simulated signal (Chapter 16). Interpolation is also applied in some builder routines (Chapter 17), but that is entirely different from interpolation of signal sequences. Interpolation (along with its complementary process *decimation*) has been probed in depth in the DSP literature. Reference [10.6] is particularly recommended for its thorough explanation and as a guide to other literature on the subject.

Interpolation in the conventional literature is accomplished by filling zero-valued samples at intervals t_s/L (where L is a positive integer) into a sequence $\{z(n)\}$ originally sampled at intervals t_s and then filtering the upsampled sequence to smooth out the zero-valued gaps. Although zero filling may be used in a simulation, that is not helpful when simulating fractional delays or extracting strobes at times not sampled. A different outlook is needed.

A short account of interpolation for simulation is provided below; further information may be found in [10.7] and [10.8]. As of this writing, the literature on interpolation for simulation is rather sparse; additional reports can be expected to be published as more research is undertaken.

10.4.1 Interpolator Model

The basic model for interpolation has been developed in [10.6, Chap. 2] and is recapitulated in [10.7]. Key points are set forth below.

Start with a time-continuous bandlimited signal $z_{\text{in}}(t)$ and sample it at uniform intervals t_s to form a sample sequence $\{z_{\text{in}}(mt_s)\}$. The sampling interval is assumed to be small enough to avoid aliasing; accurate interpolation is not possible on undersampled signals. Compute *interpolants* at intervals t_i according to

$$z_{\text{out}}(lt_i) = \sum_m z_{\text{in}}(mt_s) h_I(lt_i - mt_s) \tag{10.14}$$

(Time indexes m and l have been used instead of the symbol n employed elsewhere in this book. This choice is made because of the need to have two distinct indexes for the two sampling intervals.)

In many receivers the ratio t_i/t_s is irrational; the sampling rates are incommensurate. More complicated yet: t_i may be adjustable, not constant, in applications such as timing synchronizers. Standard DSP literature typically assumes that t_i/t_s is rational and constant, but those simplifications often obscure the problem. The ratio will be taken as irrational in this book (and time varying, where appropriate), except in particular circumstances where rational or integer simplifications arise naturally.

The quantity $h_I(lt_i - mt_s)$ represents samples of the time-continuous impulse response of a fictitious analog linear time-invariant *interpolating filter*. Equation (10.14) may be interpreted as a linear convolution of the signal sequence $z_{\text{in}}(m)$ with a linear time-varying digital interpolating filter. All digital-signal interpolation has this same underlying model, even though the model may not be evident in specific implementations.

The interpolants are computed at times $lt_i = (m_l + \mu_l)t_s$, where the time for the lth interpolant follows the m_lth sample of $z_{\text{in}}(m)$ by a *fractional interval* $\mu_l t_s$ ($0 \leq \mu_l < 1$). Index m_l is known as the *basepoint index*. Furthermore, define a *filter index* $i = m_l - m$ and rewrite (10.14) as

$$z_{\text{out}}(l) = z_{\text{out}}(lt_i) = z_{\text{out}}[(m_l + \mu_l)t_s]$$
$$= \sum_{i=I_1}^{I_2} z_{\text{in}}[(m_l - i)t_s] h_I[(i + \mu_l)t_s] \tag{10.15}$$

Practical interpolating filters have finite impulse response (see [10.7]), so the digital-filter response $h_I[(i + \mu_l)t_s]$ has $I = I_2 - I_1 + 1$ *coefficients* or *tap weights*. Those coefficients are completely defined for computation of the lth interpolant by knowledge of the time-continuous finite-duration impulse response $h_I(t)$ and the fractional interval μ_l. The I input samples (the *basepoint set*) used for computing the lth interpolant are completely specified by knowledge of the basepoint index m_l.

An interpolator process needs two control signals: one providing μ_l and the other identifying m_l. Notice that μ_l will change for every new interpolant unless $t_i \equiv t_s$. Production of control signals is covered later in the descriptions of processes that employ interpolation. The fractional interval μ_l will be passed as a numerical value, but no numerical value of m_l is ever likely to be generated. Instead, some other means of identifying the basepoint set is required—a means particular to each application.

10.4.2 Characteristics of Interpolating Filters

Ideal Interpolation

Corresponding to the impulse response $h_I(t)$ there is a frequency response $H_I(f)$—the Fourier transform of the impulse response. If the fictitious continuous-time interpolating filter had a zero-phase frequency response with flat amplitude and abrupt cutoff at half the sampling frequency, (10.14) would recover $z_{\text{out}}(lt_i) = z_{\text{in}}(lt_i)$ perfectly, without error, from the samples $\{z_{\text{in}}(mt_s)\}$. The ideal filter's impulse response is

$$h_I(t) = \frac{\sin \pi t/t_s}{\pi t/t_s} \qquad (10.16)$$

which is infinite in duration, noncausal, and nonrealizable.

A nonideal filter must be employed, one that can be implemented conveniently. Because a practical filter is not ideal, practical interpolation necessarily causes some signal distortion, as discussed in the next paragraphs.

Desirable Characteristics for Filters

To minimize signal distortion, a practical interpolating filter should resemble the ideal rectangular filter. Requirements on the filter are:

- Linear phase (symmetrical impulse response)
- Flatness of frequency response amplitude in the passband
- Good attenuation in the stopband

Passband flatness is not crucial in hardware interpolators, where other filters in the receiver can compensate for passband distortion caused by the interpolating filter. Simulation is more demanding unless a user is willing to invest the filter-design effort required to compensate interpolator distortion. Stopband attenuation is needed to reject the periodic images in the spectrum of $\{z_{\text{in}}(m)\}$. Residual image energy is aliased into the passband of the output signal $\{z_{\text{out}}(l)\}$. For further details, refer to [10.7] and [10.8].

Polynomial-Based Filters

If the impulse response $h_I(t)$ is a polynomial (or piecewise polynomials) in t, the filter is said to be *polynomial based*. The classical interpolating polynomials described in all textbooks on numerical analysis constitute a special case of polynomial-based interpolators. Their filter responses are piecewise polynomial and are given by the well-known Lagrange formulas. Classical interpolating polynomials are not the only polynomial-based interpolators, and not necessarily the best. See [10.8] for other examples. However, only classical polynomials are considered further here.

Reference [10.9] shows that the following restrictions should be imposed on polynomial interpolators to minimize distortion:

1. Employ an even number of basepoint samples: that is, an odd-degree interpolating polynomial.
2. Select each basepoint set so that the interpolation is performed between the central pair of the set.

The two restrictions assure a uniquely defined interpolant and linear phase. Among polynomials of odd degree, linear interpolation is the simplest to compute and cubic interpolation is the next simplest. Frequency response $H_I(f)$ and impulse response $h_I(t)$ of the cubic interpolator are plotted in Figs. 10.7 and 10.8.

Simulation results are reported in [10.8], comparing the performance of these two interpolators and others. The criterion of performance employed in [10.8] is the increase in signal-to-noise ratio E_b/N_0 needed to overcome the deleterious effects of interpolator distortion on bit-error ratio (BER). Some example results are 0.15 dB for linear interpolation at 3 samples per symbol, and 0.14 dB for cubic interpolation at 2

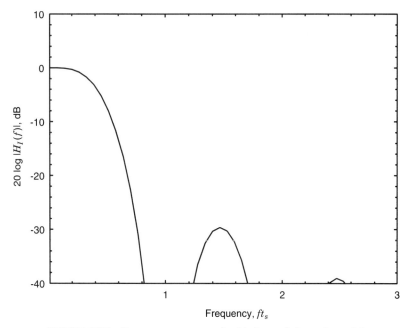

FIGURE 10.7 Frequency response of cubic interpolating polynomial.

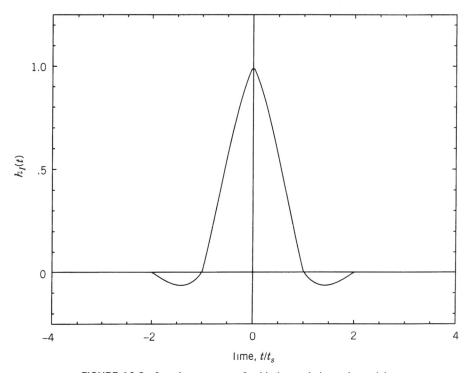

FIGURE 10.8 Impulse response of cubic interpolating polynomial.

samples per symbol. In each case the signal was bandlimited BPSK with 100% excess bandwidth, and the measurements were performed at BER = 10^{-6}. Smaller degradations are observed if the sampling density is increased, if the excess bandwidth is decreased, or if measurements are performed at larger BER. Signals with larger constellations may be more susceptible to interpolator distortion, although evidence has not been collected at this writing; additional tests are needed.

These results are for uncompensated interpolators; substantial improvement obtains with compensation. We surmise, but have not demonstrated, that the degradation is negligibly small for uncompensated cubic interpolation at four samples per symbol. Do not infer that a classical cubic polynomial is the best choice in the sense either that it has the least complexity to achieve its performance, or that its performance is the best that can be achieved with its complexity. A piecewise quadratic four-point interpolator introduced in [10.8] has a simpler configuration and was found to perform better in tests.

10.4.3 Simulation of Interpolation

Interpolation is a compound operation composed of two nearly independent suboperations:

1 Storing input samples for the basepoint set in state memory
2 Performing interpolation(s) on the basepoint set

State Memory

Regard the interpolator's state memory as an I-stage shift register for signal samples. (Software implementation as a circular buffer is more efficient, but that does not alter the external traits of the memory.) At the time of each input sample: (1) the oldest stored sample is discarded out the back of the register, (2) the surviving older samples are shifted one position down the line, and (3) the new input sample is shifted into the front of the register. State memory is shifted inexorably at each input sample time, whether or not an interpolation is performed.

Observe that an interpolator introduces a delay into the signal path; an additional $I/2$ signal samples after the m_lth sample must be collected to complete the basepoint set before an I-point central-interval interpolation can be performed. (Delay can impair system performance, particularly in a feedback loop. That could be a strong argument in favor of short interpolators.)

Immediately following a sample shift, the most recent state-memory cell contains $z_{in}(mt_s)$ and the oldest contains $z_{in}[(m - I + 1)t_s]$. Of course, m is not tallied, so a different index notation, not using m, has to be employed for the computations. In (10.16)—which is the basis for any computation—the input samples have index $m_l - i$, with i ranging from $I_1 = -I/2$ (the most recent sample) to $I_2 = I/2 - 1$ (for the oldest sample). Since m_l is not tallied either, it can be assigned any arbitrary value convenient for programming. For purposes of explanation at this juncture, m_l is set to zero, so the indexes on the state memory run from $-I_1$ on the newest sample to $-I_2$ on the oldest.

Computation

An interpolant is to be computed upon demand on whatever basepoint set is in state memory at the time a demand is made. Timing of demands is discussed below. Each demand is accompanied by a value for μ_l.

Computation is performed according to (10.16), which requires the basepoint set $\{z_{in}\}$ and I samples $\{h_I[(i + \mu_l)t_s]\}$ of the filter's impulse response $h_I(t)$. The computer program contains formulas for $h_I(t)$ and can calculate designated sample values upon demand. Formulas (Lagrange coefficients) for the classical cubic polynomial are listed below.

i	$h(i, \mu)$
-2	$\frac{1}{6}(\mu - 1)\mu(\mu + 1)$
-1	$-\frac{1}{2}(\mu - 2)\mu(\mu + 1)$
0	$\frac{1}{2}(\mu - 2)(\mu - 1)(\mu + 1)$
1	$-\frac{1}{6}(\mu - 2)(\mu - 1)\mu$

The following pseudocode indicates how the computations might proceed:

CUBIC INTERPOLATION

```
Sum = 0
FOR i = I₁ to I₂
    Compute h = h_I(i + μ)
    Product = h × z_in(-i)
    Sum = Sum + Product
NEXT i
z_out = Sum
```

Timing of Interpolator Process

Heretofore, little attention has been paid to timing of operations between processes. In most processes there is exactly one output sample for each input sample. For those few processes entailing alterations in sample rate (e.g., the symbol mapper; Sec. 7.2), the output has been synchronous with the input; an exact whole number of samples of the input (output) correspond to one sample of the output (input).

Synchronism is not a natural feature of physical signals. Receivers invariably require synchronizing operations to establish the synchronization that is crucial to digital communications. Interpolation is the simulation process employed to establish symbol synchronization. More generally, interpolation is needed to accomplish any noninteger changes in sampling rate. In this section we introduce the elements of nonsynchronous operation of interpolators, which is then covered in greater depth in later chapters.

Interpolator state memory is shifted for each new signal sample at the interpolator's signal input. Interpolation can be demanded zero or more times for each shift—for each basepoint set. Demands for interpolation are entirely independent of input-related shifts. In that sense the input and output are decoupled. It is the decoupling that allows the interpolator to perform synchronzation and rate changing.

The ordinary processes examined earlier all behave in the same manner as any line of program code; they are executed when program flow reaches them. An ordinary process has only one entrance—one event that starts execution—and just one sequence of operations in carrying out its procedures. An interpolator might be quite different; it has two separate sets of operations (state-memory shifting and interpolation itself) and two separate entrances (input signal and the execution demand). The two separate entrances are reached via separate paths in the simulation block diagram.

If fewer output samples are to be interpolated than there are input samples delivered, the interpolator is said to perform *downsampling*. Under that condition, there will be at least some input-sample intervals in which no outputs are interpolated and in no input interval will more than one interpolation be executed. If more output samples are interpolated than there are input samples, the interpolator performs *upsampling*. There will be at least some input-sample intervals in which more than one output is interpolated and in all input intervals at least one interpolation is executed. Both downsampling and upsampling are important functions for an interpolator; they

10.4.4 Variable Delay by Interpolation

Now let's return to the delay of a signal by an interpolator. For every input sample $z_{in}(n) = z_{in}(nt_s)$, there is to be an output sample

$$z_{out}(n) \simeq z_{in}[(n - I/2 + \mu_n)t_s] \quad (10.17)$$

A fixed positive delay of $t_s I/2$ is incurred, as well as a variable *negative* delay $\mu_n t_s$. That is, an increase of μ_n causes a decrease of delay. In the cubic interpolator $I = 4$, so its delay can be adjusted in the range of t_s to $2t_s$. Approximate rather than exact operation is achieved because of interpolator distortion. The fixed delay—a process latency—is of no concern in many applications, particularly if the system is open loop and has just a single path. Delay can be a problem in feedback loops, though, impairing loop stability and altering loop transfer functions.

Interpolator control signals are relatively easy to supply for simple delay operation. Exactly one output is computed for each input; no downsampling or upsampling is involved. The fractional interval μ_n is delivered for each sample; a different μ_n can be used for each n if necessary.

10.5 KEY POINTS

This chapter has treated phase shifters, frequency shifters, time shifters (delays), and interpolators (another kind of delay).

10.5.1 Phase Shifters

To shift the phase of a time-domain complex signal $z_{in}(n)$ by an angle $\theta(n)$, compute

$$z_{out}(n) = z_{in}(n)e^{j\theta(n)} \quad (10.1)$$

Phase angle $\theta(n)$ is a control signal. If θ = constant, independent of either n or k, phase shifting can be performed equally well in either the time domain or the frequency domain.

10.5.2 Frequency Shifters

Time Domain

To shift the frequency of a time-domain complex signal by f_ϵ, compute

$$z_{out}(n) = z_{in}(n)e^{j2\pi n f_\epsilon t_s} \quad (10.6)$$

This formula is a phase shifter whose phase $2\pi n f_\epsilon t_s$ changes linearly with time. Generate the phase (a control signal) in a number-controlled oscillator (NCO), realizable with the difference equations

$$p(n+1) = [p(n) + W(n)] \bmod\text{-}1$$
$$\theta(n) = 2\pi p(n)$$
$$p(0) = \theta(0)/2\pi$$
$$W(n) = f_\epsilon t_s \tag{10.7}$$

A time-domain frequency shifter is composed of a phase rotator plus an NCO.

Shift computations are dimensionless; only the product $f_\epsilon t_s$ appears in the process and not either factor alone. To avoid aliasing, it is necessary as a minimum requirement that $|f_\epsilon t_s|$ be less than 0.5. Additional margin is needed to accommodate the bandwidth of the signal being shifted.

Frequency Domain

To shift the frequency of a frequency-domain complex N-point signal block by f_ϵ compute

$$Z_{\text{out}}(k) = Z_{\text{in}}(k - Nf_\epsilon t_s) \tag{10.8}$$

Fractional values of $Nf_\epsilon t_s$ require interpolation to determine $Z_{\text{out}}(k)$.

Real Signals

To shift frequency or phase of a real signal $s_{\text{in}}(n)$, simulate the analytic signal $s_{\text{in}}^+(n)$ instead. Then perform frequency or phase shifts on the analytic signal according to (10.1) or (10.6) and find the shifted real signal $s_{\text{out}}^+(n)$ as the real part of the shifted analytic signal $s_{\text{out}}(n)$.

10.5.3 Time Shifters

Discrete Delay

A circular N-block can be shifted by any whole number of samples in either direction. A linear-block or single-point time-domain signal is delayed by a whole number of samples n_d according to

$$z_{\text{out}}(n) = z_{\text{in}}(n - n_d) \tag{10.12}$$

Continuous Shift

For a circular block $\{z_{\text{in}}(n)\}$, an arbitrary delay τ can be imposed by first using the DFT to obtain $\{Z_{\text{in}}(k')\}$ and then computing

$$Z_{\text{out}}(k') = \begin{cases} Z_{\text{in}}(k')e^{-j2\pi k'\tau/Nt_s}, & k' \neq -N/2 \\ Z_{\text{in}}(-N/2)\cos(\pi\tau/t_s), & k' = -N/2 \end{cases} \quad (10.13)$$

The delayed time-domain signal is found as $\{z_{\text{out}}(n)\} = \text{IDFT}\{Z_{\text{out}}(k')\}$.

10.5.4 Interpolation

Linear-block or single-point time-domain signals must be interpolated to effectuate noninteger delays. An interpolator can be modeled as a digital filter whose variable coefficients are chosen according to the fractional interval to be interpolated. An interpolator process needs two control signals, whose values can change for each interpolation:

1 A fractional interval μ_l that specifies the filter coefficients to use for the lth interpolation

2 A flag demanding that interpolation be performed on the current basepoint set

Continuous Delay

A variable delay can be effectuated by interpolation to give

$$z_{\text{out}}(n) = z_{\text{in}}[(n - I/2 + \mu_n)t_s] \quad (10.17)$$

where I is the number of taps in the interpolating filter. A fixed delay of $I/2$ samples is incurred as well as a *negative* variable delay of μ_n fractional samples ($0 \leq \mu_n < 1$).

Other Applications

Interpolation is also used for synchronization, strobing, and sample-rate changing, as well as simple delays.

10.5.5 STÆDT Processes

CHRT: cubic interpolation and rate changing, including control elements
DLYA: circular shift (either direction) in continuous increments
DLYC, DLYD: linear delay (toward later direction only) in discrete increments
INTP: interpolation by cubic polynomial (no control elements)
NCO: number-controlled oscillator
RTAT: phase rotator
SHFT: circular shift (either direction) in discrete increments

11

SYNCHRONIZATION

Frequency, phase, and timing of a coherent receiver have to be aligned closely with those same parameters of an incoming data signal if the receiver is to work properly. The frequency and phase alignment operations together constitute *carrier synchronization*, while the operation of timing alignment constitutes *symbol synchronization*. Carrier and symbol synchronization together constitute *signal synchronization*. Carrier synchronization is needed only for modulated-carrier signals; phase synchronization is omitted if the receiver is noncoherent or differentially coherent.

Additional synchronization at the word and frame levels (*data synchronization*) will also be required in most systems. These higher-level operations usually are purely digital and are performed on the detected digital output of the receiver. Data synchronization is not often simulated by signal-transmission programs; this book treats only signal synchronization. An excellent survey of frame synchronization may be found in [11.1].

The processes that perform the actual carrier and symbol alignment were described in Chapter 10. Each shifter accepts a control-signal input that prescribes the amount of shift to be applied. Generation of the requisite control signals is the primary topic of this chapter.

Scenarios

Two quite different synchronization scenarios commonly arise in simulations:

1 *Static Alignment: Calibration.* Timing and phase misalignments might occur solely because of fixed shifts inherent to filters and other elements in the link model—shifts that are not readily predictable a priori. Synchronization parameters otherwise would be constant over the entire duration of the simulation. If the behavior of realistic synchronizers is not being investigated, it is often suf-

ficient to determine the needed corrections by a calibration procedure and then employ constant control signals to maintain those corrections throughout the actual simulation run. This approach is the simplest, when applicable.

2 *Dynamic Alignment.* If synchronizer circuits or algorithms are to be simulated, the control signals have to be generated dynamically by synchronizer processes to be described below. Control signals can be expected to fluctuate significantly throughout the simulation run because of noise or other disturbances. This approach is more complicated than synchronization by calibration.

Both approaches are treated in the sequel.

Background

Synchronization is a broad subject that has received (and is still receiving) the attention of many researchers. Adequate coverage of the subject would require a hefty book dedicated solely to synchronizers—a book that does not now exist. The material that follows provides a brief overview of synchronizers, identifies some important structures, and examines features pertaining to simulation.

Hundreds of synchronization papers have been published over the years. Survey treatments may be found in [11.4] to [11.14]. References [11.12] and [11.13] give accounts of synchronization practices that have grown up for digital communications by satellite and [11.14] treats synchronization for CPM signals. Specialized articles of note include [11.15] to [11.28] for phase recovery, [11.29] to [11.45] for timing recovery, [11.46] to [11.51] for both phase and frequency, and [11.52] to [11.63] and [11.72] for frequency recovery. The literature on frequency recovery is sparse; the papers cited constitute a significant portion of those in existence. The literature on phase and timing recovery is huge; the papers cited constitute but a small fraction of the bulk.

11.1 SYNCHRONIZATION BY CALIBRATION

Calibration of synchronization can be approached in any of several ways. There are two broad categories of (1) *hardwire* synchronizers, whereby synchronization information is brought over from the transmitter to the receiver separately from the signal; and (2) *recovery* synchronizers, in which synchronization information is *recovered* from the receiver's input signal, without other knowledge of the transmitted information. Either way, the synchronizer produces estimates of the signal's frequency $\hat{f}$, carrier phase $\hat{\theta}$, and symbol timing $\hat{\tau}$. These estimates are delivered to a frequency/phase shifter and an interpolator that align the signal strobes correctly. Interpolators and frequency/phase shifters have been explained in Chapter 10.

11.1.1 Hardwire Synchronization

In a communications-link breadboard in the laboratory, hardwire synchronization would be accomplished by bringing timing-clock and carrier waves over wire paths from a transmitter to a receiver. These hardwire paths establish the exact frequency of each wave, but not carrier phase or symbol timing. Adjustments are needed in each path to establish correct alignment of phase and timing. An engineer would monitor some performance criterion of the receiver (such as bit error ratio) and use that to adjust phase and timing for best performance.

Simulated hardwire synchronization resembles the laboratory setup in broad outline. Carrier and clock frequencies are already known implicitly to most simulated receivers since transmitter and receiver both work from the same implied time base in the computer. Only rarely will frequency suffer change in transmission. Thus there is no need for explicit simulation of a hard wire to bring over the carrier and clock waves (which probably do not exist anyway in a simulated transmitter).

But the phase and timing adjustments are still needed; information that will permit those quantities to be adjusted must be brought over. For example, if symbol error ratio is the performance criterion to be observed, the transmitted symbols have to be brought over so that the correct message information is made known to the error evaluation process. For the hardwire schemes examined here, it is sufficient to bring over the message symbols and information on signal format.

Search

A search can be performed over all alignments to find the best delay $\hat{\tau}$ and phase $\hat{\theta}$. "Best" means that some performance criterion is optimized. The most pertinent criterion would be probability of error in detecting data, but that may call for unduly prolonged runs to gather adequate statistics. A less direct criterion (such as correlation—to be examined later) may be computable much more quickly. Employment of a less-direct criterion may entail a sacrifice in the quality of calibration. If a calibration is based on a performance criterion other than error probability, it may be prudent to verify the calibration by a limited further search using error probability as the criterion.

Search takes place over the two dimensions of delay and phase. If the data are multiamplitude, correct amplitude is needed for data detection, so the search may have to be over three dimensions. Search can be performed exhaustively over all points on a grid (or cube if in three dimensions), but much less effort is needed with a coarse/fine strategy. An initial coarse estimate can be obtained (by any convenient method) and then improved by a fine search in its near vicinity. Search methods have been treated at length in the literature on optimization; see [11.2] for an introduction and further references. Search methods can be employed for any signal format and for any transmission channels, either linear or nonlinear. Other methods may be more restricted in their applicability.

Correlation

Correlation between the receiver's output and the transmitted symbols (see Secs. 4.3.1 and 7.3.1) can be a useful criterion of synchronization performance, as a substitute for probability of detection error. Correlation is easier to compute and permits separation of phase from timing so that no more than a single one-dimensional search is needed. Signal amplitude can be ignored when correlation is used as the criterion of performance. Correlation methods are most readily applicable to linear-modulation PAM signal formats.

Analysis in Appendix 11A shows that adjusting signal delay to maximize the magnitude of correlation identifies the location of peak magnitude of the system signaling pulse $g(t)$. In many communications systems the optimum strobing point falls at or close to the pulse peak (STÆDT provides an important counterexample). Analysis also shows that the magnitude of the correlation function does not depend on phase, so timing can be calibrated without concern for phase. Once timing has been established, the optimum phase correction is given by the angle of the correlation function at its point of maximum magnitude. Synchronization by correlation has close resemblances to maximum-likelihood synchronization [11.3, 11.4].

Most QAM constellations have phase symmetry, which ordinarily leads to ambiguities of demodulation. Physical communications systems incorporate schemes to remove the ambiguity. Because it makes use of actual transmitted data, the correlation method removes the ambiguity automatically, without need for further antiambiguity provisions in the simulation.

11.1.2 Calibration by Recovery Synchronizers

Many different implementations of synchronizers are employed in physical receivers; examples are discussed later in this chapter. Any appropriate synchronizers could be simulated during a calibration to establish the proper timing and phase, and those values could then be held as constant control signals during the subsequent simulation run. This approach avoids the computational effort required for ongoing synchronization but still requires the user to model the synchronizers. It would be preferable to have a synchronization calibration method that both avoided the computational effort and relieved the user of the need to model synchronizers. For this reason, a calibration by correlation is usually more attractive, when applicable.

Modeling of recovery synchronizers is much the same whether synchronization is applied just once in a calibration or continuously throughout a simulation run. The remainder of this chapter is devoted to recovery synchronization.

11.2 RECOVERY SYNCHRONIZERS

A *recovery* synchronizer *recovers* synchronization information from the received data signal itself. In the absence of separate provisions for synchronization, all practical synchronizers are recovery synchronizers. Recovery synchronizers almost always employ nonlinear elements that are very difficult or impossible to study analytically. Simulation is often the only tool, short of prototyping, for examining the behavior of synchronizers.

Terminology

The quantities to be synchronized—phase θ, timing delay τ, and frequency f—will be called *reference parameters*. Synchronization is also called *reference recovery*.

11.2.1 Synchronizer Categories

Synchronizers can be categorized in various ways, including:

- By reference parameter: phase, timing, or frequency
- By signal character: time-continuous or time-discrete
- By their use of data information: data aided (DA), decision directed (DD), or non-data aided (NDA)
- By interactions between reference parameters: joint recovery or independent recovery
- By synchronizer configuration: feedback loops with reference-error detection, feedforward methods with reference-wave regeneration, or feedforward methods with parameter extraction

Configurations

The last category—synchronizer configuration—is considered first and most prominently, since it exposes the main elements comprising synchronizers. The other categories are treated more in passing as the elements are examined in detail. Refer to Fig. 11.1 for block diagrams of the three configurations.

Feedback synchronizers (phase-locked loops) [11.4, 11.7, 11.8] are widely used in hardware. They tend to be relatively economical to implement and they remove frequency offsets (carrier frequency or timing frequency) in addition to accomplishing the phase or timing adjustments that are their main purpose. Feedback loops are potentially unstable (an ever-present design problem) and require greater allowance for acquisition time than do feedforward methods. Feedback is used both in analog and in digital implementations, and both are often simulated.

Feedforward methods superficially appear even simpler than feedback loops, but that potential is not easily realized if frequency has to be corrected separately. Feedforward *regenerators* [11.4] have been used extensively for baseband timing recovery and for BPSK and QPSK carrier recovery. A feedforward synchronizer is always stable and inherently acquires a signal more rapidly than is possible with a phase-locked loop.

Feedforward *regenerators* have long been used in analog hardware but require excessive sampling rates (particularly for carrier recovery) that militate against their application in digital hardware. Feedforward *extractors* have appeared more recently [11.24, 11.42]; they are closely related to regenerators but are better suited to digital implementation. The principles of both are covered below.

Feedforward synchronizers recover the reference in a separate path that parallels the signal path. Stray phase and delay differences inevitably arise between paths and

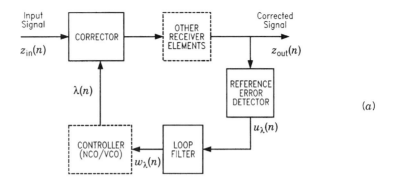

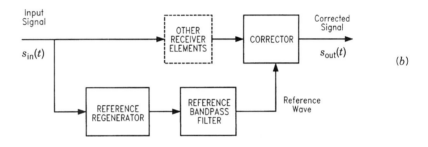

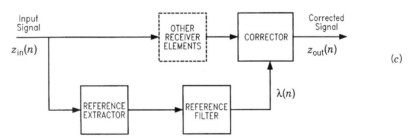

FIGURE 11.1 Synchronizer configurations: (*a*) feedback recovery; (*b*) feedforward recovery with reference regeneration; (*c*) feedforward recovery with reference extraction.

are compensated by an adjustable shifter. In analog hardware, the stray differences and compensation drift with time and temperature, requiring periodic readjustment. By contrast, a feedback synchronizer measures reference error directly on the data signal and the feedback loop drives the error to a null. No parallel paths exist; no compensating adjustment for stray shifts is needed.

Other Categorizations

In the list of categories that began this section, the first category—reference parameters—is self-explanatory.

Signal Character. All simulations on digital computers are necessarily time discrete. Sampling rate must be large enough to approximate the behavior of time-continuous synchronizers, as explained in Chapter 5. Sampling requirements are considered further below.

Use of Data. *Data aided* means that the receiver knows the transmitted symbols without error and uses that knowledge in recovery of the reference(s). Any such knowledge improves synchronizer performance. The hardwire methods described in the preceding section and Appendix 11A are data aided; it is data that the hard wire brings over. In practice, transmitted data sequences are mainly unknown to a synchronizer, so data aiding is rarely applicable in working communications systems, except in training preambles or the like.

The receiver's data decisions can be taken as error-contaminated substitutes for the transmitted symbols and used to enhance performance in a *decision-directed* synchronizer. Decision direction is not quite as effective as data aiding, but it is better than ignoring the decision information.

Non-data-aided synchronizers make no use of data information and suffer a penalty in consequence. They have been used either because they are simpler to implement (an important consideration in analog circuits, but no longer true in digital algorithms) or in conditions where data or decisions simply are not available or sufficiently reliable.

Joint/Independent Recovery. If the synchronizers (and data detection) all work together, providing information to one another, the receiver is said to engage in *joint recovery* [11.46]. Best performance is attained with the fullest use of all available information. But the prospect of all parameters being dependent upon one another has inhibited full employment of joint recovery. After all, if timing recovery is dependent on prior recovery of phase and phase is dependent on prior recovery of timing, how will synchronization ever be acquired? In fact, anecdotal evidence suggests that joint acquisition does indeed take place and rather reliably at that. At this writing, much remains to be learned about acquisition in joint synchronizers.

Fear of potential acquisition mishaps has motivated the widespread use of *independent synchronizers*, in which recovery of at least one parameter does not depend in any way upon the others. For example, non-data-aided synchronizers are independent of data detection. Some NDA timing recovery schemes are independent of carrier phase and some NDA carrier recovery schemes are independent of timing recovery.

11.2.2 Synchronizer Operations

In all that follows it is assumed that the references are recovered from the data signals themselves, with no pilots or other transmitted references being present. To be explicit: Assume that a modulated signal has no discrete carrier component present and that a baseband signal has no discrete component present related to the clock frequency. Transmitted references would expend power that could be devoted to the data instead and so are not power efficient. Some signals do incorporate pilots, but those are not treated here. Refer to Fig. 11.1; each configuration (one feedback and two feedforward) and its terminology are explained in this section.

Feedback Configuration (Fig. 11.1*a*)

Although the input signal $z_{in}(n)$ is shown as complex and sampled, the same configuration is also used with real signals or time-continuous signals; the same principles apply to all.

Corrector. The signal is first applied to a *corrector*, which adjusts the reference parameter being synchronized. Correctors and all other synchronizer elements are described in further detail in the next section. The reference parameter, whether it be frequency, phase, or timing, is symbolized here by $\lambda(n)$ applied as a control signal to the corrector.

Error Detector. The corrected signal $z_{out}(n)$, which may have been passed through other receiver elements, such as filters, downsamplers, or strobers, is delivered for further processing to other operations outside the synchronizer. Also, the output signal is applied to a *reference-error detector*, which measures some characteristic of the signal and from that measurement estimates the synchronization error with respect to the correct reference condition. A phase detector in a phase-locked loop is one kind of reference-error detector.

Output of the error detector is labeled $u_\lambda(n)$ in Fig. 11.1. Time-continuous systems are likely to produce time-continuous error signals; they should be simulated by a process that generates one error sample per signal sample. On the other hand, error detectors in digital hardware are likely to generate just one error sample per symbol interval, not per signal sample. (Besides being computationally more efficient, one error sample per symbol provides better synchronization performance.) Examples of error detectors are provided later in this chapter; details are very application dependent.

Loop Filter. The error signal u_λ—either time continuous in analog hardware or sampled in a simulation—is passed to a loop filter. Filtering in a control loop tends to be rather simple, consisting typically of a proportional path plus a small number (zero to three) of integrators. Filter elements for control signals are treated in Chapter 14. Output of the loop filter is shown in Fig. 11.1 as a *control signal* denoted $W_\lambda(n)$, to be applied to the reference controller.

Controller. The controller's task is to produce the control signal $\lambda(n)$ that drives the corrector. An analog controller typically is a voltage-controlled oscillator (VCO), whereas a controller in digital hardware or simulation may be a number-controlled oscillator (NCO). Section 10.2.1 introduced NCOs; additional features pertinent to synchronizers are presented later in this chapter. When a VCO/NCO is used as a controller, the same loop corrects for both phase and frequency errors, with no special provision needed for frequency. This is an inherent feature of phase-locked loops that contributes to simplicity of implementation. In some configurations (e.g., phase shifting alone, with no frequency-shift capability), a separate controller may be omitted and the filter output then would be $\lambda(n)$ itself.

Feedforward Regenerator (Fig. 11.1b)

A regenerator is a nonlinear device that regenerates a periodic carrier or clock wave from a data signal that contains neither. One can think of a regenerator as cross-multiplying continuous-spectrum sideband components of the data signal; the resulting products have discrete spectral lines at the symbol frequency or at harmonics of the carrier frequency.

Raw output of the regenerator is passed through a narrow bandpass filter (or, equivalently, a phase-locked loop) to remove noise and other disturbances. Cleaned-up output of the filter is the reference wave used for synchronizing the signal; reference-wave phase or timing is the same as the phase or timing of the incoming signal. A fixed (factory) adjustment is commonly included to compensate for phase or delay errors in hardware.

In carrier synchronizers, the frequency of the reference wave has to be divided back down to the frequency of the original signal and then applied to the receiver's demodulator. In timing synchronizers, the reference wave is shaped into a clock waveform and used to strobe the receiver's output. Additional provisions have to be made if nonnegligible frequency offset must be compensated. When needed, these additional features vitiate the simplicity that seems to be promised by the lean block diagram of Fig. 11.1b. Models for simulating regenerators are shown in Sec. 11.3.

Feedforward Extractor (Fig. 11.1c)

Regenerators in digital hardware would require high sampling rates, as explained in the next section. The problem is particularly severe with carrier regenerators. The signals in Fig. 11.1b have been designated as time continuous to emphasize this characteristic.

It is possible to *extract* the reference information from samples of a data signal without regenerating a reference wave and without an excessive sampling density. The extracted and filtered reference *parameter* $\lambda(n)$ is applied as a control signal to a corrector. Extractors have been implemented only digitally; they would be much too complicated in analog circuits. Accordingly, terminology in Fig. 11.1c denotes sampled signals. Principles of extractors are treated later.

It will gradually become evident that regenerators often are modeled much like extractors for simulation in a digital computer, particularly for carrier synchronizers. There may be only subtle differences, mainly in sampling rate, to distinguish regenerator models from extractor models.

11.3 SYNCHRONIZER ELEMENTS

Four different kinds of synchronization elements can be distinguished in Fig. 11.1:

- Correctors
- Nonlinear reference-measuring elements: error detectors, regenerators, and extractors
- Filters
- Controllers: VCOs, NCOs (not always present)

Characteristics of these elements and their simulation models are explored in the following pages. Let it be said from the outset that an enormous variety of nonlinear reference-measuring elements are found in practice and in the literature. Only a few representative examples are portrayed in these pages.

11.3.1 Correctors

The actual adjustment of frequency, phase, or timing is performed by a corrector.

Phase and Frequency Corrections

In analog hardware, phase and frequency shifting (correction) is accomplished in frequency mixers or demodulators. These are modeled by phase rotators in digital hardware or simulations (see Sec. 10.1). A VCO or NCO (Sec. 10.2.1) is needed as a controller to produce the steadily advancing phase that constitutes a frequency shift. Simulation of phase shifting alone needs no NCO. Phase and frequency shifting have been covered in Chapter 10 and receive no further attention here.

Timing Corrections

Phase/frequency-shifting models are quite similar for continuous- and discrete-time representations. By contrast, timing correctors for continuous-time implementation are likely to be fundamentally different from those used for discrete time. The differences exert a strong influence on the computer simulation of time shifters.

To see the nature of the differences, first consider an analog-implemented data receiver. All signals in the receiver are time continuous up until the output end, where the signal is strobed at one sample per symbol. Strobe timing in such a receiver is adjusted by shifting the phase of a local clock; no timing adjustments are performed on the signal. The most prevalent methods allow for infinitely fine resolution of clock phase so that the strobe can be selected at any instant on the signal waveform.

Now consider a time-discrete signal, either in a digitally implemented receiver or in a computer simulation. Timing adjustment is still tantamount to selection of strobe points, but now only the discrete signal sample points exist, not a time-continuous signal. How is the strobe selection to be performed without sacrificing timing resolution?

Strobing by Sample Selection. One approach is to employ such high sampling density that the sampled signal is a good approximation to a time-continuous signal, and then select one of the existing samples as the strobe. A random timing error is inflicted upon the signal with this approach. If the sample interval is t_s and the timing error can be assumed to be uniformly distributed, the rms timing jitter caused by the time quantization is $\sigma_\tau = t_s/\sqrt{12}$. A rule of thumb for choosing t_s might be that σ_τ is small compared to other sources of timing jitter, or that σ_τ have negligible effect upon system performance.

The high-density approach is similar to analog practice in that a signal point—an existing sample point—is selected as the strobe. Phase shifting of the local clock is performed in discrete steps, corresponding to the sample intervals of the signal, and not time continuously. High-density sampling may be the only method suitable for simulation of wideband signals; reasons for this statement are given below. High-density sampling is undesirable if it is not needed. It compels digital hardware to work much faster than otherwise necessary and requires that many additional signal points be computed in a simulation. Low-density sampling is preferred, when permissible.

Chapter 5 treated the sampling of simulated signals. If the signals are strictly limited to a bandwidth B, a sampling rate of $1/t_s > 2B$ is sufficient to represent the underlying time-continuous signal perfectly. Therefore, when simulating bandlimited signals it is common practice to employ relatively few samples per symbol. As few as two samples per symbol may be feasible in some instances, whereas four or eight are more usual.

Such coarse quantization is unacceptable for strobe timing in most applications; a quantization interval of, for example, $\frac{1}{4}$ symbol would have a disastrous impact on the performance of nearly any communications system. Finer timing resolution is essential for good performance, yet one does not want the burden of high-density sampling of the signals.

Strobing by Interpolation. Interpolation of the signal samples is the way to obtain fine timing resolution and yet represent the signal efficiently with low-density sampling. Any point whatever on a bandlimited, adequately sampled signal can be retrieved by interpolation. The interpolated strobe point is not one of the signal samples, except rarely and by chance.

In this manner, the signal value strobed in a simulation can be exactly the same as the strobe of the time-continuous signal being simulated. Interpolation—needed in the simulation—does not exist in the analog hardware. Interpolation, uniquely among all of the synchronizer elements, has no counterpart in the analog system under study. (Interpolators do not have to be situated immediately at the strobing point; they can be placed earlier in the receiver model and can be followed by other elements, such as additional signal filters. The principles are the same; the interpolator is to be adjusted such that one sample per symbol at the receiver's output—not the interpolator's output—falls at exactly the desired strobe instant.)

Even with interpolation, timing is still adjusted by shifting the phase of a local clock, but that local clock does not necessarily appear with a clearly recognizable identity in the simulation. Furthermore, clock-phase information has to be communicated to the interpolator in a form that the interpolator can use, not the form appearing in analog strobe circuits. That information is delivered to the interpolator by a *timing controller*, the topic of Sec. 11.3.2.

First, though, let's return to wideband signals: Why should their timing be recovered through computationally inefficient high-density sampling and not by interpolation? The answer is that a wideband signal is not bandlimited and is not well represented by samples. A nonbandlimited signal cannot be interpolated exactly, not even theoretically. If you try to interpolate among the samples of a wideband signal, the interpolants will be those of a bandlimited signal that has the same samples as the wideband signal but which may be a bad match to the wideband signal between samples.

11.3.2 Timing Controllers (Feedback)

Control of phase and frequency is relatively straightforward and has been covered in Chapter 10. Only timing control is treated here. Three different versions are considered: two related to interpolators and one to selection from among dense samples.

Interpolator Control Parameters

An interpolator computes the lth interpolant $z_{out}(l)$ from an input sequence $\{z_{in}(n)\}$ according to the convolution in (10.16):

$$z_{out}(l) = \sum_{i=I_1}^{I_2} z_{in}(n_l - i) h_I(i + \mu_l)$$

where the interpolant index n_l identifies the input basepoint set, i is the index of coefficients on samples of the FIR interpolating filter with impulse response $h_I(t)$, and μ_l is the fractional interval that identifies the sampled coefficients of the filter. To perform an interpolation, the interpolator needs the control parameters n_l and μ_l. In general, a new μ_l is needed for each interpolant. Furthermore, n_l would not be provided explicitly; some other means identifies the correct basepoint set.

Two methods for feedback control of interpolators were proposed in [10.7]: one based on the period of interpolation, and the other based on frequency. The period method might be preferable for many simulation purposes, but the frequency method has certain advantages in hardware (see [10.7]) and so also needs to be simulated upon occasion. An interpolator does not know the source of its control parameters; it simply obeys whatever parameters are delivered to it.

Period Control. Let t_s be the uniform time interval between input samples and T_i be the time interval between interpolants. The interval T_i can be dynamically varying, but that possible variation is neglected temporarily. Two successive interpolants will have the control-parameter relations

$$lT_i = (n_l + \mu_l)t_s$$
$$(l+1)T_i = (n_{l+1} + \mu_{l+1})t_s \tag{11.1}$$

From their difference one can deduce [10.7] the recursion pair

$$\Delta n_l = n_{l+1} - n_l = IP[\mu_l + T_i/t_s]$$
$$\mu_{l+1} = FP[\mu_l + T_i/t_s] \tag{11.2}$$

where FP[x] signifies the fractional part and IP[x] the integer part of the number x. Only the increment Δn_l is ever needed, never n_l itself.

COMMENT: Observe that this recursion is satisfactory because the overall feedback loop corrects for the small numerical errors that are incurred on each computation. Feedback assures that errors cannot accumulate.

Equations (11.2) determine the next interpolator control parameters from knowledge of the current parameters and the ratio $\xi = T_i/t_s$ of the interpolator and input-sampling periods. The recursion may be regarded as a sort of NCO, with ξ as its period-control signal and μ and Δn as its state variables. A varying ξ can be applied from the loop filter.

Regard Δn as the number of input samples to be shifted into the interpolator filter until the next interpolant is computed. Upon each input shift, Δn is decremented by 1. A new interpolant, a new Δn, and a new μ are computed whenever Δn decrements to zero.

Frequency Control. A more-conventional NCO can be used to control the interpolator according to the difference equation

$$p(n) = [p(n-1) - W(n)] \bmod 1 \tag{11.3}$$

where $p(n)$ is the NCO phase state variable (in units of cycles) and $W(n)$ is the NCO frequency-control signal delivered by the loop filter. Index n implies that the NCO is clocked at frequency $1/t_s$—the same frequency as the incoming samples.* Output frequency of the NCO (rate of cycling of the phase register) is

$$\frac{1}{T_i} = W \frac{1}{t_s} \tag{11.4}$$

Each cycling (underflow) of the NCO phase register indicates that a new interpolant should be computed from the basepoint set currently in the interpolator's reg-

---

*That NCO sampling frequency is sufficient for a downsampling interpolator, but not upsampling. A higher NCO sampling frequency is needed to accommodate upsampling. See Chapter 16 for more information on sample-rate changes.

ister. Furthermore, it can be shown [10.7] that the fractional interval to use for the l th interpolation is

$$\mu_l = \frac{p(n_l)}{W(n_l)} \tag{11.5}$$

The division indicated in (11.5) is too cumbersome to undertake in high-speed digital hardware. Designating the nominal interpolation frequency as $1/T_{io}$, practical implementations might employ the nominal frequency ratio

$$\xi_0 = \frac{1/t_s}{1/T_{io}} = T_{io}/t_s \tag{11.6}$$

from which the fractional interval can be approximated as

$$\mu_l \simeq \xi_0 p(n_l) \tag{11.7}$$

This expedient avoids the division at the cost of small errors in interpolation. It is common for the frequency ratio ξ_0 to be close to the actual ratio T_i/t_s within 1 part per 1000, or even much closer, so the interpolating error [10.7, 10.8] can be negligible.

Selection Control

Let the input-sample interval be t_s and let there be M_s input samples per symbol; M_s need not be an integer. Under conditions where sample selection is needed, M_s tends to be a number in the tens or hundreds, or even more. The choice for M_s is determined by the timing jitter arising from quantization of the strobe instants. Jitter has to be acceptably small; simulation is a useful way to determine what maximum interval is tolerable.

Suppose that there are to be K output samples taken per symbol; one sample out of K is taken as the symbol strobe. Typically, K is a small integer—probably just 1 or 2. An NCO makes a feasible controller for this application; clock the NCO at the input-signal sampling rate $1/t_s$, updating it according to the difference equation

$$p(n) = [p(n-1) + W(n)] \bmod\text{-}1 \tag{11.8}$$

The NCO control signal $W(n)$ is adjusted by the feedback loop such that the average of W is equal to K/M_s, whereupon the NCO output frequency is equal to the desired output sampling rate Kt_s/M_s, on average. The current sample of the input signal is selected as an output sample upon each overflow of the NCO phase register.

Integrate and Dump

Receivers for wideband signals often incorporate *integrate and dump* (I&D) circuits in place of pulse-shaping filters. An I&D is a particular form of *correlation detector*;

it is exactly equivalent to a matched filter for a rectangular pulse of one-symbol duration T.

Operation of an analog I&D is as follows. At $t = 0$, with integrator voltage starting from zero, the circuit integrates the baseband signal applied to it. At $t = T$, the voltage on the integrator is transfered to the next circuit downstream and the integrator is discharged ("dumped") back to zero, ready to start the next integration. The transfer and dump command constitutes the strobe of the integrated signal.

In a simulation, the integration is replaced by summation of samples. There are two inputs to the I&D: a data signal $z_{in}(n)$ and a control signal that initiates a strobe and dump of the I&D. The strobe can be derived from the NCO of (11.8) or from feedforward timing methods to be described later.

11.3.3 Carrier Regenerators

A nonlinear regenerator produces a periodic carrier wave from a data signal with a continuous spectrum. The regenerated carrier wave is filtered and employed as the local carrier reference in the receiver. Many issues arise in the study of regenerators; here the emphasis is on simulation topics. Although you may never have occasion to simulate an analog carrier regenerator (since digital implementations are replacing analog), the simulation issues addressed here are important in other applications as well. Don't skip over this material just because of the "analog" label.

Square-Law Example

Many different regenerator implementations are possible and several of the more common types are examined in the following pages. One widely used type is the *square-law device*; it will be taken as an example to illustrate the characteristics of regenerators as a class.

Consider a time-continuous real carrier-modulated signal

$$s_{in}(t) = x(t) \cos(2\pi f_c t + \theta_c) \tag{11.9}$$

where $x(t)$ is a real baseband signal (to be specialized soon to a baseband PAM signal), f_c is carrier frequency, and θ_c is carrier phase. Regard θ_c as a misalignment that is to be corrected in the receiver's demodulator.

Apply the signal of (11.9) to a square-law device (a square-law rectifier) which delivers a raw output

$$s_{raw}(t) = s_{in}^2(t) = x^2(t) \cos(2\pi f_c t + \theta_c)$$
$$= \tfrac{1}{2} x^2(t) + \tfrac{1}{2} x^2(t) \cos(4\pi f_c t + 2\theta_c) \tag{11.10}$$

Raw output consists of a baseband component around zero frequency and a double-frequency component around $2f_c$. Spectra of s_{in} and s_{raw} are sketched in Fig. 11.2.

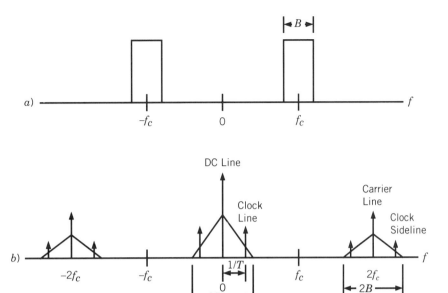

FIGURE 11.2 Idealized spectra associated with square-law regenerator: (*a*) spectrum of input signal $s_{\text{in}}(t)$; (*b*) spectrum of output $[s_{\text{in}}(t)]^2$ of square-law device.

Notice that phase information is preserved in the double-frequency term (as $2\theta_c$) and lost in the baseband term.

Carrier Wave. The rectified modulation $x^2(t)$ has a dc component, which is always real and greater than zero for any real $x(t)$, not just PAM signals. This dc component in $x^2(t)$ gives rise to a dc term in the baseband component of $s_{\text{raw}}(t)$ and also to discrete carrier lines in the spectrum at frequencies $\pm 2f_c$. These lines are shown in Fig. 11.2. The dc line can be used for signal-strength measurement in an AGC loop and the regenerated carrier lines can be used for producing a local carrier for coherent demodulation of the input signal.

Simulation Processing

Refer to Fig. 11.3 for the simulation model of a typical carrier-regenerating synchronizer. Raw output from the nonlinear regenerator is passed through a narrow bandpass filter that is tuned to the desired frequency: $2f_c$ in the square-law example. The filter suppresses disturbances such as additive noise, self-noise [11.37], or interference. A narrowband phase-locked loop is often substituted for the bandpass filter. A limiter follows the filter, to remove amplitude fluctuations from the carrier reference wave. Output from the limiter is at the wrong frequency; it must be divided (by a factor of 2 for the square-law regenerator) before it is applied to the receiver's demodulator. Frequency division also entails phase division by the same factor.

The recovered carrier wave at frequency f_c is applied to the receiver's demod-

312 □ SYNCHRONIZATION

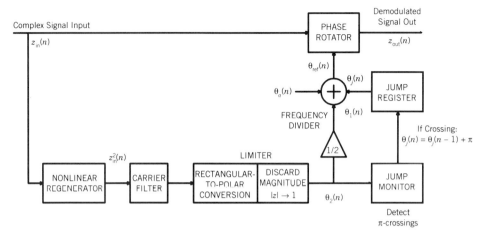

FIGURE 11.3 Simulation processes for carrier regeneration with square-law nonlinearity.

ulator to downconvert the incoming signal to baseband, to provide for coherent demodulation, and to correct for the phase misalignment θ_c. The last operation is the only one of the three that is carried out in most simulations; the other two are just implied, because a passband signal usually is simulated by its complex envelope, which already has zero center frequency.

Complex Envelope. Signal bandwidth is usually much smaller than the carrier frequency in conditions where regenerators can be applied for carrier recovery. Signals of this kind are best simulated by their complex envelopes and not as real signals that include the high-frequency carrier. (Sampling rate requirements for reproducing the carrier are much too burdensome; see Sec. 2.2.) The sampled complex envelope of the carrier signal of (11.9) is given by

$$z_{in}(n) = x(n)e^{j\theta_c} = x(n)(\cos\theta_c + j\sin\theta_c)$$
$$= I(n) + jQ(n) \qquad (11.11)$$

Raw output of a square-law carrier regenerator has a complex-envelope representation

$$z_{2\text{raw}}(n) = z_{in}^2(n) = x^2(n)e^{j2\theta_c} = I^2(n) - Q^2(n) + 2jI(n)Q(n) \qquad (11.12)$$

where subscript 2 indicates selection of the double-frequency term. Raw output of the simulated carrier regenerator is a complex signal since the square of a complex quantity is also complex.

Although a real square-law device produces both baseband and double-frequency components at its output, the simulated complex-envelope operations produce only one or the other, depending on fine details of the model. Compare the model for square-law carrier regeneration of (11.12) with square-law clock regeneration of (11.18).

The raw complex-envelope carrier is passed through a complex-envelope filter (see Chapter 3) with impulse response $h_{car}(n)$ to produce a filtered carrier envelope

$$z_2(n) = z_{2raw}(n) \otimes h_{car}(n) = I_2(n) + jQ_2(n) \qquad (11.13)$$

where $\otimes$ indicates convolution. A carrier filter is likely to be quite simple: typically just a single-section resonant circuit, tuned to the carrier frequency. Envelope response is easily simulated by a one-pole complex filter. Complex output of the filter exhibits both amplitude and phase fluctuations. Amplitude fluctuations are removed by a limiter. For this application, the limiter is conveniently simulated by converting the filtered signal to polar coordinates:

$$z_2(n) = R_2(n) \exp[j\theta_2(n)] \qquad (11.14)$$

and then simply setting $R_2 \equiv 1$, leaving the unit-amplitude exponential

$$z_{2\lim}(n) = \exp[j\theta_2(n)] \qquad (11.15)$$

A polar-coordinates complex sample $z = Re^{j\theta}$ is stored in the computer as magnitude R and angle θ—not $\exp(j\theta)$; the exponential is implied. Thus discarding the polar-coordinate magnitude leaves only the angle $\theta_2(n)$; the limited signal itself is never simulated in this representation. Frequency division by 2 is simulated by halving the phase of the limiter's output, according to

$$\theta_1(n) = \tfrac{1}{2}\theta_2(n) \qquad (11.16)$$

Phase Jumps. Conversion to polar coordinates, combined with subsequent phase halving, introduces unacceptable phase discontinuities unless corrective actions are undertaken. These phase jumps originate because angle $\theta_2(n)$ is calculated in the range $-\pi < \theta \le \pi$ and will exhibit a discontinuity whenever θ_2 passes through $\pm\pi$. Suppose that θ_2 is 179° at one sample time and 181° the next sample time. These angles will be reported as 179° and −179° (or, actually, their equivalent in radians). Farther downstream these angles are divided by 2 to ±89.5°. An event that had originally been a phase advance of just 2° now looks like a phase jump of nearly 180°, which is clearly wrong. This is a particular manifestation of the *phase-unwrapping problem* [11.65], which afflicts many different signal-processing operations carried out in polar coordinates. Something must be done to eliminate the unwrapping jumps.

Phase-unwrapping algorithms [11.65–11.67] become complicated if large sample-to-sample phase variations occur naturally and have to be taken into account. It is then difficult to distinguish between a true phase change and an unwrapping artifact. In carrier-recovery applications, large rapid phase changes in the filtered regenerated carrier occur only if amplitude of the filtered regenerated carrier wave passes nearly through zero. This is a rare event whose probability of occurrence ordinarily is small in most practical situations. On that premise, the phase unwrapping for carrier recovery with a square-law regenerator becomes rather simple:

1. Add a phase $\theta_j(n)$ to $\theta_1(n)$, starting with $\theta_j(0) = 0$.
2. Monitor the phase sequence $\{\theta_2(n)\}$.
3. Whenever $\theta_2(n)$ crosses π (in either direction), add π to $\theta_j(n)$.
4. To keep phases from growing without bound, subtract (add) 2π if phase becomes greater than 2π (less than -2π).

Phase Adjustments. In the absence of disturbances and extraneous phase shifts in the filter, the recovered angle $\theta_1(n)$ would be equal to the input phase θ_c (with an inherent phase ambiguity of 0 or τ). It is standard hardware practice to provide an adjustable phase, denoted θ_a, to compensate for unpredictable extraneous phase shifts. Therefore, the reference phase to be applied to the demodulator (a phase rotator) is

$$\theta_{\text{ref}}(n) = \theta_1(n) + \theta_j(n) + \theta_a \qquad (11.17)$$

Other Carrier Regenerators

A square-law regenerator works very well for a BPSK signal and fairly well for a staggered QPSK signal. It fails entirely for any nonstaggered MPSK signal with $M > 2$. Stated briefly, a reference carrier must be regenerated at frequency $f = Mf_c$ for any signal whose constellation has M-fold angular symmetry. Generation of the Mth harmonic from the input signal requires a nonlinearity of at least Mth degree. Regenerators that produce the Mth harmonic are known as $\times M$ *multipliers*; they have been applied extensively for $M = 4$. Two examples of $\times M$ multipliers are treated in Appendix 11B. They are a fourth-law device (applicable only for $M = 4$) and an absolute-value device.

Simulation of a $\times M$ regenerator is the same as described above for the square-law device, except for the following details:

- Substitute the appropriate regenerator equations for (11.12) of the square-law device, typically $z_{\text{in}}^M(n)$.
- In hardware, the carrier filter is tuned to Mf_c, but that has no effect on the representation of a complex-envelope filter.
- The limiter is unchanged from Fig. 11.2.
- Phase division, after filtering, is by a factor of M, instead of 2.
- Phase unwrapping is required whenever phase θ_M delivered by the limiter crosses π. At each CCW (CW) crossing, add $2\pi/M$ ($-2\pi/M$) to θ_j.
- Reference phase delivered to the phase rotator is $\theta_{\text{ref}}(n) = \theta_M(n)/M + \theta_j(n) + \theta_a$.

Sampling Rate

The objective of regenerator simulation is to model the time-continuous operations. To that end the sampling must be dense enough that the signals generated in the simulation suffer little aliasing distortion (see Chapter 5). Nonlinearities cause spectral spreading. If the input complex envelope is bandlimited to $|f| < B$, aliasing of the input itself is avoided if sampling is performed at a rate $1/t_s > 2B$. A nonlinearity of Mth degree will spread the signal's bandwith to $|f| < MB$, so a sampling rate in excess of $2MB$ is needed to prevent aliasing in the nonlinear operation. Note that the larger sampling rate has to be applied before the nonlinear operation; afterward is too late.

Figure 11.4 illustrates spectral spreading caused by a ×4 multiplier. The spectrum of the original bandlimited QPSK signal is shown in Fig. 11.4a, and the spectrum of the raw output of the ×4 multiplier appears in Fig. 11.4b, with obvious spreading. Because the input spectrum was bandlimited and because the nonlinearity is a fourth-power device (i.e., a *soft* nonlinearity), the spread spectrum is also bandlimited.

The regenerated carrier appears as the discrete line at $fT = 0$; the phase of that dc component is the reference information that is recovered by the synchronizer. Notice also the two side lines at $fT = \pm 1$; these lines are spaced by the symbol rate away from the regenerated carrier and arise because the ×4 multiplier also regenerates a clock wave.

Not all nonlinearities have finite polynomial representations or even infinite-series representations; they cannot necessarily be characterized by a specific degree. The absolute-value device of Appendix 11B is one example of a *harder* nonlinearity that cannot be expanded in a power series. Another example is the rectangular-to-polar conversion, whose amplitude is an absolute value and whose phase is an arctangent. These hard nonlinearities spread a bandlimited input spectrum into a non-bandlimited output. With them, there exists no sampling rate—no matter how large—that avoids aliasing entirely.

When faced with a nonbandlimited signal, a simulationist has to employ a sampling rate sufficient to hold the aliasing errors within tolerable bounds. Aliasing error can be quantified when dealing with signals of known spectra (Sec. 5.1), but the spectra of signals delivered from hard nonlinearities are not easy to evaluate analytically. Computer experiments with different sample rates and engineering judgment are the only tools that can be recommended in difficult cases.

11.3.4 Clock Regenerators

Hardware implementation for a clock-regenerating synchronizer closely resembles that for carrier regenerators. It has a nonlinear regenerator followed by a narrowband filter that delivers a near sine wave at the clock frequency. A limiter removes amplitude variations. A recovered timing reference wave is used directly for symbol strobing, without frequency division. Block diagrams for clock regeneration are nearly identical to those for carrier, except for the absence of a frequency divider. Despite these close similarities of the physical hardware, simulation practices for carrier and clock regeneration differ significantly, as detailed in the succeeding paragraphs.

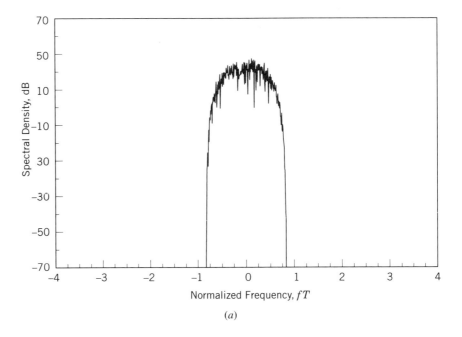

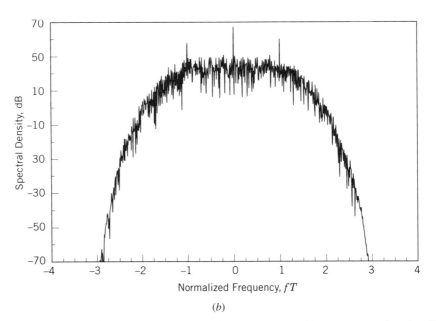

FIGURE 11.4 Spectrum spreading caused by nonlinear regenerators: (*a*) spectrum out of receiver filter. (*b*) spectrum out of fourth-law nonlinearity. Notice regenerated carrier and clock lines. Power spectra from STÆDT simulation of bandlimited QPSK signal: sampling density $M_s = 8$ samples/symbol.

Let the complex envelope of the input signal be as given in (11.11). For purposes of the immediate example, this representation can be regarded as the complex envelope of the carrier-modulated bandpass signal, or can equally well be regarded as the complex baseband signal, after demodulation. Timing is more commonly recovered from the demodulated baseband signal, so that is the viewpoint taken in the discussion of clock regeneration.

Square-law nonlinearities are used extensively in clock regenerators. Raw output from a square-law clock regenerator is

$$z_{\text{raw}}(n) = |z_{\text{in}}(n)|^2 = z_{\text{in}}(n) z_{\text{in}}^*(n) = I^2(n) + Q^2(n) \tag{11.18}$$

Compare this expression to (11.12) for square-law carrier regeneration. A simulated clock regenerator produces a real clock wave (in sampled, computer representation), whereas a simulated carrier regenerator produces the complex envelope of the carrier wave.

Figure 11.5 shows the baseband spectrum of a raw clock resulting from applying the quadratic regenerator nonlinearity of (11.18) to a QPSK signal whose spectrum was shown in Fig. 11.4a. Clock lines are evident at $fT = \pm 1$, as is the approximate doubling of spectral width. The raw clock has the following properties:

- It is a real signal in the simulation, despite being generated from a complex input.
- It is independent of carrier phase θ_c.

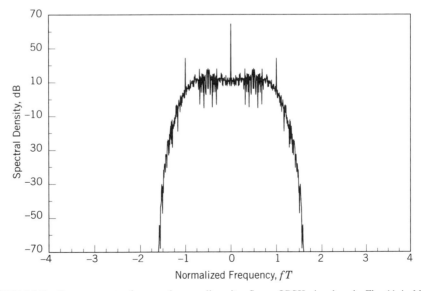

FIGURE 11.5 Spectrum out of square-law nonlinearity. Same QPSK signal as in Fig. 11.4: $M_s = 8$ samples/symbol.

The first property is inherent to all simulations of timing regenerators; the second property is unique to square-law nonlinearities.

If $z_{\text{raw}}(n)$ is real, the clock filter that operates on it has to be a real bandpass filter, too, not a complex-envelope filter. Any of the three methods of implementing real filters, described in Chapters 3 and 6, can be used for this purpose. Clock filters tend to be comparatively simple, typically containing only one or two resonant circuits tuned to the clock frequency $1/T$. Phase-locked loops might be used instead of filters and can be simulated with time-domain single-point operations. (Block mode or frequency-domain operations are incompatible with simulation of feedback loops.)

In hardware, the filtered clock wave has a phase adjustment applied to compensate for implementation deficiencies and is then passed through a limiter to remove amplitude variations. Zero crossings (either positive-going or negative-going) of the resulting square wave are used to establish the strobe instants.

Hard limiting of a real signal is easy enough to simulate; just apply the real filtered clock wave to the *sign* function:

$$\text{sgn}(x) = \begin{cases} +1, & x > 0 \\ 0, & x = 0 \\ -1, & x < 0 \end{cases} \quad (11.19)$$

[In amplitude-continuous signals, the event $\{x = 0\}$ occurs with zero probability and therefore often is not addressed separately in the definition of sgn(x). A computer signal is quantized, so $\{x = 0\}$ can occur with nonzero probability. The function has to return a value for any valid argument.]

Location of Zero Crossings. Unfortunately, the information needed for strobing is the location of the zero crossings; a limiter is just the analog-hardware method for finding the crossing locations. When a hard-limited signal is sampled, crossing locations are uncertain within the sampling interval t_s. Relatively coarse sampling is desired for efficient simulations, leading to excessively coarse timing resolution of a hard-limited timing wave. A different method for recovering the crossing locations is needed for simulation of timing regeneration. Crossings were not used in simulation of carrier recovery—the phase was extracted instead. In similar fashion, crossing locations have to be extracted for simulation of clock regeneration.

Limiting the timing wave destroys some information; it is not possible to retrieve precise crossing locations after limiting. Crossing locations should be extracted from the filtered clock wave before limiting, while full information is still present. The limiter output only provides identification of the two sample locations at either side of a crossing.

Crossings might plausibly be found by inverse interpolation in the samples of the timing wave. That is, interpolate an abscissa as a function of the ordinate values. Inverse interpolation works reliably for linear interpolation but fails for higher-order polynomials if the timing wave is not monotonic over the basepoint set. Since monotonicity cannot be assured over multiple sample points, higher-order inverse interpolation is a hazardous choice for this operation.

Error performance of linear inverse interpolation is analyzed in Appendix 11C. There it is shown that location error Δt can be as large as $0.012T$ for $M_s = 4$ samples/symbol, where T is the symbol interval. This size of error is marginally too large for some applications, but acceptable for others. Appendix 11C also shows that $|\Delta t| < 0.0013T$ for $M_s = 8$, which is more than accurate enough for all but the most demanding applications.

As one practical compromise: If sampling of the timing wave is not dense enough to permit linear inverse interpolation with adequate precision: (1) interpolate additional points into the timing wave in the vicinity of the zero crossing using a good interpolator (e.g., a cubic polynomial), and (2) perform a linear inverse interpolation between the new pair of points of opposite polarity to locate the zero crossing more precisely. As another alternative: Determine the coefficients of an L-degree polynomial that interpolates L sample points in the vicinity of the zero crossing, and solve for the approximate root of the polynomial.

Once the zero crossing has been located, by whatever technique, strobing is performed by providing the zero-crossing specifications (basepoint set identification and fractional interval) to a signal interpolator, which then computes a strobe of the data signal at the specified location. Notice that in the absence of a limiter, or conversion to polar coordinates, or a frequency division, no problem arises with phase unwrapping of the clock wave.

Timing Adjustment. Phase of the timing wave may have to be adjusted to compensate for fixed timing errors caused by filters, delays, and so on. Hardware for a mechanically adjusted phase shifter is fairly easy to build, but simulation of phase shifting of real signals is a nuisance (see Sec. 10.1.2). Rather than simulate a phase shifter of the filtered clock wave, a timing adjustment can be imposed by adding an increment to the inverse-interpolated zero-crossing location. This is just a simple addition (or subtraction) of the type already seen for phase adjustment in carrier recovery (refer to Fig. 11.3 and accompanying text).

11.3.5 Parameter Extractors

A regenerator produces a carrier or clock reference wave from the data signal applied to it. Embedded in the reference wave are the *reference parameters:* phase or timing. An analog receiver makes use of the reference wave itself to perform demodulation or strobing; the reference parameters are called upon implicitly when the reference wave is applied.

A *parameter extractor* derives a number (or a sequence of numbers) representing the reference parameter explicitly. These parameter numbers are applied to a phase rotator or timing interpolator to adjust the phase or timing of a receiver. Since the extracted parameters are represented as numbers, extractors have been realized solely in digital implementation.

In the computer simulation of regenerators, as described in Sec. 11.3.4, it is not feasible to generate reference waves in the same manner as in analog hardware. A carrier wave cannot readily be generated at all in a computer and a clock wave has to be subjected to special processes to derive its crossing locations. Regenerator simulation derives carrier phase and timing-crossing locations, which are *parameters* of their respective reference waves and are not the waves themselves. Simulation of regenerators had to be done by parameter extraction, at least in part.

Pure regenerators cannot be implemented in digital hardware either, for exactly the same reasons that exclude them from simulation. Parameter extractors have been devised to take the place of regenerators in digital implementations. Like regenerators, an extractor is a feedforward method of reference measurement. Extractors and regenerators share many similarities, differing more in detail than in broad principle. Although many different regenerators have been built, only a few extractors have yet been devised.

Carrier Extractors

The most popular phase extractors work on the strobed outputs of the receiver, one sample per symbol. Validity of strobes presupposes that timing has already been established; correct timing of the strobes is assumed throughout this section on carrier extraction. Compare the extractor to a carrier regenerator, which works on a time-continuous signal or on multiple samples per symbol sufficient to approximate time continuity. A carrier regenerator does not require prior acquisition of clock.

Prior establishment of correct timing is not a trivial matter; there are difficult questions relating to clock and carrier acquisition that have not yet been well addressed by the synchronization community. Simulation is one of the few tools available for exploring acquisition and the effects of reference errors. When simulating regenerators, it is necessary for the sampling rate to be large enough to avoid aliasing of the spectral spreading caused by the nonlinear regenerator. If only one sample per symbol is employed for an extractor, considerable aliasing is inevitable. But that aliasing is irrelevant, just as it is irrelevant to strobing itself; the purpose of carrier recovery is to synchronize the carrier, not reconstruct a signal. It can be demonstrated [11.49] that extracting carrier phase from symbol strobes is optimum; it can be accomplished with less accompanying phase jitter than that arising from any continuous-time regenerator.

Let the strobe sequence be represented as $\{z_o(m)\}$, where the subscript o will be used to indicate strobes (before they have been phase corrected) and m is the strobe (symbol) index. A strobe point is complex, resolvable into rectangular components

$$z_o(m) = I_o(m) + jQ_o(m) \qquad (11.20)$$

The best-known article on carrier extraction is by Viterbi and Viterbi [11.24]. Understanding is aided by first examining a related method proposed by Kam [11.28]. An approximation to the maximum-likelihood estimate for carrier phase $\hat{\theta}$, taken over N strobes, can be extracted from a BPSK signal by the algorithm

$$\hat{\theta} = \frac{1}{2} \arctan \frac{2 \sum_N I_o(m) Q_o(m)}{\sum_N I_o^2(m) - Q_o^2(m)} = \frac{1}{2} \arctan \frac{\sum_N \operatorname{Im}[z_o^2(m)]}{\sum_N \operatorname{Re}[z_o^2(m)]} \qquad (11.21)$$

The estimated phase is applied as a control parameter to a phase rotator that corrects the phase of the strobes delivered to the data detector.

Considerable insight can be gleaned from this formula:

- Only one phase estimate is extracted from a block of N strobes; the same estimate is used for phase correction of each strobe in the block. The N strobes are stored, the phase estimate extracted, and only then is the phase correction applied to the stored strobes. An arrangement of this kind is well suited for phase synchronization of signal bursts since no carrier-sync preamble is needed.

- The primary nonlinear operation [inside the summations of (11.21)] on the data signal is square law, much like the square-law regenerator of (11.12). The same terms appear in both. These square-law operations are similar to frequency doubling in a regenerator, except that there is no carrier frequency to be doubled in zero-frequency strobes.

- The summations act as filters of the rectangular components of the squared strobes. This filtering is equivalent to the carrier filters employed with regenerators. Filters precede the hard nonlinearities of division and arctangent, keeping most noise and interference out of these operations.

- The quotient of the sums is $\tan 2\hat{\theta}$; phase angle $2\hat{\theta}$ is extracted from the quotient by applying the arctangent. Division and arctan operations together are amplitude insensitive and so contain an implied limiter. They can be calculated in a computer but are more likely to be implemented by table lookup in hardware.

- The leading factor of $\frac{1}{2}$ is equivalent to the frequency halving needed with BPSK carrier regeneration. Phase unwrapping problems would arise here and at the arctan operation, except for the fact that only one estimate is extracted for the N-block and there can be no phase jump with just one phase. If multiple estimates were needed over contiguous blocks, provisions would have to be made for correct phase unwrapping.

- Many operations mentioned here can be performed readily in the computer or in digital hardware but are virtually impossible in analog circuits.

A comparable algorithm for QPSK is provided by

$$\hat{\theta} = \frac{1}{4} \arctan \frac{4 \sum_N I_o Q_o (I_o^2 - Q_o^2)}{\sum_N (I_o^2 - Q_o^2)^2 - 4 I_o^2 Q_o^2} = \frac{1}{4} \frac{\sum_N \operatorname{Im}[z_o^4(m)]}{\sum_N \operatorname{Re}[z_o^4(m)]} \qquad (11.22)$$

The same comments made after (11.21) apply here, too, except that this formula has a fourth-law nonlinearity.

Viterbi and Viterbi [11.24] treat the complex baseband strobe in its polar form $z_o(m) = R_o(m)e^{j\theta_o(m)}$ and subject it to the memoryless nonlinear process

$$\gamma(m) = F[R_o(m)] \exp[jM\theta_o(m)] \qquad (11.23)$$

That is, they multiply the phase by M (which is the key operation of a $\times M$ multiplier for MPSK signals) and perform an appropriate operation, symbolized by $F[R_o(m)]$, on the magnitude. Multiplied phase is reduced to the interval $-\pi$ to $+\pi$. The magnitude operation need not be R^M—the operation performed by an Mth-law nonlinearity. Different nonlinearities and their comparative performances have been analyzed in [11.71]. Simulation of (11.23) is comparatively simple after $z_o(m)$ is converted to polar coordinates as magnitude R_o and angle θ_o. In hardware, (11.23) most likely would be implemented by table lookup.

The carrier phase angle is extracted from an N-block of $\{\gamma(m)\}$ by the estimator

$$\hat{\theta} = \frac{1}{M} \arctan \frac{\sum_N \mathrm{Im}[\gamma(m)]}{\sum_N \mathrm{Re}[\gamma(m)]} \qquad (11.24)$$

which is much the same form as (11.21) and (11.22). It is necessary to convert $\gamma(m)$ back to rectangular coordinates to compute (11.24).

Clock Extractors

The only clock extractors known at this writing are really hybrids of regenerators and extractors. In these schemes a raw clock wave is regenerated first and then the needed timing parameter is extracted from the phase of the raw regenerated wave. The clock regenerator described earlier was really a hybrid extractor in this sense, where the phase was extracted by interpolating the locations of the zero crossings. A more elegant scheme has been proposed by Oerder and Meyr [11.42], as outlined in succeeding paragraphs.

Oerder and Meyr treat the square-law nonlinearity of (11.18), but any nonlinearity that regenerates a clock might be a candidate. Refer to Sec. 11.3.4 for other examples. Output of the regenerator is the raw timing wave, which can be written in the form

$$z_{\mathrm{raw}}(n) = A \sin\left(\frac{2\pi n t_s}{T} - \frac{2\pi \tau}{T}\right) + v(n) = A \sin\left(\frac{2\pi n}{M_s} - \frac{2\pi \tau}{T}\right) + v(n) \qquad (11.25)$$

where A is reference-clock amplitude (dependent on signal amplitude, pulse waveshape, data statistics, and regenerator algorithm), T is the symbol interval, τ is the unknown timing parameter whose value is to be estimated, and $v(n)$ is noise—both additive and self—accompanying the raw clock wave. Remember that the raw clock wave is real even when regenerated from a complex data signal.

Clock timing has to take into account the effects of all filters in the transmission link, so the raw clock should be regenerated after the last filter (the pulse-shaping filter) in the receiver. Additional prefilters may be placed prior to the regenerator, in the clock path, to reduce self-noise [11.36].

How many samples per symbol are needed? If the original data signal is strictly bandlimited to $|f| < 1/T$, a sampling rate of $1/t_s \geq 2/T$ would suffice to recover the data signal from its samples. But the regenerator spreads the spectrum of the raw clock; a larger sampling rate is needed to avoid (or minimize) aliasing of the raw clock. A square-law nonlinearity (just about the softest nonlinearity imaginable) doubles the spectral occupancy, and other nonlinearities spread it even wider. That seems to imply the need for a sampling rate of at least $4/T$.

On the other hand, a practical implementation only has to sample at a high enough rate to avoid significant aliasing into the bandwidth of the clock filter; it does not have to avoid aliasing completely since there is no requirement for accurate reconstruction of the raw timing wave. Therefore, a sampling rate greatly in excess of $4/T$ may not be needed for many practical regenerators. (If a hard regenerator is to be simulated, the spectrum of its raw timing wave should be scrutinized to determine the sampling rate needed for acceptable aliasing.)

If the noise sequence $\{v(n)\}$ were white and Gaussian, a maximum-likelihood estimate for timing τ would be appropriate. Self-noise is not likely to be white and no component of the noise $v(n)$ is likely to be Gaussian after the nonlinear operation in typical regenerators. Nonetheless, pretending that $v(n)$ is white and Gaussian and applying maximum likelihood anyhow leads to the estimate

$$\hat{\tau}/T = \frac{1}{2\pi} \arctan \frac{\sum_{n=0}^{N-1} z_{\text{raw}}(n) \sin(2\pi n/M_s)}{\sum_{n=0}^{N-1} z_{\text{raw}}(n) \cos(2\pi n/M_s)} \qquad (11.26)$$

where N is the size of a block of samples over which the timing estimate is made. To minimize end effects, the number of symbols in the block should be an integer.

Equation (11.26) for timing extraction is similar to (11.21), (11.22), and (11.24) for phase extraction. That is not surprising, since each estimator extracts a phase; differences lie in the the nature of the reference parameter being extracted and in the details of the regenerator/extractor nonlinearities. Comments following (11.21) also apply, in the main, to (11.26).

The summations in (11.26) will be recognized as the DFT coefficients for the clock-rate component at $f = 1/T$ contained in the raw clock wave $s_{\text{raw}}(n)$. No claim is made that the DFT is the optimum clock filter; it is only convenient. Notice that if $M_s = 4$, the sine and cosine in (11.26) take on values $\ldots, 1, 0, -1, 0, 1, \ldots$, which greatly simplifies hardware implementation.

An interpolator accepts the estimate $\hat{\tau}/T$ and interpolates one strobe per symbol from the data signal (refer to Fig. 11.1c). An interpolator requires for its control parameters the fractional interval μ_l and identification n_l of the basepoint set; it does not use $\hat{\tau}/T$ directly. One set of parameters has to be converted to the other.

To that end, suppose that interpolants are to be computed at the time instants $\{t_l\}$ when the fictitious continuous timing wave $\sin[2\pi(t - \tau)/T]$ passes through positive-going zero crossings. (In reality, a location offset from the crossings may be necessary; see below.) Those crossings occur at $t_l - \tau = lT$, l = integer, subject to $t_l \geq 0$.

Interpolator control parameters are defined by $t_l = (n_l + \mu_l)t_s$, where n_l identifies the lth basepoint set and μ_l is the fractional interval for the lth interpolant. Combining these equations and applying $T = M_s t_s$ gives

$$n_l + \mu_l = M_s \left(l + \frac{\hat{\tau}}{T} \right) \qquad (11.27)$$

which identifies the required interpolation parameters. Index l starts at $l = l_0$, where l_0 is the smallest integer such that n_{l_0} and μ_{l_0} are both nonnegative. Index n starts at $n = 0$. Notice that this algorithm does not require M_s = integer.

Once the parameters n_{l_0} and μ_{l_0} have been determined, subsequent n_l and μ_l parameters can be calculated recursively, without accumulating n and l. Define $\Delta n_l = n_{l+1} - n_l$. Subtract two successive values of (11.27) for $l+1$ and l to obtain $\Delta n_l + \mu_{l+1} = M_s + \mu_l$. Therefore,

$$\Delta n_l = \text{IP}[M_s + \mu_l]$$
$$\mu_{l+1} = \text{FP}[M_s + \mu_l] \qquad (11.27a)$$

where IP[x] means the integer part and FP[x] means the fractional part of x.

Unlimited recursion is not permissible because numerical errors will accumulate and the timing estimate would drift off from its correct value. Since the timing scheme is open loop, no feedback is present to hold the estimate close to zero error. New estimates have to be extended at frequent intervals (e.g., each data block). Compare to (11.2), which is a recursion for a feedback synchronizer.

As with any feedforward synchronizer, a timing adjustment τ_a might have to be added to $\hat{\tau}$ to compensate for stray delays in the system or to offset the strobe instant away from the zero crossings of the timing wave. This is the same kind of adjustment encountered earlier with phase θ_a in (11.16).

The foregoing scenario implies that the same timing estimate will be used for all symbols in the N-block. That is not necessary; a new estimate over an overlapping N-block could be computed for each strobe by altering details slightly. Repeated estimates require attention to phase unwrapping because of the arctangent function in (11.26).

11.3.6 Error Detectors

An error detector measures reference error on the signal itself at the location where the signal is to be used—the receiver's output. A feedback loop drives the measured error toward a null. For this reason a feedback synchronizer needs no external adjustment of stray errors of the kind seen in feedforward synchronizers. If the stray errors drift with time and temperature—as they always do in analog circuits—the feedback loop automatically compensates for the drift.

Frequency Error Detectors

Automatic frequency control (AFC) loops have had a long history in radio receivers but have been applied to data-signal reception only comparatively recently. See [11.52] to [11.64] and [11.72] for a goodly portion of the relevant literature. The approaches found in the literature are not especially difficult to simulate; methods of simulation do not differ greatly from the methods that apply to phase- or timing-error detectors, whose principles are comparatively well understood and which are treated in subsequent sections.

Principles of frequency control are under active investigation at the time of this writing (1991) and there are fundamental matters that still need to be resolved. A communications designer reading this material in the future would be well advised to consult literature current at that time instead of relying upon the incomplete knowledge of this moment.

Because frequency-error detectors are much less used, because the simulation methods are not much different, and because the principles of frequency measurement still are not nearly so well established as for phase- and timing-error detectors, the treatment of frequency is limited to these few comments and the reader is referred to the literature and the next sections, which treat phase and timing detectors.

Phase Error Detectors

In analog hardware, phase detection typically has been performed on time-continuous signals; the resulting error signal $u_\theta(t)$ has also been time continuous. In digital hardware the measurements have to be sampled, so u_θ is also sampled. It can be shown [11.49] that the best phase-error measurement is obtained on the matched-filter strobes from the receiver, one strobe per symbol. Once strobe timing has been established, any denser sampling for phase measurement actually degrades synchronization. Therefore, a typical digital carrier-sync loop will generate phase-error samples $u_\theta(m)$ once per symbol, where m is the symbol index.

Simulations, of course, have to be able to model either analog or digital implementations. To that end, simulation of time-continuous operations will compute $u_\theta(n)$, where n is the sample index and M_s samples are taken in each symbol interval. Algorithms for $u_\theta(n)$ and $u_\theta(m)$ tend to be virtually identical in their formulas, as will be seen in the examples below. Empty parentheses $u_\theta(\cdot)$ will be employed to indicate where that is true.

A phase detector computes $u_\theta(\cdot)$ as a real control signal from a complex data signal $z_o(\cdot) = I_o(\cdot) + jQ_o(\cdot)$. The best-known phase-error detector for BPSK signals, or any double-sideband suppressed-carrier signal (not necessarily a data signal) is the famous Costas detector [11.15]:

$$u_\theta(\cdot) = Q_o(\cdot)I_o(\cdot) = \tfrac{1}{2}\operatorname{Im}[z_o^2(\cdot)] \qquad (11.28)$$

This is a quadratic algorithm that has the same steady-state phase-fluctuation performance as a square-law regenerator with the same filtering [11.8]. Slightly improved tracking performance—the same as achieved with an absolute-value regenerator—is obtained from

$$u_\theta(\cdot) = Q_o(\cdot)\operatorname{sgn}[I_o(\cdot)] \qquad (11.29)$$

In a BPSK signal, the receiver's estimate of the current symbol is decided as $\hat{a}_m = \operatorname{sgn}[I_o(m)]$, so the strobed version of (11.29) may be rewritten as

$$u_\theta(m) = Q_o(m)\hat{a}_m \qquad (11.30)$$

which points up the decision-directed (DD) character of this version. Equation (11.28) is for a non-data-aided (NDA) version.

For nonstaggered QPSK signals, the quartic NDA algorithm

$$u_\theta(\cdot) = I_o(\cdot)Q_o(\cdot)(I_o^2 - Q_o^2) = \tfrac{1}{4}\operatorname{Im}[z_o^4(\cdot)] \qquad (11.31)$$

is sometimes employed. It has resemblances to the quartic regenerator of Appendix 11B. A more common implementation for nonstaggered, balanced QPSK is

$$u_\theta(\cdot) = Q_o(\cdot)\operatorname{sgn}[I_o(\cdot)] - I_o(\cdot)\operatorname{sgn}[Q_o(\cdot)] \qquad (11.32)$$

The data decision for nonstaggered QPSK is $\hat{c}_m = \hat{a}_m + j\hat{b}_m = \operatorname{sgn}[I_o(m)] + j\operatorname{sgn}[Q_o(m)]$, so the DD version of (11.32), performed on strobes, becomes

$$u_\theta(m) = \hat{a}_m Q_o(m) - \hat{b}_m I_o(m) = \operatorname{Im}[c_m^* z_o(m)] \qquad (11.33)$$

where $\operatorname{Im}[\cdot]$ indicates imaginary part and the asterisk * indicates complex conjugate. This same algorithm also applies to signals with higher-order constellations [11.46]. Furthermore, (11.33) also applies to staggered PAM-QAM if the $I_o(m)$ and $Q_o(m)$ strobes are taken at staggered time instants.

In whatever manner it may be generated, $u_\theta(\cdot)$ is applied to a loop filter that either controls an NCO or generates a phase-error estimate directly. The phase from the NCO or the filter is delived to a phase rotator, thereby closing a phase-locked feedback loop, as in Fig. 11.1a. Disturbances accompanying the error-detector sequence

$\{u_\theta(\cdot)\}$ are filtered by the narrow bandwidth of the feedback loop. See Chapter 14 for a discussion of loop filters and feedback loops.

Timing-Error Detectors

Detection of timing error historically has been performed on time-continuous waveforms in analog hardware. In simulations and in digital hardware the measurements have to be made on sample sequences. Superior performance can be obtained from sampled measurements. Simulations are obliged to model either time-continuous or sampled methods, as dictated by the specific application.

Phase-error detectors were found to be much the same for both time-continuous and time-discrete algorithms. This similarity was exploited by writing the output $u_\theta(\cdot)$ of the phase-error detector with an empty argument, signifying the applicability of the same formula to both sampled and continuous operations. That duality is not generally valid for timing-error detectors; a sampled algorithm does not necessarily have a time-continuous equivalent, and vice versa.

The phase-error detectors presented above were all memoryless devices; $u_\theta(\cdot)$ depends only upon signal conditions at the instant for which it is calculated, not earlier or later. By contrast, timing-error detectors necessarily contain memory; the calculated timing-error indication depends, in some sense, upon signal properties at earlier or later times.

For these reasons, output u_τ of the timing-error detector will not be written with empty parentheses. It will be written as $u_\tau(n)$ (where n is the sample index) when timing error calculations are performed for each sample of the signal and as $u_\tau(m)$ (where m is the symbol index) when error calculations are performed once per symbol interval. The first notation is appropriate when simulating time-continuous operations and the second when simulating typical sampled algorithms where only one error sample is computed per signal pulse.

There are only a few kinds of phase-error detectors, and most are closely related to one another, as indicated above. But there is a bewildering array of different timing-error detectors; their interrelations, if any, are difficult to recognize. Reference [11.8] lists several different kinds and many others may be found in the literature. Rather than attempting complete coverage or even a thorough summary, the paragraphs that follow just put forward some examples to explore their simulation issues. Modeling of other types has to be tailored to the specific formulations.

Maximum-Likelihood Examples. A strobe-rate sampled DD algorithm, derived [11.46] from maximum-likelihood principles, is expressed as

$$u_\tau(m) = \text{Re}[\hat{c}_m^* \dot{z}_o(m)] = \hat{a}_m \dot{I}_o(m) + \hat{b}_m \dot{Q}_o(m) \qquad (11.34)$$

where the overdot indicates differentiation with respect to time, the asterisk * indicates complex conjugate, Re[·] indicates real part, and the other notation is the same as defined above for phase-error detectors. This formula applies to all PAM–QAM signal formats, both nonstaggered or staggered channels, and either constant-amplitude or multiamplitude symbol values. For staggered operation, the $\dot{I}(m)$ and $\dot{Q}(m)$ strobes have to be time staggered.

Memory in this algorithm resides in the differentiation; it is not possible to extract the derivative of a signal from only one sample of the signal itself. Since a bandlimited signal can be reconstructed perfectly by interpolation, its derivative can also be reconstructed by interpolation. Just as perfect interpolation requires an infinite number of samples, so does perfect differentiation. In practice, an approximate differentiator is used.

For proper differentiation, sampling has to be dense enough to avoid aliasing; sampling only once per symbol interval is not adequate. If the excess bandwidth is less than 100%, the signal spectrum is confined to $|f| < 1/T$, so a sampling rate of $2/T$ avoids aliasing (where T is the symbol interval). Sampling at $2/T$ is enough to permit differentiation of a bandlimited signal and, in consequence, to perform a measurement of timing error.

The simplest approximation to differentiation of a sample sequence $\{s(n)\}$ is

$$\dot{s}(n) \simeq \tfrac{1}{2}[s(n+1) - s(n-1)] \qquad (11.35)$$

where unit-time separation of samples has been assumed tacitly.

Two-Sample Error Detectors. Aliasing notwithstanding, it appears from experience that good timing recovery can be obtained on any NRZ PAM signal, irrespective of its pulse shape and bandwidth, using just two samples per symbol. It is not necessary that the signal be bandlimited or that accurate differentiation be attainable. This observation has not been proven, but no counterexample has been found either.

At two samples per symbol, one sample falls at the desired strobe time and the other falls midway between two adjacent strobe times. For convenient notation, retain the strobe index m on the strobe samples and designate the index of the midway sample just preceding the mth strobe sample as $m - \tfrac{1}{2}$. Applying the approximate differentiator (11.35) to the DD timing-error algorithm (11.34) gives

$$\begin{aligned} u_\tau(m) &= \hat{a}_m[I_o(m + \tfrac{1}{2}) - I_o(m - \tfrac{1}{2})] + \hat{b}_m[Q_o(m + \tfrac{1}{2}) - Q_o(m - \tfrac{1}{2})] \\ &= \mathrm{Re}\{\hat{c}_m^*[z_o(m + \tfrac{1}{2}) - z_o(m - \tfrac{1}{2})]\} \end{aligned} \qquad (11.36)$$

after suppressing the irrelevant factors of $\tfrac{1}{2}$ from (11.35). Memory in the algorithm is much more apparent in this approximate form.

One would scarcely expect accurate differentiation from a two-point difference formula with a span as large as $\pm T/2$, and it is a poor approximation indeed. Nonetheless, the available evidence suggests that the algorithm (11.36) performs well for timing recovery, and that is the criterion of success.

With accurate differentiation, feedback around the algorithm of (11.34) drives the strobe point to the peak of the signaling pulse $g(t)$, irrespective of pulse shape (provided that its derivative is continuous). Approximate differentiation with (11.36) does not perform quite so well; the strobe will be driven to a location midway between two equiamplitude points on the pulse shape that are separated by T. That will be off the peak if the pulse is not symmetric about its peak.

Equation (11.36) is a DD algorithm; an NDA version is created by substituting the strobe values for the decisions:

$$u_\tau(m) = I_o(m)[I_o(m + \tfrac{1}{2}) - I_o(m - \tfrac{1}{2})] + Q_o(m)[Q_o(m + \tfrac{1}{2}) - Q_o(m - \tfrac{1}{2})]$$
$$= \mathrm{Re}\{z_o^*(m)[z_o(m + \tfrac{1}{2}) - z_o(m - \tfrac{1}{2})]\} \quad (11.37)$$

This is a quadratic algorithm and can be shown to be independent of carrier phase for nonstaggered signals. Whereas the DD version works with small to vanishing excess bandwidth, the quadratic NDA version works progressively worse as excess bandwidth diminishes, failing entirely for zero excess bandwidth. Failure with small excess bandwidth is a property of all quadratic timing nonlinearities, as has been seen before with regenerators.

Yet another variation [11.41] employs the same two samples per symbol of (11.36) or (11.37) but generates less self-noise with BPSK or QPSK signals. The quadratic NDA version is

$$u_\tau(m) = I_o(m - \tfrac{1}{2})[I_o(m) - I_o(m - 1)] + Q_o(m - \tfrac{1}{2})[Q_o(m) - Q_o(m - 1)]$$
$$= \mathrm{Re}\{z_o^*(m - \tfrac{1}{2})[z_o(m) - z_o(m - 1)]\} \quad (11.38)$$

and the DD version is

$$u_\tau(m) = I_o(m - \tfrac{1}{2})[\mathrm{sgn}[I_o(m)] - \mathrm{sgn}[I_o(m - 1)]]$$
$$+ Q_o(m - \tfrac{1}{2})[\mathrm{sgn}[Q_o(m)] - \mathrm{sgn}[Q_o(m - 1)]] \quad (11.39)$$

These two algorithms find the average location of the zero crossings in transitions between symbols and place the strobe midway between the crossings. If the pulse shape is symmetric, the strobe lands on the pulse peak.

Sequential Edge Detectors. Another important class of timing-error detectors is used in enormous quantities in clock recovery from magnetic disks, among other applications. In this class, timing information is deemed to reside entirely in the locations of the zero crossings of a baseband real binary data signal. Timing error during the mth symbol interval is measured as

$$u_\tau(m) = [\tau_r(m) - \tau_v(m)]e_d(m) \quad (11.40)$$

where $\tau_r(m)$ is the time location of a zero crossing (of either polarity) of the signal in the mth symbol interval and $\tau_v(m)$ is the time of a zero crossing (say, positive-going) of the voltage-controlled oscillator (VCO) in the timing loop.

Not all symbol intervals have zero crossings of the data signal. An editing function $e_d(m)$ is essential for satisfactory operation of this class of timing-error detector. The editing function has a value 1 if a signal crossing is present and a value 0 if a crossing is not present in the mth interval. Editing is crucial, but its details—particular to each application—are omitted here.

Timing-error detectors working on this principle are almost always implemented with flip-flops and gates. For this reason they are called *digital phase detectors* in common parlance. Yet the quantity of interest $(\tau_r - \tau_v)$ is an analog quantity—a pulse width in most implementations—so the label "digital" is misleading. To be sure, the time difference is sampled once per symbol interval, but it is not quantized and it does not appear in analog circuits as a digital number.

The fine time resolution attainable in the actual circuits should be preserved in simulations; the two zero-crossing times should not be quantized to the simulation's relatively coarse sampling interval t_s. Methods for inverse interpolating a crossing location of a bandlimited signal were discussed above for clock regenerators and in Appendix 11C; the same principles apply to locating τ_r.

Locating τ_v, the VCO crossing time, is done somewhat differently. A VCO is simulated by an NCO, whose output samples occur at time increments t_s. Output of the NCO has to be interpolated somehow to finer timing resolution. An NCO indicates the passage of another cycle (another symbol interval) by an overflow flag. The NCO is controlled by its control word $W(n)$ so that it overflows once per symbol interval while accumulating a phase increment $W(n)$ on each sample interval. Designate the contents of the NCO phase register as $p(n)$—phase measured in cycles. The NCO difference equation is

$$p(n) = [p(n-1) + W(n)] \bmod\text{-}1 \qquad (11.41)$$

Both p and W are fractional numbers between 0 and 1.

Phase $p(n)$ may be regarded as the t_s-spaced samples of a time-continuous phase $p(t)$. An overflow occurs when $p(t)$ reaches one cycle exactly and jumps back instantaneously to zero. This event is never captured exactly in the sampled version but can be deduced from inverse interpolation among the samples.

Let the NCO crossing take place at $t = (n - 1 + \epsilon_v)t_s$ just before the nth sample time, where ϵ_v is a fractional interval $0 \leq \epsilon_v < 1$. If $p(t)$ can be approximated as a linear function of time, inverse linear interpolation provides

$$\epsilon_v = 1 - \frac{p(n)}{W(n)} \qquad (11.42)$$

Both $p(n)$ and $W(n)$ are numbers that exist in the NCO module at the nth sample time and so are accessible for the crossing interpolation.

To complete the calculation of the timing difference: Suppose that the signal's zero crossing has been located by the method of Appendix 11C and Fig. 11C.1 at $\tau_r(m) = [n-1+i+\epsilon_r(m)]t_s$ and the NCO's overflow located at $\tau_v(m) = [n-1+\epsilon_v(m)]t_s$. Then the timing difference is

$$\frac{\tau_r(m) - \tau_v(m)}{t_s} = i + \epsilon_r(m) - \epsilon_v(m) \qquad (11.43)$$

As always, the computer has no absolute knowledge of t_s and therefore works with time ratios.

11.4 SYNCHRONIZER PROCESSES

The synchronizers described above have all been examples of devices encountered in practice. Many more examples could be given, each differing in some degree from those included in these pages, and each requiring its own special processes for simulation.

No simulation program could incorporate every possible synchronizer configuration; the number and variety are much too large for that to be feasible. Instead, a well-equipped simulation facility needs a good library of elementary processes that can be put together in different combinations to simulate just about any model that anyone could imagine.

From the examples of this chapter, those elementary processes should include such things as:

- Nonlinearities: power law, absolute value, sign, and so on.
- Rectangular ⟷ polar conversions
- Select real or imaginary parts
- Multiply, divide, add, and subtract
- Sum
- Filter
- Arctangent

11.5 KEY POINTS

This has been a long chapter, crammed full of small details and consequently not easy to grasp. Despite its length, it presents only a few high points of its subject; major issues have been passed over without even being mentioned. Synchronization is too large a subject to be covered thoroughly in one chapter. This summary of the chapter covers fundamental ideas and skips the details. Models in the book should be regarded as examples to be used as a guide in developing other models for specific applications.

Synchronization arises in simulations in two different ways:

1. Suppose that the user does not wish to investigate synchronization. Nonetheless, synchronization of some sort is almost always compulsory because the model of the communications link introduces delays and phase shifts that have to be compensated in the receiver. Stray delays and shifts are invariant over the course of most simulations, so the necessary compensation can be determined just once, in a calibration procedure before the main simulation run. The correlation method of Appendix 11A is attractively simple when applicable.

2. If the user wishes to investigate the dynamic behavior of synchronizers that recover their references from the received signal, the synchronizer has to be modeled fully. Modeling of recovery synchronizers constitutes the greater part of this chapter.

11.5.1 Recovery Configurations

Two basic configurations arise: feedforward and feedback. The feedforward configuration is further subdivided into regenerators and extractors. The feedforward configuration recovers a reference in a path that is parallel to the data signal's path; a recovered reference wave (from a regenerator) or reference parameter (from an extractor) is applied to the signal to correct the phase or timing (see Fig. 11.1b and c). A feedback configuration measures reference error directly from the data signal, filters that error, generates an estimate of the reference parameter, and feeds that back to correct the signal prior to the error detector (see Fig. 11.1a). Synchronizers might be non-data-aided (NDA) or decision-directed (DD). During a preamble with known data pattern, a synchronizer might be data-aided (DA).

11.5.2 Synchronizer Elements

Correctors

Phase correction is performed by phase rotators; frequency correction is performed by phase rotators plus NCOs (see Secs. 10.1 and 10.2). Timing correction is performed by interpolators (Secs. 10.3 and 11.3.1) or by sample selection (Sec. 11.3.1).

Controllers

An NCO can be regarded as a controller for a frequency shifter. Phase shifters do not necessarily have distinct controllers. Timing of an interpolator or selector is controlled by an NCO [refer to equations (11.3) to (11.7)] based on strobe frequency, or by a pair of recursive difference equations (11.2) based on strobe period.

Regenerators and Extractors

A regenerator (Secs. 11.3.3 and 11.3.4) is a nonlinear device that regenerates a carrier or clock wave from a data signal that lacks either. The wave is filtered to remove disturbances and applied to the receiver's demodulator or interpolator, as appropriate, to demodulate the signal with correct phase or to sample the strobes at the correct times.

An extractor (Sec. 11.3.5) is a nonlinear device that extracts a reference parameter—carrier phase or clock timing—from a data signal. The extracted parameter is filtered to remove disturbances and applied to the phase or clock corrector, as appropriate, to correct the signal's phase or timing.

Extractors and regenerators are closely related. Simulation of analog regenerators involves extraction of parameters since full-fledged regeneration is not feasible on a computer. The only known methods of timing extraction involve prior regeneration of a raw clock, whose phase is extracted as the timing parameter.

Error Detectors

A reference error detector (Sec. 11.3.6) is a nonlinear device whose output is an indication of the reference error in the data signal at its input. Frequency error detectors are still being researched; refer to [11.52] to [11.64] for models of algorithms that have been applied. Phase-error detectors [(11.28) to (11.33)] are well established and closely interrelated. A great profusion of timing-error detectors exists. The text provides a few examples [(11.34) to (11.43)], but many others are found in practical synchronizers.

Conclusion

Because so many different implementations of synchronizer elements are encountered, it is not possible for a simulation program to furnish full models of all versions. Instead, the program should contain a versatile set of elementary processes that the user can combine into models for the synchronizer elements. This need for combinations will arise in other areas besides synchronizers; any good simulation program has provision for ready combining of processes.

11.5.3 STÆDT Processes

Useful for Hardwire Synchronization

    CORF: correlation (circular blocks, using FFT)

    CORT: correlation (linear blocks, in time domain)

    SRCH: search for extremum

Useful for Models of Recovery Synchronizers

    CHRT: interpolator and rate changer; control elements included

DECM: decimates sample sequence (selects strobes)

I&D: integrate and dump

INTG: integrator of control signals

INTP: interpolator with cubic polynomial; control elements omitted

LET, LETS: various operations on control and data signals; some operations useful for synchronizer elements

NCO: number-controlled oscillator

RTAT: phase rotator

APPENDIX 11A: SYNCHRONIZATION BY CORRELATION

First, an argument is presented to justify the use of correlation to establish synchronization, and then the simulation model is described.

Derivation

Let the receiver output PAM signal, after filtering and before strobing, be represented as

$$r(t) = A_0 \sum_m c_m g(t - \tau_0 - mT) e^{j\theta_0} \qquad (11A.1)$$

and denote the transmitter weighted-impulse signal as (see Sec. 7.3.1)

$$d(t) = \sum_l c_l \delta(t - lT) \qquad (11A.2)$$

Both signals are complex and time continuous. Sampling of signals is treated below; a time-continuous model is used at first to simplify explanations. This particular model is nonoffset, but offset signals are handled similarly. Integers m and l both refer to the symbol index, the symbol sequence is $\{c_m\}$, and complex symbols are transmitted at intervals T. Impulse reponse of the entire link (which is assumed linear) is $g(t)$. Summations are taken over the length of a data block.

In transmission, the signal experiences an attenuation A_0, a phase shift θ_0, and a delay τ_0. These parameters are not readily determined a priori; it is the task of the synchronizer to estimate τ_0 and θ_0 so that the simulated receiver can properly demodulate and detect the signal.

The cross correlation between $r(t)$ and $d(t)$ is

APPENDIX 11A: SYNCHRONIZATION BY CORRELATION □ 335

$$\psi_{rd}(\tau, t) = E[r(t)d^*(t - \tau)]$$

$$= A_0 e^{j\theta_0} E\left[\sum_m \sum_l c_m c_l^* g(t - \tau_0 - mT)\delta(t - \tau - lT)\right] \quad (11A.3)$$

where * indicates complex conjugate. Cross correlation is time dependent (cyclostationary). Time-averaged cross correlation is

$$\overline{\psi}_{rd}(\tau, T_0) = \frac{1}{T_0}\int_0^{T_0} \psi_{rd}(\tau, t)\, dt$$

$$= \frac{A_0 e^{j\theta_0}}{T_0}\int_0^{T_0} E\left[\sum_m \sum_l c_m c_l^* g(t - \tau_0 - mT)\delta(t - \tau - lT)\right] dt$$

$$= \frac{A_0 e^{j\theta_0}}{T_0} E\left[\sum_m \sum_l c_m c_l^* g[\tau - \tau_0 + (l - m)T]\right]$$

$$= \frac{A_0 e^{j\theta_0}}{T_0} \sum_m \sum_l g[\tau - \tau_0 + (l - m)T]E[c_m c_l^*] \quad (11A.4)$$

where T_0 is the observation interval of the signals. The third line comes about because of the sifting property of the delta function $\delta(t)$.

Now assume that the random data sequence has the properties

$$\begin{aligned} E[c_m c_l^*] &= E[|c_m|^2] = C^2, & l &= m \\ E[c_m c_l^*] &\simeq 0, & l &\neq m \end{aligned} \quad (11A.5)$$

whereupon the time-averaged cross correlation reduces to

$$\overline{\psi}_{rd}(\tau, T_0) \simeq \frac{A_0 C^2 e^{j\theta_0}}{T_0} \sum_m g(\tau - \tau_0) = \frac{A_0 C^2 e^{j\theta_0}}{T} g(\tau - \tau_0) \quad (11A.6)$$

This result comes about because there are T_0/T symbols in the time interval T_0 and thus T_0/T equal summands [equal because $g(\tau - \tau_0)$ does not depend upon index m]. The sum in (11A.6) reduces to $(T_0/T)g(\tau - \tau_0)$, so the magnitude of the cross correlation is

$$|\overline{\psi}_{rd}(\tau, T_0)| \simeq \frac{A_0 C^2}{T} |g(\tau - \tau_0)| \qquad (11\text{A}.7)$$

Clearly, $|\overline{\psi}_{rd}|$ is maximum for that value of $\tau = \hat{\tau}$ corresponding to the peak magnitude of the signaling pulse $|g(t)|$. Furthermore, if $g(t)$ is real, the angle of the maximum correlation is the phase shift θ_0.

COMMENT: Correlation is an excellent method for hardwire synchronization if the pulse peak is the optimum instant for symbol strobing. The peak is always optimum if the pulse is even symmetric about the peak. The pulse is always symmetric if the receiver filter is matched to the incoming pulse shape. But practical receivers are not necessarily matched and filtered pulses are not necessarily even, so maximum correlation may not necessarily mark the optimum strobe location.

Computer Operations

Coarse Estimates. A simulated signal is sampled, not time continuous. Only one realization is available, not the whole ensemble. Only the time averaging can be performed; the statistical expectation cannot be determined from one realization. A satisfactorily representative data sequence is needed with good autocorrelation (see Sec. 7.1).

The sampled cross correlation is defined as

$$\psi_{rd}(p) = \sum_n r(n) d^*(n - p) \qquad (11\text{A}.8)$$

where $r(n)$ and $d(n)$ are the sampled versions of $r(t)$ and $d(t)$ as above, with n being the time index. Note that most samples of $d(n)$ are zero since $d(t)$ is an impulse train. The weighted impulses, of course, are replaced by weighted unit samples.

If the processes causing delay of $r(n)$ relative to $d(n)$ are all causal (as will be true in any time-domain simulation), only positive values for p need be included in the correlation; negative p corresponds to negative delay, which is impossible in a causal operation. However, frequency-domain simulations can be noncausal, so a correlation process also ought to include provision for negative delays.

Fine Estimates. Suppose that the maximum of $|\psi_{rd}(p)|$ is found at $p = p_0$. You then know that the true maximum lies in the interval between $(p_0 - 1)t_s$ and $(p_0 + 1)t_s$; a fine search of this region is needed to determine the location of the true maximum. Designate the location of the true maximum as

$$\hat{\tau} = t_s(p_0 - 1 + \hat{\epsilon} + \hat{\mu}) \qquad (11\text{A}.9)$$

where $\hat{\epsilon} = 0$ or 1, depending upon whether the true maximum precedes or follows

p_0, and $\hat{\mu}$ ($0 \le \hat{\mu} < 1$) is the fractional interval. An interpolator or fractional delay element requires $\hat{\mu}$ for correction of timing, and p_0 is needed for strobing. In terms of the notation of Sec. 7.4, μ is the fractional-interval control signal value delivered to an interpolator. The correlation ψ_{rd} itself is used only to discover the optimum p_0, μ, and ϵ and is not communicated to other synchronization processes.

Correlation is performed on the delayed signals by conjugating the impulse train, multiplying it sample by sample against the received signal, and summing the complex products. Correlation is a complex scaler for any particular choice of p_0, ϵ, and μ.

Fine search is accomplished by evaluating ψ_{rd} for different trial values of μ and ϵ, and determining the location $(p_0, \hat{\mu}, \hat{\epsilon})$ that maximizes $|\psi_{rd}|$. An exhaustive search over fine divisions of the interval is not exceptionally burdensome, but much faster convergence can be achieved with cleverer methods; for examples, see [11.2].

Frequency-Domain Methods. The foregoing schemes are appropriate for time-domain simulations. If frequency-domain, circular blocks are used instead, some elegant modifications can be applied to simplify the calibration procedures. To see how the frequency-domain methods work, let the signals be in N-point circular blocks and let $R(k)$ and $D(k)$ be the DFTs of $r(n)$ and $d(n)$, respectively. The inverse DFTs are

$$r(n) = \frac{1}{N} \sum_{k=0}^{N-1} R(k) e^{j2\pi kn/N}$$

$$d^*(n-p) = \frac{1}{N} \sum_{l=0}^{N-1} D^*(l) e^{-j2\pi l(n-p)/N} \quad (11\text{A}.10)$$

Substitute these into (11A.8) to obtain

$$\psi_{rd}(p) = \frac{1}{N^2} \sum_n \sum_k R(k) e^{j2\pi kn/N} \sum_l D^*(l) e^{-j2\pi l(n-p)/N}$$

$$= \frac{1}{N^2} \sum_k \sum_l R(k) D^*(l) e^{j2\pi lp/N} \sum_n e^{j2\pi n(k-l)/N} \quad (11\text{A}.11)$$

All sums have limits of 0 and $N-1$; all indexes are taken modulo-N. Sequences $r(n)$, $d(n)$, $R(k)$, and $D(k)$ all have the same length N. Overlap of all N elements in both blocks is assured for all shifts p because the DFT imposes circularity.

The sum on n simplifies according to

$$\sum_{n=0}^{N-1} e^{j2\pi n(k-l)/N} = \begin{cases} N, & l = k \\ 0, & l \ne k \end{cases} \quad (11\text{A}.12)$$

so the cross correlation becomes

$$\psi_{rd}(p) = \frac{1}{N} \sum_{k=0}^{N-1} R(k) D^*(k) e^{j2\pi kp/N} \qquad (11\text{A}.13)$$

The transform blocks $\{R(k)\}$ and $\{D(k)\}$ need be computed just once. Indeed, if the main simulation is in the frequency domain, these transforms are probably needed anyhow. Then $\psi_{rd}(p)$ is computed as an N-point circular block instead of one point at a time as in the time-domain method, by means of the inverse FFT applied to (11A.13).

The index p_0 is picked out from a scan of the magnitudes of all points of the correlation and then a fine search is instituted. Trial delays $\tilde{\tau}$ can be applied to the signal by multiplying $R(k)$ by $\exp(-j2\pi k\tilde{\tau}/Nt_s)$, as explained in Sec. 10.3.2. Search is over the range $-1 < \tilde{\tau}/t_s < 1$.

From the standpoint of the search procedure, the only difference between time- and frequency-domain operations lies in the method of evaluating the correlation function. The search routine is oblivious to the means of computing correlation, so the same (efficient) fine-search strategy should be used for correlation operations in either domain.

APPENDIX 11B: CARRIER REGENERATORS

This appendix develops the modeling equations for two varieties of carrier regenerator: a fourth-law (quartic) device and an absolute-value (linear rectifier) device. Both are full-wave rectifiers, operating equally on positive and negative half cycles of the carrier-modulated signal.

Signal Representation

Let the carrier-modulated signal have the real format

$$s(t) = x(t) \cos(2\pi f_c t + \theta_c) - y(t) \sin(2\pi f_c t + \theta_c) \qquad (11\text{B}.1)$$

In Chapter 2 it was shown that any data signal modulated onto a sinusoidal carrier can be represented in this format.

The quantities $x(t)$ and $y(t)$ are the baseband waveforms that have been modulated onto the carrier in the transmitter and are the quantities that are to be recovered at the demodulator in the receiver. Angle θ_c is a phase angle, best regarded as a carrier misalignment. To recover $x(t)$ without cross-channel interference from $y(t)$, and vice versa, requires that knowledge of θ_c be made available to the demodulator.

The complex envelope of the real signal can be written as

$$z(t) = [x(t) + jy(t)]e^{j\theta_c} \qquad (11\text{B}.2)$$

That representation is convenient for analytical purposes, but it does not display the characteristics of the misaligned signal particularly well since θ_c is unknown. Instead, the complex envelope can also be written as

$$z(t) = I(t) + jQ(t) \qquad (11\text{B}.3)$$

where $I(t)$ and $Q(t)$ are in-phase and quadrature components of the complex envelope $z(t)$, related to modulation components x and y and phase θ_c according to

$$\begin{aligned} I(t) &= x(t)\cos\theta_c - y(t)\sin\theta_c \\ Q(t) &= x(t)\sin\theta_c + y(t)\cos\theta_c \end{aligned} \qquad (11\text{B}.4)$$

so that the real signal can also be represented by

$$s(t) = I(t)\cos 2\pi f_c t - Q(t)\sin 2\pi f_c t \qquad (11\text{B}.5)$$

That is, the phase angle θ_c is buried inside I and Q. These are the components that are accessible for computer manipulations, not the modulation components. The models are developed in terms of I and Q so that the equations can be applied directly in simulations. For economy of notation, the time arguments will be omitted or abbreviated to: $I = I(t)$, $Q = Q(t)$ and $\phi = \phi(t) = 2\pi f_c t$.

Fourth-Law Regenerator

A fourth-law device is used for carrier regeneration of four-phase signals; it develops the fourth power of the input

$$\begin{aligned} s_{\text{raw}}(t) = s^4(t) &= I^4\cos^4\phi + Q^4\sin^4\phi + 6I^2Q^2\sin^2\phi\cos^2\phi \\ &\quad - 4I^3Q\sin\phi\cos^3\phi - 4IQ^3\sin^3\phi\cos\phi \end{aligned} \qquad (11\text{B}.6)$$

Expand the powers of trigonometric functions into multiple angles and collect like terms to obtain

$$\begin{aligned} s_{\text{raw}}(t) &= \tfrac{3}{8}(I^4 + Q^4 + 2I^2Q^2) + \tfrac{1}{2}(I^4 - Q^4)\cos 4\pi f_c t - (I^3 Q + IQ^3)\sin 4\pi f_c t \\ &\quad + \tfrac{1}{8}[(I^2 - Q^2)^2 - 4I^2Q^2]\cos 8\pi f_c t - \tfrac{1}{2}IQ(I^2 - Q^2)\sin 8\pi f_c t \end{aligned} \qquad (11\text{B}.7)$$

The first line of (11B.7) contains the baseband output of the rectifier centered on zero frequency and a double-frequency component at a frequency of $2f_c$. The second line contains the desired quadruple-frequency component at $4f_c$. For synchronizing a four-phase signal, the quadruple-frequency component would be selected in a carrier filter tuned to $4f_c$ and the other two components would be rejected. In a simulation, the complex envelope of the $4f_c$-term is represented as

$$z_{4\text{raw}} = \tfrac{1}{8}[(I^2 - Q^2)^2 - 4I^2Q^2] + j\tfrac{1}{2}IQ(I^2 - Q^2)$$
$$= \tfrac{1}{8}z^4 \qquad (11\text{B}.8)$$

Absolute-Value Device

A "linear" rectifier or absolute-value device performs the operation

$$s_{\text{raw}}(t) = |s_{\text{in}}(t)| \qquad (11\text{B}.9)$$

and generates all even harmonics of its input frequency f_c. The mth harmonic is selected by a narrowband filter tuned to mf_c to recover a carrier reference.

In the literature on statistical analysis of noise, the absolute-value device is known as a full-wave νth-law device, with $\nu = 1$. (The square-law device in the chapter text and the fourth-law device above are other examples of full-wave νth-law devices, with $\nu = 2$ and 4, respectively.) A detailed treatment of νth-law devices may be found in [11.68].

To develop a computer model, first convert the input signal to polar format:

$$s_{\text{in}}(t) = R(t)\cos[2\pi f_c t + \theta(t)] \qquad (11\text{B}.10)$$

where both of the following definitions for $R(t)$ and $\theta(t)$ are valid:

Definition A:

$$R(t) = \sqrt{x^2 + y^2}, \qquad \theta_a(t) = \arctan\frac{y(t)}{x(t)} + \theta_c \qquad (11\text{B}.11a)$$

or definition B:

$$R(t) = \sqrt{I^2 + Q^2}, \qquad \theta_b(t) = \arctan\frac{Q(t)}{I(t)} \qquad (11\text{B}.11b)$$

The second definition is the more useful for simulation purposes, for the same reasons as explained above.

From [11.68, p. 343], the output of the absolute-value device in response to the signal of (11B.10) is

$$s_{\text{raw}}(t) = |s_{\text{in}}(t)| = \sum_{m=0}^{\infty} 2F_m \epsilon_m R(t)\cos m[2\pi f_c t + \theta(t)] \qquad (11\text{B}.12)$$

where m is even, $\epsilon_m = 1$ for $m = 0$, and $\epsilon_m = 2$ for $m > 0$. The coefficient F_m is derived in [11.68] as

$$F_m = \frac{1}{4\Gamma(\frac{3}{2} - m/2)\Gamma(\frac{3}{2} + m/2)} \qquad (11\text{B}.13)$$

and $\Gamma(\cdot)$ is the gamma function [11.69]. Selected values for F_m are listed below.

m	πF_m
0	1
2	$\frac{1}{3}$
4	$-\frac{1}{15}$
8	$-\frac{1}{63}$

These values will be recognized as the coefficients of the Fourier series expansion of $|\cos x|$.

Polar-Coordinate Model. One possible simulation approach is to compute the desired term in polar coordinates directly from (11B.12). The complex envelope of the mth harmonic ($m \neq 0$) is

$$z_{m\text{raw}}(t) = 4F_m R(t) e^{jm\theta(t)} \qquad (11\text{B}.14)$$

Polar coordinates $R(t)$ and $\theta(t)$, as defined in (11B.11b), are determined from I and Q by a rectangular-to-polar conversion, and then the phase is scaled by m and reduced to the phase interval $-\pi$ to $+\pi$.

To filter the regenerated carrier wave requires that $z_{m\text{raw}}(t)$ be converted back into rectangular coordinates (see Chapter 6). Then the filtered wave has to be converted again to polar coordinates for amplitude limiting and phase division by m. Refer to the chapter text and Fig. 11.3.

Rectangular-Coordinate Model. Although the polar-coordinate approach is conceptually straightforward, the rectangular-to-polar conversions are burdensome because of their repeated invocations of trigonometric functions. Phase unwrapping has to be addressed upon both conversions to polar coordinates.

An alternative approach is to compute the raw-regenerated complex envelope entirely in rectangular coordinates, thereby avoiding one pair of coordinate conversions and one phase unwrapping. Derivation of the formulas for the raw complex envelope in rectangular coordinates is outlined below.

Expand the cosine factor in (11B.12) as

$$\cos m[2\pi f_c t + \theta(t)] = \cos(2\pi m f_c t)\cos m\theta(t) - \sin(2\pi m f_c t)\sin m\theta(t) \quad (11\text{B}.15)$$

whereupon the mth harmonic ($m \neq 0$) in the complex envelope of the raw carrier may be identified as

$$z_{m\text{raw}}(t) = 4F_m R(t)[\cos m\theta(t) + j\sin m\theta(t)] \quad (11\text{B}.16)$$

From (11B.11b) the definition of $\theta(t)$ leads to

$$\cos\theta(t) = \frac{I(t)}{\sqrt{I^2(t) + Q^2(t)}}$$

$$\sin\theta(t) = \frac{Q(t)}{\sqrt{I^2(t) + Q^2(t)}} \quad (11\text{B}.17)$$

The multiple-angle functions $\cos m\theta(t)$ and $\sin m\theta(t)$ are expanded as sums of powers of $\cos\theta(t)$ and $\sin\theta(t)$ [11.70]. Substituting (11B.17) for the trigonometric powers leads to purely algebraic expressions for the raw carrier, entirely in terms of the rectangular coordinates I and Q. Selected results are tabulated below.

m	$z_{m\text{raw}}$
0	$\dfrac{2}{\pi}\sqrt{I^2 + Q^2} = \dfrac{2\lvert z\rvert}{\pi}$
2	$\dfrac{4}{3\pi}\sqrt{I^2 + Q^2}\left(\dfrac{I^2 - Q^2}{I^2 + Q^2} + j\dfrac{2IQ}{I^2 + Q^2}\right) = \dfrac{4z^2}{3\pi\lvert z\rvert}$
4	$-\dfrac{4}{15\pi}\sqrt{I^2 + Q^2}\left[\dfrac{(I^2 - Q^2)^2 - 4I^2Q^2}{(I^2 + Q^2)^2} + j\dfrac{4IQ(I^2 - Q^2)}{(I^2 + Q^2)^2}\right] = -\dfrac{4z^4}{15\pi\lvert z\rvert^3}$
8	$-\dfrac{4}{63\pi}\sqrt{I^2 + Q^2}$ $\times\left\{\left[1 - \dfrac{32Q^2}{I^2 + Q^2} + \dfrac{160Q^4}{(I^2 + Q^2)^2} - \dfrac{256Q^6}{(I^2 + Q^2)^3} + \dfrac{128Q^8}{(I^2 + Q^2)^4}\right]\right.$ $\left. + j\dfrac{8IQ}{I^2 + Q^2}\left[1 - 10\dfrac{Q^2}{I^2 + Q^2} + 24\dfrac{Q^4}{(I^2 + Q^2)^2} - 16\dfrac{Q^6}{(I^2 + Q^2)^3}\right]\right\}$ $= -\dfrac{4z^8}{63\pi\lvert z\rvert^7}$

APPENDIX 11C: LOCATING ZERO CROSSINGS BY INVERSE LINEAR INTERPOLATION

In this appendix we derive the timing error incurred in locating the zero crossing of a timing wave by inverse linear interpolation between two adjacent samples. Since a well-filtered timing wave is nearly sinusoidal, the analysis is performed on a sine wave and the results taken as a good approximation for an authentic timing wave.

Refer to Fig. 11C.1 for an illustration of the problem. The real timing wave $s(t)$ is sampled at intervals t_s. Polarity of the sampled values changes from negative at $t = 0$ to positive at $t = t_s$, assuring that at least one positive-going zero crossing lies between these two points. Assume that the timing wave is a perfect sinusoid and that its crossing falls at $t = t_0$:

$$s(t) = A_o \sin \frac{2\pi(t - t_0)}{T} \qquad (11\text{C}.1)$$

where $0 \leq t_0 < t_s$, T is the period of the sine wave, and A_o is the amplitude of the sine wave. Define a *fractional interval*

$$\nu_0 = \frac{t_0}{t_s} = \frac{M_s t_0}{T} \qquad (11\text{C}.2)$$

such that $0 \leq \nu_0 < 1$ and M_s is the number of samples per symbol.

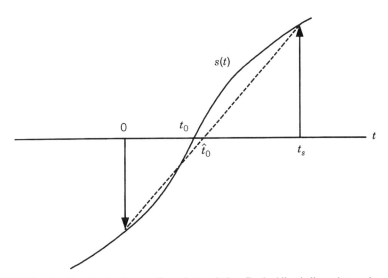

FIGURE 11C.1 Zero location by inverse linear interpolation. Dashed line is linear interpolator between sample points.

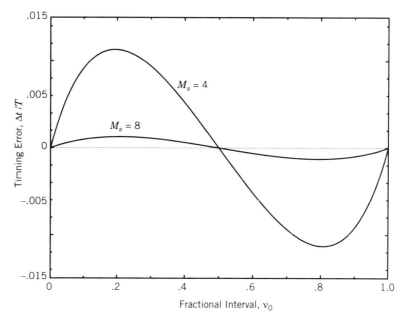

FIGURE 11C.2 Timing error incurred in locating zero crossing by inverse linear interpolation.

Sample values of the signal are

$$s(0) = A_o \sin\left(-\frac{2\pi v_0}{M_s}\right)$$

$$s(t_s) = A_o \sin\frac{2\pi(1 - v_0)}{M_s} \qquad (11C.3)$$

The straight line passing through the two sample points intersects the abscissa at

$$\frac{\hat{t}_0}{t_s} = \frac{-s(0)}{s(t_s) - s(0)} \qquad (11C.4)$$

Combining these expressions yields the relative timing error

$$\frac{\Delta t}{T} = \frac{\hat{t}_0 - t_0}{T} = \frac{1}{M_s}\left\{\frac{A_o \sin(2\pi v_0/M_s)}{A_o \sin[2\pi(1-v_0)/M_s] + A_o \sin(2\pi v_0/M_s)} - v_0\right\} \qquad (11C.5)$$

which is plotted for $M_s = 4$ and 8 in Fig. 11C.2. Notice that amplitude A_o cancels out from the estimate of (11C.4) and from the estimate's error in (11C.5).

12

SYMBOL DETECTION AND DEMAPPING

Data values are *estimated* from the strobes delivered by the receiver, one per symbol interval. The estimation process is variously known also as *symbol detection* or *symbol decision*; all three names will be used interchangeably. This chapter deals with symbol-by-symbol or "hard" decisions. That is, a decision is made on each strobe individually, with little or no attention paid to dependencies between strobes. Differential detection that involves two strobes is also treated as if it were symbol-by-symbol detection.

If the strobes are indeed independent (as they will be in PAM-QAM signals if the data stream has not been encoded and if intersymbol interference is absent), symbol-by-symbol detection yields optimum performance. If strobes are not independent, symbol-by-symbol detection is suboptimum; symbol decisions should be based on the entire sequence of strobes, not just one. Maximum-likelihood sequence estimation (MLSE), typically employing the Viterbi algorithm [12.1–12.3], affords superior performance for convolutional- or trellis-encoded signals, for CPM signals, or for any strobe sequences containing intersymbol interference.

This book concentrates on simulation of symbol-by-symbol detection. Simulations of sequence estimation and other decoding processes are also important to a communications engineer; often there is no way other than simulation to evaluate performance of a coded system. Principles for simulating the constituent processes of soft decisions, deinterleavers, and decoders are not vastly different from methods presented in this book for other processes. Once the simulation of signal processes has been mastered, methods for simulation of decoding operations can be developed from the pertinent literature; decoding methods are not included here.

The strobe sequence applied to a decision device is a signal, albeit greatly altered from the signal applied to the transmission or storage medium. In analog hardware, the strobes are still unquantized analog signals, even though they are time discrete. Output of the decision device is a sequence of digital numbers that are estimates of the symbol values; these numbers no longer constitute a signal. The decision device converts from signals to data sequences. In analog hardware, the decision device also serves as the analog-to-digital convertor that is implicit in all data receivers.

Frequently, particularly in binary receivers, the strober and symbol detector are combined into a single hardware element so that the strobe has no distinct existence in the receiver. For conceptual purposes, and for simulation, it is usually preferable to separate the strober and the symbol detector and to represent distinct strobes in the simulation model. Also, PAM symbol detectors are nonlinear, zero-memory devices (except for differential detection, which entails memory); the order of strobing and decision operations can be exchanged in these circumstances and often is exchanged in practice. This account assumes that strobing occurs first, followed by decision; the opposite order would demand greater sampling density out of the simulated receiver.

Ambiguity

A user should be alert to the fact that the detection and demapping operations to be described do not necessarily give back the original data sequence that was generated in the transmitter, not even in the complete absence of decision errors. Many signaling formats exhibit ambiguities (e.g., 0° cannot be distinguished from 180° in a received BPSK signal); there must be provision in the data stream to resolve the ambiguity. Most commonly, the transmitted signal is differentially encoded, or else a known sequence is inserted in a known location in the transmitted stream. Differential decoding after symbol detection recovers the original message from a differentially encoded stream. Observation of the known sequence after detection and demapping reveals the ambiguity state of the system and permits ambiguity-resolving interpretation of the detected stream.

Ambiguity is a property mainly of the signal format, not the detection method. With the exception of differential detection, all detection methods for the same format have the same ambiguity characteristics.

12.1 PAM–QAM DECISIONS

Assume that the signal has a balanced PAM–QAM format (Chapter 2) that can be described by an M-point constellation, examples of which are shown in Fig. 2.12. Let the mth strobe of the signal be represented by

$$z_o(m) = I_o(m) + jQ_o(m) \qquad (12.1)$$

which is presumed to be delivered in rectangular coordinates. There is exactly one strobe per symbol. If modulation is staggered, the strobes for I and Q are taken at staggered time instants.

Assume further that all amplitude,* frequency, phase, and timing corrections have already been performed in the receiver, so $z_o(m)$ is the final product of these and other signal processes. If equalization or echo cancellation are employed (not treated in this book), $z_o(m)$ incorporates their effects.

12.1.1 Minimum-Distance Decisions

Let the coordinates of the ith point in the constellation ($i = 0$ to $M - 1$) be denoted by

$$c^i = a^i + jb^i \tag{12.2}$$

where superscript i identifies the ith point in the constellation.

Decision Rule

Define the squared distance from the mth strobe to the ith constellation point as

$$\begin{aligned} d_i^2(m) &= |z_o(m) - c^i|^2 \\ &= [I_o(m) - a^i]^2 + [Q_o(m) - b^i]^2 \end{aligned} \tag{12.3}$$

To decide the mth symbol, follow these steps:

1 Compute distances $d_i^2(m)$ for each $i = 0$ to $M - 1$.
2 Decide $\hat{i}(m)$ such that $d_{\hat{i}}^2(m)$ is the smallest of the M distances.† (The caret $\hat{\;}$ indicates an estimate.)

These steps deliver $\hat{i}(m)$—the *label* of the estimated constellation point. As will be seen presently, that is the only information needed by subsequent processes. Equivalently, the detection process might deliver $\hat{a}(m) + j\hat{b}(m)$—the *coordinates* of the estimated constellation point.

Detection Tables

What defines the constellation points $\{c^i\}$? As one candidate, a formula might be used to describe a regular constellation. For broader applicability, a table can be used to define any finite constellation, no matter how irregular. Each entry in the table consists of the point's label i and the rectangular coordinates a^i and b^i.

---

*Amplitude normalization is essential for proper detection of multiamplitude constellations.

† Rarely, two or more computed distances may be equal. Any arbitrary tie-breaking expedient can be applied to force a decision.

348 □ SYMBOL DETECTION AND DEMAPPING

But this table has arisen before. It is the same table as that described in Sec. 7.2.1 for the symbol mapper in the transmitter. Therefore, provided that the constellation has not been altered in transmission, the same table can be employed for both the symbol mapper and the symbol detector. (Some signaling formats—most notably partial-response pulse shaping—do change the constellation, so that separate tables would be required for their original mapping and their symbol-by-symbol detection.) The table employed is taken from the program's library or is constructed by the user.

If the detector delivers $\hat{a}(m) + j\hat{b}(m)$, subsequent demapping is needed to obtain $\hat{i}(m)$: demapping would be accomplished as another table lookup. The sequence $\{\hat{i}(m)\}$ may be treated as a sequence of *M*-ary numbers or could be degrouped into a binary stream in opposite fashion to the bit grouping process of (7.7).

12.1.2 Quantizer Detection

Although a user-furnished table affords great flexibility for simulation of arbitrary signal constellations, computation of *M* distances for each decision can be tedious and time consuming. That method is not likely to be employed in operating hardware, and a user might prefer to use a faster scheme in simulations. Rather than computing distances, detection on regular constellations is more commonly performed by *quantizing* each strobe. The method is explained by examples that follow.

Binary Decisions

Suppose that the signal has a BPSK format and that the receiver is phased to deliver the data in the *I* channel. Then the data can be decided by

$$\hat{a}(m) = \text{sgn}_0[I_o(m)] \qquad (12.4)$$

where the function $\text{sgn}_0[\cdot]$ is a variation on the *sign* function:

$$\text{sgn}_0[x] = \begin{cases} +1, & x \geq 0 \\ -1, & x < 0 \end{cases} \qquad (12.5)$$

(In a decision context, $x = 0$ is a tie, which has been broken here by arbitrarily setting $\text{sgn}_0[0] = +1$.)

For a QPSK format, the rule would be

$$\hat{a}(m) = \text{sgn}_0[I_o(m)]$$
$$\hat{b}(m) = \text{sgn}_0[Q_o(m)] \qquad (12.6)$$

In either of these schemes or in any other quantizer detection method, the quantity delivered by the decision process is the symbol value $\hat{a}(m) + j\hat{b}(m)$, not the label $\hat{i}(m)$. Demapping with a table lookup provides $\hat{i}(m)$.

Multiamplitude Decisions

Let the constellation be situated on a $K \times K$ square grid with distance 2 between adjacent grid points, where K is a positive integer. Independent decisions can be taken separately on the I and Q channels, and then combined afterward. To be able to regard the I and Q components as independent, all K^2 points in the grid should be part of the constellation. As a counterexample, the 32-point cross constellation of Fig. 2.2e is missing the corners of its 6×6 grid, so the I and Q values could not have been assigned independently in the transmitter. Nonetheless, if one is willing to label a decision for a corner location as an erasure (presumably rare), separate I and Q decisions can be made and the pertinent constellation point identified therefrom. A minimum-distance decision would return estimated symbol values, not erasures, from the empty corner points.

The constellation values for I_o (or Q_o) are $\pm 1, \pm 3, \ldots \pm(K-1)$ if K is even, or $0, \pm 2, \pm 4, \ldots \pm(K-1)$ if K is odd. Pseudocode for computing $\hat{a}(m)$ from $I_o(m)$ is listed below. Identical operations are performed for deciding $\hat{b}(m)$ from $Q_o(m)$. The notation INT[x] means the largest integer less than or equal to x. Amplitude of $\{z_o(m)\}$ is assumed to have been normalized correctly, as is necessary for proper detection of any multiamplitude signal. Normalization is accomplished by adaptive scaling, as described in Sec. 7.5.1.

```
Multiamplitude Decisions
    IF |I_o| > K - 1
        THEN â = (K - 1) sgn[I_o]
    ELSE IF K is even
        THEN â = 2 INT[I_o/2] + 1
    ELSE IF K is odd
        THEN â = 2 INT[(I_o + 1)/2]
```

For further coverage of uniform quantizers, refer to Chapter 15.

Notice that independent detection of I and Q can produce symbol estimates that are not legitimate constellation points if disturbances push the strobe over to unused grid points. For example, an erroneous decision could yield the point $\hat{c}(m) = 5 + j5$ in the constellation of Fig. 2.12e. Although $\hat{a} = 5$ and $\hat{b} = 5$ are permissible values individually, their simultaneous occurence indicates wrongly that the received symbol was one of the missing corner points. If a symbol detector employs independent detection of I and Q, there will have to be provision for handling admissible decisions that have no corresponding symbol in the message alphabet. Provisions are specific to the application. Notice also that this problem does not arise with minimum-distance decisions, because the closest constellation point is always selected, even if a tie-breaking decision has to be forced artificially.

Independent detection can be adapted to a rectangular grid instead of square (just scale one channel to make the grid square) but any other grid pattern would seem to require joint decision with both I and Q. Independent decisions are well suited to detection of staggered signals, by strobing I and Q at staggered instants.

MPSK Detection

At least two different methods can be employed to detect MPSK signals: a polar method and a binary-decision method.

Polar Method. First calculate

$$\theta_o(m) = \arctan \frac{Q_o(m)}{I_o(m)} \tag{12.7}$$

and then quantize $\theta_o(m)$ into M uniform sectors with constellation values $\theta^i = \pm\pi/M$, $\pm 3\pi/M, \ldots, \pm(M-1)\pi/M$. Details are similar to the rectangular quantization of the preceding section. The coefficients $\pm 1, \pm 3, \ldots, \pm(M-1)$ provide symbol identification that is sufficient for demapping.

Binary Decision Method. Symbols can be detected in any MPSK signal (if M is an integer power of 2) by means of $\log_2 M$ *binary* decisions. For 8PSK the decision formulas are

$$\hat{a}_1(m) = \text{sgn}_0[I_o(m)]$$
$$\hat{a}_2(m) = \text{sgn}_0[Q_o(m)]$$
$$\hat{a}_3(m) = \text{sgn}_0[|I_o(m)| - |Q_o(m)|] \tag{12.8}$$

These three bits are Gray-encoded in the sense that points from adjoining sectors differ in only one bit position. Proofs are left to the reader.

Threshold Detection

In some instances the optimum decision boundary does not lie midway between the constellation points. One example is noncoherent detection of binary on–off keying [12.4], where the magnitude of a strobe $|z_o(m)|$ is compared to a threshold V_{th} to decide the symbol estimate $\hat{a}(m)$. A simple model for OOK detection is

$$\hat{a}(m) = \text{sgn}_0[I_o^2(m) + Q_o^2(m) - V_{\text{th}}^2] \tag{12.9}$$

This modified form saves the computer time needed to take a square root and gives the same symbol decision as $\text{sgn}_0[|z_o| - V_{\text{th}}]$.

12.1.3 Differential Detection of MPSK

Differential Reception

Recovery of carrier phase may be inconvenient, costly, or even impossible in some systems. To accomplish demodulation on MPSK signals, phase of the preceding symbol can be used as a phase reference for detection of the current symbol. Data are differentially encoded so that the information is contained in the phase changes, not the phase itself. (Note that differential encoding also might be used with coherent reception as an antiambiguity measure. Differential detection and differential encoding are two separate matters.)

Differential reception incurs a power penalty: about 0.5 dB for BPSK, 2.3 dB for QPSK, and nearly 3 dB for larger M. The penalty is regarded as tolerable in some applications.

The receiver for differential detection is identical to one for coherent reception, from the front end right through the strober. There is just one exception; no attempt is made in a differential receiver to correct the phase of the signal. Otherwise, the differential detector operates on strobes $\{z_o(m)\}$ just like those described at length previously for coherent reception.

Differential encoding takes place at the symbol mapper in the transmitter according to the rule

$$\theta(m) = \theta(m-1) + i(m)\frac{2\pi}{M} \tag{12.10}$$

where $\{i(m)\}$ ($i = 0$ to $M - 1$) is the M-ary data sequence.

Differential Detectors

As a means of retrieving the phase difference, a differential detector forms the product of strobes:

$$z_d(m) = z_o(m)z_o^*(m-1) \tag{12.11}$$

To see how that works, write the strobe expressions in polar coordinates. Since $z_o = R_o e^{j\theta_o}$, the product of strobes is

$$z_d(m) = R_o(m)R_o(m-1)\exp j[\theta_o(m) - \theta_o(m-1)] \tag{12.12}$$

The phase difference being sought is embedded in the angle of the product. Phase can be measured by converting either z_d or each z_o to polar coordinates and determining $\theta_o(m) - \theta_o(m-1)$. Phase difference is then quantized in M equal sectors around the circle to decide the data estimate. Alternatively, the strobe product could be rotated by π/M and the decisions extracted in rectangular coordinates. The rotated decision product is represented by

$$z_{\text{dr}}(m) = z_o(m)z_o^*(m-1)e^{j\pi/M} \tag{12.13}$$

Define

$$X_d(m) = I_o(m)I_o(m-1) + Q_o(m)Q_o(m-1) = \text{Re}[z_d(m)]$$
$$Y_d(m) = Q_o(m)I_o(m-1) - Q_o(m-1)I_o(m) = \text{Im}[z_d(m)] \quad (12.14)$$

In rectangular coordinates, the rotated product takes the form

$$z_{dr}(m) = X_d(m)\cos\frac{\pi}{M} - Y_d(m)\sin\frac{\pi}{M} + jX_d(m)\sin\frac{\pi}{M} + jY_d(m)\cos\frac{\pi}{M} \quad (12.15)$$

For BPSK signals ($M = 2$), the cosine terms vanish and it can be shown that the signal contribution to $Y_d(m)$ is zero, so that differential bit detection is provided by

$$\hat{a}(m) = \text{sgn}_0\{\text{Im}[z_{dr}(m)]\} = \text{sgn}_0[X_d(m)] \quad (12.16)$$

For QPSK signals ($M = 4$), multiply $z_d(m)$ by $\sqrt{2}\,e^{j\pi/4} = (1+j1)$ to perform the rotation. Bit decisions then reduce to

$$\hat{a}(m) = \text{sgn}_0\{\text{Re}[z_{dr}(m)]\} = \text{sgn}_0[X_d(m) - Y_d(m)]$$
$$\hat{b}(m) = \text{sgn}_0\{\text{Im}[z_{dr}(m)]\} = \text{sgn}_0[X_d(m) + Y_d(m)] \quad (12.17)$$

Rotation by $\pi/4$ and multiplication by $\sqrt{2}$ are effectuated in (12.17) simply by sums and differences of the real and imaginary parts of $z_d(m)$. There is no need to compute z_{dr} at all. Rotation is a conceptual aide, not a required operation, at least for $M = 2$ and 4.

π/M-PSK

Rather than performing the rotation by π/M in the receiver on the strobe product, it could be performed in the symbol mapper in the transmitter, on each symbol to be transmitted. The rotated strobe product in the receiver then will be exactly the same as in (12.13) or (12.15) and the differential decisions are taken in the same manner.

But now the transmitted signal can have any of $2M$ phases instead of M. The encoding rule is to force a phase change of an odd multiple of π/M at every symbol interval. The symbol-error performance of this format can be shown to be identical to that for regular differential detection of MPSK. These π/M-shifted signals are rich in data transitions—one occurs at each symbol. High density of transitions is favorable for clock recovery. Also, phase reversals are avoided in the modulation; that prevents nulls from occurring in the envelope of the signal and thereby reduces amplitude variations of the transmitted signal.

Multisymbol Differential Detection

In the foregoing discussions it has been assumed that differential encoding and differential detection are carried out on pairs of adjoining symbols. It is sometimes advantageous to perform the differential operations over more than two symbols. See [1.7, Chaps. 7 and 8] for further information.

Differential Detection of CPM

Continuous phase modulation (CPM; see Sec. 2.2.3) does not have a PAM–QAM format. Moreover, optimum coherent reception of CPM requires sequence estimation, not symbol-by-symbol detection. Nonetheless, CPM is often processed suboptimally by differential detection, for the same reasons that motivate differential detection of MPSK. Information in a CPM signal is contained in the phase changes, so differential detection might be carried out in much the same manner as described above for MPSK. It is also possible, though, that differential detection of CPM might be carried out over more than two adjacent symbols; exact details depend upon the specific signal format.

12.2 MULTIPLE-STROBE DECISIONS

Just one strobe per symbol is needed for a PAM signal; the simplicity of implementation made possible thereby is one of several reasons for the popularity of PAM signaling. In other signal formats, multiple strobes are produced for each symbol. As one example, there are M simultaneous strobes, one from each matched filter, for each received symbol in an MFSK signal. As another example, a PPM signal with M slots per symbol requires M strobes to be taken sequentially at intervals T/M in each symbol interval T.

No matter what the signal characteristics may be or how the strobes are taken, the detection problem is essentially the same for all multiple strobe situations: The symbol detector must identify the largest strobe. In a simulation, the largest strobe is found by storing the magnitudes of all M strobes in an array with index $i = 0$ to $M-1$ and then searching the array for the strobe with greatest magnitude. A tie-breaking rule forces an arbitrary choice in those rare instances when two or more strobes have equal magnitude. Output of the detector is the M-ary number $\hat{i}(m)$ that identifies the index of the largest strobe. Subsequent demapping, in the same manner as for other signals, recovers the original data sequence.

12.3 KEY POINTS

In this chapter we have explored models for symbol-by-symbol data detection, also known as hard decisions. The method is optimum for reception of noncoded PAM–QAM signals that are free of intersymbol interference. Sequence estimation yields better performance for coded PAM signals, for CPM signals, or for signals with appreciable intersymbol interference. Symbol detection is performed on the strobes from the receiver. With PAM signals, there is just one strobe per symbol.

12.3.1 Minimum-Distance Detection

For those signals describable by a constellation (signals with PAM–QAM formats), detection can be accomplished by computing the squared distance from the complex strobe value to each point in the constellation. The constellation point corresponding to the smallest distance is declared to be the estimate of the symbol. By placing the constellation coordinates in a table, any arbitrary constellation, no matter how irregular, can be represented without difficulty. This same table is employed in the transmitter for the symbol mapper (see Chapter 7).

12.3.2 Quantizer Detection

A regular constellation, especially one laid out on a square grid, is conducive to detection by quantizing of the strobes. This is the method most likely to be employed in hardware since it requires less computation than the minimum-distance determination. Rules have been provided for quantizing simulated strobes.

Signals with binary basebands (BPSK, QPSK) are particularly simple because the quantizer reduces to determining the signs of the I and Q components of the strobes. Signals in which all constellation points have the same amplitude—MPSK signals—can be detected with binary comparisons (if M is an integer power of 2). An example is given for 8PSK. Signal amplitude is irrelevant to the detection. Signals with multiamplitude constellations have to be normalized to a reference amplitude before detection. (This is as true for minimum-distance detection as it is for quantizer detection.)

12.3.3 Differential Detection

Phase-change information can be retrieved from both MPSK and from CPM signals by means of differential detection. The current strobe is multiplied by the complex conjugate of the preceding strobe and the angle of the resulting product is the phase change from one symbol to the next. Differential encoding in the transmitter assures that the phase changes represent the message data. Differential detection incurs a power penalty compared to coherent detection.

12.3.4 Largest-of Detection

In some signals—such as MFSK or PPM—there will be M strobes per symbol, not just one. Optimum detection of the symbol is accomplished by selecting the strobe with the largest amplitude.

12.3.5 Demapping

Several of the detection schemes deliver an M-ary number instead of the original data bits. Demapping is needed for recovery of the original data stream.

12.3.6 STÆDT Processes

DDET: differential detector
DGRP: bit degrouper
DIFD: differential decoder
DMAP: symbol demapper
EXTR: extreme value (for largest-of detector)
MDET: M-ary minimum-distance detector

13

EVALUATION OF PERFORMANCE

A data communications link (or storage system) is judged largely upon its ability to deliver correct message data in the presence of transmission disturbances. Probability of data error is the main criterion of performance. Probability of error can be estimated by experiment, or it can be predicted by analysis. Both approaches are applied in simulations. A simulation could monitor either the symbol errors or the bit errors, as best meets the needs of the user. When bit errors are monitored, the observed ratio is known as the *bit-error ratio* (BER).

In the purely experimental approach, the entire communications system is simulated, including all noise and other disturbance sources. Data outputs from the receiver's symbol detector (or decoder) are compared against data furnished to the transmitter by the message source. Transmission errors show up as disagreements between symbols (or bits) in these two streams. The probability of error is estimated as the ratio of errors observed to the total number of symbols (bits) transmitted. This approach is known as the *error-counting* or *Monte Carlo* method; it is the primary method of performance evaluation by simulation and is explained in this chapter.

When links perform well, large numbers of symbols have to be simulated to induce just a small number of errors. Enough errors must be collected to establish reliable estimates; how much is "enough" is an important question to be examined below. Computer effort required for collecting "enough" errors can easily become intolerably burdensome.

There have been many attempts to relieve the computing burden inherent in Monte Carlo methods. One modification—known as *importance sampling*—will be mentioned briefly; it offers future promise, but still has barriers that have inhibited practical application.

A completely different approach turns away from the Monte Carlo method entirely and applies analysis instead. In the *quasianalytic* or *semianalytic* method, the additive noise at the receiver is not simulated at all. Instead, the system is simulated noise-free and a strobe value is collected for each symbol. Then the probability of detection error at each strobe is computed analytically from knowledge of the strobe value, the noise statistics that would arise if noise were present, and the decision boundaries of the data detector. Overall error probability is estimated as the average of the calculated probabilities of all of the strobes that are simulated. A quasianalytic simulation need only be long enough to provide a sufficiently representative signal sequence (see Sec. 7.1.2); there is no need for the prolonged runs of the Monte Carlo approach.

Application of the quasianalytic method is severely constrained to special cases where all necessary conditions are satisfied (e.g., Gaussian noise, a linear time-invariant receiver, and a regular constellation). Nonetheless, the special conditions are satisfied for many situations of great practical importance, so quasianalytic performance evaluation is widely used. Its principles are explained in detail in this chapter.

13.1 MONTE CARLO ERROR COUNTING

Brute-force error counting can be applied to any simulatable model. The method is not restricted to noise disturbances alone; the disturbances need not be additive nor Gaussian nor white nor stationary; the system need not be linear. Because of the absence of restrictions, error counting is a highly versatile tool for performance evaluation. Not only is it applicable in simulations, but it is also the most-used method for evaluation of communications hardware. Much the same principles apply equally to evaluations in either software or hardware.

13.1.1 Monte Carlo Configuration

An arrangement for error counting is depicted in Fig. 13.1. Data are generated in a message source as described in Sec. 7.1 and then processed through a simulated transmission link. The simulated link can be as complex as may be desired and can have any signal format whatever, within the capabilities of the program. Disturbances of any kind may be applied to the link simulation.

None of these details matter in the least to the error-counting process, provided that the data recovered from the simulated link are in the same format as the original transmitted data. The recovered bit stream will have been delayed by the transmission processes, and some recovered bits will be in error. Otherwise, the transmitted and received bit streams are the same.

> COMMENT 1: Instead of *bits* of data, it is equally feasible to deal with *symbols* or *words*. Wherever *bits* are mentioned in this section, *symbols*, *words*, or whatever data unit is appropriate may be substituted. The comparator to be described may have to be modified if the data unit is not individual bits or symbols.

COMMENT 2: Some transmission links may transform the data to a different format that does not agree with the transmitted data, even in the absence of errors. In that case the same transformation should be applied to the transmitted bit stream where it is brought into the error-counting process. Then the recovered stream and the transformed transmitted stream will agree, except for errors and delay.

Comparison

The recovered bit stream is compared, bit by bit, against the delayed transmitted bit stream. For each bit pair, the comparator delivers a 0 if the two bits agree and a 1 if they disagree (if there has been a transmission error). If the two bit streams are in Boolean 0–1 format, the comparator could be an exclusive-OR. Otherwise, a different test for equality is needed.

STÆDT employs a symbol-equality test that treats each individual bit or symbol as a number, irrespective of format. If the two numbers of a comparison pair are equal, the test outputs a 0; if the two numbers are unequal, the test outputs a 1. This plan allows any arbitrary symbols, real or complex, to be compared without need for special arrangements for different formats.

Counting

The two counters in Fig. 13.1 are set to zero at the start of a simulation run. The error counter then counts the 1's (i.e., detected errors) delivered by the comparator, while the bit counter counts all bits (or symbols) delivered to the comparator. Running contents of the two counters are designated in the figure as CNE and CNB, respectively. Error probability is estimated simply as the ratio BER = CNE/CNB.

Delay and Phase

The simulated transmission link and associated recovery processes almost invariably impose a delay on the recovered bit stream. However, the individual bits of the transmitted and recovered streams have to be time aligned to permit correct bit-by-bit comparisons to be performed. Therefore, the transmitted stream has to be delayed to compensate for the delay of the simulated link.

Delay of the recovered stream will always be an integer number of bits (or symbols), so that a relatively straightforward discrete delay process might be thought to be sufficient. In fact, there are several complicating factors that make the delay problem nontrivial, as summarized below. For further detail, refer to Appendix 13D.

There is the question of how to determine the correct delay to apply. A user is not likely to know beforehand the delay of a simulation; delay is a parameter that is best measured in a calibration procedure. Once measured, the calibration delay can often be employed for the entire duration of a simulation run. (But see below for further complications.)

Measurement of delay is closely related to timing synchronization, as treated in Chapter 11. Each of the two bit streams in a calibration would be contained in a block of N_b bits. The transmitted block is delayed by a trial delay of n_d bits ($0 \leq n_d \leq N_d$)

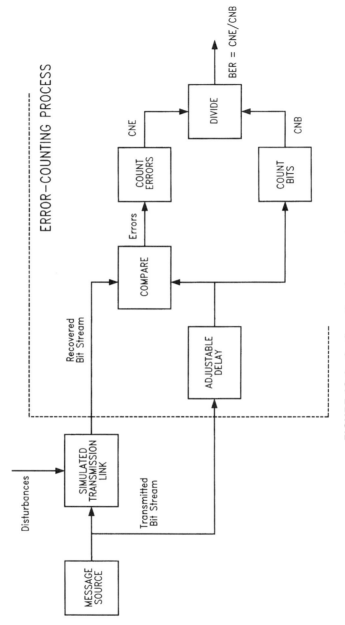

FIGURE 13.1 Configuration of error counter process.

and then the bits in each block are compared, bit by bit. For practical situations, $N_b \gg N_d$, where N_d is an upper bound on delay.

Comparison errors are counted and stored for each trial value of n_d. The number of errors will be small for the correct value of delay (especially if the calibration is performed without transmission disturbances) and comparatively large for all other values of delay. The value of delay providing the smallest number of comparison errors is selected as the calibration value to be used in the subsequent simulation run. (The correlation function of the data sequence used for setting delay should have small sidelobes, so there is little risk of mistaking a sidelobe for the main lobe.)

Phase Ambiguity. So far, it has been assumed tacitly that signal phase is correct—that the phase ambiguity has been resolved before the bit comparison is performed. If that is not true—if the bits or symbols to be compared are still phase ambiguous—the one-dimensional search for correct delay has to be expanded into a two-dimensional search for correct delay and correct phase. If the phase is incorrect, a signal has to be rotated to bring the two symbol streams into phase alignment with each other. Some simulations will include phase-ambiguity resolution in the main simulation while others will have to include it in the evaluation routines. A flexible simulation program makes provision for either eventuality.

Cycle Slips. An overstressed clock synchronizer can undergo a *cycle slip* where one or more clock cycles are gained or lost. A clock slip shows up as missing or extra bits in the recovered stream. Similarly, phase slips occur in carrier synchronizers; their effect is somewhat different from that of timing slips. A slip can arise only in a recovery synchronizer of the kinds described in Sec. 11.3. There can be no slips if the receiver is hardwire synchronized in a prerun calibration, as described in Sec. 11.1, and held to constant timing and phase thereafter.

When a timing slip occurs during a simulation run, the previously established value of delay is lost and the data streams will no longer be aligned correctly. Anomalously large numbers of apparent errors will be detected by the comparator. When a phase slip occurs, the effect depends on the method employed for ambiguity resolution. If differential encoding and decoding are employed, there will be a burst of errors in the vicinity of the slip and then the link will settle down to its normal behavior. If phase ambiguity was resolved instead from a synchronization preamble, the resolution is wrong after a phase slip has occurred and large numbers of errors will be detected thereafter. These slip behaviors closely mimic those of typical communications hardware.

Slips tend to be rare events, except under highly stressed conditions (such as low signal-to-noise ratio). They are sufficiently rare in well-designed communications systems that simulation has not been used much for investigating their statistics, because of the inordinate amount of computing time required. But they do happen and a user should be prepared for them.

Data-error statistics that are taken after a slip has occurred are worthless if the slip is not corrected promptly. Time spent gathering those statistics is wasted. To recover from a slip and return to counting only the ordinary errors, it is necessary to change the adjustable delay and/or phase rotation in compensation for the slip. Changing the delay or phase is tantamount to recalibration. To know that recalibration is needed requires that the existence of a slip and its approximate location be detected. A user might wish to excise the data lying in the near neighborhood of a slip. These matters are examined more carefully in Appendix 13D.

Auxiliary Message Generator. An alternative data-alignment configuration is often employed, with the following characteristics:

- Delay and phase alignment are handled entirely by the receiver synchronizer.
- Message bits or symbols are applied to the comparator from a separate auxiliary message generator.
- The auxiliary generator produces an exact replica of the transmitted sequence.
- Output of the auxiliary generator is blocked to match the blocks (or single points) of the received bit stream.

Commercial error-counting instruments ("BERC meters") usually incorporate auxiliary message generators of this kind.

13.1.2 Monte Carlo Statistics

Suppose that a Monte Carlo simulation is performed in which N_b recovered bits (or symbols) are examined and n_e errors are found. The error ratio is estimated as n_e/N_b. A different n_e is likely to be found if the simulation were to be repeated with a different realization of the disturbances, leading to a different value for the estimated error ratio.

The observed error count is a random variable, subject to statistical scatter arising from different realizations of the disturbances. A central question is: How many bits must be examined so that the scatter is a tolerably small fraction of the estimated ratio? This question has been addressed in [13.1, Sec. II]. Simplified results were distilled from [13.1] and are presented below.

If the Monte Carlo simulation is repeated many times, with different realizations of the disturbances for each repetition, the average of the observed error counts n_e will approach a value $\bar{n}$. The ratio $\bar{n}/N_b$ is the true error ratio; in statistical terminology, the estimator is *unbiased*.

Confidence Limits

However, only one simulation will be performed, from which only n_e and N_b will be obtained. The count $\bar{n}$ is always unknown. Given the known information n_e and N_b, one would like to set bounds on $\bar{n}$. Bounds can be expressed probabilistically as

$$\text{Prob}\{n_L \leq \bar{n} \leq n_U | n_e, N_b\} = \Theta \qquad (0 < \Theta < 1) \qquad (13.1)$$

where Θ is the *confidence level* and n_L to n_U is the *confidence interval*. Equation (13.1) states that the average count $\bar{n}$ lies in the interval n_L to n_U with probability Θ, where n_L and n_U are functions of the observables n_e and N_b and of the value assigned to Θ.

If certain reasonable conditions are fulfilled (effects of disturbances are independent for each error and error-count statistics can be approximated as Gaussian distributed), n_L and n_U can be expressed in closed form (see equation (8) of [13.1]). Further simplification is afforded if $N_b \gg 1$, $n_e \ll N_b$, and Θ is assigned the value 0.954. These conditions lead to the numerically simple formula

$$n_L = n_e + 2 - 2\sqrt{n_e + 1}$$
$$n_U = n_e + 2 + 2\sqrt{n_e + 1} \qquad (13.2)$$

which is independent of N_b (provided that N_b is large). The confidence interval depends only on n_e and Θ. The influence of Θ is concealed in the numerical elements of (13.2), which will be altered for different choices of Θ. See [13.1] for the effects of different Θ. Values from (13.2) of n_L, n_U, and their ratio are tabulated below for several values of n_e and confidence level $\Theta = 0.954$.

n_e	n_L	n_U	n_U/n_L
5	2.1	11.9	5.7
10	5.4	18.6	3.5
20	12.8	31.2	2.4
50	37.7	66.3	1.8
100	81.9	122	1.5
200	174	230	1.3

Confidence limits diminish slowly as n_e increases. Large numbers of errors, needing large amounts of computing time, must be observed if narrow confidence limits are to be achieved.

Rule of Thumb. Collecting 10 errors provides an estimate of $\bar{n}/N_b$ within a factor of 2 either way ($n_U = 20$, $n_L = 5$), with a 97.5% confidence level. Tighter bounds are usually needed, so 10 is a minimal error count, barely acceptable only if loose bounds are tolerable or if collecting many more errors would be inordinately time consuming.

Independence

The foregoing results are based upon the assumption that errors are independent, an assumption that is not always valid. Anything that causes errors to occur in bursts violates the assumption of independence. Bursts can be caused, for example, by cycle slips (timing or phase), by error propagation in decoding of encoded data streams, or by error propagation in decision-feedback equalizers.

Confidence limits established for independent errors are too optimistic if the errors are dependent. In general, it is difficult to establish the nature of dependency and doubly so to devise modified confidence limits. Reference [13.1] considers a special case where an error in one symbol is virtually certain to induce errors in the next m following symbols, and concludes that the confidence interval, for that special case, should be expanded by the factor $\sqrt{2m+1}$.

Another example—of bursty errors in a digital tape recorder—is presented in [13.2], which gives a method of establishing confidence intervals and levels for this dependent model. Experimental data (on hardware, not a simulation) reveal that uncritical acceptance of the assumption of independence would have led to seriously wrong estimates of confidence levels.

Other dependencies will have different effects and have to be considered individually when they arise. A user of Monte Carlo simulation should always be alert for error dependencies; laboriously collected data may be invalid if dependencies are overlooked.

13.1.3 Importance Sampling

Although Monte Carlo simulation is a powerful method of performance evaluation, it is exceedingly computation intensive. Estimation of moderately low error rates easily might require simulation of many millions of symbols, consuming days or weeks of time on any but the fastest computers. Researchers have searched assiduously for shortcuts that would reduce the computing burden; several methods are outlined in [13.1].

Importance sampling is one of those methods. It gives promise of major improvement and accordingly has received abundant attention [13.1, 13.3–13.12]. For additional literature, see the reference lists of the cited papers. To understand the premise of importance sampling, regard the link disturbances as a series of individual events. Most of these events do not cause errors; only the largest of the disturbances—the most important ones—lead to data errors. Simulation time conventionally is taken up mostly with simulating the less important events.

If the probability distribution of the simulated disturbances could be modified—could be *biased*—to favor the more important events, many more data errors would be induced in a relatively short simulation. Afterward, the large error ratio obtained with a biased disturbance is corrected for the bias to obtain an estimate of the much smaller error rate obtaining from an unbiased disturbance. Reduction of computing effort by orders of magnitude appears, at first glance, to be attainable. See the references for details; much of the literature is devoted to different methods of biasing.

Unfortunately, the presence of memory in the simulation model—in the form of filters, for example—tends to vitiate the potential benefits of importance sampling [13.5, 13.6]. For that reason the topic is not pursued further in this book. However, importance sampling is being investigated by many researchers; a dedicated student of simulation would do well to keep abreast of importance-sampling developments in case its potential is realized more fully in the future.

Optical transmission is one area of particular relevance for importance sampling. Optical channels typically have wide bandwidths (compared to the symbol rate) and so have small to negligible memory. Also, because required error probabilities tend to be very small ($<10^{-9}$), brute-force Monte Carlo simulations take unspeakably long times to run. Applications of importance sampling to optical links are exemplified in [13.24] and [13.25].

13.2 QUASIANALYTIC METHOD

In Monte Carlo simulation, all noise disturbances must be simulated and added to the signal, sample by sample. In quasianalytic (QA) simulation, only the signal is simulated, not the noise. The probability of decision error is computed analytically at each symbol strobe, based upon known statistics of the noise. A QA simulation need only be long enough to generate a sufficiently representative signal* (see Sec. 7.1.2) whose duration is typically far shorter than that needed for noise in Monte Carlo evaluations. Often, QA requirements are satisfied with a few hundred symbols where Monte Carlo might need millions. Furthermore, all points on a BER curve for different signal-to-noise ratios can be calculated from one QA signal simulation, whereas each individual point requires a separate run in Monte Carlo simulation. Where applicable, QA provides enormous savings in computing time as compared to Monte Carlo.

Unfortunately, QA is applicable only under narrowly restricted conditions (to be described presently); it cannot be used when the conditions are not met. However, the restrictive conditions are satisfied for several important classes of signals and receivers that are used extensively in many practical systems. Therefore, QA evaluation is widely applied in simulations despite its restrictions; where it can be used, it will almost always be preferred to Monte Carlo.

Quasianalytic evaluation is well established in the simulation community [13.1, 13.13, 13.14], but published accounts have been sketchy. This section is devoted to the practical implementation details of QA simulation.

13.2.1 Overview of QA Methods

Quasianalytic evaluation makes use of the sequence of strobes delivered from a simulated receiver, one strobe per symbol. In a Monte Carlo simulation, those strobes include noise and would go to a decision device for symbol-by-symbol detection of signal plus noise (or to a decoder, after "soft" decisions). In QA simulation the strobes do not include noise and are not sent on to decision devices.

---

*A sufficiently representative signal incorporates all significant intersymbol interference combinations. If the signal is not sufficiently representative, the QA evaluation may not be reliable.

Instead, each strobe value is sent to a computation routine that has knowledge of the decision boundaries of the signal constellation (Sec. 4.4.2) and of the statistics of the noise to be modeled. From this information the computation routine calculates the probability that the observed strobe plus added noise would fall outside the correct decision boundaries and cause a decision error for that strobe. Error probability for the entire simulation run is estimated as the average of the error probabilities associated with each individual strobe.

Because this method deals with ensemble statistics of the noise instead of a particular noise realization, there is no statistical scatter originating from the noise. The QA method computes the actual probability of detection error for the given noise statistics and collection of strobe values; it does not estimate the probability from an experimental count of errors. Scatter still might arise from other disturbances that are present or from an insufficiently representative signal realization, but not from the noise.

Restrictive Conditions

The following conditions are implied or explicit:

- Only symbol-by-symbol detection can be evaluated—not sequence estimation or other decoding schemes.
- Noise output of the receiver is modeled as additive to the strobes; superposition applies.
- Noise statistics are known at the receiver output.
- The decision boundaries are known and expressible in reasonably compact format.
- The transmitted symbol corresponding to each strobe is known, so that the correct decision region can be selected for each evaluation.
- Evaluation procedures are computationally feasible.
- The receiver is time invariant.

These conditions are not absolutely rigid—exceptional cases can be contrived—but they do apply to the great majority of QA-amenable problems.

The requirement that noise be additive to the strobes implies that the communications link be linear from the entry point of the noise through to the strobe output of the receiver. Since noise often is applied at the receiver input, the requirement typically devolves to a need for a linear receiver.

The requirement that the noise statistics be known at the receiver output usually implies that the noise be Gaussian at its entry point. Gaussian noise has the property that it remains Gaussian after passage through linear processes, such as the filters of the receiver. If system noise were non-Gaussian, it might be very difficult or impossible to know the noise statistics after passage through the receiver.

In a system design, the noise typically is specified at the receiver's input, not its output. A relation between noise input and output has to be established, usually by a noise calibration of the receiver. The principles of noise calibration are presented subsequently.

The requirements for compact expressions for the decision boundaries and computationally feasible evaluation procedures tends to restrict the method to coherent reception of PAM–QAM signals with regular constellations. That is the configuration given the greatest attention in the paragraphs to follow. However, extensions to other signals have been devised and will be mentioned briefly. The requirement for time invariance of the receiver will be examined further after noise calibration has been explained.

QA Configuration

Figure 13.2 shows the main elements of QA procedures. A noise-free signal $\{z_{in}(n)\}$ is generated by the simulation program and delivered to the simulated receiver. System noise typically (but not always) is referred to the receiver input. But no noise is generated; only its effects are to be calculated from its specified characteristics. Because the noise is not actually present in the simulation, we have dubbed it *phantom noise*.

The simulated receiver operates on the simulated signal and delivers a sequence of strobes $\{z_o(m)\}$, one per symbol. The simulated receiver includes all necessary processes described in the preceding chapters, excluding symbol detection and any digital operations lying farther downstream. Provided that the restrictions are satisfied and the noise calibration is possible, the QA method is not concerned with receiver details.

Strobes are delivered to the computation routine, along with the corresponding

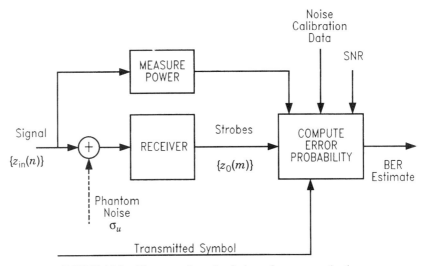

FIGURE 13.2 Elements of quasianalytic performance evaluation.

transmitted symbols. Because of system delay, there has to be a compensating delay on the transmitted sequence, in the same manner as described for Monte Carlo runs, as shown in Fig. 13.1. (Alternatively, a properly synchronized auxiliary message generator would supply the transmitted symbols to the comparison routine.) Furthermore, since phase ambiguity most often is resolved after symbol detection, and since postdetection processes are not included in QA evaluation, an ambiguity-compensating phase rotation also has to be included. Delay and phase calibration are needed in addition to noise calibration; these are treated in Appendix 13D.

Calibrations are performed prior to a simulation run and the calibration data are supplied to the computation routine. Information on decision boundaries (specific to the particular signal format) is provided to the routine. Error probability usually is computed for a range of signal-to-noise ratios (SNR); the user specifies the SNR values to be used in the computations.

13.2.2 QA Computation of Error Probability

Represent the *m*th noise-free strobe delivered by the receiver of Fig. 13.2 as

$$z_o(m) = x_m + jy_m \tag{13.3}$$

Figure 13.3 illustrates the geometry of the strobe components. Observe that even though $z_o(m)$ is noise-free, it does not necessarily coincide with the constellation point $a_m + jb_m$. Other disturbances and distortion may cause the noise-free strobe to deviate from the ideal location.

COMMENT: Representation of the signal by a constellation point and by a single strobe for each symbol presupposes a PAM–QAM format. The remainder of this section is limited to particular PAM–QAM signals.

Now imagine that complex noise sample $w_m = u_m + jv_m$ were added to the noise-free strobe to produce a noisy strobe:

$$\begin{aligned}\bar{z}_o(m) &= z_o(m) + w_m = x_m + jy_m + u_m + jv_m \\ &= x_o(m) + jy_o(m)\end{aligned} \tag{13.4}$$

The effect of added noise is demonstrated by dashed lines in Fig. 13.3. Noise provides another component of deviation from the ideal constellation point.

Probability of Decision Error

Designate the decision region (see Sec. 4.4.2) associated with the *m*th symbol as $\beta_m(x, y) = \beta_m$. The decision region surrounds the *m*th constellation point $(a_m + jb_m)$. If the *m*th strobe falls inside β_m, the symbol detector will provide the correct symbol estimate $(\hat{a}_m + j\hat{b}_m) = (a_m + jb_m)$. If the *m*th strobe falls outside β_m (in the comple-

368 ☐ EVALUATION OF PERFORMANCE

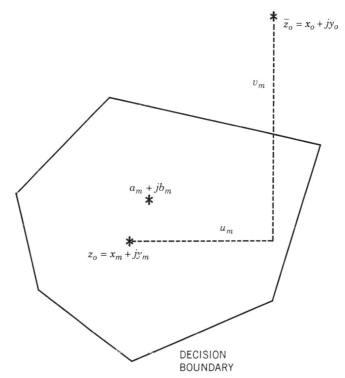

FIGURE 13.3 Geometry of signal and noise components of strobe. Region outside decision boundary is $\beta'_i(x, y)$.

mentary region denoted β'_m), a decision error will be incurred. A conceptual decision boundary has been included in Fig. 13.3; the illustrated noise-free strobe z_o falls inside β_m, while the illustrated noisy strobe $\bar{z}_o$ falls in β'_m.

Erroneous decisions can arise from additive noise, from ISI, from interference, from synchronization errors, and from many other disturbances too numerous to itemize. Commonly, but not universally, additive noise is the dominant cause of decision error.

Probability of decision error on the mth strobe is expressed by

$$\Pr\{\epsilon_s\}_m = \Pr\{\text{symbol error on } m\text{th strobe}\} = \Pr\{\bar{z}_o(m) \text{ lies outside } \beta_m\}$$

$$= \iint_{\beta'_m} \text{pr}\{x_o, y_o\} \, dx_o \, dy_o \tag{13.5}$$

where $\Pr\{\cdot\}$ denotes probability of the indicated event and $\text{pr}\{\cdot\}$ denotes the probability density function (pdf) of the indicated random variable(s). That is, the probability of symbol error on the mth strobe is found by integrating the pdf of $\bar{z}_o(m) =$

$x_o(m) + jy_o(m)$ over that portion of the x–y plane lying outside the decision region β_m for the constellation point $a_m + jb_m$ corresponding to the mth transmitted symbol.

As an important special case, assume (1) that the noise components u_m and v_m are independent, (2) that they are zero-mean Gaussian, and (3) that their variances are equal. Stated formally, these conditions are

1. $\text{pr}\{u_m, v_m\} = \text{pr}\{u_m\}\text{pr}\{v_m\}$

2. $\text{pr}\{u_m\} = \dfrac{1}{\sqrt{2\pi}\,\sigma_{um}} \exp\left(-\dfrac{u_m^2}{2\sigma_{um}^2}\right)$; $\text{pr}\{v_m\} = \dfrac{1}{\sqrt{2\pi}\,\sigma_{vm}} \exp\left(-\dfrac{v_m^2}{2\sigma_{vm}^2}\right)$

3. $\sigma_{um} = \sigma_{vm} = \sigma_o$ \hfill (13.6)

In consequence of these special assumptions and because $u_m = x_o - x_m$ and $v_m = y_o - y_m$, the pdf in (13.5) becomes

$$\text{pr}\{x_o, y_o\} = \text{pr}\{x_o\}\text{pr}\{y_o\}$$

$$\text{pr}\{x_o\} = \dfrac{1}{\sqrt{2\pi}\,\sigma_{um}} \exp\left[-\dfrac{(x_o - x_m)^2}{2\sigma_{um}^2}\right];$$

$$\text{pr}\{y_o\} = \dfrac{1}{\sqrt{2\pi}\,\sigma_{vm}} \exp\left[-\dfrac{(y_o - y_m)^2}{2\sigma_{vm}^2}\right] \quad (13.7)$$

so the integral (13.5) transforms to

$$\Pr\{\epsilon_s\}_m = \dfrac{1}{2\pi\sigma_o^2} \iint_{\beta_m'} \exp\left[-\dfrac{(x_o - x_m)^2 + (y_o - y_m)^2}{2\sigma_o^2}\right] dx_o\, dy_o \quad (13.8)$$

In general, (13.8) cannot be evaluated analytically for arbitrary decision regions β_m; numerical evaluation usually is needed. Fortunately, the decision regions for several widely used classes of constellations permit analytical evaluation or an analytical approximation. Simple evaluation is possible for linear constellations, rectangular constellations, or MPSK constellations, as developed in Appendix 13A and summarized below.

QA Decision Formulas

The formulas given below are listed here for purposes of reference. Consult Appendix 13A for their derivation and context. Temporarily assume that the noise-free strobe $z_o(m)$ falls inside the decision region β_m—that there is no decision error in the absence of noise. The procedure to be followed when that assumption is violated is treated in a later section.

Notation. In the subsequent formulas, δ is a generic notation indicating a distance from the noise-free strobe to a decision boundary. The function

$$Q(X) = \frac{1}{\sqrt{2\pi}} \int_X^\infty e^{-t^2/2} \, dt \qquad (13.9)$$

is the complementary probability integral, found in numerous tables, such as those of [13.15, Chap. 26]. Computable approximations for $Q(X)$ are discussed in Appendix 13C. Recollect that $Q(0) = 0.5$, $Q(\infty) = 0$, and $Q(-\infty) = 1$. The probability of symbol error on the mth strobe is denoted $\Pr\{\epsilon_s\}_m$, and the probability of bit error is denoted $\Pr\{\epsilon_b\}_m$.

Baseband Binary Decisions. Let $b_m = 0$, $y_m = 0$, and $v_m = 0$, as occurs in a real baseband signal. Furthermore, let a_m take on the values ± 1 so that the decision level is at $x = 0$. Probability of bit error, which is the same as probability of symbol error since each symbol is one bit, is given by

$$\Pr\{\epsilon_b\}_m = \Pr\{\epsilon_s\}_m = Q\left(\frac{x_m}{\sigma_o}\right) \qquad (13.10)$$

Baseband Multilevel Decisions. Let a_m take on more than two values so that some constellation points will be sandwiched between two decision levels. If the distance from the strobe to one level is designated δ_1 and the distance to the other level is designated δ_2, the probability of symbol error is

$$\Pr\{\epsilon_s\}_m = Q\left(\frac{\delta_1}{\sigma_o}\right) + Q\left(\frac{\delta_2}{\sigma_o}\right) \qquad (13.11)$$

Specification of the probability of bit error $\Pr\{\epsilon_b\}_m$ will be covered later.

All points of a baseband constellation lie on the same straight line. Suppose that the constellation points and therefore the decision boundaries are equally spaced and balanced about zero. Decision boundaries would lie midway between constellation points. This is the predominant choice for multilevel signaling.

Then it is convenient to scale the constellation so that the distance between adjacent points is 2 and distance from any constellation point to its closest decision boundaries is 1. The receiver is required to normalize the amplitude of the signal so that the strobe amplitudes fit the decision regions properly. Significant misadjustment of signal amplitude will lead to excessive decision errors.

Designate the mth transmitted symbol as a_m. If a_m is an inner point of the constellation, the distances in (13.11) are

$$\delta_1 = x_m - (a_m - 1), \qquad \delta_2 = (a_m + 1) - x_m \qquad (13.12)$$

for the lower and upper boundaries, respectively. These distances both are positive if the noise-free strobe x_m falls between the two decision boundaries. If a_m is an outer point, one of the distances is infinite and only the other (either lower or upper, as appropriate) is given by (13.12).

In a binary baseband system, all points are outer points. Taking that into account, the formulation of (13.11) and (13.12) can be applied to binary systems as well as to multilevel.

Two-Dimensional Constellations. Let the constellation be situated on a square grid, with distance 2 between adjacent constellation points in horizontal and vertical directions. Decision boundaries lie at a distance 1 from the closest points. See Sec. 12.1 for additional discussion of two-dimensional symbol detection.

Decomposing the decision into independent operations on x and y separately, the probabilities of erroneous decisions are

$$\Pr\{\epsilon_x\}_m = Q\left(\frac{\delta_{x1}}{\sigma_o}\right) + Q\left(\frac{\delta_{x2}}{\sigma_o}\right)$$

$$\Pr\{\epsilon_y\}_m = Q\left(\frac{\delta_{y1}}{\sigma_o}\right) + Q\left(\frac{\delta_{y2}}{\sigma_o}\right) \quad (13.13)$$

where the distances for an interior point of the constellation are (see Fig. 13.4)

$$\delta_{x1} = x_m - (a_m - 1); \quad \delta_{x2} = (a_m + 1) - x_m$$
$$\delta_{y1} = y_m - (b_m - 1); \quad \delta_{y2} = (b_m + 1) - y_m \quad (13.14)$$

Appropriate modifications are made for an exterior point by setting one of the distances to infinity.

Equations (13.13) and (13.14) relate to neither symbol error nor bit error (except for QPSK with its binary baseband). Appendix 13A shows that the probability of symbol error is given by

$$\Pr\{\epsilon_s\}_m = \Pr\{\epsilon_x\}_m + \Pr\{\epsilon_y\}_m - \Pr\{\epsilon_x\}_m \Pr\{\epsilon_y\}_m \quad (13.15)$$

MPSK. Decision boundaries for MPSK are rays emanating from the origin of the complex plane (see Fig. 13A.4). The probability of error cannot be expressed exactly in closed form. A good upper bound on symbol error probability is set by

$$\Pr\{\epsilon_s\}_m < Q\left(\frac{\delta_1}{\sigma_o}\right) + Q\left(\frac{\delta_2}{\sigma_o}\right) \quad (13.16)$$

where δ_1 is the distance to the clockwise boundary and δ_2 is the distance to the counterclockwise boundary of the decision region.

372 ☐ EVALUATION OF PERFORMANCE

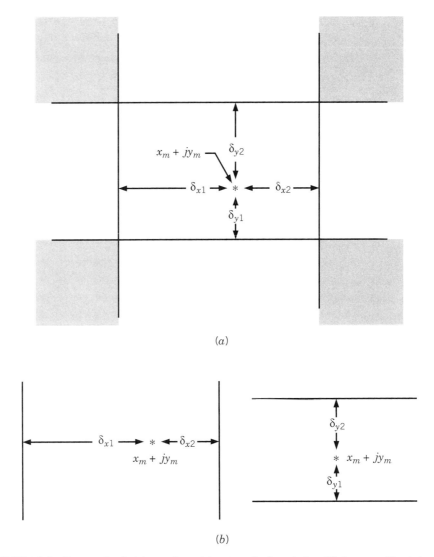

FIGURE 13.4 Rectangular decision regions: (*a*) rectangular boundaries; (*b*) decomposition into separate straight-line *x* and *y* boundaries.

Let the mth symbol have the value i_m ($i = 0$ to $M - 1$) and let the constellation phases be

$$\phi_m = \frac{(2i_m + 1)\pi}{M} \tag{13.17}$$

Then the distances are found from trigonometry to be

$$\delta_1 = y_m \cos \frac{2\pi i_m}{M} - x_m \sin \frac{2\pi i_m}{M}$$

$$\delta_2 = x_m \sin \frac{2\pi(i_m + 1)}{M} - y_m \cos \frac{2\pi(i_m + 1)}{M} \qquad (13.18)$$

Both distances are defined to be positive if the strobe value falls in the correct decision region.

Use of the Formulas

The formulas given above are specific to particular signal formats. There are other aspects of the method that are more general; these are covered below.

Noise-Free Decision Errors. It is possible that nonnoise disturbances will cause a strobe to fall outside the correct decision region, so that a decision error would happen in the absence of noise. Such an event would be considered a severe design fault in many systems and a simulation engineer might want to make note of its occurence. To that end, simulation runs often include provision to check that all noise-free symbols were received correctly and to identify any that were received in error.

Some programs stop a QA evaluation when noise-free errors are discovered, while others carry on with the evaluation. If the evaluation is to be continued, the errors cannot be permitted to disrupt the run (the program must include provision for dealing with decision errors) or to deliver faulty probability estimates (the program must compute the proper effect of the errored strobes). The distance formulations given here satisfy both requirements. If the strobe falls in the correct decision region, the distances will be positive; if the strobe falls in the wrong decision region, one distance will be negative. Finding a negative distance is an unequivocal indication of a noise-free decision error.

The probability of error computed with a negative distance will be correct; the formula will indicate that the probability exceeds 0.5. The function $Q(X)$ is defined for negative X as well as positive, so it could accept negative distances without trouble. However, the computer uses an approximation to $Q(X)$—call it $\tilde{Q}(X)$. See Appendix 13C for a short account of approximations. An approximation might be excellent for positive X but increasingly less satisfactory as X goes negative. To use an approximation $\tilde{Q}(X)$ that is not accurate for negative X, employ the following expedient:

$$\text{For } X \geq 0, \text{ use } \tilde{Q}(X)$$
$$\text{For } X < 0, \text{ use } 1 - \tilde{Q}(|X|) \qquad (13.19)$$

This expedient is based on the relation for the normal probability integral that $Q(-X) = 1 - Q(X)$.

Average Probability of Error. The formulas give the probability of symbol error for the mth strobe. Unless all decision regions are identical for all points of the constellation and strobes fall in identical locations in each decision region (these are highly unlikely occurences), the formulas will give a different probability for each strobe. An estimate of the true error probability is obtained by averaging the probabilities determined for the individual strobes.

Let N_Q strobes be included in one QA evaluation. Assume that all points in the constellation are represented in the N_Q strobes in proportion to their probability of being selected for transmission. If all points are equally likely to be transmitted, the strobes should include all constellation points in nearly equal numbers. With that proviso satisfied, the average probability of symbol error is calculated as

$$\overline{\Pr}\{\epsilon_s\} = \frac{1}{N_Q} \sum_{m=1}^{N_Q} \Pr\{\epsilon_s\}_m \qquad (13.20)$$

where the overbar denotes average.

Probability of Bit Error. Although the probability of symbol error is a valuable datum, an engineer more often wants to know the probability of bit error. The latter frequently can be approximated from the former by the following reasoning. Assume that the transmitted symbols have been Gray-encoded so that adjacent symbols in the constellation differ by just one information bit. Assume further that decision errors are made predominantly to adjacent constellation points, to the extent that the probability of errors to greater distances can be neglected. If there are b bits in each symbol and the average probability of symbol error is $\overline{\Pr}\{e_s\}$, the average probability of error per bit is approximately

$$\overline{\Pr}\{\epsilon_b\} \simeq \frac{1}{b} \overline{\Pr}\{\epsilon_s\} \qquad (13.21)$$

Theoretical Probability of Error. All of the QA formulas above are applied to imperfect strobes that, although noise free, have been perturbed by other disturbances and distortion. If the other disturbances and distortion were absent, the noise-free strobes should fall exactly on the ideal constellation points; errors would then be caused solely by the additive noise. Error probability attained under these ideal conditions is the best performance that can be expected for the signal format being investigated; that is what is meant by the term *theoretical performance*.

Theoretical performance can be derived from the QA formulas by calculating error probability using the distances from the ideal constellation points instead of the perturbed strobes. Results are developed in Appendix 13B for the signal formats considered above. As a convenience to the user, a simulation program should be capable of displaying theoretical performance to compare to simulation results. Theoretical performance is unrelated to the method of performance evaluation; it could also be compared against results of Monte Carlo simulations.

Ambiguity Resolution

Any complex signal with p-fold symmetry in the I–Q plane will have p-fold phase ambiguity in its reception. For example, QPSK has fourfold phase ambiguity. In like manner, absolute polarity of baseband signals can be lost in transmission. Examples arise in magnetic recording or balanced wire lines, where a polarity inversion occurs with the reversal of wire connections. Such signals have a twofold ambiguity if the connection sense is not controlled.

To recover the data, these ambiguities must be resolved. There is nothing in the signal format itself to aid the resolution of ambiguity; some antiambiguity pattern must be embedded in the data. Two different antiambiguity methods are often encountered:

1 Embed a known word in the data stream at periodic intervals. Antiambiguity circuits in the receiver watch for the known word and its $(p-1)$ ambiguous translates. Once the word or one of its translates is found in the data stream, the receiver knows the ambiguity state and can interpret all data accordingly.

2 Apply differential encoding to the signal, wherein the information is transmitted as changes between symbols rather than as the symbols themselves. Differential decoding in the receiver then recovers the original information.

All of the decision-error formulas above were based on absolute decisions on individual symbols and an assumption that no further error mechanisms were at work. However, if differential encoding is employed, a decision error in one strobe most often will cause two decoded symbol errors; a decision error on one strobe propagates to decoding of the following strobe as well. To take account of the effect on decision performance of error propagation with differential decoding, it is usually sufficient to double the probability of error calculated from absolute errors that ignore the differential encoding.

Absolute decisions require unambiguous detection of the symbols. That is possible in a simulation because the transmitted symbols are known and the received strobes can be rotated to agree with the transmitted stream (see Appendix 13D).

Other Signal Formats

Only a narrow range of signal formats have been included in the foregoing list of QA formulas. The square-constellation formulas depend upon the separability of x and y decisions for their simplicity; the separation permits the error probability to be evaluated by two individual one-dimensional integrals. If the constellation is not based on a rectangular grid, the decisions no longer are separable and two-dimensional integrations become necessary for evaluation of the probability of error. Moreover, the region of integration may become so complicated that the integrals can no longer be evaluated analytically, not even approximately. Numerical integration can be invoked, but that is much more time consuming than analytical evaluations.

Nonetheless, it is possible to apply the QA method to other signal and receiver formats, provided that the noise is additive to the strobes and the noise statistics are known. Reference [13.1] takes a brief look at a general approach, [13.16] describes QA evaluation of differentially coherent demodulation (the methods explained above all relate to coherent demodulation), [13.17] treats QA evaluation of FSK demodulation, and [13.18] introduces a nonlinearity into the transmission path.

With sufficient effort, QA methods undoubtedly can be adapted to other signal formats as well. When deciding whether to attempt QA evaluation of a system for which no QA model yet exists, a user has to trade off the analytical and programming effort of developing the model against the possibly extravagant consumption of computer time required by brute-force Monte Carlo operations.

13.2.3 Phantom Noise Calibration

Terms of the form $Q(\delta/\sigma_o)$ arise repeatedly in all of the error-probability formulas of the preceding section. Distance δ is determined from the strobes, one by one, of the simulated signal and from knowledge of the transmitted symbol and the decision boundaries of the constellation. The rms phantom noise σ_o that would accompany the strobes is calculated from a noise calibration and from specification of input signal-to-noise ratio (SNR). This section contains an outline of a calibration procedure whose foundations are developed in Appendix 13E.

Noise Transmission

Let the receiver be linear and time invariant. Implications of time invariance are explored below. For purposes of the immediate discussion, assume also that the receiver's I and Q channels are balanced; this last assumption simplifies the presentation but is not essential. Appendix 13E broadens the treatment to include unbalanced I and Q channels.

If the signals are represented by their complex envelopes and if the foregoing assumptions are met, the receiver can be modeled as a filter with complex impulse response $h_R(n)$, followed by a downsampler (decimator) that selects the strobes. Figure 13E.1 illustrates the receiver model but in the greater detail needed for unbalanced I and Q channels.

Complex phantom noise at the receiver's input is assumed to be white, Gaussian, and stationary, with independent rectangular components of equal variances σ_u^2. It can be demonstrated (see Appendix 13E) that the resulting noise variance of each component of the output strobes is

$$\sigma_o^2 = \sigma_u^2 \sum_n |h_R(n)|^2 = \sigma_u^2 \xi_R^2 \qquad (13.22)$$

Since the noise is stationary and the receiver is time invariant, all strobes have the same variance. The summation in (13.22) is over all nonzero samples of $h_R(n)$; the summation will have infinite limits for an IIR filter.

This formulation is in the time domain, but a similar frequency-domain formulation could also be developed. If the transfer function of the receiver filter is $H_R(k)$ [furnished as an N-element block], the noise calibration is

$$\xi_R^2 = \frac{1}{N} \sum_k |H_R(k)|^2 \tag{13.23}$$

which is closely related to the noise bandwidth of the receiver.

Receiver Calibration

There are two parts in establishing σ_o^2: (1) specify σ_u^2 and (2) measure $\xi_R^2 = \Sigma |h_R|^2$. The latter task will be considered first.

A brute-force technique could be employed. Just apply white Gaussian noise of known variance to the receiver's input and measure the output variance. Because this approach uses a particular noise realization, the receiver calibration obtained thereby is subject to statistical scatter. Time is needed to collect enough samples for good estimates of the variance. Undecimated filtered-signal samples may not be accessible, so measurements would have to be performed on the strobes, thereby further increasing the time required to gather valid statistics.

A more elegant approach uses a deterministic calibration signal, in the form of unit samples ("impulses") to measure $h_R(n)$ directly. By definition, $h_R(n)$ is the receiver's response to a unit sample. It is a simple matter to form $\Sigma |h_R(n)|^2$ as the samples of $h_R(n)$ are generated. Because $h_R(n)$ is deterministic, no statistical scatter occurs in its measurement. As a result, there is no need to accumulate large numbers of output samples, in contrast to the requirement when calibration is attempted on a noise input.

Practical Matters. If the receiver is I–Q balanced, it is sufficient to apply a single calibrating impulse $z_1(n) = \delta(n)(1 + j0)$ to the I-input alone. If the receiver is I–Q unbalanced, separate calibrations are needed of I and Q individually. That can be accomplished by applying a second impulse $z_2(n) = \delta(n)(0 + j1)$ to the Q-input. Separate calibrations for I and Q are explained in Appendix 13E.

The formulation presented so far has been based on equal sampling rates at input and output of the receiver. There will be many simulations in which sampling rate will not be preserved. More commonly, downsampling (or *decimation*) will be performed in the receiver and the sampling density at the output will be substantially reduced from that at the input. For many receivers, only the symbol strobes—one sample per symbol—are delivered at the receiver output, while M_s samples per symbol are present at the receiver's input.* If only downsampled strobes are available at the output, it is not possible to determine $h_R(n)$ from a single impulse calibration signal on each input channel.

---

*Elsewhere in this book, M_s has been the designation for the number of samples per symbol. For this one application, expand its meaning so that M_s is defined as the receiver's decimation ratio. If receiver output samples are taken only once per symbol interval, the two meanings coincide.

This obstacle is overcome by applying M_s impulses to each input channel, in the form $z_1(n,p) = \delta(n-p)(1+j0)$, with $p = 0$ to $M_s - 1$, and similarly for $z_2(n,p)$. The required $\Sigma |h_R(n)|^2$ is calculated entirely from the externally delivered strobe responses to these multiple inputs, with no need to have internal access into the receiver. Appendix 13E provides mathematical details.

Internal states of the filters must be set to zero before applying each of the calibrating impulses. Other than that, the calibration operates entirely externally to the receiver; it is oblivious to details of the receiver implementation, provided that the receiver is linear and time invariant. Receiver simulation could be single-point or linear block or circular block, or a mixed implementation. Simulation could be in the time domain or in the frequency domain. None of these matter to the calibration or to the QA evaluation.

If the receiver filter has finite impulse response, only a finite number of terms need be summed in the calibration of $\Sigma |h_R(n)|^2$. There should be no difficulty in encompassing all nonzero terms in the sum. A fundamental problem arises, though, if the filter has infinite impulse response; it is not possible to measure and sum an infinite number of terms. In practice, the impulse response has to be truncated to finite length and only a finite number of terms can be summed. The resulting calibration always yields a value that is slightly smaller than the correct value, so the ensuing QA evaluation will be slightly optimistic. It is the user's responsibility to set the truncation long enough that the calibration error is acceptably small. Some guidance on this matter is provided in Appendix 13F.

Noise Specification

The objective of the calibration procedure is to specify the output noise variance σ_o^2 in terms of the signal-to-noise ratio (SNR)—either E_b/N_0 or E_s/N_0—at the receiver's input. Appendix 9A and equation (9.11) tell how to relate the receiver's input-noise variance σ_u^2 to a measurement of the signal power in a calibration block and to the desired SNR.

To recapitulate: First measure the power P_z of a sequence of N signal samples. Power is defined in the time domain as

$$P_z = \frac{1}{N} \sum_n |z_{\text{in}}(n)|^2 \qquad (13.24)$$

or in the frequency domain as

$$P_z = \frac{1}{N^2} \sum_k |Z_{\text{in}}(k)|^2 \qquad (13.25)$$

where $\{z_{\text{in}}(n)\}$ and $\{Z_{\text{in}}(k)\}$ are discrete Fourier transforms of one another.

Next, specify the desired value (or a series of values) for either E_s/N_0 or E_b/N_0.

Then from (9A.18), the required noise variance is

$$\sigma_u^2 = \frac{M_s}{2(E_s/N_0)} P_z = \frac{M_s}{2b(E_b/N_0)} P_z \qquad (13.26)$$

where M_s is the number of samples per symbol and b is the number of bits per symbol.

In Chapter 9 the calculated value for σ_u was used to set the scaling on a noise generator. Identical signal-power measurement and SNR calculations are used for phantom-noise specification, but the resulting value of σ_u is not used for scaling a noise generator. Instead, the calculated σ_u is taken as the rms phantom noise at the receiver input. The noise quantity needed for QA evaluation is σ_o, the rms phantom noise in the receiver's output strobes. That is determined from the receiver calibration simply as

$$\sigma_o^2 = \xi_R^2 \sigma_u^2 = \sigma_u^2 \sum_n |h_R(n)|^2 \qquad (13.27)$$

for balanced I and Q channels, or

$$\sigma_{ou} = \xi_I \sigma_u, \qquad \sigma_{ov} = \xi_Q \sigma_u \qquad (13.28)$$

for unbalanced channels. See Appendix 13E for definitions of ξ_I and ξ_Q.

13.2.4 Implications of Time Invariance

If the requirement for time invariance of the receiver is to be observed rigorously, the following restrictions apply:

- Interpolators cannot be included in the receiver, because interpolators are time-varying filters (see Chapter 10).
- The decimation factor M_s must be an integer. A noninteger value implies interpolation or other time-varying processes inside the receiver.
- Gating or other time windowing of the signal is not permitted, since those are time-varying processes.
- Adaptive operations, if present, have to settle to their equilibrium states and be frozen before calibration and evaluation can be performed. Some examples of time-varying adaptive operations include phase rotation, automatic gain control, adaptive equalization, and echo cancelling.

Expedients sometimes can be found to circumvent restrictions. An example for interpolators is reported briefly in [10.8]. If M_s is an integer—that is, the symbol rate is commensurate with the sample rate—the optimum fractional interval μ needed in the interpolator is a constant. Optimum μ depends on the phase relation (*decimation phase*) between the samples and the symbols.

Receiver calibration and QA-evaluated performance will both be dependent on μ. Average performance—as encountered for systems with M_s not an integer and thus with time-varying optimum μ—can be estimated by averaging the performance of an integer-M_s system over different decimation phases, each with a fixed optimum value of μ ($0 \leq \mu < 1$).

13.3 KEY POINTS

A prime reason for running simulations is to evaluate the data-error performance of the system under study. Two different techniques for performance evaluation have been employed extensively:

1 Monte Carlo simulations, in which detection errors are counted directly and all transmission disturbances, including additive noise, are incorporated into the simulation

2 Quasianalytic (QA) simulations, in which the signals are simulated but additive noise is treated analytically

Monte Carlo methods are extremely versatile; in principle, any system whatsoever can be evaluated by Monte Carlo methods, provided that the system can be simulated. The main drawback of Monte Carlo simulations lies in their inordinate expenditure of computer time when error probabilities are small.

Quasianalytic methods have highly restricted applications. The receiver must be linear and time invariant, the signal has to have a regular constellation with readily computable error probabilities, and the additive noise needs to be Gaussian in most instances. Fortunately, many communications links of great practical importance satisfy these restrictions, so QA evaluation is often applicable. It has an enormous advantage over Monte Carlo methods in the time needed for computing.

13.3.1 Monte Carlo Method

A Monte Carlo simulation models the entire transmission chain, from data source at the transmitter through symbol (or bit) detector or decoder in the receiver. All disturbances are also modeled. Because of the disturbances, some symbols will be detected in error. Errors are revealed by comparison of the detected symbols with the transmitted data stream. Error events are counted over the course of the simulation run. At the end of the run, the probability of detection error is estimated as the ratio of the number of errors counted divided by the number of symbols (bits) transmitted.

A Monte Carlo simulation is a statistical experiment; the observed error ratio is an *estimate* of the true probability of error and is subject to statistical scatter. Enough errors have to be collected so that the estimate is adequately reliable. Confidence levels and intervals are discussed in the text. The engineers' standby of 10 errors is usually not enough; 100 is better and some experienced simulationists try to collect 1000. System performance commonly is presented as a plot of error probability versus signal-to-noise ratio (SNR). Each point on such a curve requires another complete Monte Carlo simulation to be run with the appropriate SNR.

13.3.2 QA Method

Quasianalytic simulations generate noise-free strobes out of the receiver. Symbol detectors and other elements farther downstream are omitted. Additive noise is not simulated; only noise statistics are involved. Because no actual noise is present, we have called this the *phantom noise* method. On each strobe the evaluation procedure computes the probability of decision error that would be caused by noise of specified statistics. Average probability of error is determined by averaging the individual error probabilities over all of the strobes in the simulation.

Since ensemble statistics of the noise are used in the calculations instead of a particular noise realization, there is no statistical scatter arising from the phantom noise. As a result, the necessary evaluation can be performed with a comparatively short simulation. Moreover, all points on an error probability versus SNR curve can be calculated with just one short simulation. It is vitally important that the signal be sufficiently representative—that it incorporates all significant intersymbol interference patterns.

13.3.3 Alignment

Both methods require that the transmitted data symbol or bit be known to the evaluation procedure. In Monte Carlo simulations the transmitted symbol is compared against the detected symbol; any discrepancies are counted as errors. In QA simulations, the transmitted symbol is needed so that each received strobe can be placed in the correct decision region for calculation of the probability of detection error.

A communications link imposes time delay and phase-ambiguity rotation upon its signals. Transmitted data have to be delayed to compensate for the system delay and any phase ambiguity has to be counter-rotated out, so that the receiver-recovered data can be compared correctly with the proper transmitted data.

The amount of compensating delay and phase rotation are not known a priori; they have to be determined in a calibration procedure that is undertaken before the main simulation run. Techniques for accomplishing the alignment are treated in Appendix 13D. These are closely related to synchronization techniques treated in Chapter 11.

13.3.4 Signal and Noise Calibration

Performance is evaluated as a function of SNR at the receiver's input. To set up the simulation, signal power is established by measurements on a block of signal, and then the noise is specified to set the desired SNR. In Monte Carlo simulations, the noise specification is applied to a noise generator (see Chapter 9). In QA simulations, no noise is actually generated; the noise specification is used to establish the value of rms phantom noise to be used for calculations on the strobes.

Translation of user-specified input SNR to the corresponding rms noise on the strobes is accomplished through a calibration of the receiver's noise transmission. If σ_u^2 is the variance of the input phantom noise, then the phantom-noise variance at the strobe is

$$\sigma_o^2 = \sigma_u^2 \sum_n |h_R(n)|^2$$

where $h_R(n)$ is the impulse response of the receiver. See Appendix 13E for details of the calibration procedure. Input phantom noise variance is given by

$$\sigma_u^2 = \frac{M_s}{2(E_s/N_0)} P_z = \frac{M_s}{2b(E_b/N_0)} P_z$$

where P_z is the measured power of the receiver's input signal, M_s is the number of samples per symbol, b is the number of bits per symbol, and E_s/N_0 and E_b/N_0 are user-specified signal-to-noise ratios.

13.3.5 STÆDT Processes

    BERC: error counter

    CMPR: comparison of two sample sequences (to detect errors)

    LET: various operations, including evaluation of $Q(X)$

    NCALI, NCALO: noise calibration of receiver

    PLER, PRER: plot or print error ratio versus signal-to-noise ratio

    QA: quasianalytic evaluation of signal strobes

    TBER: theoretical error ratio

    TDER: theoretical error ratio for differentially detected or decoded data

APPENDIX 13A: ERROR PROBABILITIES FOR QUASIANALYTICAL EVALUATIONS

This appendix contains a short derivation of the QA error-probability formulas that were presented in the chapter text. Although results are given only for simple constellations and additive, Gaussian noise, these conditions are of immense practical importance. Therefore, the results are applicable to a great many simulations that will be encountered.

APPENDIX 13A: ERROR PROBABILITIES FOR QUASIANALYTICAL EVALUATIONS □ **383**

Binary Decisions

First consider real baseband binary antipodal decisions, whose results will be extended also to BPSK and (O)QPSK passband signals. Moreover, the same principles can be applied, with some modification, to any PAM–QAM signals with linear or rectangular constellations. The strobe model for real binary signals consists only of the real parts of (13.4); all contributions from b_m and v_m disappear. Equation (13.4) reduces to

$$\overline{z_o}(m) = x_o(m) = x_m + u_m \tag{13A.1}$$

where the noise-free x_m is the desired signal information plus ISI resulting from imperfect timing synchronization and non-Nyquist pulse shaping, and u_m is the added noise. Assume that the data symbol $a_m = +1$ or -1, with equal probability.

A symbol detector for antipodal binary signals examines the strobe and decides $\hat{a}_m = +1$ if $\overline{z_o}(m) \geq 0$, or $\hat{a}_m = -1$ if $\overline{z_o}(m) < 0$. Assume that x_m has the correct polarity and then calculate the probability that the addition of a sample of zero-mean Gaussian noise u_m of known variance σ_o would cause $x_o(m) = x_m + u_m$ to have the wrong polarity. The error integration of (13.8) reduces to

$$\Pr\{\epsilon_s\}_m = \frac{1}{\sqrt{2\pi}\,\sigma_o} \int_{\beta'_m} \exp\left[-\frac{(x_o - x_m)^2}{2\sigma_o^2}\right] dx_o \tag{13A.2}$$

Since transmitted symbols are binary for this class of signals, the symbol error probability is the same as the bit-error probability.

If $a_m = +1$, the error region β'_m is defined by $x_o < 0$ and the error integration can be manipulated as follows:

$$\Pr\{\epsilon_s\}_m = \frac{1}{\sqrt{2\pi}\,\sigma_o} \int_{-\infty}^{0} \exp\left[-\frac{(x_o - x_m)^2}{2\sigma_o^2}\right] dx_o$$

$$= \frac{1}{\sqrt{2\pi}\,\sigma_o} \int_{-\infty}^{-x_m} \exp\left(-\frac{u^2}{2\sigma_o^2}\right) du$$

$$= \frac{1}{\sqrt{2\pi}} \int_{-\infty}^{-x_m/\sigma_o} \exp\left(-\frac{\mu^2}{2}\right) d\mu$$

$$= \frac{1}{\sqrt{2\pi}} \int_{x_m/\sigma_o}^{\infty} \exp\left(-\frac{\mu^2}{2}\right) d\mu = Q\left(\frac{x_m}{\sigma_o}\right) \tag{13A.3}$$

where $u = x_o - x_m$, $\mu = u/\sigma_o$, and the last line derives from its preceding expression because of the even symmetry of the Gaussian pdf. The function $Q(\cdot)$ is the

384 □ EVALUATION OF PERFORMANCE

complementary probability integral of (13.9) and may be found tabulated in many publications (e.g., [13.15, Chap. 26]).

[An alternative notation for the same result is often encountered. Let $\mu = u/\sqrt{2}\sigma_o$, so that the error probability becomes

$$\Pr\{\epsilon_s\}_m = \frac{1}{\sqrt{\pi}} \int_{x_m/\sqrt{2}\sigma_o}^{\infty} e^{-\mu^2}\, d\mu \triangleq \frac{1}{2}\operatorname{erfc} \frac{x_m}{\sqrt{2}\sigma_o} \qquad (13A.4)$$

The function erfc($\cdot$) is known as the complementary error function and is also found in many tables, such as [13.15, Chap. 7]. Clearly, $Q(\cdot)$ and erfc($\cdot$) are the same function with slightly different scaling, namely $Q(X) = \frac{1}{2}\operatorname{erfc}(X/\sqrt{2})$.]

A graphical illustration of the meaning of the error integral for binary decisions is shown in Fig. 13A.1. Added noise would cause a noisy strobe value x_o to depart from the noise-free value x_m. Since noise is random, the noisy strobe value is also random and has a pdf determined by the noise statistics. The pdf of x_o is shown as a Gaussian curve, centered at x_m. An error is committed if signal plus noise causes the composite strobe value to fall on the wrong side of the decision boundary, which is $x_o = 0$ for antipodal binary signals. Probability of error is the integral of the pdf in the error region—the shaded area in Fig. 13A.1.

Multilevel Decisions

Consider a real baseband PAM signal with more than two amplitude levels. Each inner level will have two decision thresholds, one above the constellation point and the other below. Figure 13A.2 illustrates the decision problem. One decision threshold is located at a distance δ_1 below the noise-free strobe point x_m, and the other is located δ_2 above x_m. Following the same reasoning as for binary decisions, the probability of error given by the shaded areas in Fig. 13A.2 is

$$\Pr\{\epsilon_s\}_m = Q\left(\frac{\delta_1}{\sigma_o}\right) + Q\left(\frac{\delta_2}{\sigma_o}\right) \qquad (13A.5)$$

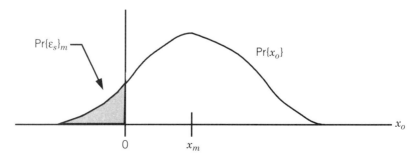

FIGURE 13A.1 Error integration for binary decisions.

APPENDIX 13A: ERROR PROBABILITIES FOR QUASIANALYTICAL EVALUATIONS □ **385**

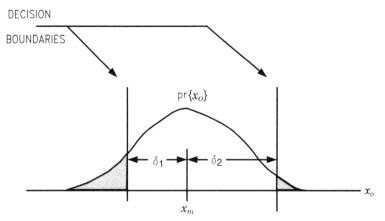

FIGURE 13A.2 Error integration for multilevel real decisions. Probability of error is represented by total shaded area.

Extension to Complex Plane

Now consider a complex noise-free strobe point $x_m + jy_m$ at a distance δ from a straight-line decision boundary, as exemplified by Fig. 13A.3. Probability of error is the integral of the pdf of $\overline{z}_o(m)$ over the half-plane to the right of the vertical decision boundary. The integral (13.8) may be segmented for a vertical boundary as follows:

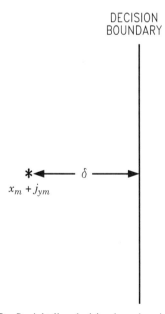

FIGURE 13A.3 Straight-line decision boundary in complex plane.

$$\Pr\{\epsilon_s\}_m = \frac{1}{2\pi\sigma_o^2} \int_{x_m+\delta}^{\infty} \exp\left[-\frac{(x_o - x_m)^2}{2\sigma_o^2}\right] \int_{-\infty}^{\infty} \exp\left[-\frac{(y_o - y_m)^2}{2\sigma_o^2}\right] dy_o \, dx_o$$
(13A.6)

But the inner integral evaluates as $\sqrt{2\pi}\,\sigma_o$, irrespective of x_m, y_m, or δ, leaving only the outer integration over x_o to be performed. That is accomplished by substituting $\mu = (x_o - x_m)/\sigma_o$ to obtain

$$\Pr\{\epsilon_s\}_m = \frac{1}{\sqrt{2\pi}} \int_{\delta/\sigma_o}^{\infty} \exp\left(-\frac{\mu^2}{2}\right) d\mu = Q\left(\frac{\delta}{\sigma_o}\right) \quad (13A.7)$$

which is exactly the same result obtained for baseband binary decisions in (13A.4).

Figure 13A.3 shows a vertical decision boundary, and (13A.7) was derived on that basis. Since the noise pdf is circularly symmetric, the same result would obtain for a straight-line decision boundary at any angle whatever; working with a vertical (or horizontal) boundary merely simplifies the explanation.

Now augment the geometry in Fig. 13A.3 so that the noise-free strobe point $x_m + jy_m$ lies between two parallel straight-line decision boundaries in the complex plane, at distances δ_1 and δ_2. The derivation leading to (13A.7) can be repeated, integrating over two disjoint error regions. The result is identical to (13A.5), which was derived for multilevel real baseband decisions.

Next, consider a rectangular decision region, as was shown in Fig. 13.4a. The probability of symbol error is given by the integral of the noise pdf outside the decision rectangle. The problem may be decomposed into independent decisions on x and y separately, as shown in Fig. 13.4b. Probabilities of erroneous decisions on x and y are, from (13A.5),

$$\Pr\{\epsilon_x\}_m = Q\left(\frac{\delta_{x1}}{\sigma_o}\right) + Q\left(\frac{\delta_{x2}}{\sigma_o}\right)$$

$$\Pr\{\epsilon_y\}_m = Q\left(\frac{\delta_{y1}}{\sigma_o}\right) + Q\left(\frac{\delta_{y2}}{\sigma_o}\right) \quad (13A.8)$$

Under some circumstances, the separate x and y error probabilities may be the quantities of interest, and then (13A.8) is exactly the right expression to be used. More commonly, an engineer needs the probability of symbol error, which is found from a combination of the two separate constituent probabilities of (13A.8). To form that combination, recognize that symbol errors occur in three circumstances: error in x only, error in y only, or errors in both x and y. Since the errors in x and y are independent, the probability of symbol error in a rectangular decision region is given by

APPENDIX 13A: ERROR PROBABILITIES FOR QUASIANALYTICAL EVALUATIONS □ 387

$$\Pr\{\epsilon_s\}_m = \Pr\{\epsilon_x\}_m[1 - \Pr\{\epsilon_y\}_m] + \Pr\{\epsilon_y\}_m[1 - \Pr\{\epsilon_x\}_m] + \Pr\{\epsilon_x\}_m\Pr\{\epsilon_y\}_m$$
$$= \Pr\{\epsilon_x\}_m + \Pr\{\epsilon_y\}_m - \Pr\{\epsilon_x\}_m\Pr\{\epsilon_y\}_m \qquad (13A.9)$$

Simple addition of the separate x and y probabilities would double count the overlaps of the x and y error regions, the shaded areas in Fig. 13.4a. Subtracting the product of the two probabilities from their sum eliminates double counting.

MPSK

Exact analytical evaluation of error probability is rarely possible when decision boundaries are not rectangular. Analysts are forced to fall back on analytical approximations or numerical integration. A simple approximation has been found to provide a good upper bound on error probability for the important case of MPSK constellations.

Figure 13A.4 shows the decision geometry for MPSK. (The constellation has been rotated so that the decision sector is centered on the positive x axis.) Decision regions are bounded by a pair of rays emanating from the origin of the x–y plane, with angle $2\pi/M$ between them. The noise-free strobe $x_m + jy_m$ lies at distances δ_1 and δ_2 from the two boundary rays. Only if the noisy strobe $\overline{z}_o(m)$ falls inside the correct sector between the rays will the mth decision be correct. If the strobe falls outside the decision sector, the decision will be in error.

A closed-form expression cannot be found for the integral of (13.8) over the error region of Fig. 13A.4. To obtain an approximate upper bound on error probability, extend each ray of the decision boundary to infinity in the back directions (dashed lines in Fig. 13A.4), forming two straight-line boundaries that intersect at the origin. It is easy to integrate the strobe pdf over the two half planes lying to the far side of each extended ray, with results $Q(\delta_1/\sigma_o)$ and $Q(\delta_2/\sigma_o)$. An upper-bound approximation for the probability of MPSK decision error is

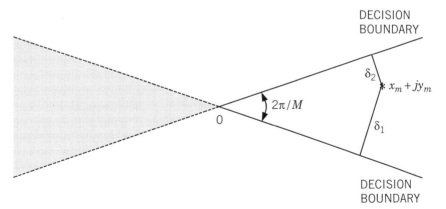

FIGURE 13A.4 Decision geometry of MPSK. Shaded area is double-counted by MPSK approximate formula.

$$\Pr\{\epsilon_s\}_m < Q\left(\frac{\delta_1}{\sigma_o}\right) + Q\left(\frac{\delta_2}{\sigma_o}\right) \qquad (13A.10)$$

where the distances δ_1 and δ_2 in Fig. 13A.4 are

$$\delta_1, \delta_2 = y_m \cos(\pi/M) \pm x_m \sin(\pi/M) \qquad (13A.11)$$

Reference [1.2, Sec. 4.2.6] arrives at the same approximation by very different arguments.

Approximation (13A.10) is pessimistic because it double counts the area of the sector indicated by shading in Fig. 13A.4. If signal-to-noise ratio is relatively high, and if M is sufficiently large, the double-count contribution will be small. Bounds on the error in (13A.4) can be estimated by examining QPSK, for which the exact result is known from (13A.8) and (13A.9). The approximation of (13A.10) is equivalent to neglecting the effect of double counting—the third term of (13A.9). It is evident that double counting is inconsequential at reasonable error probabilities for QPSK. The effect is even less for $M > 4$.

APPENDIX 13B: THEORETICAL ERROR PERFORMANCE

Traditional formulas for average probability of decision error can be derived from the per strobe expressions of Appendix 13A. First, assume that all signal-related disturbances disappear: namely, (1) ISI is absent (the overall impulse response of the link is Nyquist), (2) $\Delta\tau = 0$ (timing error is absent), (3) $\Delta\theta = 0$ (phase error is absent), and (4) amplitude has been normalized exactly correctly. Thus $x_m + j y_m = a_m + j b_m$, so additive Gaussian noise is the only contributor to incorrect decisions. Second, assume that the receiver filter is matched to the incoming pulse shape so that the signal-to-noise ratio is maximized at $\rho^2 = 2E_p/N_0$, in accordance with (4.32). The probability of error depends only on ρ and the distances from the constellation points to the decision boundaries. Third, assume that each pulse conveys exactly one symbol; the data are not encoded.

Binary Decisions

If the signal is binary, the constellation points are $a_m = \pm 1$, $b_m = 0$. Assuming that $|x_m| = 1$ for all strobes, the signal-to-noise ratio is $\rho = 1/\sigma_o^2$, and the distance from the constellation point to the decision boundary is 1. From (13A.3), the probability of decision error is

$$\Pr\{\epsilon_s\} = Q\left(\frac{1}{\sigma_o}\right) = Q\left(\sqrt{\frac{2E_p}{N_0}}\right) \qquad (13B.1)$$

Note that the subscript m that appeared in (13A.3) has been removed; the probability of decision error is the same for each strobe in this ideal case.

In a binary system, each symbol contains just one bit, and the pulse energy E_p is the same as the energy per bit E_b. Thus the probability of bit error in terms of the energy per bit to noise density ratio E_b/N_0 is

$$\Pr\{\epsilon_b\} = Q\left(\sqrt{\frac{2E_b}{N_0}}\right) \qquad (13\text{B}.2)$$

This is the so-called "theoretical" bit-error probability, or *bit-error ratio* (BER) for an antipodal, binary signal in white Gaussian noise.

Multilevel Baseband Decisions

Let a_m take on L (where L is even) equally spaced centered levels $a_m = \pm 1$, $\pm 3, \ldots, \pm(L-1)$ with equal probability, and let $b_m = 0$. Distance from any level to the closest decision boundary is 1. The $L - 2$ inner levels lie equidistant between two decision boundaries, while the two outer levels are adjacent to only one decision boundary.

From (13A.5), the probability of decision error at an inner level is $2Q(1/\sigma_o)$ and the probability at an outer level is $Q(1/\sigma_o)$. Average probability of error over all L levels, assuming equal probability of occurrence of all levels, is

$$\Pr\{\epsilon_s\} = \frac{1}{L}\left[(L-2) \times 2Q\left(\frac{1}{\sigma_o}\right) + 2 \times Q\left(\frac{1}{\sigma_o}\right)\right]$$

$$= \frac{2(L-1)}{L} Q\left(\frac{1}{\sigma_o}\right) \qquad (13\text{B}.3)$$

To relate this formula to the bit energy ratio E_b/N_0, invoke the matched-filter condition $2E_{pm}/N_0 = a_m^2/\sigma_o^2$ for the mth strobe. Solve for E_{pm} and then average the pulse energy over the symbol alphabet to obtain

$$\overline{E}_p = E_s = \frac{1}{L} \sum_{l=-L/2}^{L/2} E_{pl} = \frac{N_0}{2\sigma_o^2 L} \sum_{l=-L/2}^{L/2} a_l^2 = \frac{N_0}{\sigma_o^2 L} \sum_{l=1}^{L/2} (2l-1)^2$$

$$= \frac{N_0(L-1)(L+1)}{6\sigma_o^2} \qquad (13\text{B}.4)$$

where E_s is the average energy per symbol and l is a character index in the symbol alphabet.

There are $b = \log_2 L$ bits per symbol, so the energy per bit is

$$E_b = \frac{E_s}{b} = \frac{N_0(L^2 - 1)}{6b\sigma_o^2} \qquad (13\text{B}.5)$$

Solve (13B.4) or (13B.5) for $1/\sigma_o$ and substitute into (13B.3) to obtain the probability of symbol decision error for L-level PAM as

$$\Pr\{\epsilon_s\} = 2\left(1 - \frac{1}{L}\right) Q\left(\sqrt{\frac{6bE_b}{N_0(L^2 - 1)}}\right) = 2\left(1 - \frac{1}{L}\right) Q\left(\sqrt{\frac{6E_s}{N_0(L^2 - 1)}}\right) \qquad (13\text{B}.6)$$

[Although (13B.6) was derived for even L, it also applies to an odd number of equiprobable, equispaced levels with $a_m = 0, \pm 2, \pm 4, \ldots, \pm(L - 1)$.]

A close approximation to the probability of bit error (BER) can be obtained if two assumptions are valid:

1 Error events occur predominantly as crossings of just one decision boundary.
2 The symbols have been *Gray encoded*. That is, mapping from the binary information bits is such that symbols at adjoining levels differ in just one information bit.

If these conditions are satisfied, the BER is related to the probability of symbol error by

$$\Pr\{\epsilon_b\} \simeq \frac{\Pr\{\epsilon_s\}}{b} \qquad (13\text{B}.7)$$

This is the "theoretical" BER for L-level PAM, including $L = 2$—the binary case. Figure 13B.1 plots BER for several choices of L.

Square PAM–QAM Constellations

Now consider a PAM–QAM signal whose constellation points lie on an $M = L \times L$ square grid. As shown in Appendix 13A, decisions may be partitioned into separate operations on the x and y components. Ideal reception places all noise-free strobes at the center of the correct square decision region, and the square geometry means that x decisions have the same statistics as the y decisions. Furthermore, these separate decisions have average error probabilities given by (13B.3), so that

$$\Pr\{\epsilon_x\} = \Pr\{\epsilon_y\} = 2\left(1 - \frac{1}{L}\right) Q(1/\sigma_o) \qquad (13\text{B}.8)$$

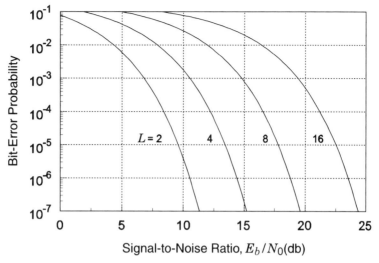

FIGURE 13B.1 Theoretical BER for L-level PAM, or for square MQAM with $M = L^2$.

The average probability of symbol error is then given by (13A.9) as

$$\Pr\{\epsilon_s\} = \Pr\{\epsilon_x\}[2 - \Pr\{\epsilon_x\}]$$

$$= 4\left(1 - \frac{1}{L}\right) Q(1/\sigma_o)\left[1 - \left(1 - \frac{1}{L}\right) Q(1/\sigma_o)\right] \quad (13\text{B}.9)$$

Following the same steps that led to (13B.4) to (13B.7), the average energy per symbol is

$$\overline{E}_p = E_s = \frac{N_0(L^2 - 1)}{3\sigma_o^2} = \frac{N_0(M - 1)}{3\sigma_o^2} \quad (13\text{B}.10)$$

which is double that for the L-level baseband signal. There are now $b = \log_2 M = 2 \log_2 L$ bits per symbol, so the energy per bit is unchanged from (13B.5). Solving (13B.5) or (13B.10) for $1/\sigma_o$ and substituting into (13B.9) gives the theoretical probability of symbol error of an M-ary square constellation as

$$\Pr\{\sigma_s\} = 4\left(1 - \frac{1}{\sqrt{M}}\right) Q\left(\sqrt{\frac{3E_s}{N_0(M - 1)}}\right)$$

$$\cdot \left[1 - \left(1 - \frac{1}{\sqrt{M}}\right) Q\left(\sqrt{\frac{3E_s}{N_0(M - 1)}}\right)\right] \quad (13\text{B}.11)$$

or just about double (13B.6)—the average probability of symbol error for an L-level baseband constellation.

The approximate BER for an M-ary square PAM–QAM constellation (if data are Gray encoded and if most decision errors are to adjacent decision regions) is just

$$\Pr\{\epsilon_b\} \simeq \frac{\Pr\{\epsilon_s\}}{b}$$

$$= \frac{4}{b}\left(1 - \frac{1}{\sqrt{M}}\right) Q\left(\sqrt{\frac{3bE_b}{N_0(M-1)}}\right)$$

$$\cdot \left[1 - \left(1 - \frac{1}{\sqrt{M}}\right) Q\left(\sqrt{\frac{3bE_b}{N_0(M-1)}}\right)\right] \quad (13\text{B}.12)$$

Since $Q(\cdot) \ll 1$ for most useful values of E_b/N_0, the bracketed term in (13B.12) is very nearly 1, so the BER shown in Fig. 13B.1 for L-level baseband signals also applies to $M = L \times L$ square-constellation passband signals as well. The contribution of the bracketed term of (13B.12) would be invisible to the eye in the plots of Fig. 13B.1.

QPSK

Various small approximations have been employed in developing these BER expressions for MQAM. It is instructive to examine QPSK, for which an exact BER expression can be derived, with the objective of evaluating the degree of approximation error. As a particular form of square MQAM, a QPSK strobe can be detected separately on its x and y components, with probabilities of erroneous decisions of $\Pr\{\epsilon_x\}$ and $\Pr\{\epsilon_y\}$. The probability in any one decision of deciding exactly one bit in error is

$$\Pr\{\epsilon_1\} = \Pr\{\epsilon_x\}[1 - \Pr\{\epsilon_y\}] + \Pr\{\epsilon_y\}[1 - \Pr\{\epsilon_x\}]$$

and the probability of committing exactly two bit errors in the decision is

$$\Pr\{\epsilon_2\} = \Pr\{\epsilon_x\}\Pr\{\epsilon_y\}$$

There are two bits per strobe, so the probability per bit of a bit error is

$$\Pr\{\epsilon_b\} = \tfrac{1}{2}[\Pr\{\epsilon_1\} + 2\Pr\{\epsilon_2\}] = \tfrac{1}{2}[\Pr\{\epsilon_x\} + \Pr\{\epsilon_y\}]$$

For the ideal signal assumed for theoretical performance, the separate probabilities are equal:

$$\Pr\{\epsilon_x\} = \Pr\{\epsilon_y\} = Q(1/\sigma_o)$$

and from (13B.5) the Q-function argument is $1/\sigma_o = \sqrt{2E_b/N_0}$, whereupon the theoretical BER for QPSK is given by

$$\Pr\{\epsilon_b\} = Q\left(\sqrt{\frac{2E_b}{N_0}}\right) \quad (13\text{B}.13)$$

which is identical to (13B.2) for binary signals.

The exact (13B.13) may be compared to the approximate (13B.12) with $M = 4$. These differ only by the bracketed factor in (13B.12), which is negligibly different from 1 for most reasonable values of E_b/N_0.

MPSK

Refer to the approximation for MPSK symbol-error probability (13A.10) and the distance expression (13A.11). If the noise-free signal strobe is ideal, the strobe falls at $x_m + jy_m = 1 + j0$ and the distances are equal, with $\delta = \sin \pi/M$. The probability of symbol error is

$$\Pr\{\epsilon_s\} \simeq 2Q\left(\frac{\sin \pi/M}{\sigma_o}\right) \quad (13\text{B}.14)$$

Energy per symbol is found from the matched-filter condition $2E_p/N_0 = 1/\sigma_o^2$. Solving for $1/\sigma_o$ and substituting it into (13B.14) gives

$$\Pr(\epsilon_s) \simeq 2Q\left(\sin\frac{\pi}{M}\sqrt{\frac{2E_s}{N_0}}\right) \quad (13\text{B}.15)$$

Dividing $\Pr\{\epsilon_s\}$ by the number of bits per symbol and substituting bE_b for E_s, the BER for MPSK is found as

$$\Pr\{\epsilon_b\} \simeq \frac{2}{b}Q\left(\sin\frac{\pi}{M}\sqrt{\frac{2bE_b}{N_0}}\right) \quad (13\text{B}.16)$$

where Gray encoding and predominant error events to adjacent decision sectors have been assumed.

Surprisingly, in light of all the approximations that have been made, (13B.16) reduces to the exact (13B.13) for $M = 4$. Not so surprisingly, the approximation (13B.16) fails for $M = 2$, yielding twice the correct result of (13B.2). Equation

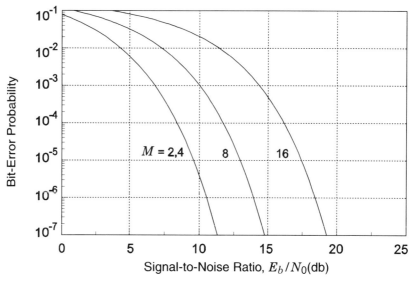

FIGURE 13B.2 Theoretical BER for MPSK.

(13B.16) is plotted in Fig. 13B.2 for several values of M. [The curve for $M = 2$ is from the correct expression (13B.2).]

Differential Detection. Equations (13B.15) and (13B.16) are valid for coherent detection of MPSK signals. For the theoretical performance of differential detection or differential decoding, refer to the documentation for the TDER command in Chapter 5 of Baker & Gardner, *Simulation Techniques: The STÆDT Program*.

Non-PAM Signals

The foregoing material applies only to PAM–QAM signals. Symbol-detector rules and theoretical performance for other signal formats and other demodulation methods may be found in the textbooks, such as [1.2, Chap. 4].

APPENDIX 13C: EVALUATION OF $Q(X)$

From the foregoing material on symbol detection, it is clear that the complementary probability integral $Q(X)$ plays a vital role in evaluation of the effects of Gaussian noise and that a simulation program should have facilities for ready calculation of the function. The fact that $Q(X)$ may be found in many published tables is of little help to a simulationist, unless someone is willing to copy printed tables into computer memory.

More compactly, it would be preferable to have an approximation to $Q(X)$ that the computer could evaluate upon each invocation of the function. Reference [13.15, Chap. 26] contains series expansions of $Q(X)$ and $P(X) = 1 - Q(X)$. Reference [13.19] proposes a Fourier series expansion with potential for highly accurate evaluation. Reference [13.20] has collected many approximations to $Q(X)$, along with specifications of accuracy for each approximation.

For our program STÆDT we have selected from [13.20] the approximation

$$Q(X) \simeq \frac{1}{(1-a)X + a\sqrt{X^2 + b}} \frac{1}{\sqrt{2\pi}} e^{-X^2/2} \qquad (13\text{C}.1)$$

Different choices for the parameters a and b are discussed in [13.20]. (Do not confuse these parameters with the signal data values a_m and b_m; there is no connection.) From among the several offerings, we have selected $a = 0.339$ and $b = 5.510$ for use in STÆDT. With this choice, the approximation (13C.1) departs from the true value of $Q(X)$ by no more than 0.27% for $X \geq 0$. Moreover, relative error decreases as X grows larger. This approximation error would be invisible on any graphical plot of $Q(X)$.

Sometimes it is necessary to evaluate $Q(X)$ for negative X. Accuracy of the approximation (13C.1) has not been evaluated for $X < 0$. To obtain a reliable approximation for negative X, employ the complementary relation $Q(-X) = 1 - Q(X)$. With this expedient, (13C.1) never needs to be evaluated for $X < 0$. [The complementary relation is built into all STÆDT evaluations of $Q(X)$; no user intervention is needed.]

APPENDIX 13D: DELAY AND PHASE ALIGNMENT FOR PERFORMANCE EVALUATIONS

Two methods for performance evaluation have been identified in the text: Monte Carlo error counting and quasianalytic (QA) computations on simulated strobes. Both methods require knowledge of the transmitted data symbol corresponding to each individual strobe or detected data symbol. Because of delays and phase ambiguities in the simulated transmission link, the recovered symbols or strobes will not be aligned in time or phase with the transmitted symbols. The delay and phase differences have to be measured and corrected; the transmitted and recovered streams have to be brought into alignment before performance evaluation itself can begin.

Alignment is closely related to synchronization. Indeed, the correlation method of synchronization of Appendix 11A also furnishes data alignment and phase-ambiguity resolution as a side benefit. When that method can be applied, little or no additional alignment is likely to be needed. Other synchronizers—the recovery synchronizers of Sec. 11.3—perform fine phase and timing adjustments, but they do not remove the phase ambiguities nor do they align the transmitted and received data symbols. This appendix delves into the mechanics of measurement and data handling needed to accomplish the data alignment and ambiguity resolution left undone by synchronizers.

Block Structure

Receiver outputs for a Monte Carlo evaluation are detected symbols or bits. Receiver outputs for a QA evaluation are symbol strobes. To avoid clumsiness in the ensuing narrative, the words *symbols* or *data* will be used to mean all three entities: symbols, strobes, or bits. Wherever the generic terms *symbols* or *data* occur in the following paragraphs, substitute the term appropriate to your application.

It will be assumed that the transmitted and received symbols are presented to the evaluation process in block format, either linear or circular (see Sec. 5.3). The intervening simulation might employ single-point operations, but it is assumed that the symbols are collected into block format before being delivered to the evaluation process. Serial single-point operations to align the symbols would be wasteful of computation time.

Blocks can be circular or linear, depending upon how the simulation is arranged. Block structure is not related to the evaluation scheme; either type of block can be used with either method of evaluation. If all blocks in the simulation are circular, then:

1 The simulated conditions are steady state, so there is no initial transient that has to die away before evaluation can begin.

2 Delay is realized by a discrete circular shift; phase is corrected by a rotation. The circular shift is confined within the given block, with no overlap to other blocks.

3 Proper delay and phase can be calibrated readily using the frequency-domain correlation method described in Appendix 11A.

4 Slips usually will be absent, whereupon delay and phase calibration remains the same for all blocks.

The first three properties are inherent to circular-block simulations, while the last comes about because of the typical ways in which circular blocks are employed.

Aligning circular blocks is relatively simple. If all data for evaluation were contained in circular blocks, there would be no need for this appendix. Linear-block simulations are not handled nearly as easily as circular. Allowance must be made for the initial transient. Delay-aligning shifts overlap into adjoining blocks. The frequency-domain method of correlation cannot be applied quite so simply. Provision must be made for slips. These matters, and others, are addressed in succeeding sections.

Delivery of the recovered symbols in concatenated linear blocks implies that at least some operations in the simulation were performed on concatenated linear blocks or in single-point format. Alignment processes can be arranged, for the most part, to ignore the formats used in the intervening signal simulations and to work solely with the two sets of data blocks—transmitted and recovered. Independence of the alignment procedure from the system details is a desirable feature in a simulation program but is not always possible.

Initial Transient

Linear-block or single-point operations exhibit transients upon startup of a simulation, closely mimicking the behavior of physical systems. Performance evaluation ordinarily is concerned with steady-state conditions, established after transients have died out. To evaluate steady-state performance, a simulationist must wait until the system has settled before accepting any recovered data. Some of the early portion of the recovered data has to be discarded since it is contaminated with transients.

Be aware that transients can be long lasting in some systems; durations into the thousands of symbols are not unusual. Other systems have much shorter transients, on the order of tens of symbols. The amount of initial data to be discarded is determined by the user, who makes an engineering judgment of the discard quantity from observations on relatively short trial simulations.

You may not want to discard the initial blocks; your intention may be to determine error performance during transient conditions. That task is even more complicated than evaluation of steady-state performance and will not be considered here.

Block Overlaps

Simulations with concatenated blocks are run with one transmitted block per concatenation. In response to that one transmitted block and to the state information left over from previous blocks, the simulation generates one block of recovered symbols. These two blocks do not necessarily have the same length, because of incompletely processed samples remaining within state memory of the link model. Unless special provisions are made, no earlier blocks are likely to have been saved (computer memory would soon be exhausted if all simulated data were preserved) and no later block yet exists.

Unless a compensating delay has been included in the main simulation, these two blocks are not aligned; the transmitted block has to be delayed within the evaluation procedure to bring its data into alignment with the recovered data. The delay process incorporates state memory to accommodate overlaps; the symbols needed for a correlation point come partly from the current block and partly from the preceding block. All simulation programs have provisions for delays.

Phase Alignment

If a constellation is symmetric about M_p different angles, a signal might be synchronized in any of M_p different phases, leading to M_p-ambiguous interpretation of data. Correct interpretation requires either that the data format be oblivious to phase ambiguity (e.g., via differential encoding and decoding) or that the correct phase be identified from special patterns embedded in the data stream.

Differential decoding or pattern identification is carried out after received symbols have been detected. A Monte Carlo simulation might or might not include the ambiguity-resolution process, but a quasianalytic (QA) simulation surely does not, since it deals with strobes delivered prior to the symbol detector.

If ambiguity is not resolved otherwise, the alignment process is required to (1) measure the ambiguity state, and (2) rotate a signal to establish the correct phase. Measurement is considered in the next section, while certain peculiarities of rotation are brought to light in this section. Phase rotators were introduced in Chapter 10.

In a QA simulation, both the transmitted data and the recovered strobes are generally complex quantities; either may be rotated (by $2k\pi/M_p$, where k is the integer that yields correct phase alignment) to align their phases. The ambiguity-removing rotation is applied to data delivered to the alignment procedure and does not have to be applied to signals in the main simulation; the rotation can be made part of the evaluation procedure and need not be incorporated into the model of the communications link.

Monte Carlo simulations could be performed under any of three different conditions:

1 Data blocks (transmitted and recovered) contain complex symbols, and ambiguity resolution is omitted from the simulation. Rotation is possible and is required. Either data block can be rotated to resolve the phase ambiguity.

2 Data blocks contain message bits—taken before symbol mapping in the transmitter and after symbol detection and demapping in the receiver. If ambiguity resolution is included in the simulation, no additional phase rotation is needed.

3 Data blocks contain bits, and ambiguity resolution is not simulated. A phase rotation is necessary to resolve ambiguities but cannot be applied to bits in the data blocks (except for polarity reversal). A rotator must be inserted in the stream ahead of the alignment process, at some point where rotation is meaningful.

Rotating the signal externally to the alignment process makes for extra computing effort compared to rotating a data block locally. The extra effort is minimized by inserting the rotator as far downstream as possible, as will be shown in the next section.

Alignment Measurement

Various different methods are available for testing alignment of two data blocks. Several are outlined in this section.

Frequency-Domain Correlation. Circular blocks can be aligned using the frequency-domain correlation method of Appendix 11A. Indeed, the problem is even simpler than indicated earlier since only coarse estimates are needed; no fine estimates or adjustments have to be performed. Delay is quantized to the closest symbol interval, and phase rotations are quantized to the M_p ambiguity positions.

The frequency-domain correlation method of Appendix 11A can be applied directly to two blocks of equal length. The two blocks are Fourier transformed into the frequency domain, the product of one block multiplied by the conjugate of the other is calculated and a correlation function is determined from the inverse Fourier transform as explained in (11A.10) to (11A.13).

If the blocks are really linear blocks, the results of circular correlation are not the same as would be obtained from correlation of circular blocks or from linear correlation of the received block against the entire transmitted data stream. Nonetheless, the peak magnitude of correlation will still be very large, provided that the shift $\hat{n}_d$ for peak correlation is small compared to the block length N_b.

COMMENT: A correlator can be fooled by tricky patterns that produce multiple high peaks in the correlation function. Existence of such a pattern is ignored here, but a simulationist should be alert to their potential threat.

The phase rotation needed to resolve ambiguity is approximately the negative of the angle of the correlation at $\hat{n}_d$, rounded to the closest $2\pi/M_p$ increment (provided that $\hat{n}_d \ll N_b$).

Time-Domain Correlation. Alternatively, correlations might be computed point by point, using a time-domain shift-and-multiply routine based upon (11A.8). Each correlation point is computed over all overlapping points of the two blocks. To achieve good correlation values, it is necessary that the number of overlapping points greatly exceed the shift amount. The calibration delay $\hat{n}_d$ is the delay that yields the largest magnitude of correlation, and the calibration phase rotation is the negative of the angle of that correlation.

If the number of correlation points is sufficiently small, correlation by time-domain methods might entail less computing effort than that involved in frequency-domain methods. As an example, if 30 correlation points are to be calculated from data blocks of length 512, rough calculations indicate that the two methods have approximately equal computational burden. Larger numbers of correlation points favor the FFT method, while smaller numbers favor the time-domain correlation. The FFT would be vastly superior if all points of a circular correlation function were to be computed.

Detect and Compare. As an alternative to correlating the two data blocks, you could send the receiver strobes into a symbol detector and compare the detected symbols against the transmitted symbols. Very few symbol errors will be detected if the two blocks are aligned properly. If the blocks are misaligned, there will be large numbers of errors.

This approach might be particularly applicable to a Monte Carlo evaluation, where detection, comparison, and error counting are necessary parts of the simulation anyhow. Refer to Chapter 12 for a discussion of symbol detection and Sec. 13.1.1 for comparison and error counting. Some details of this method have interesting features, as explained in the next few paragraphs.

If the signal constellation has multiple amplitudes, the amplitudes must be normalized before the strobes are applied to the symbol detector. Amplitude becomes a third dimension of alignment, in addition to timing and phase. By contrast, amplitude is inconsequential in a correlation measurement, since the coordinates (timing and phase) of maximum correlation are not influenced by amplitude.

Phase alignment might be more burdensome in a detect-and-compare measurement. Unlike a complex correlation, an error count is a real number with no phase information; the necessary phase rotation cannot be inferred from the comparisons. Instead, at each timing position tested for alignment, it is also necessary to test all M_p phase-ambiguity positions. The search is expanded by a factor of M_p as compared to correlation searches.

Suppose that the comparison is performed on complex detected symbols, which themselves retain phase information of the constellation. Then the data block containing these symbols (or the corresponding transmitted block) can be rotated in the M_p-position search for the correct phase. But the comparison might be performed instead on demapped real numbers—either M-ary or binary. (See Chapter 12 for demapping.) These numbers individually retain no phase information and cannot be rotated to resolve ambiguities. Any rotation has to be performed farther upstream at a location where phase information is preserved. A phase rotation alters the detected symbols; these must be demapped again into a different sequence of real data for another trial comparison.

The crucial point here is that any operations lying between the rotation and the comparison have to be repeated for each phase position tried, at each timing position tried. To minimize repetition of operations, perform the rotation as far downstream as possible. If ambiguities are resolved by means of differential encoding and if differential decoding precedes the symbol comparator, no search over phase positions is needed and no corrective rotations are needed either.

Cycle Slips

Synchronizers are subject to cycle slips wherein the recovered carrier or clock reference gains or loses one or more cycles with respect to the transmitted carrier or clock. Slips most often are caused by noise disruptions of normal operation of the synchronizer, but signal dynamics can also play a part. Slips have been studied in phase-locked loops (see [11.7], [11.8], [11.23], [13.21], and [13.22]). However, slips also occur in feedforward synchronizers that use passive tuned filters. The latter slips are closely related to Rice's FM clicks [13.23].

Slips occur in synchronizers that continually update the estimate of local phase or timing; the extra or missing cycles grow out of the variations of the estimates. A slip cannot occur in one-shot calibrations in which just one timing or phase estimate is applied to an entire simulation. A one-shot estimate might be a poor estimate, but it cannot contain slips since it is a single, fixed value. The synchronization by correlation of Appendix 11A is slip-free, whereas the recovery synchronizers of Sec. 11.2 are all subject to slipping.

Phase and timing slips show up differently in the data blocks to be evaluated. An isolated timing slip causes an extra symbol to be added to or deleted from a received data sequence. An isolated phase slip causes a rotation to a different phase ambiguity and does not affect the number of symbols in a sequence. Alignment of transmitted and received blocks is destroyed during and following a slip. For that reason, large numbers of errors will be observed by an error comparator. Similarly, correlation will be greatly reduced. Either of these events is an indication that a slip has occurred.

Slips are not instantaneous events; they take place over a time interval determined by the bandwidth of the synchronizer. With typical synchronizers, slips could have durations of tens to hundreds or even thousands of symbols. A slip might be spread over multiple data blocks, making it difficult to detect. Because a slip is not instantaneous, its time of occurrence cannot be specified precisely.

Excision of Slips. Communications systems are normally designed so that slips are very rare events. Probability of errors caused by slips ought to be small compared to the probability of errors caused directly by noise; the two probabilities typically would be evaluated separately. Simulations have not been particularly useful for evaluating slip probabilities, because slips are so rare and therefore because of the excessive computing time ordinarily needed to gather good statistics. Accounts of slip simulations may be found in [11.23] and [13.22].

The vast majority of simulation runs are free of slips, but occasionally one does crop up. Because typical runs do not accumulate large numbers of errors, and because a slip introduces a burst of many errors, occurrence of a slip in a run will badly skew the estimated error probability. To avoid contamination of the simulation results, it is advisable to excise slip-containing blocks from the evaluation (unless, of course, you are trying to evaluate the influence of slips).

A slip is detected when a wrong number of elements is found in a received block, when the timing or phase alignment is found to have changed (determined by correlation or by detect and compare), or when undue numbers of errors are discovered in a block. Since the exact location of a slip cannot be determined, the block in which a slip is detected should be dropped from the compilation of error statistics. Moreover, the slip effects might spill over to adjacent blocks; as a minimum, the blocks immediately preceding and immediately following the slipped block should be also dropped.

It is easy enough to drop the error contributions of the current or future excised blocks from error statistics; just never add them into the running count of errors and symbols or into the average of QA computations. Contributions of previous blocks are more trouble to handle; either defer their contribution until enough succeeding blocks are found to be slip-free, or remove their contribution from the running count (or average) if a slipped block is found too soon afterward. Whichever plan is followed, extra storage is needed to hold the contributions of individual earlier blocks, and a excision procedure conditional on slip detection has to be invoked.

After a slip has been detected and the contaminated blocks excised, it is necessary to realign the data blocks, either in timing or phase or both. Realignment is accomplished in exactly the same manner as initial alignment, except that any search might best be initiated from the positions of the previous alignment and the search might be a zigzag from that position. Most slips do not deviate by many cycles, so the new position most commonly will lie close to the old.

You could omit any provision for block excision and take your chances that no slips would occur during a run. Then it is not necessary to bother to simulate the

402 □ EVALUATION OF PERFORMANCE

detection of slips or the data-handling procedures that accompany excision. In most runs there will be no slips, particularly if the signal-to-noise ratio is adequate. If a slip did occur (suspected because the run indicated unreasonably large probability of error), you could just excise the entire run and do it over again (with different signal and/or noise realizations so that the same slips were not simply repeated). That would not be a happy solution if each run consumed large amounts of computer time.

APPENDIX 13E: PHANTOM NOISE CALIBRATION

Assume that the input signal to the receiver to be calibrated is represented by its complex envelope and that the receiver is linear and time invariant. Under these assumptions, the receiver can be modeled as a complex filter followed by a sampler for taking strobes, as in Fig. 13E.1. Receiver details might be far more complicated than this simple model, but those details are irrelevant to the calibration if the assumed conditions of linearity and time invariance are met.

Receiver Response

The receiver's filtering is represented by the four subfilters with "impulse" responses (more correctly, unit-sample responses) $h_{II}(n)$, $h_{QQ}(n)$, $h_{QI}(n)$, and $h_{IQ}(n)$. Each of these impulse responses is real. Balance between the individual subfilter responses

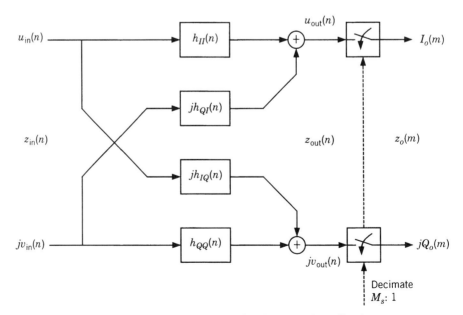

FIGURE 13E.1 Receiver model for phantom noise calibration.

APPENDIX 13E: PHANTOM NOISE CALIBRATION 403

is not necessary, although balance in the form $h_{II} = h_{QQ}$ and $h_{IQ} = h_{QI}$ is the most common condition. (The complex filter of Sec. 3.2.3 and Fig. 3.1 is the balanced case of Fig. 13E.1.)

Represent the input to the filter as

$$z_{\text{in}}(n) = u_{\text{in}}(n) + jv_{\text{in}}(n) \tag{13E.1}$$

and the filter output, prior to strobing or other decimation, as

$$z_{\text{out}}(n) = u_{\text{out}}(n) + jv_{\text{out}}(n) \tag{13E.2}$$

It is possible that the undecimated output $z_{\text{out}}(n)$ will not actually be computed in many simulations. That does not matter; the undecimated output is used only in derivation of the results, not in the calibration itself.

Outputs are formed by convolving the inputs with the subfilter responses:

$$u_{\text{out}}(n) = u_{\text{in}} \otimes h_{II} - v_{\text{in}} \otimes h_{QI}$$
$$= \sum_{l} u_{\text{in}}(l) h_{II}(n-l) - \sum_{l} v_{\text{in}}(l) h_{QI}(n-l)$$

$$v_{\text{out}}(n) = v_{\text{in}} \otimes h_{QQ} + u_{\text{in}} \otimes h_{IQ}$$
$$= \sum_{l} v_{\text{in}}(l) h_{QQ}(n-l) + u_{\text{in}}(l) h_{IQ}(n-l) \tag{13E.3}$$

where $\otimes$ denotes convolution. Unless stated otherwise, all summations can have infinite limits.

Noise variance is to be calibrated, so the squared outputs are the quantities of interest:

$$u_{\text{out}}^2(n) = \sum_{l}\sum_{i} u_{\text{in}}(l) u_{\text{in}}(i) h_{II}(n-l) h_{II}(n-i)$$
$$+ \sum_{l}\sum_{i} v_{\text{in}}(l) v_{\text{in}}(i) h_{QI}(n-l) h_{QI}(n-i)$$
$$- \sum_{l}\sum_{i} u_{\text{in}}(l) v_{\text{in}}(i) h_{II}(n-l) h_{QI}(n-i)$$
$$- \sum_{l}\sum_{i} u_{\text{in}}(i) v_{\text{in}}(l) h_{II}(n-i) h_{QI}(n-l)$$

$$v_{\text{out}}^2(n) = \sum_l \sum_i v_{\text{in}}(l)v_{\text{in}}(i)h_{QQ}(n-l)h_{QQ}(n-i)$$

$$+ \sum_l \sum_i u_{\text{in}}(l)u_{\text{in}}(i)h_{IQ}(n-l)h_{IQ}(n-i)$$

$$+ \sum_l \sum_i v_{\text{in}}(l)u_{\text{in}}(i)h_{QQ}(n-l)h_{IQ}(n-i)$$

$$+ \sum_l \sum_i v_{\text{in}}(i)u_{\text{in}}(l)h_{QQ}(n-i)h_{IQ}(n-l) \qquad (13\text{E}.4)$$

Noise Statistics

Let the input be complex white stationary zero-mean Gaussian noise with the following properties:

$$E[u_{\text{in}}(n)] = 0 = E[v_{\text{in}}(n)]$$
$$E[u_{\text{in}}^2(n)] = E[v_{\text{in}}^2(n)] = \sigma_u^2$$
$$E[u_{\text{in}}(l)u_{\text{in}}(i)] = \sigma_u^2 \delta_{il} = E[v_{\text{in}}(l)v_{\text{in}}(i)]$$
$$\delta_{il} = \begin{cases} 1, & i = l \\ 0, & i \neq l \end{cases}$$
$$E[u_{\text{in}}(l)v_{\text{in}}(i)] = 0 \qquad (13\text{E}.5)$$

Noise variances of the two outputs are found by taking the expectations of (13E.4) using the statistical properties of (13E.5), to obtain

$$E[u_{\text{out}}^2(n)] = \sigma_u^2 \sum_l [h_{II}^2(n-l) + h_{QI}^2(n-l)] \qquad (13\text{E}.6)$$

and similarly for $E[v_{\text{out}}^2(n)]$. Because the noise is stationary and the receiver is time invariant, each output sample has the same noise statistics—in particular, the same variance. Therefore, the variance calculated for $n = 0$ has the same value as for any other n. Set $n = 0$ to get

$$\sigma_{ou}^2 = E[u_{\text{out}}^2] = \sigma_u^2 \xi_I^2$$
$$\sigma_{ov}^2 = E[v_{\text{out}}^2] = \sigma_u^2 \xi_Q^2 \qquad (13\text{E}.7)$$

where, substituting $l' = -l$ yields

$$\xi_I^2 = \sum_{l'} h_{II}^2(l') + h_{QI}^2(l')$$

$$\xi_Q^2 = \sum_{l'} h_{QQ}^2(l') + h_{IQ}^2(l') \tag{13E.8}$$

These are the noise calibration factors needed for QA evaluation procedures.

Calibration Signals

The noise calibration factors only involve the impulse response components of the receiver; they can be measured directly by exciting the receiver with impulse (really unit-sample) inputs and observing the resulting outputs. Details are laid out below.

A total of $2M_s$ different unit-sample sequences will be used for calibration; they are defined as

$$z_1(n,p) = \delta(n-p)(1+j0)$$
$$z_2(n,p) = \delta(n-p)(0+j1)$$
$$0 \le p < M_s$$
$$\delta(n-p) = \begin{cases} 1, & n = p \\ 0, & n \ne p \end{cases} \tag{13E.9}$$

Outputs in response to these inputs are simply

$$u_{\text{out},1}(n,p) = h_{II}(n-p)$$
$$v_{\text{out},1}(n,p) = h_{IQ}(n-p)$$
$$u_{\text{out},2}(n,p) = -h_{QI}(n-p)$$
$$v_{\text{out},2}(n,p) = h_{QQ}(n-p) \tag{13E.10}$$

where the subscripts 1 and 2 indicate responses to $z_1(n)$ or $z_2(n)$, respectively. To repeat an earlier statement: These sequences (13E.10) may not actually be computed in many simulations.

The quantities that are always computed are the strobes $z_o(m) = I_o(m) + jQ_o(m)$, according to the rule

$$z_o(m) = z_{\text{out}}(n)|_{n=mM_s} \tag{13E.11}$$

The signal is decimated by the ratio $M_s:1$. The requirement for time invariance implies that M_s must be an integer. Decimated outputs are

406 ☐ EVALUATION OF PERFORMANCE

$$I_{o,1}(m,p) = h_{II}(mM_s - p)$$
$$Q_{o,1}(m,p) = h_{IQ}(mM_s - p)$$
$$I_{o,2}(m,p) = -h_{QI}(mM_s - p)$$
$$Q_{o,2}(m,p) = h_{QQ}(mM_s - p) \qquad (13\text{E}.12)$$

Take sums of squares to form

$$I_{o,1}^2(m,p) + I_{o,2}^2(m,p) = h_{II}^2(mM_s - p) + h_{QI}^2(mM_s - p)$$
$$Q_{o,1}^2(m,p) + Q_{o,2}^2(m,p) = h_{QQ}^2(mM_s - p) + h_{IQ}^2(mM_s - p) \qquad (13\text{E}.13)$$

and then sum over all m and all p as follows:

$$\sum_{m=-\infty}^{\infty} \sum_{p=0}^{M_s-1} I_{o,1}^2(m,p) + I_{o,2}^2(m,p) = \sum_m \sum_{r=mM_s}^{mM_s-(M_s-1)} h_{II}^2(r) + h_{QI}^2(r)$$

$$= \sum_{r=-\infty}^{\infty} h_{II}^2(r) + h_{QI}^2(r)$$

$$= \xi_I^2$$

$$\sum_{m=-\infty}^{\infty} \sum_{p=0}^{M_s-1} Q_{o,1}^2(m,p) + Q_{o,2}^2(m,p) = \sum_m \sum_{r=mM_s}^{mM_s-(M_s-1)} h_{QQ}^2(r) + h_{IQ}^2(r)$$

$$= \sum_{r=-\infty}^{\infty} h_{QQ}^2(r) + h_{IQ}^2(r)$$

$$= \xi_Q^2 \qquad (13\text{E}.14)$$

where $r = mM_s - p$ was substituted.

These last equations provide the receiver calibration that has been sought. Sum-of-squares responses ξ^2 are obtained entirely from the strobes, and the desired output noise variance is calculated from (13E.7). Because calibration is performed on the strobes (and not the more densely sampled, undecimated output of the receiver's filter), and because the receiver model is complex, a total of $2M_s$ individual unit-sample sequences have to be applied as calibrating inputs.

APPENDIX 13F: TRUNCATION OF INFINITE IMPULSE RESPONSES

Noise transmission of a simulated receiver is calibrated by measuring the sum $\sum_n |h_R(n)|^2$, where $h_R(n)$ is the impulse response of the receiver. If $h_R(n)$ has finite

APPENDIX 13F: TRUNCATION OF INFINITE IMPULSE RESPONSES

duration, the sum can be formed readily. But if $h_R(n)$ is infinite in duration, it is not possible to measure and sum an infinite number of terms. Instead, $h_R(n)$ has to be truncated so that a finite number of points can be processed. In this appendix we examine the effects of truncation on one particular example of impulse response, but the results can be generalized much more broadly.

Consider a filter with causal exponential impulse response

$$h(n) = \begin{cases} e^{-nt_s/\tau}, & n \geq 0 \\ 0, & n < 0 \end{cases} \quad (13F.1)$$

where t_s is the sampling interval and τ is the filter's time constant. The sum of the squared response over n_p terms is

$$\Sigma(n_p) = \sum_{n=0}^{n_p-1} e^{-2nt_s/\tau} = \frac{1 - e^{-2n_p t_s/\tau}}{1 - e^{-2t_s/\tau}} \quad (13F.2)$$

For $n_p = \infty$ the sum becomes

$$\Sigma(\infty) = \frac{1}{1 - e^{-2t_s/t}} \quad (13F.3)$$

The relative calibration error caused by truncation at n_p terms is

$$\epsilon_r = \frac{\Sigma(n_p)}{\Sigma(\infty)} = 1 - e^{-2n_p t_s/\tau} \quad (13F.4)$$

Expressed in decibels, the relative error is (for $e^{-2n_p t_s/\tau} \ll 1$)

$$10 \log_{10} \epsilon_r \simeq -4.343 e^{-2n_p t_s/\tau} \quad (13F.5)$$

Some numerical examples are tabulated below.

n_p	$h(n_p) = e^{-n_p t_s/\tau}$	$10 \log_{10} \epsilon_r$ (dB)
$2\tau/t_s$	0.135	−0.08
$2.3\tau/t_s$	0.1	−0.043
$4.6\tau/t_s$	0.01	−0.00044

Comments

Infinite impulse responses of lumped filters can be represented as the sum of exponentially weighted terms with different values for τ. If complex poles are present, two conjugate poles combine as the product of a real exponential multiplied by a sinusoid, whereupon the exponential describes the envelope of the response. One time constant τ will exceed all others, so the term containing the largest τ dominates for large n. When that dominant term has died out sufficiently, all other terms have also died out. Therefore, it often will be sufficient to consider just the term with the largest time constant.

In simulations, the ratio τ/t_s might be as small as 2 to 4 and possibly as large as 100 or so. Thus small calibration errors can be attained even when the response is truncated at quite modest lengths n_p. Moreover, since the relative error in dB is exponentially dependent upon n_p, the error falls off extremely rapidly as n_p increases beyond some minimally acceptable value.

14

FEEDBACK LOOPS AND CONTROL SIGNALS

Numerous communications links employ feedback loops for functions such as automatic gain control, automatic frequency control, timing or carrier synchronization, adaptive equalization, echo cancellation, and other tasks. A communications engineer frequently needs to simulate these functions when exploring the larger system. Simulation of feedback loops imposes special requirements that do not arise in simulation of feedforward processes—the great majority of processes that have been described so far. One portion of this chapter is devoted to the special simulation requirements of feedback loops.

The loops that are most commonly employed in communications systems often have a particular structure that is a small subclass of all feedback loops. In brief, data signals (i.e., the signals discussed at length in preceding chapters) are processed in the forward portions of the loop, but possibly nothing but *control signals* might be processed in the feedback portions of the loop. Characteristics of control signals and control processes are examined in a second portion of this chapter.

14.1 SIMULATION OF FEEDBACK

Various special conditions govern the simulation of feedback loops. Some of those conditions are:

- Single-point data format (Sec. 5.3.2) is usually necessary*; data blocks with multiple data points cannot be used inside the loop.

---

*This statement has been oversimplified but is valid for feedback loops in which all processes run at the same sampling rate. Simulations with multiple sampling rates are introduced in Chapter 16.

- Simulation is performed module by module on an opened loop. Any arbitrary break-open point can be selected.
- A computable feedback loop contains a delay, at least implicitly.

These conditions, and others, are explained in the pages that follow. In addition, always remember that feedback loops can be unstable, even in simulations. You might want to use a simulation to explore stability, or you might unexpectedly find unstable behavior in a feedback simulation. Stability is not considered further in this book, but that does not mean that it can be ignored.

14.1.1 Single-Point Format

In any feedback loop, there is always a process—call it the *closure process*—that combines an input signal and a feedback signal to generate a processed signal. Denote these signals as $z_{in}(n)$, $z_f(n)$, and $z_p(n)$, respectively, as in Fig. 14.1. The closure process computes

$$z_p(n) = F[z_{in}(n), z_f(n)] \tag{14.1}$$

where $F[\cdot]$ indicates that the process's output is a function of the two inputs. At its simplest, the closure function might be just the difference of the two inputs, but other functions arise much more often in communications simulations. Closure processes that have been introduced in earlier chapters include phase and frequency rotators, timing interpolators, and gain controllers.

Block Failure

Both $z_{in}(n)$ and $z_f(n)$ have to exist before the process can compute $z_p(n)$. With that in mind, consider what would happen if one attempted to compute (14.1) with data blocks. A data block is an array of data samples, such as

$$\{z(n)\} = \{z(0), z(1), z(2), \ldots, z(N-1)\} \tag{14.2}$$

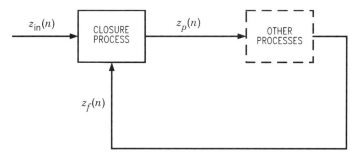

FIGURE 14.1 Nomenclature of feedback closure.

A two-input block process would accept both input blocks $\{z_{in}(n)\}$ and $\{z_f(n)\}$ and then compute the entire output block $\{z_p(n)\}$. No other processes in the simulation would be invoked until the entire output block had been computed. As noted in Secs. 5.3.2 and 5.3.3, block format is computationally efficient.

But in a feedback loop, the feedback signal z_f is an output as well as an input. The entire feedback signal is not known when the computations begin; only $\{z_{in}(n)\}$ is available. Successive values of z_p and z_f have to be computed from z_{in}. It is not possible to start the computations with an already existing feedback block $\{z_f(n)\}$, since none yet exists. Instead, along with $\{z_{in}(n)\}$, there will be an initial value for z_f. (Initiation of loop computation is explained further below.) The closure process computes the first value for z_p from the initial z_f and the first point of $\{z_{in}(n)\}$. No further computations can be performed by the closure process until a new value for z_f becomes available.

A new z_f can only be produced by the succeeding processes in the loop, operating on the current z_p, which is just one data point. Each succeeding process computes one data point and passes it to the next process in the loop. Eventually, a new z_f emerges, whereupon the closure process is invoked once again to repeat the cycle with the next input data point.

This description is the essence of single-point operation, wherein each data point is taken alone and processed through each module before the next data point can be consumed. Single-point processing is inherent to simulation of feedback loops; no other course is possible. For reasons discussed in Sec. 5.3.3, single-point format is much less efficient computationally than block format. In STÆDT we recommend that block format be used whenever possible and that single-point format be used only when unavoidable, primarily in the simulation of feedback loops.

Exception

A strictly linear time-invariant feedback loop has a closed-loop frequency-domain transfer function or an equivalent time-domain function. In principle, a feedback loop of this kind could be treated as a feedforward filter and simulated with block operations. Substantial improvements in computation speed could be gained. However, the feedback loops of greatest interest in communications links—the synchronizers, the adaptive filters, the gain and frequency controls—are not linear time-invariant networks and cannot be treated as ordinary filters; they cannot be simulated in block format.

14.1.2 Opened-Loop Simulation

A feedback loop is a closed ring. It may have inputs and outputs, but it does not have a beginning or an end. A computer program is quite different; it is an in-line list of instructions. There is a first instruction in the list and a last instruction. Program execution begins at the first instruction and proceeds (barring internal jumps) to the last, one instruction at a time. A feedback loop is simulated by going back to the head of the list and starting over again after the last instruction has been executed.

Program looping notwithstanding, the program is constructed as a linear sequence that must have a definite beginning and a definite end. The closed feedback loop has to be opened up to be modeled in a sequential computer.

Opening Location

Where should the loop be opened? To answer that question, consider a generic feedback loop as illustrated in Fig. 14.2a. The loop contains a closure process that accepts external input z_{in} and delivers output z_p. The loop also contains other processes, labeled A to F. Signal outputs of the individual processes are labeled z_a to z_f, with z_f being the feedback signal to the closure process. No time indexes are yet assigned to these signals; it will be seen shortly that there is not yet sufficient information to assign indexes.

An external output is shown as being taken from process C. That is nonrestrictive; any of the internal signals could be taken as an external output. One might intuitively want to open the loop at the closure process. That is certainly a legitimate choice, but it is not required. Instead, the loop can be opened at any point whatsoever, as most convenient. In Fig. 14.2a the loop has been opened arbitrarily between processes B and C, as indicated by the X mark. The opened loop is shown in Fig. 14.2b.

It does not matter that the external input is inserted well down on the list of loop processes. All that is needed is that the external-input data value be available when program execution reaches the closure process; this same requirement on input applies to any process. It does not matter that the external output in the opened loop is taken off before the external input is inserted. The opened loop is causal despite any appearance of noncausal structure; initial external output is computed from initial conditions of the loop processes.

Time Indexing

One pass through the opened-loop process list is called an *iteration*. In most simulations, a loop will be iterated many times. Assign an index n to each iteration. All signal points computed in the nth iteration are assigned index n. In addition, the external input point inserted during the nth iteration is also assigned index n. These indexes are illustrated on Fig. 14.2b. In execution of that example, the signal points $z_c(n)$ to $z_f(n)$ are computed; $z_c(n)$ is delivered as external output; an external-input point $z_{in}(n)$ is inserted; and signal points $z_p(n)$, $z_a(n)$, and $z_b(n)$ are computed. All processes have well-defined outputs once their inputs have been established, and all inputs are well-defined once the output of the preceding process has been established.

Loop Delay and Initialization

All signals are well defined except for one: the input to the first process on the execution list. In the closed feedback loop, process C obtains its input from the output of process B. In the opened loop of Fig. 14.2b, the output of process B on the nth iteration is $z_b(n)$. But that cannot possibly be the input to process C on the nth iteration, since C is executed before $z_b(n)$ exists.

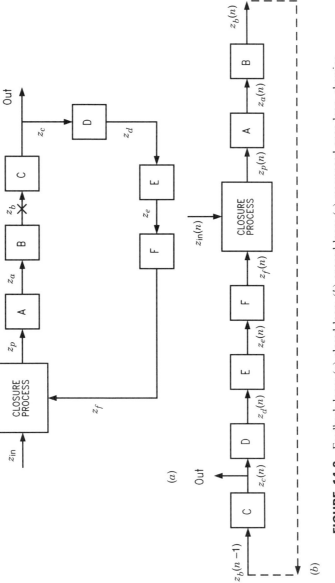

FIGURE 14.2 Feedback loop: (a) closed loop; (b) opened loop; (c) rearranged open loop, showing execution order and initialization switch.

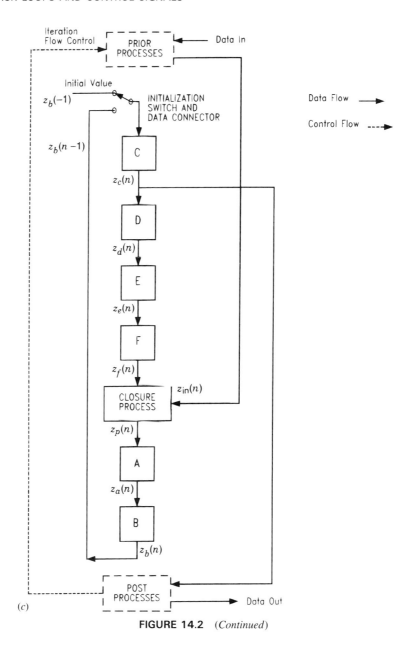

(c)

FIGURE 14.2 (*Continued*)

To provide an input to process C, use the latest-known output of process B, which is $z_b(n-1)$, as computed on the $(n-1)$th iteration. In other words, to simulate a feedback loop, a one-sample delay has to be inserted into the loop, at least implicitly. Otherwise, the loop is not computable. This requirement for a delay applies to all sampled loops and is not a vagary of simulations.

Suppose that the first iteration of the loop begins with $n = 0$. Where does the starting value $z_b(-1)$ come from, particularly since there have not yet been any preceding iterations? The answer is: from the program user. Starting value $z_b(-1)$ is an initial condition, determined by the user from knowledge that is outside the simulation program. This initial condition is unusual since it is indispensable even if process B does not otherwise have ordinary state variables to be assigned initial conditions. The loop starting value is a state variable of the loop, not necessarily of the opened loop's end process. Data for the first process in an opened loop comes from either the last process or the loop initial condition. The simulation loop needs *data connectors* to establish the data paths, and an *initialization switch* to select between the two sources.

Figure 14.2c shows the same opened loop as in Fig. 14.2b, but rearranged vertically to better emphasize the order of execution of the modules. In addition, an initialization switch and data connector have been inserted into the simulation, as have generic processes to be executed before and after the feedback loop. Modules in Fig. 14.2c are executed strictly from top to bottom. There are no jumps in execution until the last module is reached. Execution starts again at the very first module on a new iteration. By contrast, data flow (solid connecting lines) skips forward or backward, as required; data flow and program flow are entirely separate. This important feature is well understood by programmers but may not be quite so obvious to communications engineers.

For the first iteration of a simulation, the initialization switch is set to provide the loop initial value as the first input to process C. Once C has been executed for the first time, the switch is reset (by a software flag) so that output from B is input to C upon later iterations. Other programs might employ mechanisms other than an initialization switch (such as a one-sample delay module) to accomplish the same purpose, but some mechanism is indispensable.

14.1.3 Multiple Loops

Only simple loops have been considered so far, but many communications systems contain multiple feedback loops. Separated loops are simulated independently without taking special measures. However, multiple loops may be interlinked in some manner; what measures are needed for simulation of interlinked multiple loops?

Examples of double loops are shown in Figs. 14.3 and 14.4. Both loops have to be broken open, so that the simulation is converted to a feedforward (open-loop) configuration. Both examples demonstrate that it may be possible to break open all loops at a single point, a potential convenience. Structures of practical multiple loops often are amenable to single-point breaking. If multiple loops cannot be opened at a single point, more than one point will have to be opened. No feedback loops can be left unbroken in a simulation.

Figure 14.3 shows a pair of nested loops; one loop is contained entirely inside the other. Figure 14.4 shows two crossed loops that overlap one another. Conventional programming languages have *program* looping structures that permit nested program loops but prohibit crossed loops. That prohibition does not apply to simulation of

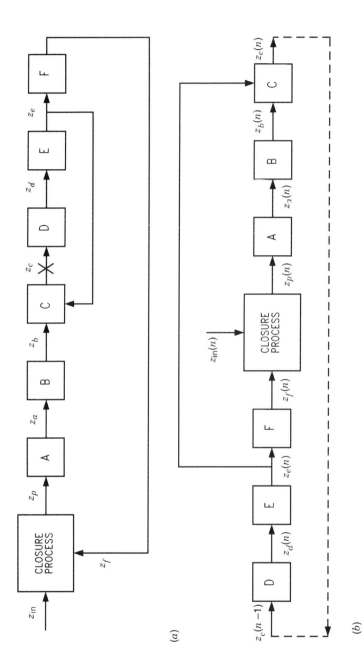

FIGURE 14.3 Nested multiple loops.

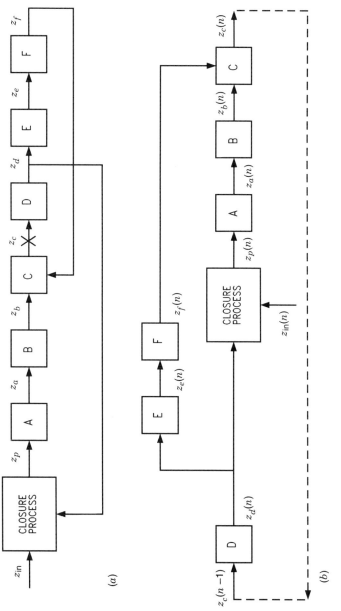

FIGURE 14.4 Crossed multiple loops.

feedback loops when the feedback loop has been broken open. As far as the program is concerned, there is only one program loop in either of Fig. 14.3 or 14.4, as indicated by the dashed lines. A program loop is not the same thing as a feedback loop.

14.2 CONTROL SIGNALS

Processes and signals in the feedback portions of many feedback loops have specialized properties that are distinct from most of the processes and *data signals* that have been scrutinized in the preceding chapters. We have named these *control processes* and *control signals* (not an established nomenclature); their nature will be explored in the pages to follow. Control signals need not be distinguished from data signals; simulations could be carried out perfectly well with just one type of signal that included both categories. It is solely because of the authors' preferences that STÆDT distinguishes two types.

14.2.1 Configuration and Properties

Figure 14.5 shows a block diagram of a typical feedback loop, to serve as an example for explaining control signals. The particular example happens to be a phase-locked carrier synchronizer, but timing synchronizers, frequency-control loops, and gain-control loops would not differ in any important manner. Adaptive equalizers and echo cancelers are substantially more complicated, but the same control-signal principles could be applied to them also.

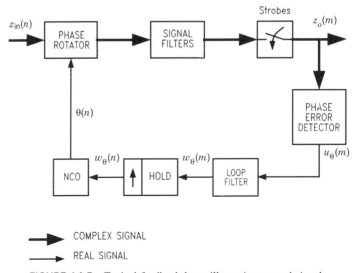

FIGURE 14.5 Typical feedback loop, illustrating control signals.

Forward-Path Operations

In the upper part of Fig. 14.5 there are several processes of the kinds that have been explained earlier: a phase rotator, signal filters, and a strober (downsampler). These processes operate on complex data signals; the potential existence of complex signals is indicated in the figure by heavy connecting lines and arrows. Input to the forward path is $\{z_{\text{in}}(n)\}$, with M_s samples per symbol. External output consists of strobes $\{z_o(m)\}$, with one sample per symbol. Changes of sampling rate are commonplace within communications feedback loops. See Chapter 16 for more on sample-rate changes.

A reduced sampling rate at the output is characteristic of digital implementation of receivers; digital implementation is covered in Chapter 15. Analog implementation has time-continuous signals throughout the loop; a simulation of an analog system would maintain the same sampling rate of M_s samples per symbol throughout the model, instead of downsampling. The following explanations are in terms of symbol-rate-sampled feedback operations, but a reader should understand that the sampling rate need not be reduced from that established on the data signal. Like any process in a feedback loop, the forward-path processes operate with single-point data format. Otherwise, they are exactly the same as the nonfeedback processes of previous chapters.

Feedback-Path Operations

The feedback path consists of a phase-error detector, a loop filter, a hold process, and a number-controlled oscillator (NCO). The phase-error detector generates a real sample $u_\theta(m)$ from the complex strobe $z_o(m)$. Examples of error-detector algorithms have been given in Sec. 11.3.6. The error samples are processed by the loop filter to produce a real control word $w_\theta(m)$, once per symbol interval (or maybe even less frequently). The hold function upsamples back to the input sampling rate with M_s samples per symbol. (See Chapter 16 for more on hold functions and sample-rate changing.) The NCO delivers a phase $\theta(n)$ (a real number) to the phase rotator; phase is advanced from sample to sample by an increment determined by the control word w_θ.

All of these feedback-path processes produce real output signals, not complex signals. Real signals in Fig. 14.5 are indicated by lightweight lines and arrows. The real signals in the example are $u_\theta(m)$, $w_\theta(m)$, $w_\theta(n)$, and $\theta(n)$. In the context of this discourse, they are the control signals, and the processes that generate them are the control processes.

Properties of Control Signals

Generalizing from Fig. 14.5, and anticipating later matters, control signals and processes have the following properties:

- Control signals are real in most feedback loops.
- Single-point format must be used (as is true for any signal in a feedback loop).

- Loop bandwidths almost invariably are much smaller than the data-signal bandwidth. Accordingly, control signals vary much more slowly than do data signals. Longer simulation times often are needed to produce control signals with sufficiently representative statistics.
- Loop filters are much simpler and configured differently from the data-signal filters of earlier chapters (to be treated below).
- Some control processes require greater numerical precision than has been encountered heretofore (to be treated below).

These special properties suggest that control signals and processes might be implemented in special ways, as will be illustrated presently.

14.2.2 Loop Filters for Control Signals

Among the control processes in Fig. 14.5, the phase-error detector has been covered in Chapter 11 and NCOs have appeared in Chapters 7, 10, and 11. Hold functions will be deferred to Chapter 16. That leaves loop filters to be introduced now.

Typical Loop Filters

Loop filters for analog phase-locked loops (PLLs) often have transfer functions of the form

$$F(s) = \frac{s\tau_2 + 1}{s\tau_1(s\tau_3 + 1)} \tag{14.3}$$

which may be regarded as a proportional-plus-integral portion

$$\frac{s\tau_2 + 1}{s\tau_1} = \frac{\tau_2}{\tau_1} + \frac{1}{s\tau_1} \tag{14.4}$$

in cascade with a single-pole lowpass filter

$$\frac{1}{s\tau_3 + 1} \tag{14.5}$$

where the τ_i, $i = 1$ to 3, represent time constants of simple filters. These simple forms, perhaps with several sections connected in cascade, serve very well to model the vast majority of loop filters found in practical feedback loops. Almost all realistic loop filters can be modeled by gain scaling and adders, integrators, and one-pole lowpass filters. There is no need to use the diverse and complicated models provided for data filters. Furthermore, as will become evident, the data-filter models may not serve well for loop filters.

Simulation Models for Loop Filters

Scaling and adding are trivial; they need no further explication. That leaves integrators and single-pole lowpass filters to be modeled. Section 6.2.2 described impulse-invariant and step-invariant methods for transforming an analog filter to an equivalent digital filter. Those transformations for the integrator and for the one-pole lowpass filter are tabulated below. (The impulse-invariant transformation has been scaled by t_s to obtain a nondimensional digital transfer function and to force the dc gain of the digital lowpass filter to be near unity [1.9, Chap. 5].)

Operation	Analog	Transfer Functions Digital Impulse-Invariant	Step-Invariant
Integration	$\dfrac{1}{s\tau_1}$	$\dfrac{t_s/\tau_1}{1-z^{-1}}$	$\dfrac{z^{-1}t_s/\tau_1}{1-z^{-1}}$
One-pole lowpass	$\dfrac{1}{s\tau_3+1}$	$\dfrac{t_s/\tau_3}{1-z^{-1}e^{-t_s/\tau_3}}$	$\dfrac{z^{-1}(1-e^{-t_s/\tau_3})}{1-z^{-1}e^{-t_s/\tau_3}}$

Model Selection. The two integrator models differ only in the z^{-1} (a one-sample delay) in the numerator of the step-invariant version. The two lowpass models differ both in the delay and in the gain scaling. If $t_s/\tau_3 \ll 1$, the gains are nearly equal. To preserve flexibility, the delay could be omitted from the step-invariant models. Delay can be cascaded separately if the system being simulated requires it, but cannot be removed easily for delay-free simulations if the models include the delays.

Based on these transformations, useful delay-free difference equations for the two models are

$$x_{out}(i) = ax_{in}(i) + x_{out}(i-1) \tag{14.6}$$

for the integrator, and

$$x_{out}(i) = ax_{in}(i) + (1-a)x_{out}(i-1) \tag{14.7}$$

for the one-pole filter. This formulation models the one-pole filter as a leaky integrator with exactly unity dc gain. Variables x_{in} and x_{out} are the real input and output control signals for each filter, and a is a user-specified gain parameter. These models are neither strictly impulse invariant nor step invariant, but partake of some of the features of each.

Initial Values. Observe that x_{out} is a state variable whose initial value has to be specified at the start of a simulation and propagated along as the simulation proceeds.

All difference-equation filters have state variables, but the initial values for STÆDT data-signal filters are arbitrarily set to zero, on the premise that the transient time for these filters is relatively short and that the user does not want to be bothered specifying large numbers of initial states in complicated filters.

Control-signal filters typically are much slower than data-signal filters and the user may not want to wait out the transient. Alternatively, the transient behavior of control loops often has to be examined from particular nonzero starting conditions; the convergence of the feedback loop to equilibrium is the behavior of interest. Furthermore, feedback loops are likely to have only small numbers of initial states contained in relatively simple filters. For these reasons, control-signal processes with state variables should have provision for user specification of the initial values.

Numerical Precision. Integrators and NCOs—and to a lesser extent, lowpass filters—are capable of developing comparatively large values for their state variables, while the input-signal variations might be comparatively small. That can be a problem since the computer has only finite numerical resolution, even with floating-point number representation. A representation capable of holding the largest value of the state variable might be unable to accommodate the smallest inputs of importance. To reduce the likelihood of such occurrences, double-precision numbers (or long integers) ought to be employed for control-signal integrators, NCOs, and one-pole control-signal filters.

This underflow/overflow problem is much less likely to arise in the state variables of data-signal filters, so STÆDT employs single-precision floating-point numbers for them. Further attention is given to overflow and underflow and to precision in Chapter 15.

Accumulators

Many digitally implemented feedback loops employ downsampling accumulators, devices that sum L input samples to produce one output sample. Such devices relate to digital implementations as described in Chapter 15, and to sample-rate changing, covered in Chapter 16. They are mentioned at this point, somewhat prematurely, since they are also control-signal processes.

An accumulator forms the sum

$$x_{\text{out}}(i) = \sum_{l=(i-1)L}^{iL-1} x_{\text{in}}(l) \qquad (14.8)$$

which is similar to the integrate and dump (I&D) of Sec.11.3.2. There is a difference in that the I&D is dumped in response to an external Strobe command, while the accumulator process needs an internal L-counter to establish its timing. Refer to Chapter 16 for timing considerations in processes that change the sampling rate.

14.3 KEY POINTS

14.3.1 Feedback Loops

Special conditions apply to simulation of feedback loops:

- Single-point data format must be used inside the loop; block format is not possible (at least not in loops in which all processes run at the same sampling rate).
- The loop must be opened for simulation.
- The loop can be opened at any convenient point; opening need not be at the location of the loop input.
- A computable feedback loop contains an implicit delay.
- An initial input must be provided to the opened loop to start the loop simulation.
- Simulated feedback loops can be unstable.

14.3.2 Control Signals

In many feedback loops encountered in communications systems, the forward portion of the loop handles *data signals* of the kinds described in the earlier (nonfeedback) chapters. However, the feedback portion of the loop is likely to process *control signals*, which are quite different from data signals. Special properties of control signals include:

- Control signals will usually be real, not complex.
- Control signals tend to vary much more slowly than data signals. Longer simulation times are needed to collect valid statistics.
- Control signals typically will be sampled less densely than associated data signals.
- Single-point format is usually mandatory for control signals.
- Loop filters—an element found in most feedback loops with control signals—are simple structures, composed of small combinations of very simple building blocks, such as adders, gain scalers, integrators, one-pole filters (leaky integrators), and accumulators.
- Some control signals might require enhanced numerical precision.
- Provision is needed for user specification of initial values in control processes that have state variables. This is a separate matter from the initial input specification needed to start simulation in the opened loop.

14.3.3 STÆDT Processes

Feedback Processes

    FBC, FBD: Data connectors/initialization switches for marking bounds of feedback loops

    SP .. SPND: Delimiters for establishing single-point data mode

Control-Signal Processes

    ACCM: accumulator

    DLYC: delay (discrete increments)

    HOLD: zero-order hold and upsampler

    INTG: integrator

    LET: various operations, most of which can be applied to control signals

    LPFC: one-pole lowpass filter

    NCO: number-controlled oscillator

    QUNC: quantizer

    SWC: switch

15

SIMULATION OF DIGITAL PROCESSES

Digital signal processing is rapidly replacing functions that previously have been performed by analog circuits. Although a few digital methods crept into the earlier chapters, most of the preceding material related to simulation of analog systems. Novel features of digital implementation impose additional tasks upon simulations, beyond those arising in analog implementations. This chapter is devoted to simulation issues particular to digital processes.

Overview

As might be anticipated, digital simulations make full use of most of the techniques already developed for analog systems. Indeed, since digital processes are already sampled, the distortion inherent to time-discrete approximation of time-continuous processes is absent. Also, many elements—filters foremost among them—are implemented in digital hardware in close conformity to the simulation model. However, certain features of digital implementations raise issues that do not appear in analog. The main complicating feature, and the one given the most attention in this chapter, is the use of fixed-point numbers with comparatively few bits. Numerical precision is a prominent issue in planning of digital processors. How many bits should be employed in the numbers? What are the effects of different choices of word lengths? To explore these questions, a simulator of digital processes has to be able to *quantize* signals or parameters to a specified precision. Quantization is the dominant topic of this chapter.

Along with quantization, there arises a question of how numbers should be represented. Although decimal numbers are most familiar, numbers in digital hardware will be in some kind of binary notation (e.g., two's complement). Numbers inside the computer are almost surely also represented in a binary notation, probably two's complement. You might think that a binary representation in the simulation would also be feasible and perhaps even easy to provide. However, it would be a severe burden on a user to have to deal directly with binary (or octal, or hexadecimal) numbers, particularly for a nonprogrammer user who was not proficient with nondecimal numbers. Furthermore, high-level programming languages require binary or hexadecimal numbers to be entered as integers, but parameters more generally have fractional parts as well. These matters and other number representation issues will be discussed from the user's point of view.

Floating-point operations adjust scaling automatically; it would be unusual for registers to overflow in the course of floating-point simulations. That is not true for fixed-point processors, where the designer is obliged to perform explicit scaling in each process to avoid overflow. Exactly the same scaling is needed in fixed-point simulations. Scaling has been studied intensively in the DSP literature [1.11, Chap. 5], so is covered only briefly here. Overflow behavior is a critical matter and is given somewhat more attention. Digital hardware often is pipelined to permit multiple processors to operate simultaneously. Pipelining introduces throughput delay. A simulation might be required to model the delay; that question, too, is examined in this chapter.

15.1 QUANTIZATION

A quantizer is a zero-memory nonlinear device whose output $x_{\text{out}}(n)$ is related to its input $x_{\text{in}}(n)$ according to

$$x_{\text{out}}(n) = q_i, \qquad x_i \leq x_{\text{in}}(n) < x_{i+1} \tag{15.1}$$

where q_i is an output number that identifies the input interval $[x_i, x_{i+1})$. Inputs and outputs of quantizers are treated as real throughout this discussion. A complex quantizer consists of two real quantizers.

Outputs $\{q_i\}$ can be assigned any arbitrary numbers whatsoever, provided only that subsequent processes know how to interpret the numbers. Think of the outputs as labels rather than as numbers with intrinsic meaning. This viewpoint is helpful in distinguishing between the different quantizer laws that might be encountered.

Quantization arises in two different situations: (1) in analog-to-digital convertors (ADCs), where a sampled analog quantity of infinitesimally fine resolution is converted to a digital number of finite resolution, and (2) in fixed-point digital processes, where the result of an operation contains more digits than can be handled downstream, so the result must be shortened.

15.1.1 Quantizer Laws

Uniform Quantizing

Uniform quantization is the most widely used law in data signal processing and the only one considered here. The uniform law is defined by

$$q_{i+1} = q_i + \delta q$$
$$x_{i+1} = x_i + \Delta x \qquad (15.2)$$

where δq and Δx are constants. With equal increments δq for the output numbers and equal increments Δx on the inputs, a uniform quantizer characteristic plots as a staircase with equal risers and equal step widths. All uniform quantizers are based on the uniform staircase of Fig. 15.1. Uniform quantizers differ in the number of levels included, limits of operation, scale factors, and location of origins. These are all described in succeeding paragraphs.

Other, nonuniform, laws are important in special applications [15.1]. Worthy of passing note are the millions of logarithmic quantizers [15.2] that have been built for use in speech codecs in the telephone systems of the world.

Over-Range Limits. Every quantizer characteristic has a finite range; the staircase extends between the input limits $x_{\min}$ to $x_{\max}$. Outside those limits, some other behavior intervenes. In this chapter and in the QUNC and QUND commands of the STÆDT library, the output *saturates* at the highest and lowest levels of q_i. Stated formally:

$$x_{\text{out}}(n) = q_{\min}, \qquad x(n) \leq x_{\min}$$
$$x_{\text{out}}(n) = q_{\max}, \qquad x(n) \geq x_{\max} \qquad (15.3)$$

Saturation is illustrated in Fig. 15.1.

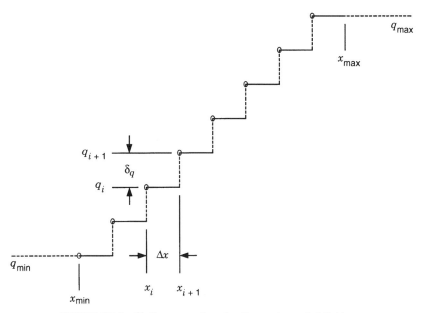

FIGURE 15.1 Uniform quantizer, $L = 8$; notation and definitions.

COMMENT: Over-range behaviors other than saturation are also possible. One that occurs in some computer arithmetics is *recycling*; an example of recycling is provided later in this chapter.

A quantizer exhibits small-scale nonlinearity in its individual steps and large-scale nonlinearity in its over-range behavior. The amplitude of the input has to be set carefully to avoid severe distortion of the signal from either nonlinearity [15.1]. This amplitude setting is known variously as *loading adjustment* in ADCs, as *scaling* in DSP operations, or as *normalization* in many data-signal applications.

Levels. A finite quantizer has a finite number L of output levels and, equivalently, L input intervals of width Δx. It is commonplace to employ $L = 2^b$ or $L = 2^b - 1$ levels, where b is a positive integer. The operation is then said to be a *b-bit* quantization. With these definitions, the following relations apply in a uniform quantizer:

$$\Delta x = \frac{x_{max} - x_{min}}{L}$$

$$\delta q = \frac{q_{max} - q_{min}}{L - 1} \quad (15.4)$$

Scales. It is often convenient to define

$$q_i = Q_i \, \delta q$$
$$x_i = X_i \, \Delta x \quad (15.5)$$

whereupon

$$Q_{i+1} = Q_i + 1$$
$$X_{i+1} = X_i + 1 \quad (15.6)$$

Notice that Q_i and X_i need not be integers, despite the unit increments between adjoining values. Noninteger examples are shown below.

Frequently, particularly in ADCs, the meaning of δq is arbitrary and flexible. For example, if the output of the ADC is a b-bit number q_i, one might interpret q_i equally well as an integer ($\delta q = 1$) or as a binary fraction ($\delta q = 2^{-b}$). In such cases, Q_i is an unequivocal specification and may be preferred for describing the ADC output. The ratio $\delta q / \Delta x$ defines a large-scale gain of the quantizer.

Axis Translations

The quantizer staircase can be translated to any location in the (x_{out}, x_{in}) plane. Various seemingly different quantizer characteristics may simply be translations of one another, as will be illustrated by several examples.

Polarity. A quantizer might be unipolar or bipolar on its input and, independently, on its output. Many ADCs work satisfactorily only with unipolar inputs; a bipolar input signal has to be biased so that the net signal applied to the ADC has just the single polarity allowed.

Signals of interest usually are bipolar; for that reason, the examples considered below are mostly bipolar, on input and output both. However, quantizer models ought to be adaptable enough to simulate unipolar as well as bipolar inputs and outputs.

Bipolar quantizers are categorized as possessing *midstep* or *midriser* staircases, according to their properties in the vicinity of zero input. Each has its own advantages and drawbacks, and each is encountered extensively in practice. Examples of both kinds are presented below.

Midstep Characteristics. Output of a midstep quantizer is constant over a region on both sides of $x_{in} = 0$. A usual specification is

$$x_{out}(n) = 0 \quad \text{for} \quad -\frac{\Delta x}{2} \leq x_{in}(n) < \frac{\Delta x}{2} \tag{15.7}$$

Two similar midstep quantizer staircases are shown in Fig. 15.2a and b. They differ by one in the number of levels employed. A midstep quantizer needs an odd number of levels if it is to be symmetric about its center. The extra level of an even-level quantizer can be suppressed, if necessary.

COMMENT: An asymmetric quantizer potentially acts as a rectifier of large-amplitude input signals. If the quantizer is not symmetric and if a signal saturates in both directions, the quantizer output will not have zero mean because of the extra level on one side, even if the input signal has zero mean.

A dead zone lies at the center of a midstep quantizer. One might fear that a small-amplitude signal could disappear into the dead zone without a trace under some circumstances. This would come about from improper scaling, but it is not always possible to achieve proper scaling. As a more subtle concern, feedback loops tend toward instability when their error-detector characteristics have dead zones. A designer of feedback systems might want to avoid dead zones by avoiding midstep quantizers.

Notice the dashed straight line representing the unquantized $X_{out}(n) = X_{in}(n)$ in Fig. 15.2. Subtract the straight line from the staircase to obtain the *quantizing error* or *quantizing noise*. Quantizing error of a centered midstep quantizer is the error incurred by *rounding* the input [1.9, Chap. 9; 15.1]. Although the quantizing error is uniformly distributed with zero mean, the energy of the quantized output exceeds the energy of the unquantized input (even for unity gain, $\delta q/\Delta x = 1$). This energy gain is conducive to limit cycle instabilities in recursive digital filters [15.3]. A different quantizing rule, discussed below, can help to suppress limit cycles.

430 □ SIMULATION OF DIGITAL PROCESSES

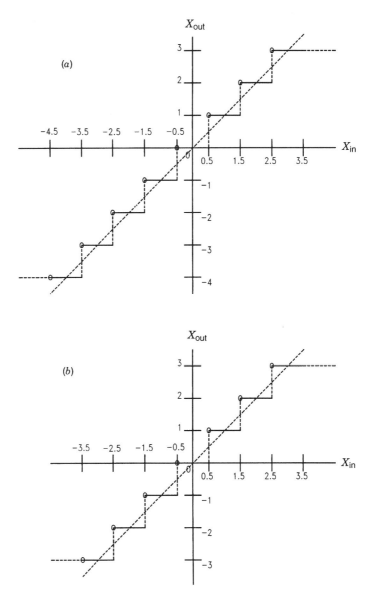

FIGURE 15.2 Midstep quantizers: (a) $L = 8$; (b) $L = 7$.

Midriser Characteristics. A midriser quantizer is discontinuous at $x_{in}(n) = 0$. Two examples are shown in Fig. 15.3. A centered midriser, as in Fig. 15.3a, will never lose a zero-mean input signal entirely, no matter how small its amplitude. If nothing else, the centered midriser acts as a hard limiter and preserves the polarity of the input.

The centered midriser has no dead zone around zero input. A feedback loop with a discontinuous step at zero in its error detector will dither about the step and can

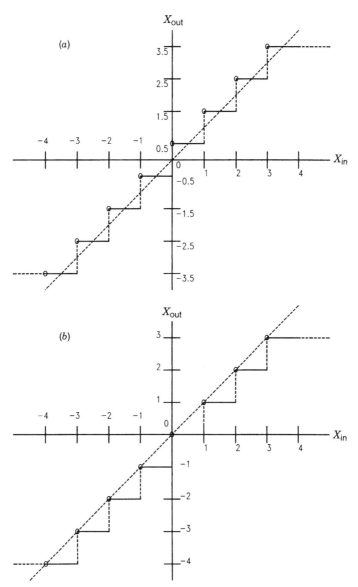

FIGURE 15.3 Midriser quantizers: $L = 8$: (a) symmetric placement; (b) offset placement.

never settle to zero error. Dithering is often preferable to a dead zone. Absence of a zero level is unsatisfactory in some applications, where zero output might be desired in response to zero input.

Observe that the outputs X_{out} are not integers; all Q_i are odd-integer multiples of 0.5 for the centered midriser characteristic. In a binary system, that would mean that the least-significant bit (LSB) is always the same, so there would be no need to produce it in a physical ADC. That extra bit could be provided in the digital processes after the ADC and would not be needed in the ADC itself.

Suppression of the LSB leads to an *offset* quantizer, as illustrated in Fig. 15.3*b*. This is identical to Fig. 15.3*a* with the staircase dropped vertically down by 0.5. Output levels now are integers. If the half-integer is added back in the digital processes, operation of the offset midriser is identical to that of the centered midriser. If the half-integer is discarded, there will be different characteristics. Since the offset quantizer is no longer symmetric, it will cause some rectification of input signals, even of small-amplitude signals. That can be disconcerting when the input is zero mean and a zero-mean output is expected or needed.

An offset quantizer has a dead zone to one side of zero input, which may induce peculiar behavior in feedback loops. In other applications, a zero-output level is required and the dead zone may be necessary. The offset characteristic of Fig. 15.3*b* results from *truncation* of two's-complement numbers [1.9, Chap. 9] or from truncation of decimal numbers. When the numbers are represented in two's-complement, the characteristic of Fig. 15.3*b* is sometimes known as *offset two's-complement* notation. (Truncation is represented by the function IP[x]—the integer part of x—which means the largest integer less than or equal to x. As examples: IP[3.2] = 3, IP[3.0] = 3, but IP[−3.2] = −4.)

Notice the relation of the midriser characteristics to the line $X_{\text{out}} = X_{\text{in}}$. The quantizing error for the centered midriser is zero mean, whereas there is a negative mean for the offset midriser (a manifestation of the rectification mentioned above). With bipolar signals, the output of either will have more energy than the input, in a manner similar to that noted for the midstep quantizer.

Magnitude Quantization. It is advantageous to employ a *sign-magnitude* number representation [1.9, Chap. 9] in some digital operations. Sign reversal then becomes an exceedingly simple operation, while multiplication is no more difficult than for other representations. Addition and subtraction are more cumbersome, though, so sign-magnitude is not employed as often as two's complement.

Truncation of the magnitude of an input signal leads to the quantizer characteristic of Fig. 15.4. This is no longer a strictly uniform quantizer; the center interval (a dead zone) has double width. Observe that the line $X_{\text{out}} = X_{\text{in}}$ has magnitude greater than the staircase at most points and no less than the staircase at isolated intersections. Thus the quantized output has less energy than the unquantized input. This feature is helpful in suppressing limit cycles in IIR filters [15.4].

Unipolar Quantizers. Several examples of unipolar quantizers are shown in Fig. 15.5. There are slight differences among these characteristics; all those shown are variations on offset midriser quantizers.

Noisy Inputs

In many instances the input to a quantizer consists of a desired signal plus unwanted noise. It is common practice to adjust the input level such that the noise fluctuations

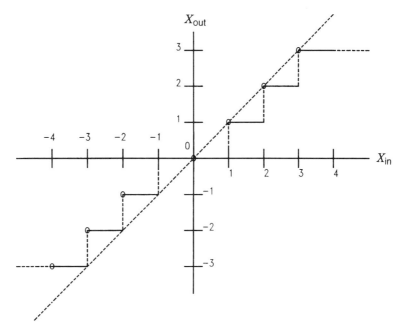

FIGURE 15.4 Magnitude quantizer, $L = 7$ (nonuniform; there are eight input increments).

extend over several quantizer increments, assuring thereby that the additive noise dominates strongly over the quantizing noise. Issues of midriser versus midstep characteristics become irrelevant when sufficient noise (or signal alone, for that matter) is present.

15.1.2 Simulation of Quantizers

In the foregoing section we provided background on quantizers; now let's turn to particulars of their simulation.

Models

A saturating, uniform quantizer is modeled by

$$x_{out}(n) = \left\{ \text{IP}\left[\frac{x_{in}(n)}{\Delta x} - B_X\right] + B_Q \right\} \delta q, \quad x_{min} \leq x_{in}(n) < x_{max}$$
$$= q_{max}, \quad x_{in}(n) \geq x_{max}$$
$$= q_{min}, \quad x_{in}(n) < x_{min} \qquad (15.8)$$

where B_X and B_Q are horizontal and vertical offsets, respectively, in units of the horizontal and vertical increments Δx and δq. Furthermore, observe that

434 □ SIMULATION OF DIGITAL PROCESSES

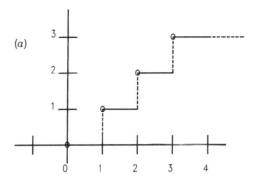

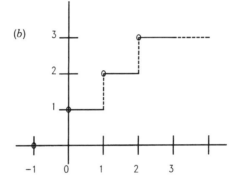

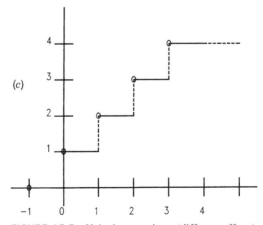

FIGURE 15.5 Unipolar quantizers (different offsets).

$$q_{\max} = \left\{ \text{IP}\left[\frac{x_{\max} - \Delta x}{\Delta x} - B_X \right] + B_Q \right\} \delta q = \left\{ \text{IP}\left[\frac{x_{\max}}{\Delta x} - B_X \right] - 1 + B_Q \right\} \delta q$$

$$q_{\min} = \left\{ \text{IP}\left[\frac{x_{\min}}{\Delta x} - B_X \right] + B_Q \right\} \delta q \tag{15.9}$$

The slightly different formats for $q_{\max}$ and $q_{\min}$ come about because of the different treatments of the left and right endpoints of each input interval.

Simulation programs might include only uniform quantization in their libraries. To perform nonuniform magnitude quantization as in Fig. 15.4, first save the sign of the input, then quantize its magnitude, and finally, multiply the quantized output by the previously stored sign. In brief:

$$\begin{aligned} x_{\text{out}}(n) &= \left\{ \text{IP}\left[\frac{|x_{\text{in}}(n)|}{\Delta x} \right] \right\} \delta q \, \text{sgn}[x_{\text{in}}(n)], & |x_{\text{in}}| &< x_{\max} \\ &= q_{\max}, & |x_{\text{in}}| &\geq x_{\max} \end{aligned} \tag{15.10}$$

Most programs will have the sign and absolute-value functions available.

Specification of Quantizer Parameters

Uniform saturating quantization is modeled by (15.8); a user need only furnish the model parameters when incorporating the model into a simulation. How are the parameters to be specified? You will find that (15.8) is substantially overdetermined; the quantizer is completely specified by the five primary parameters $L, x_{\min}, x_{\max}, q_{\min}$, and $q_{\max}$. The other four parameters $\Delta x, \delta q, B_X$, and B_Q are determined entirely by the primary five.

First, the increments Δx and δq were specified in (15.4). Next, it turns out that the offsets B_X and B_Q are not unique. To see how this comes about, consider that

$$\begin{aligned} \text{IP}[X - B_X] + B_Q &= \text{IP}[X - \text{IP}[B_X] - \text{FP}[B_X]] + B_Q \\ &= \text{IP}[X - \text{FP}[B_X]] - \text{IP}[B_X] + B_Q \end{aligned} \tag{15.11}$$

where $\text{FP}[\cdot]$ means "fractional part of." It is clear that either one of B_X or B_Q can be assigned an arbitrary integer part and the effect of that assignment can be absorbed in the other offset parameter. To obtain a unique specification on the offsets, it is reasonable to require $\text{IP}[B_X] = 0$. That is,

$$0 \leq B_X < 1 \tag{15.12}$$

Having imposed the constraint upon B_X, now consider the quantizer output in the close vicinity of one of its discontinuities, say at $x_{\text{in}} = x_{\min} + k \Delta x \pm \epsilon$, where $0 < k \leq L - 1$, k an integer, and $0 < \epsilon \ll \Delta x$. The quantizer delivers

SIMULATION OF DIGITAL PROCESSES

$$q_{\min} + k\delta q = \delta q \left\{ \mathrm{IP}\left[\frac{x_{\min} + k\Delta x + \epsilon}{\Delta x} - B_X \right] + B_Q \right\}$$

$$= \delta q \left\{ \mathrm{IP}\left[\frac{x_{\min} + \epsilon}{\Delta x} - B_X \right] + k + B_Q \right\} \quad (15.13)$$

at the right side of the discontinuity and

$$q_{\min} + (k-1)\delta q = \delta q \left\{ \mathrm{IP}\left[\frac{x_{\min} - \epsilon}{\Delta x} - B_X \right] + k + B_Q \right\} \quad (15.14)$$

Take the difference between these two expressions to obtain

$$1 = \mathrm{IP}\left[\frac{x_{\min} + \epsilon}{\Delta x} - B_X \right] - \mathrm{IP}\left[\frac{x_{\min} - \epsilon}{\Delta x} - B_X \right] \quad (15.15)$$

from which can be inferred

$$B_X = \mathrm{FP}\left[\frac{x_{\min}}{\Delta x} \right] \quad (15.16a)$$

By similar arguments based on $x_{\max}$ instead of $x_{\min}$, it can be shown that B_X is also given by

$$B_X = \mathrm{FP}\left[\frac{x_{\max}}{\Delta x} \right] \quad (15.16b)$$

To find B_Q, solve either form of (15.9) to obtain

$$B_Q = \frac{q_{\min}}{\delta q} - \frac{x_{\min}}{\Delta x} + B_X$$

$$= \frac{q_{\max}}{\delta q} + 1 - \frac{x_{\max}}{\Delta x} + B_X \quad (15.17)$$

using the fact that $x_{\min}/\Delta x - B_X$ and $x_{\max}/\Delta x - B_X$ are integers by definition.

Examples. Parameter values are tabulated below for the quantizers shown in Figs. 15.2, 15.3, and 15.5. All figures were drawn with Δx and δq of unit size.

Figure	B_X	B_Q	L	x_{max}	x_{min}	q_{max}	q_{min}
15.2a	0.5	1	8	3.5	−4.5	3	−4
15.2b	0.5	1	7	3.5	−3.5	3	−3
15.3a	0	0.5	8	4	−4	3.5	−3.5
15.3b	0	0	8	4	−4	3	−4
15.5a	0	0	4	4	0	3	0
15.5b	0	1	4	3	−1	3	0
15.5c	0	1	5	4	−1	4	0

15.1.3 Quantizer Applications

Quantization is applied in ADCs and in fixed-point DSP operations.

A/D Convertors

The term *A/D convertor* usually means the combination of a sampler (actually a *sample-and-hold* or *track-and-hold* circuit), followed by a quantizer. Each of these devices is afflicted by imperfections that impair its performance. Some sampler impairments are:

- Finite aperture (noninstantaneous sampling)
- Aperture jitter
- Switching-charge leakage
- Hold sag
- Slew limiting
- Bias

Some quantizer impairments are:

- Differential and integral nonlinearity (nonuniform increments)
- Missing codes (increments with zero span)
- Nonmonotonicity (codes out of order)
- dc offset
- Dynamic nonlinearities

Unfortunately, most of these imperfections are extremely difficult to simulate at the block diagram hierarchical level. (See Sec. 1.2.1 for a discussion of hierarchy.) Exceptions are dc offset, which can easily be simulated by adding a constant to the input signal, or integral nonlinearity, which can be handled separately as a nonlinear process of the kinds treated in Secs. 4.5 and 7.5.2. The other impairments usually have to be simulated at a lower level in the hierarchy.

Fixed-Point DSP

Time-domain filtering (see Secs. 6.1.2 and 6.1.3) is an example of a digital process needing repeated quantizations. Digital filters perform numerous multiply and add operations. The two factors of each product consist of a filter coefficient and a sample of the signal. Coefficient quantization can be handled differently from that of signals, at least for time-invariant filters.

Coefficient Quantization. Coefficients typically are specified prior to a simulation. They can be quantized at the time of specification (as they would be in realistic design of hardware) rather than at run time. This approach allows the effect of quantization of coefficients to be investigated even if the signals themselves are not quantized.

If you specify coefficients directly, remember that the computer can only represent exactly those numbers that are sums of positive or negative powers of 2; any other numbers are only approximated. As an example, the expression $13{,}107/131{,}072$ [$= (2^{13} + 2^{12} + 2^9 + 2^8 + 2^5 + 2^4 + 2 + 1)/2^{17}$] will be evaluated and the quotient stored exactly in binary, single-precision format in the computer, but display of the quotient will be rounded to seven decimal digits. You cannot enter this 18-bit number (17 bits plus sign) as a single-precision (seven-digit) decimal; the exact decimal representation needs 11 digits. Instead, enter it as a rational expression whose denominator is an integer power of 2. Because the internal storage and operations are binary, the exact binary quotient will be employed for calculations, even though the numbers in the entry expression are decimal integers.

As another example, you can enter the expression $1/10$ or the decimal 0.1 into the computer, but the result is stored as a binary approximation. When converted back to decimal for display, the number reappears as 0.1, even though internal operations use the approximate value.

Signal Quantization. An isolated quantizer, such as needed for an A-D convertor or a finite-precision digital integrator, is readily incorporated into a simulation. It is not different from other signal processes in any important respect and is handled in the same manner. Dilemmas arise when large numbers of quantizers are needed. Fixed-point digital operations require that word size be examined after virtually every operation; requantization is needed each time if the word has developed too many digits.

Digital filters are prime examples. A digital filter consists of many repetitions of multiply-and-accumulate operations. Each multiplication forms the product of a signal sample with b_s bits by a coefficient value with b_c bits. The product—an internal signal inside the filter—will have $b_s + b_c$ bits. Each product is added to (accumulated with) the result of another multiplication or multiply–accumulate. Each addition can increase the signal's word size by one more bit, for a total of $b_s + b_c + 1$ bits for one stage of multiply–accumulate. The next stage of multiply–accumulate only accepts a signal word size of b_s bits; requantization is needed to reduce the word size from the preceding stage.

COMMENT: It is usual for hardware multiplier-accumulators to have sufficient register length so that the quantization need not be performed until after the accumulation step has been performed. If the register were shorter, quantization would have to be performed after each multiplication and perhaps after each addition—a major increase in the number of quantizers to be simulated.

Different programs might offer various options with regard to quantization. Options include:

1 No quantization processes whatever are included in the command library; only analog systems can be simulated.

2 Simple quantizer processes are included in the command library. These can be used in conjunction with inputs and outputs of other processes.

3 All (or most) processes in the library include options for control of internal word sizes.

In the second option, intricate digital processes can be simulated by combining quantizers with simple processes. This avenue becomes overly tortuous when simulating truly elaborate processes such as filters. Elaborate processes are developed in the first place because they are too complicated to be constructed readily from simple constituents; one does not want to have to abandon them and perform the elaborate constructions whenever internal quantization is needed.

Therefore, the most complete and satisfying choice would seem to be option 3. Unfortunately, that places the greatest burden of foresight on the program developer, who must predict all of the various quantization schemes that will be needed by future users, and make provision for the user to specify them when they are needed. These provisions add to the number of parameters to be specified for the process and annoy those users who do not need them. The extra operations also tend to make the processes run slower, even when they are not invoked (they need bypass code).

An underlying problem here is that the details of quantization are really a step down in the simulation hierarchy. (The notion of hierarchy was introduced in Chapter 1.) Transmission simulation is usually intended to be at the block-diagram, black-box level. Quantization detail (or any other fine detail of digital implementation) is at the next lower level in the hierarchy—the numeric operations themselves. At one level lower might be digital gates, while circuits would be one level lower yet.

Whenever simulation hierarchies are mixed together, the lower level almost invariably impedes operations at the higher. The lower level introduces details that the higher level tends to ignore. The user and programmer are forced to consider these additional details, so the simulation model grows more complicated and subject to error. More options have to be included in the models, thereby inflating the individual processes. More operations have to be performed in each process, so the simulation runs more slowly.

Despite all of these difficulties, quantization is a vital factor in the design of DSP systems and its effect frequently needs to be known. For this reason, quantization according to option 3 is included in some simulation programs, particularly those specifically intended for DSP simulation.

Quantization of Complex Signals. A real signal is quantized by one real quantizer. A complex signal is quantized by two real quantizers. If the signal is in rectangular coordinates (the usual case), one quantizer is for the I channel and the other is for the Q channel. Less commonly, signals might be quantized in polar coordinates.

An emerging trend in hardware digital receivers is to sample and quantize (A-to-D convert) a real passband signal rather than a real or complex baseband signal. To simulate quantization in this configuration, the simulated signal also has to be real and to be quantized by a real quantizer. Complex quantization of the complex envelope of the real signal does not give the same results, not even if (as is usual) the sampled real signal is immediately converted to an I–Q complex signal.

Computer Precision. Numbers by which signals are represented in the computer (usually floating point, although long integers also have been employed) can often be regarded as having effectively infinite precision. Nonetheless, it must always be remembered that these numbers have only finite precision. A single-precision floating-point number might have 24 bits in its mantissa, and a double-precision number might have 53 bits. Don't try to model quantizers so fine-grained that they approach or surpass the resolution of the numbers furnished as signal input.

15.2 OTHER DIGITAL ISSUES

In this section we touch briefly on other digital issues that were raised at the beginning of the chapter.

15.2.1 Number Representation

Many specialized digital processors work with binary numbers in two's-complement representation. The host computer probably also represents numbers internally in two's-complement format. One might expect that direct simulation of a two's-complement processor would be a straightforward matter. However, a high-level programming language intervenes between the computer and the simulated processor. The programming language probably will not support user access to the number arrangement internal to the computer; that detail is hidden. If the language permits binary representations (or, equivalently, hexadecimal representations) at all, that probably is limited strictly to integers.

There are really two issues intertwined here. One is how a number is represented and the other is how the representation might influence arithmetic. If the arithmetic issue can be ignored, the representation does not matter; a particular number preserves its identity no matter what label (i.e., representation) is assigned to it. That is, +27 (decimal) = 011011 (two's complement) = 1B (hexadecimal).

Upon accepting this conclusion, one simply employs whatever number representation is offered by the programming language and does not try to duplicate the representation of the processor to be simulated. Most simulation programs at the black-box hierarchical level (including STÆDT) adopt this reasoning. But arithmetic cannot be ignored entirely; the different representations do not work by quite the same rules. Something has to be done to accommodate varying treatments of details such as quantization and overflow/underflow. Two approaches might be visualized:

1. Map the two's-complement numbers onto numbers that are provided by the programming language (e.g., the positive integers) and write special arithmetic functions (add, multiply) that obey the two's-complement rules of arithmetic [15.5]. Different mappings and different rules may be needed for other number representations.
2. Use the ordinary numbers and arithmetic provided by the programming language. Then, in addition, separate out the arithmetic features that may differ among different number representations (e.g., quantization, overflow/underflow) and furnish these as explicit processes that operate on results obtained from ordinary arithmetic.

The second approach was already chosen here tacitly when quantization was presented as a separate process.

15.2.2 Overflow and Underflow

A computer stores its numbers in a finite number of bits, so the range of storable numbers is limited. If the results of a calculation fall outside the limited range, the storage register is said to over- or underflow. (*Over*flow means a number larger than the largest number that can be supported; *under*flow means a number smaller than the smallest number that can be supported. For the remainder of this section, *overflow* will be used to designate both situations.)

Floating-point numbers employ an exponential representation so that the range is extremely large. It would be unusual in ordinary simulations to encounter floating-point overflow. When overflow does occur, the computer detects it, stops execution of the simulation, and exits with an error message. It is then the user's responsibility to determine why overflow took place and to alter the simulation so as to avoid it.

Fixed-point representations have much smaller number ranges, so overflow is much more likely. Indeed, simulations sometimes are arranged to force fixed-point overflows deliberately as a means of investigating their effect on signal processing. Thus explicit handling of overflows has to be an integral part of fixed-point simulations.

Overflow Types

Although various rules could be devised, the predominant forms of overflow behavior are *saturation* and *recycling*. In saturating arithmetic, the result of a calculation takes on the largest number in the allowable range in the event of overflow (smallest number in case of underflow). The quantizers of Sec. 15.1 illustrated saturation. The numbers in a saturating arithmetic lie on a finite straight-line segment.

In recycling, an overflow result cycles around to a number within the allowable range. For example, consider an 8-bit system in which one bit designates the sign. This system can represent the numbers −128 to +127. The expression 56 + 75 (= 131) gives a recycled result −125 (= 131 − 256) because the overflowed sum cycles around to the other end of the range. This is addition modulo-256. Think of the numbers in a recycling arithmetic as lying on a circle, rather than in a straight line. Unless there is intervention to prevent it, two's-complement overflows ordinarily will recycle. (Recycling overflows of partial sums are not necessarily harmful, provided that compensating overflows in the opposite direction later occur to cancel the previous recycles from the final sum.)

Both behaviors—recycling and saturation—are needed in practical simulations; both could probably be encountered in the same simulation. A number-controlled oscillator (see Sec. 7.4) is an example of recycling; an NCO must recycle to operate as intended. By contrast, a digital integrator in a control loop almost invariably would be constrained to saturate; recycling upon overflow would be disastrous to the stability of most control loops. Operations in IIR digital filters ordinarily are made saturating; recycling leads to limit cycles upon overflow in recursive filters [15.6].

Simulation of Overflow

In a simulation, the overflow treatment (and the quantization) conveniently can be separated from the main arithmetic, as long as the host computer's number system does not overflow. Separate treatment is accomplished by first examining the result of a standard operation after it has been produced and then modifying the result to emulate any desired overflow rule. In the example of recycling arithmetic above, the ordinary sum 56 + 75 = 131 would be formed first. That sum would then be examined and found to exceed 127 (the overflow limit). Then 256 would be subtracted to emulate recycling modulo-256. Alternatively, to emulate saturation, set the result equal to 127.

Rules of this kind are easily implemented in high-level programming languages. They might be accomplished quite differently at the bit level in DSP hardware, but the net result is the same irrespective of the implementation. By separating the overflow treatment from the arithmetic, a simulation program can make use of the standard arithmetic of the programming language and also be capable of highly flexible implementation of overflow rules.

Scaling

Provision must always be made for overflows if for no other reason than the processor or simulation program has to know what to do under all conditions that might arise. Processor designers are likely to expend considerable effort, though, to *scale* the signals so that overflow does not occur in the processes; consequences of overflow usually are deleterious in most applications.

Scaling consists of multiplying the signal by a fixed constant (a fixed gain) to adjust its amplitude level. Any simulation program will have ample facility for scaling signals at any designated point in the simulated system. Scaling factors are chosen by means outside the simulation program and then specified as parameters for the model to be simulated. The simulation program is concerned only with the implementation of scaling and not the design considerations that lead to the scaling values. A reader is referred to the literature for the design considerations.

15.2.3 Processing Delays

Consider a simulation program operating in single-point mode (Sec. 5.3.2). Each process in the single-point model is invoked once for each input data point before operations begin on the next input point. If there are no transport delays within the model, the output due to each input will be delivered prior to starting on the next input. Although one round of processing might take a long time to complete, *simulated delay* is not involved. If delay is desired, it must be included explicitly in the model.

One can imagine a hardware processor that is also delay free; it accomplishes all of its operations in a time no longer than the interval between input samples. A simulation can always be arranged delay free in this sense if the system being simulated is delay free.

A delay-free processor has to run very fast indeed if the sample interval is small or if a large number of operations have to be performed in each interval. At higher data rates, a *pipelined* architecture is often more feasible. A pipeline consists of several stages in cascade, each performing just one of the processes to which the data signal is subjected. Each stage performs its own operation on an input and delivers its output to the next stage only once per sample interval. If there are L stages in the pipeline, a point at the input cannot reach the output before L sample intervals have elapsed. The pipeline has a transport delay of at least L sample intervals; that delay has to be included in the simulation. Techniques for delay modeling have been presented in Sec. 10.3.

15.3 KEY POINTS

When simulating digital processes, concerns over distortions that arise in digital simulation of analog processes can be set aside. However, digital processes raise new concerns relating to:

- Quantization
- Number representation
- Fixed-point operations
- Overflows and underflows

- Scaling
- Process delays

The greatest portion of this chapter has been devoted to quantization.

15.3.1 Quantization

A quantizer is a nonlinear process that delivers a comparatively small number of discrete output values. Most interest in simulation of data communications lies in *uniform* quantizers, in which the output levels are equally spaced and the input range is divided into equal increments. The input–output characteristic of a uniform quantizer plots as a staircase function.

Staircases are finite in extent; those examined here *saturate* at their ends. That is, if the input signal falls outside specified range limits, the corresponding outputs take on one of two specified extreme values. A staircase can be described by the input and output step sizes, axis offsets, minimum and maximum output levels, input saturation limits, and number of levels. These features are interdependent; if the quantizer is uniform and saturating, it can be specified completely by the number of levels and the extreme values on input and output. Examples and formulas are provided in the text.

Various characteristics—such as midriser or midstep—are desirable in different circumstances; a quantizer model has to be flexible enough to simulate any reasonable uniform characteristic. The different characteristics implement truncation, rounding, or other rules for shortening overlong numbers. A versatile quantizer is easy enough to provide as a stand-alone process, but it becomes more challenging when quantization capabilities have to be embedded into other, more-intricate processes, such as filters. The finer detail needed to embed quantizers is really at a lower level in the simulation hierarchy. Hierarchy mixing is a broader problem that almost invariably leads to complications in the model and increases in computing burden.

15.3.2 Other Digital Issues

Number Representation

Hardware processors to be simulated are not likely to represent their numbers in the same convention used by the programming language in which a simulation is written. Moreover, the rules of arithmetic are also different in some degree. A simulation program could be written to emulate the hardware number representation closely, with different rules for each possible representation. Or the main calculations could be carried out in the representation and arithmetic of the programming language, with the result then being modified to conform to the hardware's rules on quantization and overflow. STÆDT is implemented according to the second approach.

Overflow

The result of a calculation sometimes exceeds the range capabilities of the processor and an over- or underflow occurs (generically, both are *overflow*). A simulation program for digital processors needs a capability for modeling the effects of overflow. The two most important overflow rules are those of *saturation* and *recycling*. Both might be applied at different locations in the same simulation.

15.3.3 STÆDT Processes

    GAIN: adjustable gain (for scaling)

    LET, LETS: various operations for control signals (and variables) and for data signals, respectively; include absolute value and sign, among many others

    QUNC, QUND: quantizers for control and data signals, respectively

16

SAMPLE RATE CONVERSION

Sampling rates are changed repeatedly within simulations and within digitally implemented equipment. Examples of rate changing have emerged in earlier chapters, but the task has not been examined closely so far, except to note that interpolation lies at the heart of any nontrivial rate conversion. In this chapter we treat sample rate conversion in a simulation context, identifying structural features needed to support rate changes. For clarity of explanation, block mode will be covered first, followed by the additional considerations imposed by single-point and feedback-loop operations.

Rate conversion in simulations has received scant attention in the literature, mainly in articles that treat it as just one of several issues [16.1–16.4]. Many simulation packages provide only rudimentary facilities for rate changing, or omit the capability entirely. (To be fair, it should be recognized that the great majority of simulations do not have to include any but the simplest of rate changes.) The material described here is exploratory in nature; there is substantial room for further work on the subject. Rate changing in DSP applications has been treated exhaustively in [16.5] and other DSP literature, but the problems to be considered for simulation are quite different. The DSP principles do not provide much guidance to simulation practices.

Rate conversion might refer to a change of sampling rate on an underlying time-continuous signal that otherwise is supposed to remain unaltered. Or it might equally well refer to a time-base dilation of the underlying signal (e.g., due to Doppler shift), with the sampling rate remaining unchanged. Since computer operations do not have an intrinsic time scale, a user may ascribe either interpretation to any rate change. Most of the sections to follow speak in terms of sampling-rate changes but the techniques developed are also applicable to simulating time-base dilations. Also treated here are miscellaneous operations that are performed on data blocks: decimation, reblocking, zero padding, and others. These are distantly related to rate conversions and so have been incorporated into the same chapter.

16.1 BLOCK-MODE CONVERSIONS

A data block is just a group of data elements in memory. A block-mode module reads input from one data block, processes that data, and writes output to another data block. A sampling-rate change is manifested by the number of points N_{out} of the output block being different from the number of points N_{in} of the input block. In block mode the rate-change operations tend to be isolated locally to individual modules and are largely independent of other modules in the simulation. In single-point mode (to be treated after block mode has been described), the operations are not isolated; program flow control has to be coordinated among several modules, which is why block-mode conversion is easier to explain and to implement.

16.1.1 Resampling Definitions

Integer Ratios

An integer-ratio module might *upsample* with L_u output points for each input point or might *downsample* with one output point for each M_d input points, where L_u and M_d are both integers. In the standard DSP literature [16.5], integer upsampling is called *interpolation* and integer downsampling is called *decimation*. Integer up- and down-sampling are relatively straightforward and are treated as block alteration processes later in this chapter.

The term *interpolation* is employed here more broadly for the computation of new samples from among existing samples and can be applied to either up- or downsampling, or even to a process (such as a fractional-sample delay) that does not change the net sampling rate at all. The generic term *resampling* will be used to include upsampling, downsampling, or interpolation of new samples without rate change.

Upsampling of a signal always requires interpolation. An interpolating filter (see Chapter 10) suppresses the spectral images that cause distortion. Downsampling can be accomplished without interpolation if the output samples are composed exclusively of selected input samples. If the output samples are not selected from the input samples—if new samples are computed in the downsampling—interpolation is required here, too. Aliasing can be avoided in downsampling only if the output sampling rate exceeds twice the bandwidth of the input signal or the interpolating filter, whichever is smaller.

Block operations for integer ratios are conceptually straightforward. It is only necessary to reserve memory adequate for both the input and output blocks. (For the most part, STÆDT takes care of block sizes without much user intervention, as long as a specified maximum size is not exceeded.)

Rational Ratios

Suppose that a module generates L_u output points for each M_d signal-input points, where L_u and M_d are integers with no common factors. Then the rate change is *rational* instead of integer. How might this rate change be accomplished in the computer?

Method 1: First downsample by selecting only one out of M_d input samples and discarding the rest. Then upsample by a factor of L_u; upsampling requires interpolation.

Method 2: First upsample by a factor of L_u (requires interpolation of $L_u - 1$ new points for each input point) and then select one out of M_d points for the output, discarding the other interpolants.

Method 3: Interpolate the output points directly; an average of $\xi = M_d/L_u$ input points are needed for computation of each output point.

Method 1 seems appealingly direct, but it causes unacceptable aliasing unless the input signal has been oversampled by a factor of at least M_d. Such oversampling is not likely to be encountered except for small values of M_d. Method 2 better avoids aliasing and assures that exactly the desired rate change is applied. It wastes an inordinate amount of computing effort and can require large amounts of intermediate storage if L_u is large. Method 3 is computationally the most efficient. Interpolators and their control are examined later.

COMMENT: The multirate DSP literature [16.5] describes various structures, such as polyphase filters and DFT filter banks, that are extremely valuable in hardware implementations. Because it does only one operation at a time, a computer is subject to different rules from hardware DSP, which might perform many simultaneous operations in parallel. There may be occasions when the hardware-efficient structures have to be simulated, but they are not necessarily the preferred ways of rate changing within a simulation.

Irrational Ratios

Designate the input-to-output sampling-rate ratio as ξ. Upsampling is performed if $\xi < 1$ and downsampling if $\xi > 1$. Now suppose that ξ is irrational—there exist no integers L_u and M_d such that $\xi = M_d/L_u$. This is the situation in most hardware systems, where the receiver clock is not inherently synchronous with the transmitted clock frequency. A generally applicable rate-changing scheme, as described below, has to be able to cope with irrational rate ratios. Rational or integer ratios then become special cases of irrational ratios.

16.1.2 Block-Mode Issues

Number Representation

From a practical standpoint, simulation programs can perform arithmetic on floating-point or integer numbers and are not designed to do rational arithmetic. Thus ξ must be represented as a floating-point number even if it is rational. In floating-point representation, no distinction is made between rational and irrational numbers. Of course, all floating-point numbers have finite length and so all are rational; they can approximate irrationals because they are so closely spaced.

Note that a particular rational number—even the ratio of small integers—may not be representable exactly in a binary-based floating-point number; 3/7 is a simple example. If exact rational rate changing is essential, one should consider method 2 above or deal only with ratios that can be represented exactly in the computer's number system.

Memory Management

If there are N_{in} points in the input block, the output block needs $N_{out} = N_{in}/\xi$ points, on average. But if ξ is irrational,* there is no integer N_{in} such that N_{out} can be an integer. Yet both blocks can only have integer numbers of data points. How is this discrepancy to be handled?

Fixed Input-Block Size. To gain insight, first assume that N_{in} is a prespecified, fixed integer. The irrational number N_{in}/ξ consists of the sum $I_{out} + \phi_{out}$, where I_{out} is the integer part of N_{in}/ξ and ϕ_{out} is the fractional part ($0 \leq \phi < 1$). Block invocations of a module with irrational ξ and exactly N_{in} input points will sometimes generate I_{out} output points and sometimes $I_{out} + 1$ output points. A user has no easy way to predict $N_{out} = I_{out}$ or $N_{out} = I_{out} + 1$ in any particular invocation; provision must be made to accommodate either.

Fixed Output-Block Size. In an attempt to avoid downstream propagation of variable block size, suppose that N_{out} in the rate-changing module is specified as a fixed number. Then ξN_{out} is an irrational number with integer part I_{in} and fractional part ϕ_{in}. Generation of exactly N_{out} output points sometimes requires $N_{in} = I_{in}$ and sometimes $N_{in} = I_{in} + 1$. Just as with a variable output size, the user has no easy way to predict the exact N_{in} required on any one invocation.

But now the problem is even more insidious; the required value of N_{in} would have to be communicated to the upstream module that generates the input data. However, N_{in} for any one invocation cannot be determined until the rate-changing module has operated on the input data and generated the specified N_{out} points. By then, of course, it is too late. This feedback arrangement for communicating block size is inherently unworkable in block mode.

Block-Size Specification. It is apparent that irrational rate changing is incompatible with fixed specifications on block size. Two ways of specifying block sizes may be distinguished:

1 *Static Specification.* Size of any one block is fixed and may be specified before the simulation run begins. It is not necessary that all blocks for every module have the same size—only that any one block not change size.

2 *Dynamic Specification.* Size of any one block can vary in the course of the simulation. Specification prior to the run is not posssible.

---

*Since N_{in} cannot take on arbitrarily large values, the same problem arises in practice with rational ξ.

The terms *static* and *dynamic* have been adapted from software terminology commonly used to describe dimensioning of arrays.

Block-mode rate changing can be accommodated by dynamic specification of block sizes. Each module must be capable of determining the size of a block presented as input, and working with that size, whatever it might be. No prespecified size is imposed on output blocks.

In STÆDT, each block is tagged with its size, and any block-mode module takes input block size into account. Most block-mode modules in STÆDT will work with any arbitrary block size (within constraints imposed by available memory). A different approach may be found in [16.1], [16.3], and [16.4].

16.1.3 Reblocking

Some modules demand a fixed block size for their inputs. For example, in STÆDT the FFT block size is restricted to be an integer power of 2. These necessarily statically assigned block-size modules can be connected with dynamically varying block sizes from other modules through a *reblocking* process that absorbs the variations and delivers specified fixed block sizes.

A reblocker is an *elastic buffer* that does not itself change rates or otherwise generate or delete signal points. It collects input blocks of possibly varying sizes and delivers outputs of a fixed size. Depending on the relative sizes of input and output blocks, reblocking will require more invocations of upstream modules than downstream, or vice versa. If, for example, each input block is small compared to the specified output-block size, many executions of the upstream (input-side) modules will be needed to collect enough points to deliver one full-sized output block to downstream modules.

Scheduling. In the absence of rate changes and reblocking, the modules in a simple chain would be invoked one at a time in the exact order in which they appeared in the chain. All block interface specifications would be static. To concatenate signal blocks, the chain is just invoked again, starting at the top. Scheduling (program flow control) in such a chain is rudimentary. When reblocking is in the chain, invocations become conditional on the state of the reblocker's memory fill. Underfill requires looping back to the upstream side to bring the output block size up to specification, while overfill requires repeated invocations of the downstream modules to clear out the memory until its fill is below the size specified. Introduction of reblocking therefore has an interaction on module scheduling.

Circular Blocks. Reblocking works best with concatenated linear blocks; caution must be exercised with circular blocks. Even if input blocks are circular, the reblocked outputs cannot be circular except under special circumstances; circular operations should not be applied to them without careful thought.

16.1.4 Block-Mode Interpolation

An interpolator is needed in any resampling operation that produces new samples, so interpolation lies at the heart of generalized sample-rate conversion. Interpolators were introduced in Sec. 10.4 and their controllers in Sec. 11.3.2. A time-domain interpolator is a finite impulse response filter that processes the signal according to (10.15), repeated here as

$$z_{\text{out}}(l) = \sum_{i=I_1}^{I_2} z_{\text{in}}(m_l - i) h_l(i + \mu_l) \qquad (10.15)$$

where l is the index for the output samples, m the index for the input samples, i the index for samples of the filter's time-continuous impulse response $h_l(t)$, and μ_l the fractional interval associated with the lth output sample, which occurs at time $t = (m_l + \mu_l) t_s$.

Interpolator Control

To work properly, the interpolator must be provided with its *control signals* μ_l and m_l for each output point. (Only μ_l is provided as an explicit number; m_l is provided implicitly as a flag that identifies a particular basepoint set as one upon which an interpolation is to be performed.)

Two control methods were presented in Sec. 11.3.2. One employs an NCO and is based on the frequency of interpolation. It is useful for simulation of many timing synchronizers. The other method is based on the period between interpolations and is the more appropriate choice for self-contained rate-change simulations.

Control by the period method uses the difference equations

$$\Delta m_{l+1} = m_{l+1} - m_l = \text{IP}[\mu_l + \xi]$$
$$\mu_{l+1} = \text{FP}[\mu_l + \xi] \qquad (16.1)$$

where IP[·] means the integer part, FP[·] the fractional part, ξ the input-to-output rate-change ratio, related to the interpolation period T_i by

$$\xi = T_i / t_s \qquad (16.2)$$

and t_s is the input sampling period. (Note that ξ need not be a constant unless the rate change ratio is supposed to be constant. However, ξ can change only from one block to the next, not within any one block.)

The interpolating filter has $I = I_2 - I_1 + 1$ cells of signal state memory, holding the past and current input-signal samples. This memory can be visualized as a shift register, although implementation as a circular buffer is more efficient in software. Contents of this state memory constitute the basepoint set from which an interpolant is computed. Other memory cells contain Δm (the number of samples to be shifted into the filter before the next interpolant is to be computed), μ (the fractional interval to be used for the next interpolant), and ξ.

The number of output samples per input sample is a fluctuating quantity. If the interpolator only downsamples, there will always be either zero or one output sample per input. If the interpolator only upsamples, there will always be one or more output samples per input. The control method must allow for all possibilities. The following pseudocode indicates one method of accomplishing block-mode interpolation. (Initialization and termination details are omitted.)

BLOCK INTERPOLATE

```
FOR p = p₁ to p₂ (Input block index runs from p₁ to p₂)
    SHIFT signal data by one position in register,
        discarding oldest sample and placing
        new input sample into input end.
    Δm = Δm - 1
    WHILE Δm = 0
        COMPUTE interpolant, using current basepoint
            set and most recent μ, according to (10.15).
        Δm = IP[μ + ξ]
        μ = FP[μ + ξ]
    WEND
NEXT p
```

Constraints on Interpolation

Block-mode interpolation is comparatively simple to implement; the effects of rate changing appear as a change in the number of points in the output block as compared to the input block. If varying block sizes can be tolerated by other modules, these block-size effects are entirely local to the interpolator; there is little or no need to alter the scheduling of other modules because of the rate change.

Interpolation of the kind described may be applied readily to linear blocks but not so readily to circular blocks. Only exactly rational resampling ratios $\xi = M_d/L_u$ are possible and M_d must divide N_{in} to assure an integer number of elements in the output circular block. Furthermore, provision has to be made for wrapping around the input circular block when loading the shift register. The time-domain interpolators in STÆDT will not handle circular blocks correctly.

16.2 SINGLE-POINT CONVERSIONS

A single-point module might be supposed to take in one input sample, process it, and deliver one output sample to the next module, which expects just one sample for its input. That is the simplified model. If a module changes the sampling rate, there could be more or less than one output sample: How shall a variable number of samples be handled in single-point mode? That is the question addressed in the paragraphs that follow. Rate-changing definitions and terminology are the same for single-point operations as presented above for block-mode operations.

Scheduling

The simplest model of a single-point simulation is that of a string of modules, connected in an open chain, all running at the same sampling rate. Scheduling of such a *homogeneous-rate* chain is straightforward. Starting at the module farthest upstream, compute a processed sample and successively invoke the next module downstream until every module in the chain has been executed. To process the next signal point, start again at the upstream end of the chain.

Single-point operation is especially important in simulation of feedback loops. Each module in a loop is invoked once per pass-around (*round* for short) if the sampling rate is homogeneous. Upon completing the loop, another pass is started to process the next signal point. Feedback has been explained in Chapter 14.

Homogeneous scheduling does not suffice when rate changes occur. A different arrangement is needed—one that invokes each module as it is needed, not necessarily in the order it appears in the chain or loop. Furthermore, scheduling must respond to the varying number of output samples delivered by a rate-changing module. Responsibility for variable scheduling may be placed upon the user, or it may be built into the simulation program. Both approaches are considered in the following sections, and both are incorporated into STÆDT.

16.2.1 Flag-Mediated Control

Variable scheduling can be implemented with the aid of control flags, as explored in this section.

Bypass

If no rate-changing module is *ever* called upon to perform upsampling—never deliver more than one output sample per input sample—rescheduling can be handled with *bypass flags*. Bypass works as follows. Upon each invocation, a rate-changing module decides whether one or zero output points are to be produced. That decision is used to set a binary flag whose state tells downstream modules whether or not to respond to invocation on the current round. Any processes that would need the missing output point are simply bypassed until the next round.

All modules in STÆDT have a bypass provision. In the absence of a flag (the usual condition) the module will respond whenever invoked. If a flag has been specified, the module responds only if the flag is set correctly; execution flow skips the module if the flag is set for bypass. Several rate-changing commands in STÆDT generate flags that indicate, for example, the presence or absence of a new output.

These flags can be used for bypass or for any other purpose. Alternatively, flags can be generated by combinations of housekeeping commands; provisions of this sort will be found in most simulation programs.

Other simulation programs also have bypass capability. Bypass can be described alternatively in terms of its logical complement, which might be called EXECUTE, ENABLE, or other similar term; the principle is the same. We have opted for the BYPASS model over ENABLE because most modules that are included in most simulations will be enabled on most rounds. Bypassing is an event that is out of the ordinary and will occur much less frequently than not-bypassed.

Clocked Bypass

If more than one output sample per input is *ever* required from a rate-changing module (an upsampling process), the rate-changing module itself and its downstream successors have to be invoked more often than its upstream predecessors. The simplistic bypass arrangement described above cannot handle upsampling unless the scheme is modified.

In one modification, all modules in a simulation might be regarded as being driven from a common clock. Each tick of the clock demands that an entire round of modules be invoked. The clock rate must be at least as high as the sampling rate of the fastest module in the simulation. Slower-rate modules are bypassed if they are not to be invoked on any one round. Counters, or other timing devices, control the bypass flags. With this viewpoint, there is no such thing as upsampling of invocation frequency—there is only downsampling (through bypassing) from a maximum frequency.

Clocking can be simulated implicitly by visiting each module once in each round in an order specified by the user. User-specified control logic determines whether a module is enabled or bypassed in the current round. Hardware processors normally clock their operations. A clocked model therefore might bear close resemblance to the system being investigated.

Conditional Jumps

In place of using bypasses, one could introduce jump commands into the simulation. A jump directs execution around one or more modules; flag conditions indicate whether the jump is to be taken, or if execution should continue without a jump. Preliminary examination suggests that jump commands and bypasses are nearly the same from a logical standpoint. If anything, bypasses might impose slightly less burden on a user since it is then not necessary to bother with jump commands as well as the flag conditions. On the other hand, jumps might be more efficient computationally since it is then unnecessary to visit modules that will not be executed. Jump commands are essential building blocks of conventional programming. Most simulation programs, STÆDT among them, include a variety of jump commands for users who might want to employ them.

16.2.2 Internal Scheduling

Some programs offer flag methods, with jump commands and perhaps bypass, as their only means of variable scheduling. Other programs have internal provisions that handle much of the scheduling without flags and with less need for user intervention. In this section we examine several aspects of internal scheduling.

Drawbacks to Flags

The methods of Sec. 16.2.1 can be awkward and confusing if more than just a few flags have to be generated. Their burden on the user quickly becomes excessive; the likelihood of user error is high. Two features tacitly implied in the preceding section exacerbate the problem:

1. Strict single-point processes cannot accept or deliver more than one signal point per invocation, even if a process naturally could handle more than one. This restriction creates a demand for extensive rescheduling whenever sampling rate is to be changed.

2. Conditional execution of modules is controlled by flags generated in other modules. Applying these flags correctly requires intensive logic design and planning by the user.

The burden would be reduced substantially if (1) processes were flexible, so that they could accept or deliver any reasonable number of elements on each invocation of a module, and (2) control logic could be written into the modules by the programmer and not displaced onto the user. The following paragraphs explore methods to attain these ends.

Buffers

Data connections between modules are provided by means of *buffers*. A buffer is a memory location for storing one or more data elements. Output from a module is written to a buffer; the module reads its input from another buffer. The output buffer for one module is the input buffer for another module. Buffers are employed in block mode and in single-point mode. A block-mode buffer contains multiple elements; a strict single-point buffer would contain only one element.

Flexible Buffers

Rather than restricting all buffers for single-point mode to just one element, a program would be much more versatile if module-output buffers could be expanded flexibly to accommodate as many elements as might be needed. The next module in the stream would then be required to process multiple data points on one invocation instead of just one point. This section examines some of the rules that might be applied to implement flexible buffers. Other implementations with other rules are also possible, so the ideas to follow should be considered as examples.

COMMENT 1: In an earlier section of this chapter it was stated that dynamic assignment of block memory was essential for successful rate changing. *Dynamic assignment* is the same as *flexible buffers*. Thus rate changing in block mode demands flexible buffers. The distinction between block mode and single-point mode begin to blur if multiple elements are allowed in a "single-point" buffer.

COMMENT 2: Not all buffers can be made flexible; some processes might demand a fixed number of elements. In STÆDT, the buffers for control signals are inherently restricted to single elements. Even so, there is benefit in implementing flexible buffers for as many other processes as possible.

With the advent of flexible buffers, the previous notion of strict single-point operation is no longer fully applicable. Single point operation was found to be necessary for simulation of feedback loops (see Chapter 14). That rule remains rigorously applicable for a single-rate feedback loop but can be modified for a loop with multiple sampling rates, as addressed further in the sequel. STÆDT buffers are flexible in both block and single-point mode.

Buffer Rules and Fill Tags

For purposes of explanation, consider two processes: a generating process (denoted GP) that places its output into a buffer, and an accepting process (denoted AP) that takes its input from GP's output buffer. Of course, AP also has its own output buffer and GP takes its input from the output buffer of yet another module.

Generating Process. It is assumed in this paragraph that each module has only a single output buffer; a multiple-buffer alternative is considered later. Any number of accepting processes can read their inputs from one buffer. Assume that every accepting process connected to GP's buffer reads all the points in the buffer before GP is invoked again. This is a crucial restriction in what follows (and is incorporated into STÆDT) but not a necessary one, as will be discussed in a later section. With this restriction, GP can blindly write data into the buffer overwriting anything that was placed there on the previous invocation. Because of the overwriting, it is necessary to make certain that the accepting processes indeed do consume all points, with none left over.

Accepting processes must know how many points are in the buffer if all points are to be consumed. To this end, GP attaches an *output fill tag* to the buffer stating how many points were written on the most-recent execution.

Accepting Process. The accepting process reads the fill tag attached to its input buffer and acts accordingly. The process is activated if it has any useful inputs to work upon and is automatically bypassed if there are no inputs. The mechanism is built into the process itself and is hidden from the user. No user-mediated flags or control logic are required.

Some processes will be capable of performing useful actions—such as initialization of state variables or incrementing a counter—even if the input buffer were empty.

Such actions would be contained in the code for the particular processes. Buffer reading has to be nondestructive if more than one AP is to use the same buffer. Therefore, no AP in a single-buffer architecture can modify an input buffer or its fill tag.

Single-Element Buffers

Buffers for some processes will be restricted to just one element, even if other buffers are flexible for single-point operations. Control-signal buffers in STÆDT are so restricted. Such processes cannot generate any more than one output point per invocation and may not be able to accept more than one input point per invocation. These restrictions take away some of the benefits that are available with flexible buffers. However, a single-element buffer can have a fill tag, in exactly the same manner as described above for flexible buffers. The value of the tag would be either 0 or 1, never larger. Process flow could be regulated in the same manner as described for flexible buffers. This arrangement is implemented in STÆDT.

Multiple Buffers

In the foregoing material on buffers it has been assumed that each generating process will have only one output buffer. That constraint means that every accepting process that takes its input from that buffer must read the entire contents of the buffer on each invocation, because the next invocation of the generating process will overwrite the contents of the buffer. Greater versatility can be attained if a separate flexible buffer is provided for every connection from a generating process to each of its accepting processes. A buffer manager can keep track of which elements have been read and which should be preserved in each individual buffer. If any earlier elements have not yet been consumed, the new elements can simply be appended after them while increasing the buffer size if necessary. Separate buffers for the same process can evolve to different sizes and different contents as a simulation progresses.

Some of the burden left on the user by single-buffer schemes (see the next section) can be mitigated by multiple buffers. If ample memory is available, multiple buffers are surely more versatile than single buffers. For STÆDT, memory usage was a major constraint, so no consideration was given to multiple buffers for one module. BLOSIM [16.1, 16.3] is a well-reported simulation program that incorporates multiple buffers.

16.2.3 Mixed Scheduling

Flexible buffers with fill tags work very well in controlling operation of processes that execute one after the other in fixed order. Details of scheduling are hidden from the user, reducing the user's burden. Computing efficiency is likely to improve because higher-rate processes are able to deal with multiple points on a single round, an undertaking that is forbidden in strict single-point mode with only single-element buffers. Nonetheless, there are scheduling tasks that cannot be handled solely by flexible buffers and fill tags. Sometimes it is also necessary to employ flags and jumps, and perhaps bypass. Several examples that require mixed scheduling are considered below. The examples are all for single-point mode, but some could apply in block mode as well.

Context

Although single-point mode can certainly be used for simulation of feedforward systems, the more-challenging problems of scheduling will be encountered in feedback loops. For that reason the mixed-scheduling examples presented below all concentrate on feedback simulations, even though the same techniques can also be applied to nonfeedback models.

Overriding considerations in a feedback loop are (1) that the feedback signal be available when needed, (2) that the samples used for computing the feedback signal belong to the correct time instant, and (3) that no process be supplied with more input points than it can handle in one invocation. These considerations are readily satisfied in the strict single-point operation described in Chapter 14. However, if higher-rate modules in a loop are allowed to generate multiple points per round, a closer scrutiny by the user will be required for each individual simulation model. Each of the examples below pays particular attention to feedback considerations.

Traffic Jams

Visualize a situation in which one module (denote it as module-A), capable of generating multiple-element outputs, drives another module (module-B), which can output no more than one element per execution. Such modules are not hard to find in typical simulations: Module-A might be a signal filter (Chapter 6) and module-B might be a timing-error detector (Chapter 11).

On any one execution, module-B can accept no more inputs than are needed to generate one output element; denote that number as n_B. The flexible output buffer for module-A (assuming a single buffer per module) has no knowledge of the nature of module-B; it does not know or care how many elements module-B can accept. Module-A's buffer can accept an essentially unlimited number of elements on each execution. But if too many elements are put into module-A's buffer, they will not all be consumed by module-B and the excess will be overwritten wrongly on subsequent executions. Overwritten elements are lost from the simulation and the simulation results will be incorrect. There is nothing in the flexible buffer/fill tag arrangement to protect against *traffic jams* of this kind.

> COMMENT: Multiple buffers offer some protection in that buffer elements remaining unconsumed from one execution will not be overwritten on subsequent executions. However, if module-A were to consistently generate more elements than module-B could consume on one execution, the flexible buffer connecting the two modules would grow in size, eventually exceeding any reasonable memory limits.

Clearly, something must constrain the number of outputs produced on each execution by module-A, something that is not inherently part of the flexible buffer/fill tag combination. One answer is to apply reblocking at a suitable location so that the buffer for module-A never accumulates more than n_B elements. With that restriction, module-B will always consume all the elements that it finds in the module-A buffer and there will never be a traffic jam. Reblocking has been described in Sec. 16.1.3 as a block-mode process, but it can also be applied quite well in single-point mode.

A reblocker generates a flag indicating whether it needs more inputs to supply its specified-size output or whether it has more than enough output elements and needs to have those elements consumed. Conditional-jump and flow-control commands would use that flag to direct the scheduling. Note that it usually will not be necessary to deliver exactly n_B elements to the buffer of module-A; it should only be necessary that not more than n_B elements be delivered. The flexible buffer/fill tag architecture should be able to readily accommodate an underfill of the buffer; module-B will not generate an output if it does not have enough inputs, and subsequent modules will react in accordance with their lack of input. The next module-B output will be produced on a subsequent round. For this arrangement to work successfully, module-B usually will require state memory to accumulate partial data until an output can be generated. Module-B has to be designed to handle partial data.

This situation is similar to the downsampling clocked bypass described earlier in Sec. 16.2.1 except that here the bypassing is done internally to the modules, by reading the fill tags and reacting appropriately, instead of by external flags.

Where should the reblocker be located? If the modules in question are located inside a feedback loop, the reblocker should not also be inside the loop, because that would destroy the time integrity of the feedback. New feedback samples require contributions from the output of module-B. If the reblocker is located inside the feedback loop (e.g., in front of module-A), feedback samples without current module-B contributions could be accumulated, wrongly; or the simulation might simply collapse from disruption of the loop.

For these reasons the reblocker should be placed outside the feedback loop, where it can meter inputs into the loop, like a traffic control at an entrance to a highway roundabout. A traffic jam at module-B is prevented by restricting the data flow into the loop at a point remote from module-B.

Parallel Operations

Most of the emphasis so far has been on sequential operations—output of one module feeding input to another. Situations also arise in which two (or more) modules must operate in parallel; they both are supplied the same signal inputs but generate different outputs. An example might be a timing error detector (Sec. 11.3.6) and a symbol strober. Each of these has to deliver one sample per symbol. Both are required to derive their one output sample from the same set of multiple samples per symbol at their shared input. Both must be timed identically; the timing error computed in one process has to apply at the time of the strobe taken in the other process.

If these processes are in separate modules (as is likely), communication between those modules is necessary so that they are able to act in synchronism. One of these parallel modules will have internal timing, probably a counter to count down from the input sampling rate to the symbol rate. A flag would be generated by the counter to indicate the end of each count cycle, and the flag would be communicated to the other module to notify it of the cycle timing and to direct it to perform its function. Alternatively, the timing might originate in a separate counter and the counter's flag would be distributed to both parallel modules. A flag, or other equivalent indicator, is required in situations of this sort, irrespective of other scheduling provisions. A flexible buffer and fill tag do not provide the necessary communication between parallel modules.

In Chapter 14 it was conjectured that a feedback loop could be broken open for simulation modeling at any arbitrary point in the loop. Now it can be seen that the loop breaking should not separate parallel operations—both must be on the same side of the break to have the same timing. Don't carry a timing flag across a loop break. This restriction holds even if one of the parallel operations is outside the loop.

16.3 RATE-CHANGING PROCESSES

Various processes reduce or increase the number of samples in a sequence, even though their main purpose may not be rate changing. Examples of such processes are presented here.

16.3.1 Downsampling Processes

Decimators

An integer decimator delivers one output sample for every M_d samples supplied as input. The output sample is one of the input sample points and not a new point to be computed by the decimator. (New points would be generated by interpolators.) Strobing is an example of decimation. A decimator has to be provided with the value of the downsampling ratio M_d as a parameter and also with an identification of the first sample to be retained from the input block or sequence; the next sample to be retained is selected by counting by M_d from the first sample selected; and so on for succeeding output samples. The count is a state variable to be carried over from one invocation of a decimator to the next.

Unless the input signal is suitably bandlimited, the decimated output signal will include aliasing, so that waveform reconstruction will be impaired. That aliasing is immaterial for strobing (the most frequent application of decimation) but could be crucial in other applications where waveforms must be preserved. An appropriate antialias filter should be placed before the decimator if aliasing is to be avoided.

Sum and Dump

Need often arises for a process that sums a number M_d of samples and delivers the sum as output. A downsampling of $M_d : 1$ occurs. This process is a discrete-time approximation to the time-continuous integrate-and-dump that is used in many data receivers. The number of samples in each sum might be a specified fixed number. Or dumps of the sum might be forced by an external flag, with possibly varying numbers of samples contributing to each sum.

16.3.2 Upsampling Processes

Integer Upsampling

To upsample by an integer factor L_u, intersperse $L_u - 1$ zero-valued samples between each pair of input sample points and then pass the zero-padded sequence through a filter that interpolates a new sequence with the higher sampling density. This operation is ordinarily called *interpolation* in the DSP literature. Properties of integer-ratio interpolating filters are described at length in [16.5]. To maintain the same signal amplitude before and after upsampling, the signal should be multiplied by a scale factor L_u. Scaling could be incorporated within the interpolating filter, or provided as a separate gain process.

The zero-padded sequence has the same spectrum as the original, unpadded sequence; the spectrum is periodic, with *spectral images* spaced at frequency intervals $1/t_s$ (where t_s is the sample spacing of the input sequence). The interpolating filter is supposed to suppress the original images, without much alteration in the spectrum of the desired spectral component. Images of an ideally interpolated signal are spaced at intervals L_u/t_s.

It is wasteful of computing effort to perform filtering operations on large numbers of zero-valued input samples, which contribute nothing to the output. Efficient upsampling methods, such as polyphase filters [16.5], can be used to reduce the computing burden. Alternatively, use an interpolator directly, as described earlier in this chapter.

Hold

A hold process has a one-element output buffer whose fill tag is permanently set equal to 1 (after the buffer is first initialized). As each new input is received, the buffer content is updated to be equal to the input value. Only strict single-element inputs can be accepted. In the absence of a valid new input, the buffer holds the value of the last valid input, so a hold module always has a valid output. In sampled-data terminology, this process is called a *zero-order hold*. A hold module is often used for upsampling from a low sampling rate to a higher rate, where the higher rate accepting process demands an input for every invocation at the higher rate.

16.3.3 Block Alterations

Various editing tasks often have to be carried out on data blocks. Such editing may alter the number of samples to be processed, even though the sampling rate is not really changed, strictly speaking. Examples are listed below.

Zero Padding

Necessity sometimes demands that a contiguous group of zero-valued samples be inserted into an existing block, perhaps to lengthen the block to a prescribed size. A padding process requires input parameters that specify the number of zeros to add and the location of the start of the added group.

Extracting and Inserting

Sometimes only a portion of a block is wanted for later operations. An *extracting* process selects the desired samples and copies them to another block. Parameters specify the number of samples and the location of the first sample in the group to be extracted. On other occasions, two blocks are to be merged, one being inserted into the other. Both blocks have to be specified; the one to be inserted has to be identified, as does the location of insertion. As a special case, insertion of one block at the end of another is known as *appending*.

16.4 KEY POINTS

This chapter dealt with sample-rate conversion and with other processes that alter the number of samples.

16.4.1 Sample-Rate Conversion

Many systems incorporate sample-rate changes, due either to rate conversions in sampled systems or to time-dilation effects (e.g., from Doppler shifts) in time-continuous systems. Therefore, simulations often have to be able to handle sample-rate changes. Define ξ as the ratio of the sample rate at the input to the sample rate at the output of the process. In hardware, ξ is generally irrational; simulations must be able to mimic irrational rate changes, despite the fact that a computer employs rational numbers only.

As special cases, ξ could be rational or integer. Simplified methods of rate conversion are available for the special cases. Rate changes can involve upsampling to higher rates or downsampling to lower rates. Techniques for rate changing have to be able to accommodate both in the same simulation.

16.4.2 Interpolation

Versatile rate changing requires that new samples be interpolated from among the input samples provided. Interpolation is performed by an interpolating filter with impulse response $h_I(t)$ according to

$$z_{\text{out}}(l) = \sum_{i=I_1}^{I_2} z_{\text{in}}(m_l - i) h_I(i + \mu_l) \tag{10.15}$$

Control signals m_l and μ_l may be updated according to

$$\Delta m_{l+1} = m_{l+1} - m_l = \text{IP}[\mu_l + \xi]$$
$$\mu_{l+1} = \text{FP}[\mu_l + \xi] \qquad (16.1)$$

For each input point $z_{\text{in}}(m)$ shifted into the interpolator's data register, Δm is decremented by 1. An interpolation is performed whenever $\Delta m = 0$; the input-sample index m is not tallied explicitly. The control signals are updated immediately following each interpolation.

16.4.3 Block Mode

The average number of output points N_{out} generated by interpolating N_{in} input points at the rate ratio ξ is

$$N_{\text{out}} = N_{\text{in}}/\xi$$

which is not generally an integer. But N_{out} and N_{in} both must be integers, so N_{out} will vary slightly even if N_{in} and ξ are constants.

To accommodate varying N_{out}, the block size has to be specified *dynamically*; the simulation program cannot be restricted to blocks of prespecified, fixed lengths. With dynamic specification of block lengths, scheduling for block-mode rate changes is local to the rate-change modules. Nonetheless, some processes will demand a fixed size of input block. That demand can be met by reblocking the input stream in an elastic buffer. Complicated flow controls are needed to regulate access to and from the reblocker's buffer.

16.4.4 Single-Point Mode

Data are exchanged between simulation modules through data buffers; one module does not communicate directly with any other. In one kind of architecture (incorporated into STÆDT), each module has only a single output buffer. All downstream processes using that output are required to read the same buffer. With that constraint, a module cannot be executed again until its previous output has been read fully by every connected downstream process. The constraint has an important bearing upon *scheduling* of simulations in which rate changes occur.

For strict single-point buffers without rate changes, execution of a simulation starts at the first module and proceeds downstream, one module at a time, to the last. Each module accepts one input sample from its input buffer and delivers one output sample to its output buffer. After executing the last module in the simulation, program flow returns to the first and starts all over again. One complete execution of all modules is called a *round*. The sampling rate is one sample per round.

Bypass Flags

How should a single-point simulation be scheduled when rate changes are included in the system? One approach would be to employ jump commands to skip over those modules not yet ready for execution. Another equivalent approach is to employ clocked-bypass operation. An implicit clock is established, running at the sampling rate of the fastest process in the simulation. There is one round per sample at this highest rate.

Each module is invoked, in proper order, once per round. If the module is ready for execution, it is executed. If not, it is bypassed. Each module includes a bypass provision, controlled by a bypass flag. The flags are generated by processes within the system but generally external to the controlled modules. Thus scheduling with bypass flags is global, not local.

Flexible Buffers

Generation of bypass flags imposes a tedious, error-prone task on the user. As an alternative, single-point scheduling can be performed locally within each module if multielement flexible buffers are employed instead of single-element buffers. To each buffer is attached a fill tag that shows the number of output elements generated on the most recent execution. Upon being invoked, each module would examine its input buffers and fill tags to determine whether it could execute on the current round. If execution were feasible, it would then generate as many output points as it could. If the module cannot generate an output, it would put a zero into its output fill tag and pass execution to the next module in line. This alternative reduces the burden upon the user by placing more intelligence (more code) within the individual modules.

Not all rate changing can be governed by flexible buffers and fill tags. Control flags and jump commands are still needed for some tasks. Examples include prevention of traffic jams (which also requires reblocking) and synchronization of parallel processes.

16.4.5 STÆDT Processes

    ACCM: accumulator for control signals
    CHRT: cubic polynomial interpolator (controllers included)
    DECM: decimator
    HOLD: zero-order hold and upsampler for control signals
    I&D: integrate and dump
    IF .. EIF, FOR .. NEXT, WHIL .. WEND: flow controls
    INTP: cubic polynomial interpolator (controllers omitted)
    LETS: various operations on data signals, including padding and chopping
    RBLK: reblocker
    SIZE: reads value of fill tag of a data signal
    TMBS: upsamples by interspersing zeros. Also a hold function for data signals

17

ENTRY AND BUILDER ROUTINES

A typical simulation module consists of (1) a *process* to be applied to an input signal (e.g., difference equations for filtering), and (2) *characteristics* of the process for the application being simulated (e.g., particular numerical values for the coefficients of the difference equations). Preceding chapters have concentrated on the processes and have paid little attention to handling the characteristics. Characteristics are also known as *process parameters*, or just *parameters*. Both terms will be used interchangeably.

As one vital step in preparing a simulation, each process to be included must be supplied with its particular characteristics. If a process is employed at multiple locations within the simulation, it generally requires different parameters at each location. Characteristics ordinarily do not originate within the simulation program. Values for characteristics are a matter for design, whereas the simulation program is an analysis and evaluation tool. The program can reveal whether particular characteristics yield satisfactory performance, but by itself, it is incapable of specifying characteristics.

Because characteristics originate externally, provisions are needed for entering them into the simulation without having to write new code; there is a need for *entry routines*. Moreover, characteristics provided externally are not likely to be in the format demanded by individual processes. Format conversions and transformations that are required are performed by *builder routines*. In this chapter we examine various aspects of entry and builder routines. Since specification of characteristics is a design matter, some experienced users maintain that all builder operations should be external functions and not a part of the simulation program. Contrariwise, others point out that one may wish to alter device characteristics in a series of related simulation runs. For this purpose it is extremely convenient to be able to incorporate builder routines within the simulation command list. Our viewpoint is that entry and builder routines, either internal or external, are needed for setting up almost all simulated systems and that a simulation facility ought to incorporate a variety of them.

17.1 OVERVIEW

Device characteristics are delivered to individual processes as parameter values. Most processes have a small number of associated parameters that are best typed in by the user in the course of preparation of the simulation. Options for entry of small quantities of parameters are examined further in Chapter 19. Here the concern is with processes that need large numbers of parameters for their operation. For example, a frequency-domain filter might require several thousand complex values to specify its characteristics. In such cases, direct typing by the user is too time consuming and too prone to error. Some other method is needed.

One way to ease the user's burden of parameter entry is to store large parameter sets in named *parameter files*. Then, when preparing a command list, the user need only call for the appropriate file by name and not be bothered with generating the parameter values. It will be assumed henceforth that parameter files are employed in this manner.

Of course, introducing parameter files only shifts the task of generating parameter sets; they still have to be generated somehow. Two plans for generation can be envisaged:

1 External generation of the parameter values and direct transfer of those values into a named parameter file

2 Entry of device characteristics into *device data files* and subsequent *builder* operations on the device data to generate suitable parameter files

The first plan might be exemplified by an external filter-design program, especially for digitally implemented filters. The format of the output file from the external program must conform to the file format expected by the simulation program. Otherwise, there is but little more that can be said about the first approach in this book.

The second plan might be exemplified by measured input–output characteristics of a nonlinear amplifier, or by coefficient values for an analog filter function. Neither of these *raw* characteristics would be in a format usable directly in a parameter file. Instead, the raw characteristics would be entered into the computer and stored in device-data files. Then the raw data would be reformatted, rescaled, transformed, and converted in builder routines to produce run-time parameter files. This is the plan described in the sequel.

Data in parameter files are highly specific to a particular process and to the conditions of a particular simulation. Specifications of sampling rate, number of samples per symbol, and—in some instances—block length are all involved in establishing parameter files. If any of these conditions are altered, the parameter files have to be changed completely. For the most part it is not usually feasible to employ a parameter file generated for one simulation in another simulation with different conditions. It is not generally easy to modify one parameter file as a means of generating another parameter file suitable for changed simulation conditions. The device-data characteristics are more pliable. They can be altered in many ways by builder routines and be made to fit many different simulation conditions. Indeed, they *must* be altered by builder routines; a device-data characteristic will not normally be provided in a form that is usable as a parameter file.

Figure 17.1 shows elements related to entry and builder routines and illuminates

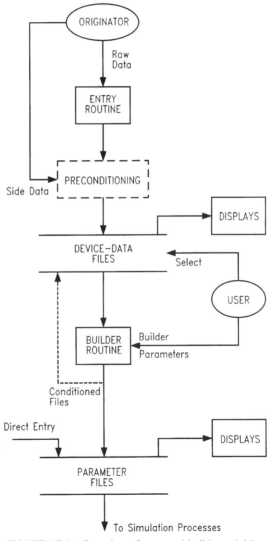

FIGURE 17.1 Overview of entry and builder activities.

several issues that arise. The material that follows is divided in accordance with the figure.

First, it is useful to distinguish between a *user* and an *originator*. A user is a person who prepares and runs a simulation. In the course of preparation, the user is likely to have to generate parameter files. To do so, the user selects device-data files and operates upon them with builder routines. An originator is a person who enters device characteristics into the computer. The originator might be the same person as the user, but need not be. Originator data can be entered at the time of simulation preparation but could also be entered at any earlier time. An originator has the duty to provide enough side information along with the raw data that a user can interpret and apply the data appropriately.

Raw data are entered with the aid of an entry routine. It is possible that some preconditioning will be performed within the entry routine; examples of preconditioning tasks are given in the next section. Preconditioning typically involves reformatting data into standardized formats. Side data, in addition to those found in typical raw data sets, are likely to be needed for preconditioning. Raw/preconditioned data are stored into device-data files.

Preconditioning could be performed either in the entry routines or in the builder routines. If in the entry routines, the builders are simplified because only a few raw data formats need be accommodated. If in the builders, the entry routines can be made very simple indeed. The preconditioning box in Fig. 17.1 is shown in dashed outline because of this uncertainty over its location.

Most likely, though, conditioning of various kinds will be performed in both locations and by either the originator or the user. Some "builder" routines might also serve as "conditioning" routines. This situation is depicted in Fig. 17.1 by the dashed-line feedback connection from the builder routine box back to the device-data files. A terminology distinction might be drawn as follows:

- A builder routine outputs a parameter file, specialized for a particular set of simulation conditions.
- A conditioning routine outputs a nonspecialized device-data file that will require further builder operations to produce a parameter file.

The input to either kind of routine is contained in a device-data file.

To generate a parameter file, a user selects a device-data file, invokes builder routines to operate upon the data, and provides builder parameters as needed. The builder routine constructs a parameter file according to its instructions. The user can then inspect the characteristics of the parameter file on the displays of the simulation facility. (Displays are treated in Chapter 18.)

Needs for entering and building have been indicated in preceding chapters for such items as filter responses, nonlinear amplifier characteristics, and mapping tables, among others. As a practical matter, the overwhelming bulk of entry and building efforts are devoted to generating filter parameters. The examples that follow reflect that preponderance of filter activities.

17.2 ENTRY ROUTINES

The main entry issues to be treated are file formats and entry tools.

17.2.1 File Formats

Raw data might be typed in by the originator, supplied by automated test equipment, or generated by some other program such as a spreadsheet. An originator or user should be able to read the file, no matter how produced, with the aid of readily available viewing tools. The requirement for readability applies to parameter files as well as device-data files. Furthermore, it is desirable to be able to accept data files from a wide variety of external sources and to process those files through a wide variety of external programs (e.g., for plotting or printing). Readability is assured by writing the data as ASCII text files. Many other programs, such as text editors, spreadsheets, and graphics programs can import and export data in ASCII format. No other format offers the same wide acceptance.

Data could be arranged within the file in many different ways. Because of the popularity of spreadsheets, it is advisable to have a row and column arrangement. A header would provide for side data and textual descriptions that did not fit naturally into rows and columns.

A descriptive header is essential for any data file that you ever hope to reuse. A header would explain the nature of the data, specify its format, summarize the characteristics, identify the originator and origination date, and so on. Header information should be transferred from file to file as the device characteristics are converted and transformed from raw data to parameter files. Without sufficient header information, a data file becomes a jumble of meaningless numbers. The simulation community has considered a standard file format [17.1] with the foregoing features, but no convention has been adopted at the time of this writing.

17.2.2 Entry Tools

A text editor is an essential tool for data entry. If an ASCII file format is adopted, any ASCII text editor or word processor can be employed for the task. A text editor offers means for entering data manually, for viewing the file contents, for printing the file, for naming and storing the file, for other file-management duties, and—most important—for altering the contents of the file easily. Also crucial is the ability to insert additional data points at any stage of the entry–conditioning–building sequence.

A spreadsheet is another valuable entry tool that greatly eases row/column data entry. Also, spreadsheets have formula-evaluation capabilities. If device properties can be expressed as a computable formula, a spreadsheet might be used to generate a numerical table of characteristics suitable for further conversions by builder routines or even for direct entry to a parameter file.

Furthermore, commercial spreadsheets typically have flexible graphics capabilities; they readily plot almost any data that can be entered into their row/column format. Plotting routines that are included in a simulation program usually are much more specialized in the data formats they will accept; they are not likely to accommodate arbitrary data files nearly so well as is done by a good spreadsheet program. Entry tools might be included in the simulation facility, or external tools may be required. STÆDT adopts the latter expedient to afford greater flexibility.

17.3 CONDITIONING

Various tasks fall under the heading of "conditioning." These tasks are best described by showing particular examples of meaningful applications. Two examples are presented here: (1) input–output characteristics of nonlinear saturating amplifiers, and (2) frequency response of a filter. These data typically result from measurements on actual devices. Conditioning operations are considered in this subchapter, and their associated builder operations are considered in the next.

Input to a conditioning session is a file of raw or preconditioned device data. Output is another device-data file. The conditioning tool might be a text editor or spreadsheet—that is, indistinguishable from an entry tool—for the simplest tasks. Or it might be a specialized conditioning process that differs from a builder routine only in the disposition of its output. A conditioning process will be preferred whenever more than trivial calculations on the data are involved. One might think that conditioning should be a straightforward matter, if perhaps somewhat tedious. To the contrary, significant modeling issues arise, as will become evident in the pages below.

17.3.1 Conditioning of Nonlinear Amplifiers

Input Data

Measured characteristics of nonlinear amplifiers typically are presented in tables with output power and phase shift as dependent variables, and input power as the independent variable. Data points are usually presented in order of increasing input power. The powers most often will be expressed in logarithmic units—dBW or dBm—and the phase in degrees.

Figure 17.2 is a conceptual plot of representative saturating characteristics. The two plots are commonly known as the AM/AM and AM/PM nonlinearities (AM = amplitude modulation; PM = phase modulation). Continuous plots are shown for illustration, but the raw data are always discrete points.

Nonlinear Functions

These data are to be conditioned and subsequently delivered to builder routines (considered later), which will construct nonlinear functions of output amplitude and phase shift versus input amplitude. Following Sec. 7.5.2 and (7.27), the nonlinear processes have the functional form

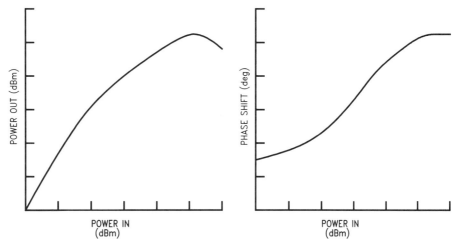

FIGURE 17.2 AM–AM and AM–PM characteristics of typical saturating amplifier.

$$R_{\text{out}}(n) = \gamma_R[R_{\text{in}}(n)]$$
$$\phi_{\text{out}}(n) = \gamma_\phi[R_{\text{in}}(n)] + \phi_{\text{in}}(n) \qquad (17.1)$$

where $\{R_{\text{in}}(n), \phi_{\text{in}}(n)\}$ and $\{R_{\text{out}}(n), \phi_{\text{out}}(n)\}$ are the complex-envelope values in polar coordinates of the nth sample of the input and output signals, and $\gamma_R[\cdot]$ and $\gamma_\phi[\cdot]$ are the nonlinear functions to be built. The format of (17.1) is broadly applicable to memoryless, complex-envelope nonlinearities whose behavior depends solely upon the input amplitude.

Conditioning Tasks

Units. Although measured powers are normally expressed in decibel units, the signal process works with the dimensionless amplitude $R(n)$ of the complex envelope of the signal; decibel powers have to be converted to amplitudes. To obtain rigorously correct amplitudes, it would be necessary to specify the input and load resistances of the amplifier and to take account of the factor of 2 that arises with in the definition of power of a complex envelope (see Sec. 2.3.2).

Rather than becoming entangled in these fine details, it is much more convenient to normalize the amplitudes, as will be described below. Most simulations are conducted with normalized quantities. The computer itself works with dimensionless numbers and is not concerned with units. All that matters is that the shapes of the nonlinear functions be maintained and that the signals be scaled properly to the shapes.

To have consistent units, convert any input powers expressed in dBm to dBW by subtracting 30 dB from each power value. Denote dBW-expressed power by the symbol PdBW. Then the amplitude corresponding to each power is given by

$$R = 10^{\text{PdBW}/20} \qquad (17.2)$$

This is a fictitious amplitude, since no account has been taken of the input and load resistances or the complex-envelope multiplication by 2. However, both the input and the output of the nonlinear device have the same confounding factors, so the ratio of output and input amplitudes—the gain—is correct.

Phase-shift measurements almost always are reported in degrees, but the signal process will require radians. All phase angles have to be converted to radians. These unit conversions clearly should be performed by the computer, not the user. Therefore, a conditioning routine is needed for unit conversions. The user should only be called upon to identify which power data, if any, are in dBm and need converting to dBW.

Normalization. Simulations could be performed perfectly well using the unnormalized amplitudes obtained from (17.2). Often, though, the system behavior relative to a saturation level of the nonlinear amplifier will be more informative. Simulations are frequently conducted with differing amounts of backoff from the saturation level. Under those circumstances, it is convenient to normalize the nonlinear characteristic to the saturation level.

The first step in normalization is for the user to specify an input reference level that marks the onset of saturation. If the AM–AM characteristic is like that of Fig. 17.2—with a distinctive maximum in the output power—the reference level is fairly easy to define; it is simply the input that produces the maximum output. However, if the AM–AM curve is monotonic, the choice of reference level becomes rather more subjective. Such criteria as the 1-dB compression point or the intersection of the small-signal and large-signal asymptotes might be applied.

Choice of a good criterion is highly dependent on the application. An AM–AM curve with a well-defined peak is typical of a traveling-wave tube amplifier, while a monotonic curve is typical of a transistor amplifier. Once the reference level has been specified, the computer can scale the input and output amplitudes (or powers) for unit input and output amplitude (0-dB relative power) at the reference level. This scaling in effect sets the normalized gain at saturation—the large-signal gain—equal to unity.

You might reasonably object that it is often more appropriate to set the small-signal gain to unity and let the large-signal normalized gain be compressed to a smaller value. That is certainly a valid approach when the small-signal behavior of the amplifier is linear, but it fails if the small-signal behavior is nonlinear—if there is crossover distortion. No matter what definition is used for gain normalization, any arbitrary gain can be set after normalization by following (not preceding) the nonlinear amplifier by a linear-gain process. Therefore, the large-signal gain-normalizing rule is not really confining.

Out-of-Range Points. Any set of raw data includes only a finite number of data points that cover only a finite range of input amplitudes. In a simulation, it is possible for the input signal to have an amplitude that is outside the range of the data set. Despite the fact that the raw data are incomplete, it is essential that the computer program respond sensibly to any input amplitude that might arise. Provisions have to be made for out-of-range points. The problem is fairly straightforward for small-amplitude inputs. The input amplitude can never be negative, and a zero input leads to a zero output amplitude. Thus a small-amplitude endpoint (0,0) can be appended automatically to the AM–AM portion of the conditioned data.

The phase shift at zero input amplitude has to be entered by the user, since no natural value exists. You might extrapolate the AM/PM curve to zero amplitude or might simply choose the phase of the smallest-measured amplitude as the zero-amplitude phase. Handling of large out-of-range amplitudes is treated as a builder task and discussed in Sec. 17.4.

Task Assignments. Tasks described above might be regarded as conditioning or might equally well have been regarded as a part of building; there is no clear boundary between the two. Output data from these tasks will consist of a table of input and output amplitudes and phase, with units converted, amplitudes normalized (perhaps), and some additional data points inserted.

17.3.2 Frequency Response of Filters

One feature that is valuable in a simulation program is an ability to use experimental data taken from measurements on physical devices, such as filters. Characteristics of conventional analog filters are almost invariably measured as a frequency response. In this section we examine the various conditioning operations that may have to be performed on the raw frequency-response data. In later sections we briefly discuss conditioning tasks for other methods of specifying filters, while filter builders are pursued in detail in Sec. 17.4.

Input Data

Measured frequency response is most commonly presented as a table, each entry of which consists of the measurement frequency f_i, the amplitude of the response (in dB), and the phase of the response (in degrees). Instead of phase, the measurements might provide delay—or even delay distortion, where a constant delay has been removed. Operations to be described subsequently will require that data be arranged in order of increasing frequency.

In addition to actual measurement data, it is sometimes necessary that other, non-measured points be inserted into the record. For example, dc response is not often measured but may be needed by later operations. In situations where dc response is needed, it is often easily inferred by the user. Also, explicit identification of transmission nulls may be required, as discussed below.

For purposes of this discussion, assume that the frequency entry is provided explicitly. If the frequencies are equally spaced, the data may provide only a base frequency and a frequency increment, leaving the frequency of each point implicit and to be calculated as needed by the software.

Transfer Functions

Eventually, the conditioning and building will produce a digital transfer function $H_d(k)$. Two different methods of arriving at $H_d(k)$ can be envisaged:

1 Perform system identification operations on the measured data to arrive at an analytical expression $H_a(s)$ for the transfer function of the physical filter. Then evaluate $H_a(s)$ for those values of $s = j2\pi f_k$ corresponding to the sample index k.
2 Scale and interpolate the measured data to produce the values for $H_d(k)$.

System identification is an elaborate topic that has inspired entire books [17.2–17.6]; it will not be considered further here. Instead, assume that the conditioned data will be interpolated by the filter builder.

Units and Formats

Frequency could be entered in any of several customary units: Hz, kHz, MHz, and so on. The units of frequency do not matter; the only requirement is that all frequency entries be in the same units. The frequency entries could be scaled by any arbitrary, positive coefficient whatever and no important information would be lost. In consequence, it is not necessary to retain the units (e.g., kHz, MHz) of the frequency data.

Communications engineers, and this book, define a transfer function as the ratio of output divided by input. With this definition, the dB-amplitude is positive for a gain and negative for a loss. Phase shift is positive for a leading angle. Sometimes, though, the data are prepared from a filter designer's definition, in which dB attenuation is positive and leading phase is negative. If the data are in the attenuation format, the signs of phase and dB amplitude must be reversed to place them into gain format. Delay has the same sign in either format, so no sign reversal should be applied to delay data.

Phase data usually are reported in degrees whereas the builder routines will demand radians (to be compatible with the trigonometric functions of the underlying programming language in which the routines have been written). Phase in degrees has to be converted to phase in radians.

Delay will be reported in time units of seconds, milliseconds, microseconds, and so on. The builder will require that the units of delay be compatible with the units of frequency: that is, microseconds with MHz, and so on. The product of delay units multiplied by frequency units should be unity. If the time and frequency units are compatible, no units need be carried over to the builder. If the units are not compatible, the delay data should be scaled to make them compatible. For example, if frequency is expressed in MHz and delay in nanoseconds, all delay values should be divided by 1000 to convert them to microseconds.

Delay Integration

For constructing the digital frequency response $H_d(k)$, the interpolating builder requires phase information, not delay. If the raw data are provided as delays, phase will have to be computed from the delay data. The relation between phase and delay is expressed by

$$\tau(f) = -\frac{1}{2\pi} \frac{d\phi(f)}{df} \qquad (17.3)$$

where $\tau(f)$ is the delay as a function of frequency and $\phi(f)$ is the phase. To find the phase, within a constant of integration, integrate the delay over all frequencies:

$$\phi(f) = -2\pi \int_{-\infty}^{f} \tau(\nu) \, d\nu \qquad (17.4)$$

where ν is a frequency dummy variable of integration.

Data are provided only over a finite range of frequencies, say f_0 to f_1. Within that range, phase can be determined as

$$\phi(f) = -2\pi \int_{f_0}^{f} \tau(\nu) \, d\nu + \phi(f_0) \qquad (17.5)$$

Of course, the phase $\phi(f_0)$ is not known, and usually cannot be known from the raw data. This matter will be taken up again presently.

Data are provided only in discrete points, not as continuous functions. True integration is not possible; an approximating summation must be performed instead. If the data points are spaced closely enough, the trapezoidal rule will serve:

$$\phi(f_i) = \phi(f_{i-1}) - 2\pi \times \tfrac{1}{2}[\tau(f_i) + \tau(f_{i-1})][f_i - f_{i-1}] \qquad (17.6)$$

where the N data points were measured at frequencies f_i, $i = 0$ to $N - 1$. It is not necessary that the frequencies be equally spaced. In fact, it is advisable to take points more densely in regions where sharp changes of delay (or amplitude or phase) occur. If a plot of the measured delay versus frequency looks smooth when the points are connected by straight lines, the trapezoidal rule probably is adequate. If the plot looks rough, a higher-order rule would improve the approximation, but additional data points would be even more helpful.

What is to be done about the unknown constant of integration $\phi(f_0)$, needed for starting the recursion of (17.6)? If the filter is to be applied to a complex-envelope signal, nothing need be done in many situations. ["Doing nothing" means setting $\phi(f_0)$ to any arbitrary value, such as zero.] An incorrect value for $\phi(f_0)$ rotates the filter response by a constant angle. If subsequent carrier synchronization corrects for phase angle, or if the signal processing is independent of phase angle, an erroneous phase angle simply does not matter.

In other situations, such as filtering a real signal with a lowpass filter, the phase angle does matter; an incorrect phase angle will cause distortion of the signal. If the conditioned phase data are to be valid, the user has to specify a reference phase at one point in the data set. Then a correct phase is computed as follows at each data point:

1. Denote the user-specified reference phase at frequency f_a as $\theta(f_a)$; f_a has to be one of the data frequencies f_i, but it is not necessary that $f_a = f_0$.

2. Set the unknown phase $\phi(f_0)$ to zero and compute $\phi(f_i)$ according to (17.6) for each data point.

3. The correct phase $\theta(f_i)$ at each frequency is found as

$$\theta(f_i) = \phi(f_i) + \theta(f_a) - \phi(f_a) \tag{17.7}$$

How is the reference phase $\theta(f_a)$ to be determined? That is a responsibility of the user; the simulation program cannot help. Side information of some kind has to be available so that the reference phase can be inferred. As one example, typical lowpass filters have zero phase at zero frequency. As another example, bandpass filters often have known phase (typically zero) at their center frequency. Absence of side information would stymie the needed determination.

Treatment of Nulls. Some filters have nulls in their frequency responses (gains of $-\infty$ dB). Nulls pose difficulties when phase is to be obtained by integrating delay data. Those difficulties are the topic of these next paragraphs.

A null of degree m at frequency f_n is caused by m coincident zeros in the frequency response $H_a(f)$ at frequency f_n. (In practical filters, m rarely exceeds unity.) At the null frequency, the amplitude $|H_a(f_n)| = 0$. Also, the phase of the frequency response will have a discontinuity of $m\pi$ radians at the null frequency.

The discontinuity will appear prominently in measurements of phase versus frequency but could easily be lost in measurements of delay versus frequency. If the null is perfect (i.e., the zeros at frequency f_n lie precisely on the imaginary axis of the s-plane), delay is an impulse with area of $m\pi$, exactly at frequency f_n. Measurement at discrete frequencies will never catch f_n exactly—that is, f_n will not ordinarily be one of the measurement frequencies f_i—so the measured data points will miss the impulse of delay. The data will appear to have a continuous delay characteristic, without singularities.

In the real world, no null is perfect. The amplitude $|H_a(f_n)|$ will not quite reach zero and the phase will not be quite discontinuous at f_n. A plot of delay will exhibit a narrow spike in the vicinity of f_n. If the null is shallow enough and if a sufficient number of data points are taken so that the delay spike is well defined, integrating the delay will lead to the correct phase. But if the delay spike is too narrow for the resolution of the data, the integration will not provide the correct phase.

How shall this problem be handled? If additional data can be collected, an increased density of measured points in the vicinity of the null will improve the situation. However, if additional points cannot be taken, the user's engineering judgment will be called upon for editing the data. Four actions will be needed:

1. Append to the data set an identification of the frequency f_n and order m of each null. (Don't overlook nulls at zero frequency.)

2 Edit the delay data to eliminate any delay spikes in the vicinity of the nulls.

3 Integrate the edited delay data according to (17.6) and add in the constant of integration of (17.7), if that is applicable. Phase discontinuities will not appear in the integrated phase.

4 For each null frequency f_n, add $m\pi$ to the phase of each data point at every frequency $f_i > f_n$.

Conjugate Extension for Real Filters

Frequency-response measurements are almost always conducted at positive frequencies only. That's just fine for filters of complex envelopes of modulated signals; omitting negative frequencies accomplishes one step in the necessary real-to-complex conversion described in Sec. 6.2.1. But filters for real signals must have negative frequencies in their responses as well as positive. A conditioning routine should fill in the negative frequencies when so ordered by the user. If the filter has a purely real impulse response, the frequency response at negative frequencies is the complex conjugate of that at positive frequencies.

17.3.3 Other Conditioning Tasks

Analytical Functions

Some kinds of data are entered to the computer as analytical expressions. For example, the transfer functions of many filters can be expressed as a rational function—the ratio of two polynomials—in the Laplace variable s. There are several different formats that could be used for the same function:

- *Expanded Format:* a single rational function expressed as the ratio of two high-order polynomials
- *Factored Format:* locations of the individual poles and zeros, and value of the flat-gain coefficient, of the transfer function
- *Biquad Format:* coefficients of biquadratic partitions of the transfer function

Considerations of numerical precision (see Sec. 3.1.2) encourage biquad representation of these filters in the signal processes and in the associated filter builders. In that case, conditioned data delivered to the filter builder should be in biquad format.

Conditioning routines might include a polynomial root solver (to reduce the expanded polynomial format to the factored pole and zero format) and a biquad expander (to collect zeros and poles in pairs and expand them to biquad format). Alternatively, a simulation package could omit these routines (the option chosen for STÆDT) and require that all rational transfer functions be entered as partitioned biquads. Conditioning of data in expanded or factored formats would then have to be performed externally to the simulation facility.

17.4 NONLINEARITY BUILDER ROUTINES

A user will be confronted by a multitude of builder activities when preparing modules for a simulation. In this section and the next we attempt to bring some order to the subject by way of examples. Continuing the approach of Sec. 17.3, examples in this section show builders for nonlinear amplifiers, and treat filters in Sec. 17.5.

17.4.1 Nonlinearity Overview

Nonlinearities have been discussed in Chapters 4 and 7. All nonlinearities considered in this book are memoryless, which means that the current output from the nonlinear signal process depends only upon the current input, not on past (or future) inputs or outputs. In essence, a memoryless, nonlinear signal process is a single-valued function for which the signal input is the independent variable and the signal output is the dependent variable. A signal sample is provided at the input; the process evaluates the nonlinear function for that input sample and delivers the function value as the corresponding output sample.

Two broad classes of nonlinear functions are common: (1) closed-form expressions, and (2) tabular data to be fitted or interpolated. Many different closed-form expressions might be needed in different applications. In general, closed-form expressions have to be programmed individually but require little or no builder activity.

Examples here concentrate on a narrow subclass of functions, which derive their characteristics from data measured on actual devices. One approach is based on the observation that many nonlinear charactoristics can be well approximated by a rational function. How might the coefficients of the rational function be determined from the data; that is, how is the rational function to be *fitted* to the measurements?

Alternatively, intuition tells us that it should be possible to interpolate measured data to determine output for an input value that lies between discrete points of the data set. What interpolation rules should be applied? What parameter information has to be supplied to the nonlinear signal process so that it can interpolate correctly?

These are the questions examined in the next few sections. All memoryless nonlinear operations take place in the time domain; frequency-domain nonlinearities have no physical meaning in most communications systems. The text to follow implies that the signal is treated as a complex envelope and that the signal process operates in polar coordinates. However, similar considerations would apply to real signals or to processes working in rectangular coordinates.

17.4.2 Function Fitting

Fitting tabulated data to a rational function usually is a problem in nonlinear estimation of parameters [17.7, Sec. 14.4]. The task is best assigned to a capable optimization program, external to the simulation facility. Nonlinear fitting is an extensive topic that is not treated further in this account. The Saleh model [17.8] of equation (7.28) is a notable exception, in which least-squares-fit parameters can be found from explicit formulas instead of nonlinear estimation. Even though quite simple, Saleh's rational functions afford good models of traveling-wave-tube amplifiers (TWTAs); they have been employed in numerous simulations of communications links that include TWTAs.

A builder for the Saleh model accepts AM–AM and AM–PM data as input, already conditioned as described in Sec. 17.3, and applies the formulas to calculate two pairs of model coefficients as output. One pair of coefficients specifies a rational function for the AM–AM nonlinearity and the other pair of coefficients specifies the AM–PM nonlinearity. Any simulation facility that is likely to deal with TWTAs ought to have a Saleh-model builder in its library.

17.4.3 Interpolation

Reasons to prefer curve fitting over interpolation were outlined in Chapter 7. Nonetheless, interpolation can be accomplished with much simpler software than needed for general curve fitting and can be applied to almost any reasonably smooth data set. Therefore, interpolation is a meritorious tool for inclusion in a simulation library.

Configuration

Visualize the conditioned data set as $D + 1$ ordered pairs (x_i, y_i), where x_i is the independent variable (e.g., input-signal amplitude) and y_i is the dependent variable (e.g., output-signal amplitude or phase shift). There are D intervals between data points. An interpolator-builder selects approximating polynomials for the conditioned data set, one for each interval in the set. Each polynomial is denoted as $y = p_i[x]$. The nonlinear signal process accepts an input point $x(n)$, determines the interval i in which it falls in the data set, and evaluates the polynomial $y(n) = p_i[x(n)]$ to find the corresponding output point. This section is concerned with the actions taken by the interpolator-builder to establish the polynomials $\{p_i[x]\}$.

Let the polynomials have degree m, and let's concentrate attention on the interval (x_{i-1}, x_i). The objective is to specify a unique polynomial $p_i[x]$ of degree m that passes through (x_{i-1}, y_{i-1}) and (x_i, y_i) and which has other desirable properties.

Conventional Polynomials

A first candidate might be the conventional interpolating polynomials, each one of which passes through $m + 1$ successive points of the data set. These polynomials are the interpolating functions that are found in the many textbooks on numerical analysis. Designate those $m + 1$ points as the *basepoint set* of the polynomial.

If $m > 1$, the interval (x_{i-1}, x_i) will be included in more than one basepoint set, each with a different polynomial. Something must be done to assign a unique polynomial to each data interval. One solution is to select an odd value for m and restrict (x_{i-1}, x_i) to be the central interval of the basepoint set for $p_i[x]$. (This approach has to be modified for intervals too close to either end of the data set; modifications will not be examined here.)

Having specified degree m, a unique definition of basepoint set, and the data values at the basepoints, a unique output $y(n)$ can be calculated for any input $x(n)$, using any appropriate mth-order interpolation formula from the many presented in textbooks. If, as is likely, m is a fixed number built into the signal process ($m = 3$ is a popular choice), there is little for a builder to do; the burden rests on the signal process.

Artifacts

To discern a greater role for a builder, consider a second candidate for the approximating polynomial $p_i[x]$. For an introduction, consider a property of the conventional polynomials. The point $x = x_i$ is at the right end of the interpolating interval for the polynomial $p_i[x]$ and at the left end for the adjoining polynomial $p_{i+1}[x]$. Indicate the derivatives of these polynomials by a prime. Although $p_i[x_i] = p_{i+1}[x_i]$, the derivative $p_i'[x_i]$ is not equal to $p_{i+1}'[x_i]$. On the contrary, derivatives of all orders 1 through m will be discontinuous at every point of handover from one polynomial to the next.

As explained in Chapter 7, discontinuities in the derivatives will cause the spectrum of the output signal to exhibit spuriously large components at far-out frequencies. The discontinuities are small-scale nonlinearities that are not present in the device being modeled but are artifacts of the model. It is desirable to reduce those artifacts to the smallest magnitude feasible.

Splines

Rather than force each polynomial to pass through $m+1$ points of the data set, relax that requirement so that the ith polynomial is only fixed to two points (x_{i-1}, y_{i-1}) and (x_i, y_i)—the boundary points of its interval of applicability. Use the remaining $m-1$ degrees of freedom to force as many derivatives as possible to be continuous at the boundary points of handover to the adjoining polynomials.

A *cubic spline* ($m = 3$) forces both the first and second derivatives of adjoining polynomials to be continuous at the boundary points [17.7, Sec. 3.3]. A substantial improvement in spectral artifacts can be expected as compared to the conventional interpolating polynomials with discontinuous low-order derivatives.

To work with a cubic spline, the interpolating signal process needs an additional quantity for each data point: the second derivative of the spline polynomials at each boundary point. These second derivatives are calculated by a *spline builder* from the conditioned $\{x, y\}$ data set to form an extended table $\{x, y, y''\}$, from which the signal process evaluates the spline polynomials. Spline building need be performed only once, before the simulation run begins. Spline interpolation in the $\{x, y, y''\}$ table will be performed repeatedly in typical simulations.

Secondary Matters

Either type of interpolation requires that certain smaller details need attention, as noted below.

By inserting the point (0, 0) into the conditioned $\{x, y\}$ data set, you will guarantee that a zero input leads to a zero interpolated output.

Suppose that $x(n)$ exceeds the largest abscissa x_D of the data set; what is the signal process supposed to do? There is a dilemma here: *Something* has to be done; the program will crash if the process does not know what to do. Moreover, whatever is done has to be consistent with the action of the device being modeled, or else the simulation results will be incorrect.

On the other hand, knowledge of the device characteristics extends only over the range of the data set; nothing is known (to the simulation program) about behavior outside the data set. It would be a very poor idea to extrapolate an output using the polynomial from the highest data interval; interpolating polynomials extrapolate badly. Additional information is needed, in the form of specific instructions to the signal process for large-amplitude inputs.

If the measured data set is to be useful, out-of-range inputs had better occur only rarely. Then the statistical behavior of a system probably would not be disrupted significantly if rare large-signal amplitudes were not treated quite accurately. With that in mind, consider these three simple rules, one of which is to be applied when $x(n) > x_D$:

- *Flat Limiting.* Set $y(n) = y_D$.
- *Rolloff.* Set $y(n) = [x_D/x(n)] y_D$.
- *Zeroing Out.* Set $y(n) = 0$.

Flat limiting might apply well to an AM–PM nonlinearity or to the AM–AM nonlinearity of a saturating transistor amplifier. Rolloff applies to typical TWTAs. Zeroing out is an escape route that can be taken if the other alternatives seem even worse. The over-range contingency has been explained in conjunction with interpolator builders, but the applicable instructions would have to be part of the interpolating signal process.

17.5 FILTER BUILDER ROUTINES

Almost every simulation will incorporate one or more filters and the characteristics of most filters will be prepared with the aid of filter builder routines. This subchapter is devoted to explanations of the arrangements and tasks of filter building.

17.5.1 Filter Builder Constituents

Figure 17.1 showed the builders' place in the larger simulation facility; a builder's task is to convert and transform conditioned device-data files into process parameter files. Figure 17.3 provides somewhat greater detail, showing categories of builders and files.

Signal Processes

Start at the bottom of Fig. 17.3 with the filter signal processes, and move upward. In most simulation programs, there will be only three kinds of filter processes, as outlined in Chapter 6:

482 ☐ ENTRY AND BUILDER ROUTINES

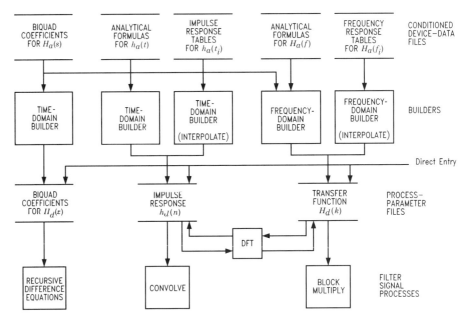

FIGURE 17.3 Filter builders, files, and processors.

1. A frequency-domain process, in which a signal block $Z(k)$ is multiplied by a filter-response block $H_d(k)$
2. A nonrecursive time-domain process, in which a signal sequence $z(n)$ is convolved with an impulse-response sequence $h_d(n)$
3. A recursive time-domain process, in which a signal sequence $z(n)$ is delivered as input to difference equations having a z-transform $H_d(z)$, a rational function

A subscript d indicates a *digital* filter, used for discrete-time simulation by the computer.

Process Parameter Files

The coefficients of $H_d(k)$, $h_d(n)$, or $H_d(z)$ are kept in process-parameter files. Coefficient values are highly specific to the type of filter involved, the characteristics of the filter, and the parameters of the particular simulation, such as symbol rate, samples per symbol, carrier frequency (when applicable), and block size (for frequency-domain filters). If any of these quantities have to be altered, it is necessary to produce new process parameter files with new coefficients.

Parameter files usually are produced by builder routines. There are two exceptions: (1) parameter files can be entered directly from external sources (e.g., filter-design programs), or (2) transformations between time-domain impulse responses $h_d(n)$ and frequency-domain responses $H_d(k)$ can be performed by the discrete

Fourier transform (DFT). Characteristics of digitally implemented filters are likely to be generated externally and then entered directly in the process parameter files. Therefore, in this section we concentrate on device data from analog filters.

Domain transformation between $h_d(n)$ and $H_d(k)$ would treat the filter parameters just like a signal sequence. Other operations, such as truncation and windowing (to obtain a finite-length impulse response), zero filling (to pad an impulse response to fit the block length required for the FFT), and delay insertion (to obtain a causal impulse response), may be needed in addition to the DFT. These operations all would be general-purpose signal processes and not specialized builder routines. To make it possible to perform such operations on parameter files, the files have to be in a format that can be interpreted as signals by signal processes. STÆDT files for $h_d(n)$ and $H_d(k)$ can be treated as signals.

The remainder of this section is devoted to builder topics; signal processes have been treated in depth elsewhere in the book.

Device-Data Files

Now skip to the top of Fig. 17.3, to the conditioned device-data files. Five kinds are shown and other kinds could also arise. Those shown contain:

- Coefficients of biquadratic-partitioned analog transfer functions $H_a(s)$
- Analytical formulas of impulse responses $h_a(t)$
- Tables of impulse responses $h_a(t_i)$, presumably obtained by measurements
- Analytical formulas of frequency response $H_a(f)$; includes any transfer functions that have not been reduced to partitioned biquads
- Tables for frequency responses $H_a(f_i)$, presumably obtained by measurements

The subscript a indicates an *analog* filter.

Conventional analog filters have not commonly been specified or measured in terms of impulse responses, so there may be little demand for their device-data files or associated builder routines. Accordingly, they are not pursued further. Some of the principles of building from impulse responses can be gleaned from the methods described below of building from frequency reponses.

Builder Categories

After excluding impulse-response files, only two broad categories of files remain: partitioned biquads that can be built into either time- or frequency-domain filters, and frequency responses (either in analytical formulas or tables) that are most readily built into frequency-domain filters. These are the categories that are examined in the sections that follow.

COMMENT: A distinction has been made in Chapter 14 between data signals and control signals. The filter builders considered here are for filters of data signals only. Control signals are also subjected to filtering, but the control-signal filters tend to be so simple that their parameters can be specified directly by the user without any need for a builder routine.

Figure 17.3 shows two ways to categorize the builder routines themselves: (1) between time domain and frequency domain, and (2) between interpolators and function evaluators. Examples from all categories are scrutinized below.

17.5.2 Standard Filters

Besides the distinctions shown at the top of Fig. 17.3, device-data files can be categorized according to their origin: either they are entered by an originator from outside the computer, or are resident permanently in the simulation library, having been provided by the program authors. Certain *standard filters* are of sufficient importance that their characteristics are often included in the program library.

Rational Functions

Some standard filters are described by rational transfer functions; these include Butterworth, Chebyshev, Bessel, and elliptic (Cauer) characteristics. Such filters are often used in communications links because their properties are well known, as are methods for their design.

Despite their frequent employment, these particular filters are not good choices for data transmission. All but the Bessel filter have significant delay distortion, and the Bessel filter has overly gentle amplitude rolloff. Better-suited filters can be built. Nonetheless, they might well be included in a simulation library because of their ubiquity in physical systems.

To call one of these filters from the library, a user specifies the filter type, the number of poles, and such other information as may be required by the particular filter. A builder routine then retrieves the necessary information from the library file and delivers coefficients of a biquadratic partitioning of the transfer function.

Butterworth and Chebyshev pole locations can be found from well-known formulas [17.9–17.11], Bessel pole locations have been tabulated [17.12] and procedures have been described for finding elliptic poles and zeros [17.9]. Pole–zero information is converted to biquad coefficients, and these coefficients are then processed further by other builder routines in exactly the same manner as any other set of analog biquad coefficients. Processing of biquad coefficient sets is described in a later section.

The standard filters have normalized frequency scales. Butterworth and Bessel filters are normalized to the 3-dB bandwidth and Chebyshev filters are normalized to the ripple bandwidth. They all have to be rescaled to fit the simulation in which they are to be used. Frequency rescaling is a builder task to be applied to nearly all filters, not just standard filters. As such, it will be described separately in the sequel.

Nyquist Filters

Ideal Nyquist filters have a vital role in establishing the theoretical capabilities of a communications system. Every communications simulation library should include a

Nyquist filter characteristic. The most popular Nyquist filter derives from the cosine rolloff Nyquist spectrum of (2.8), which describes a Nyquist pulse with Fourier transform $G(f)$. That pulse is formed by the overall filtering of the entire communications link. To achieve matched filtering in the system, the overall filtering has to be distributed equally between transmitter and receiver (see Sec. 4.3). Therefore, the simulation library would contain an expression for a root-Nyquist, amplitude-normalized, cosine-rolloff filter with frequency response:

$$H_a(f) = \begin{cases} \sqrt{G(0)/T} = 1, & |fT| \leq \frac{1-\alpha}{2} \\ \cos\left[\frac{\pi}{4\alpha}(|2fT|-1+\alpha)\right], & \frac{1-\alpha}{2} < |fT| < \frac{1+\alpha}{2} \\ 0, & |fT| \geq \frac{1+\alpha}{2} \end{cases} \quad (17.8)$$

where T is the symbol interval.

17.5.3 Frequency-Domain Builder Routines

In building a frequency-domain filter, think of the device data as constituting a frequency-response function $H_a(f)$. The building process samples the data at frequencies $f_{k'}$ and stores $H_d(k') = H_a(f_{k'})$ in the process-parameter file. The main builder tasks are: (1) determine the frequencies $f_{k'}$, and (2) evaluate $H_a(f_{k'})$ corresponding to each value of k'. Frequency determination tasks will be described first, followed by two kinds of function evaluations: interpolation of tabular data and operations on analytical formulas.

Frequency Determination Tasks

The following frequency tasks have been set forth in Sec. 6.2.1.

- Lowpass-to-bandpass (LP-BP) transformation (optional)
- Frequency translation for real-to-complex conversion (only for complex filters)
- Frequency rescaling (required)

These operations were shown in Sec. 6.2.1 to easily be combined so that only one evaluation of frequency response need be performed for each point of $H_d(k')$, according to the formula

$$H_d(k') = H_{a(+)}\left[r_p\left(\frac{\gamma k' M_s}{NT} + f_0 - \frac{f_p^2}{\gamma k' M_s/NT + f_0}\right)\right] \quad (17.9)$$

where the subscript (+) indicates setting $H_a(f) = 0$ for $k'M_s/NT < -f_0/\gamma$ if a complex conversion is included. The sample points for the process parameter array are computed as $H_d(k') = H_{a(+)}(f_{k'})$, with

$$f_{k'} = r_p \left(\frac{\gamma k' M_s}{NT} + f_0 - \frac{f_p^2}{\gamma k' M_s/NT + f_0} \right) \quad (17.10)$$

User-Specified Parameters. Parameter r_p is the bandwidth dilation ratio, f_p the center frequency of the bandpass filter resulting from a LP–BP transformation, and f_0 the translation frequency for real-to-complex conversion. These two operations are optional; they are omitted by setting $r_p = 1$, $f_p = 0$, and $f_0 = 0$. See Sec. 6.2.1 for complete details.

The quantities M_s and N are simulation parameters; they are the number of samples per symbol and the number of elements per block, respectively. These parameters must agree with the same parameters for the signal in the simulation model. If either M_s or N is altered, a new process-parameter file for $H_d(k')$ must be built, incorporating the altered simulation parameter.

The quantity $1/T$ is the symbol rate in the user's frequency scale. This quantity can have any positive value whatsoever ascribed to it; the assignment is a completely free choice. Whatever value of $1/T$ is chosen will be compensated by the concomitant choice of γ that converts from the frequency scale of the data to the frequency scale of the user. Here are two ways (others exist also*) to choose γ:

1. In the frequency scale of the user: What frequency dilation γ has to be applied to $H_a(f)$ to make it a satisfactory filter for the user's ascribed symbol rate $1/T$?

2. In the frequency scale of the data: What fictitious symbol rate $1/T' = \gamma/T$ would be suitable for the filter $H_a(f)$?

In the first way, the user might set the symbol rate to $1/T = 1$, so that all frequencies can be regarded as normalized to the symbol rate. Then the user inspects the filter data to decide how much the bandwidth should shrink ($\gamma > 1$) or expand ($\gamma < 1$) to make the filter suitable for a data rate of 1 symbol per unit time. In the second way, the user might set $\gamma = 1$ and then specify a fictitious symbol rate $1/T'$ that is appropriate for the undilated filter.

Both ways lead to the same ultimate results; the choice between them is a matter of personal taste. If the builder routine provides for independent specification of γ and $1/T$, you can select whichever way you prefer. No matter how you proceed, the value chosen for γ (or $1/T'$) is a filter-design problem that you have to determine outside the simulation.

---

*These simple ways become more convoluted if the LP–BP transformation is applied, with its nonlinear rescaling of the frequency axis.

Block Indexing. It is often convenient to picture the block index k' as running from $-N/2$ to $N/2 - 1$ for N even, or from $-(N - 1)/2$ to $(N - 1)/2$ for N odd. This is intuitively appealing for real filters or for complex-envelope filters, since it places zero frequency ($k' = 0$) at the center of the block. In reality, the block actually might be stored with index k running from 0 to $N-1$, to conform to conventions on the DFT. From 0 to IP$[(N - 1)/2]$ the labeling of k and k' are the same, but from IP$[(N + 1)/2]$ to $N - 1$, the k-packed block contains the negative-index elements of the k'-packed block with index range IP$[-(N - 1)/2]$ to -1. (*Notation:* IP$[x]$ means the integer part of x.) One block is a simple circular shift of the other. See Sec. 5.2.3 for an explanation of different indexing schemes.

Although the digital block $H_d(k')$ is regarded as periodic, the original data function $H_a(f)$ is aperiodic. Therefore, the range of k' in (17.10) cannot be regarded as periodic. A different choice of range for k' would amount to a linear translation along the frequency axis of $H_a(f)$, equivalent to a different specification of f_0.

Frequency Response Interpolation

Interpolation in frequency-response tables is another task encountered in building some frequency-domain filters. Interpolation is one kind of evaluation of the function $H_a(f_{k'})$. Assume that the tables have been conditioned as described in Sec. 17.2.2 so that their records consist of data triples of dimensionless frequency, gain amplitude in dB, and phase in radians. Assume further that conjugate response at negative frequencies has been filled in if the filter is to be real, that dc response has been (or can be) filled in if it is needed, and that any transmission nulls have been identified and specified.

Let there be L data points in the device-data table and N points in the process-parameter file. Typical values for L might range from as small as 10 to as many as 200. The number of parameter-file points is likely to be an integer power of 2, to take maximum advantage of FFT efficiencies. Values of N from 128 to 8192 are likely to be encountered; lesser or greater block sizes are more unusual. Represent the device-data set as $\{H_a(f_i)\}$; the interpolator is required to find $H_a(f_{k'})$, where $f_{k'}$ is not generally one of the frequencies f_i in the device-data set. Methods for determining the frequencies $f_{k'}$ have been presented above.

Amplitude Interpolation. We have experimented with cubic polynomial interpolation—a conventional interpolating polynomial from the textbooks. Results of the experiments have demonstrated forcefully that low-order polynomials work poorly on frequency responses that are actually rational functions of frequency. Polynomial interpolation works equally badly on the rectangular components of $H_a(f_i)$ or on polar components expressed as amplitude and phase. (There is one exception. The rectangular components interpolate very nicely in the vicinity of a null; more on this later.)

You might interpolate with rational functions instead of polynomials, as one way to circumvent the difficulty. Information on rational-function interpolation may be found in [17.7, Sec. 3.2], [17.13], and [17.14], among other sources. The impression gained from the references is that interpolation with rational functions would be burdensome to a user, so we have not pursued the matter. Further investigation might be warranted.

What we have found experimentally is that reasonable polynomial interpolation is possible for the logarithmic amplitude—that is, the gain expressed in dB. Although power gain $|H_a(f_i)|^2$ is a rational function of frequency, the logarithm of gain is not rational. The features that cause interpolation troubles with $|H_a(f_i)|$ are softened by taking the logarithm.

As an example, amplitude of a filter response typically rolls off proportionally to $1/f^m$, where m is an integer. A polynomial is hard-pressed to fit this shape; the problem comes about because $1/f^m$ cannot be approximated in a power series that can be truncated to a finite polynomial. But after conversion to dB, the amplitude is $-20m\log(f)$, which does have a power-series expansion. (No power-series expansion is needed to set up or perform an interpolation; the example only serves to illustrate the nature of the problem.)

Phase Interpolation. Polynomial interpolation of phase is better behaved than that of amplitude, except for one practical snag. Measured phase data are most often reported modulo-2π, typically in the range $-\pi$ to π. If $\phi_i = \pi - \epsilon_-$ and $\phi_{i+1} = \pi + \epsilon_+$, where ϵ_- and ϵ_+ are small positive angles, the measurement for ϕ_{i+1} would be reported as $-\pi + \epsilon_+$ which shows up in the record as an apparent phase discontinuity of almost 2π.

Never interpolate through a discontinuity; the interpolated function will be seriously incorrect in its vicinity. Something has to be done to avoid trouble from this basically fictitious discontinuity. One solution is to interpolate $\sin\phi_i$ and $\cos\phi_i$, instead of ϕ_i itself. Sine and cosine are periodic; they have the correct values no matter how the angle is expressed. That is, $\sin(\pi + \epsilon) = \sin(\pi + \epsilon)$, and similarly for cosine. Interpolation of both sine and cosine—not just one alone—is necessary to retain quadrant information.

It will turn out that both sine and cosine are needed to form the process parameter file, which has to be in rectangular coordinates. The operation immediately following interpolation will be to convert the interpolated value to rectangular coordinates.

Nulls. Interpolation of dB gain and of $\cos\phi_i$ and $\sin\phi_i$ would be disrupted severely by nulls in the frequency response. The logarithm of gain has a singularity at a null, and the sine and cosine will be discontinuous if the null has odd multiplicity. These functions should not be interpolated across a null; a different interpolation scheme is needed.

One solution in the near vicinity of a null is to interpolate the two rectangular components of $H_a(f_i)$. Although dB gain and possibly phase are ill behaved at a null, the rectangular components are continuous and well approximated by a low-order polynomial for nulls of reasonable multiplicity.

The strategy adopted for STÆDT is to interpolate the rectangular components in any basepoint set whose span contains an identified null, while interpolating dB gain and sine and cosine in basepoint sets whose spans do not contain a null. This requirement for differing interpolation schemes in different parts of the table favors interpolators with comparatively short basepoint sets. STÆDT employs a conventional cubic interpolator with just four points in each basepoint set.

Formula Evaluation

If $H_a(f)$ is expressed as an analytical formula, the digital filter coefficients $H_d(k')$ are found by substituting N values of $f_{k'}$ into the formula and performing the necessary calculations. A different builder routine is needed for each different kind of formula to be accommodated. This account has presented two different examples of formulas: root-Nyquist, cosine-rolloff forms, or partitioned biquads.

Cosine-Rolloff Formula. The cosine-rolloff characteristic $H_a(f) = \sqrt{G(f)/T}$ was given in (17.8). The user specifies the number of sample points per symbol M_s, the number of sample points per data block N, and the excess-bandwidth factor α. The Nyquist builder then computes the digital filter coefficients $H_d(k') = H_a(f)$ at $fT = k'M_s/N$ for $k' = -N/2$ to $N/2 - 1$. Notice that the frequency in $H_a(f)$ is already normalized to the symbol interval T, so frequency rescaling is not needed in this instance. Furthermore, the lowpass-to-bandpass transformation would never be applied, nor is there any need for a real-to-complex conversion. If a root-Nyquist filter approximation is needed in the time domain, obtain the filter's impulse response from the frequency-domain version via the DFT and windowing.

Partitioned Biquads. A device-data file for partitioned biquads (see Sec. 3.1.2) contains information on L sections, each with transfer functions of the form

$$H_{al}(s) = H_0 \frac{a_2 s^2 + a_1 s + a_0}{b_2 s^2 + b_1 s + b_0} \qquad (17.11)$$

where l is an index on the partitioned sections. Device data consist solely of the values of the coefficients and of L; it is the builder routine that has knowledge of the formula (17.11).

Each biquad is evaluated by the builder at the N frequencies $f_{k'}$ specified in (17.10) to produce

$$H_{al}(f_{k'}) = H_{al}(s)|_{s=j2\pi f_{k'}} \qquad (17.12)$$

Then the process-parameter file is calculated as the product of all the individual biquadratic frequency responses in cascade:

$$H_d(k') = \prod_{l=1}^{L} H_{al}(f_{k'}) \qquad (17.13)$$

Frequency Limits

The device-data filter $H_a(f)$ is specified over a frequency span from f_1 to f_2 and is not specified outside that span. If $H_a(f)$ is an analytical formula, either or both limits can be infinite. If $H_a(f)$ consists of tabular data, both limits are finite. The digital filter $H_d(k')$ always has a finite frequency span. Let the limits of that span be denoted f_3 and f_4, referred to the same frequency scale as f_1 and f_2.

Rarely will the two sets of limits coincide; in general, $f_1 \neq f_3$ and $f_2 \neq f_4$. There will be regions of one span that are not included in the other. Provisions have to be made for the regions of nonoverlap. Two rules help dispose of the out-of-limits portions:

1. Any portions of $H_a(f)$ that extend below f_3 or above f_4 are truncated.
2. Any elements of $H_d(k')$ that correspond to unspecified data frequencies below f_1 or above f_2 are set to zero.

The rules will be carried out automatically if the builder initially sets all elements of $H_d(k')$ to zero and if the builder skips any values of k' corresponding to data frequencies outside (f_1, f_2). There is no need to evaluate f_3 and f_4.

These rules work reasonably well for lowpass and bandpass filters but could give anomalous results in some instances. For example, suppose that the filter is highpass, and the digital-filter upper-frequency bound f_4 exceeds the data-frequency upper bound f_2. Then the rules would set the digital filter response to zero for all frequencies above f_2—a nonsensical outcome for a highpass filter. The user should extend the data set by extrapolation to prevent such discordances.

Check Display

After a parameter file $H_d(k')$ has been built, it is prudent to display a plot of its response characteristics. Mistakes happen easily; this is a good place to intercept them. The user should verify that the desired scaling was achieved, that truncation of the filter response occurs sufficiently far down on the filter skirts, and that the filter passband was actually included in the span of the constructed filter.

17.5.4 Time-Domain Builder Routines

For the builders considered here, device data for a time-domain filter consist of the coefficients of partitioned analog biquads in the form of (17.11). The builders convert and transform the analog functions $H_a(s)$ to yield coefficients of digital biquads in the form

$$H_d(z) = \frac{A_0 + A_1 z^{-1} + A_2 z^{-2}}{1 + B_1 z^{-1} + B_2 z^{-2}} \tag{17.14}$$

In this section we examine the different builders needed to obtain the coefficients of $H_d(z)$.

Builder Flow Sequence

Refer to Fig. 17.4, a flow diagram for the time-domain builders. The builder operations to be discussed are (1) a lowpass-to-bandpass (LP–BP) transformation, (2) a real-to-complex conversion, and (3) an analog-to-digital (A/D) transformation. Builder operations have to be performed in the order shown in the figure, although two of the three operations are optional. Optional builders are bypassed when not needed.

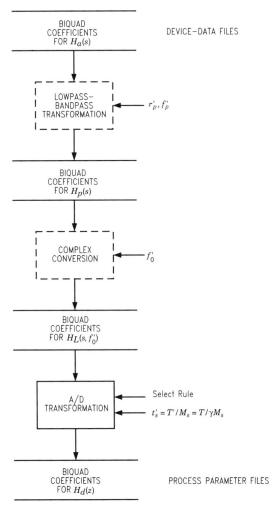

FIGURE 17.4 Time-domain filter builders.

Each builder produces coefficients of a modified transfer function. Only the last operation—the A/D transformation—produces a process parameter file for a digital filter. The intermediate steps generate biquad coefficients for converted or transformed analog filters; these intermediate coefficients can be used only as inputs to subsequent builder routines and not as process parameters. Analytical bases for each of the builder functions have been presented in Chapters 3 and 6. The calculations needed to implement the analytical bases are embedded inside the builder routines themselves and are of concern only to the programmer. The following sections are devoted to a user's view of the builders.

Frequency-domain builders of Sec. 17.5.3 were all grouped together; the various builder operations were activated simply by the manner of assigning frequency-determination parameters. The digital frequency response $H_d(k')$ was calculated directly from the analog frequency response $H_a(f)$, with no intermediate transfer functions needed. By contrast, the time-domain builders are all separate; they cannot be combined readily. Each time-domain builder produces a new set of biquad coefficients. All sets must be computed explicitly; all but the last are intermediates.

Time and Frequency Scales

The filters $H_a(s)$ contained in the device-data files include standard library filters furnished with the program, or biquad coefficient sets entered by an originator. Origins of device data were described earlier in this chapter. Device-data biquads are each represented in their own unique time and frequency scales. Only in exceptional circumstances will the device-data scale coincide with the user's scale. A rescaling will be needed between the two scales. Either of two rescaling plans could be followed:

1 Immediately rescale the device data to the user's scale by the relation $H_u(s) = H_a(\gamma s)$, where γ is the same rescaling factor introduced for frequency-domain builders.

2 Defer the rescaling to the A/D builder, where it can be incorporated without performing an extra rescaling step.

Lowpass-to-Bandpass Transformation

No time-domain LP–BP transformation is included in STÆDT. If it is needed, the transformation could be performed externally to the program and the transformed coefficients entered into a device-data file.

Real-to-Complex Conversion

The real transfer function $H_a(s)$ is to be converted to a complex transfer function $H_L(s, f'_0)$ as explained in Sec. 3.3.2 and Appendix 3A. Conversion is optional and so is shown in dashed outline in Fig. 17.4. Coefficients of the converted filter depend on the user-specified translation frequency f_0 (in the user's frequency scale) or $f'_0 = \gamma f_0$ (in the frequency scale of the device data). A different specification of f_0 leads to an entirely different set of coefficient values. In most instances, f_0 (or f'_0) will be specified equal to the carrier frequency of the signal to be filtered.

The transfer function $H_L(s, f'_0)$ consists of biquads with complex coefficients. These coefficients are written to an intermediate file, to be delivered subsequently to the A/D transformation.

A/D Transformation

The A/D builder accomplishes two actions in combination: (1) it optionally rescales the time and frequency scale, and (2) it calculates the biquad coefficients for $H_d(z)$ according to a particular A/D rule.

A/D Rules. Section 6.2.2 described four different rules that might be used for the A/D transformation:

1. Impulse-invariant transformation
2. Step-invariant transformation
3. Matched-z transformation
4. Bilinear transformation

STÆDT offers all four transformations; the user selects whichever is best for the application at hand. Other programs might offer fewer choices.

Certain restrictions must be obeyed when applying the transformation. The impulse-invariant transformation can only be used on filters for which all biquadratic or bilinear sections are proper; that is, the degree of the denominator must exceed the degree of the numerator. It is the user's responsibility to avoid selecting the impulse-invariant transformation if any section is improper.

Aliasing has to be guarded against in all but the bilinear transformation. Denote the imaginary part of a pole location (also, zero location for the matched-z transformation) as β. Then the restriction $\beta t_s < \pi$ must be observed, or else the transformed pole (zero) will be aliased. (The sampling interval t_s is in the same frequency scale as β.)

Time and Frequency Rescaling. The different transformation rules each incorporate sampling interval t_s in their defining formulas. A value of t_s (or $t'_s = t_s/\gamma$) has to be specified to permit the transformation coefficients to be calculated. Instead of the sampling interval t_s, it may be more convenient to specify the sampling rate $1/t_s$. Or, since a simulationist is likely to think in terms of symbol rates instead of sample rates, the ratio $M_s/T = 1/t_s$ could be used instead; this last option is the one offered in STÆDT.

To convert to the user's time or frequency scale from that of the device data, use the formula

$$1/t'_s = \gamma M_s/T \tag{17.15}$$

494 ▢ ENTRY AND BUILDER ROUTINES

The parameters γ, M_s, $1/T$, and $1/T' = \gamma/T$ are exactly the same parameters that arose for frequency-domain rescaling, as explained in Sec. 17.5.3. If the time-domain builder calls for γ, M_s, and $1/T$, the same ways of specifying these quantities will apply in both domains.

17.5.5 Specimens of Frequency Rescaling

Rescaling can be applied to a frequency-domain file or to a biquad file. Each is treated in its own appropriate manner, as described below. Actual calculations are performed by the builder command, not the user. A frequency-domain device-data file contains multiple data records, each record consisting of a measurement frequency f_i, a dB gain, and a phase angle. When rescaling, the gain and phase are left unchanged, while the frequency f_i is replaced by γf_i.

A partitioned biquad file consists of values for coefficients for each section in the form of (17.11). Coefficients will be scaled as follows:

$$\begin{aligned} H_0 &\to H_0 \\ a_2 &\to \gamma^2 a_2 & b_2 &\to \gamma^2 b_2 \\ a_1 &\to \gamma a_1 & b_1 &\to \gamma b_1 \\ a_0 &\to a_0 & b_0 &\to b_0 \end{aligned}$$

As an example, consider the simple one-pole section

$$H_a(s) = \frac{\alpha}{s + \alpha}$$

which has its 3-dB frequency at $f = \alpha/2\pi$. Scaling by γ results in

$$H_a(\gamma s) = \frac{\alpha}{\gamma s + \alpha}$$

which has its 3-dB frequency at $f = \alpha/2\pi\gamma$.

For another example, consider the two-pole resonant tank with transfer function

$$H_a(s) = \frac{s\omega_r/Q}{s^2 + s\omega_r/Q + \omega_r^2}$$

which has peak gain of unity at the resonant frequency $f_r = \omega_r/2\pi$ and a 3-dB band width of f_r/Q. Frequency scaling this function by γ gives

$$H_a(\gamma s) = \frac{\gamma s \omega_r/Q}{\gamma^2 s^2 + \gamma s \omega_r/Q + \omega_r^2}$$

$$= \frac{s\omega_r/\gamma Q}{s^2 + s\omega_r/\gamma Q + \omega_r^2/\gamma^2}$$

which has its resonant frequency at f_r/γ, 3-dB bandwidth of $f_r/\gamma Q$, and unchanged values for Q and for peak gain.

17.6 KEY POINTS

Some simulation modules require large numbers of *process parameters* to describe their characteristics. These parameters are usually stored in files and retrieved by file name when a user is preparing a simulation. Before the process parameter files can be called up, they have to be constructed from *device data* that have been entered into other files. To enter the data into device-data files, an *originator* needs assistance from *entry routines*, which might be nothing more than spreadsheets or text editors.

Various *conditioning* operations have to be performed upon device data files to bring them into standard formats compatible with subsequent procedures. Conditioned data are stored in modified device-data files to await being called up for a simulation. Device-data files selected for use in a simulation are converted and transformed by *builder routines* to produce process-parameter files tailored to the particular simulation module being constructed. A user specifies builder parameters and simulation-model parameters to be used in the builder operations. A device-data file may be adapted to a wide range of simulation parameters, but a process-parameter file is highly specific to its particular simulation. If you change any builder or simulation parameter, a process-parameter file must be rebuilt; it cannot be reused in a different simulation, barring exceptional circumstances.

17.6.1 Nonlinearity Builders

Many different kinds of nonlinearities might arise in simulations. Two examples were presented of builders for saturating amplifiers, a narrowly restricted but prevalent class. In one example the builder fits a rational function to experimental data, and the signal process evaluates the rational function for each signal sample.

In another example, measured data are interpolated by the signal process for each signal sample. To reduce spectral artifacts caused by discontinuous derivatives, it is advantageous to use spline interpolation instead of conventional interpolating polynomials. A spline builder is then needed to augment the device data prior to interpolation.

17.6.2 Filter Builders

The numerous filter-building tasks can be categorized according to:

- Frequency domain versus time domain
- Tabular data versus analytical formulas
- Internal origin versus external origin

Frequency rescaling can be incorporated into any filter builder.

Frequency-Domain Builders

Device data for a frequency-domain filter will be in the form of a frequency response, denoted $H_a(f)$. Tabular device data are specified at discrete frequencies $f = f_i$, while analytical formulas are specified at continuous frequencies. Building a frequency-domain digital filter $H_d(k)$ entails sampling the analog frequency response $H_a(f)$ at selected frequencies f_k to establish $H_d(k) = H_a(f_k)$. Two distinct builder tasks are involved: (1) determine the sampling frequencies f_k, and (2) evaluate $H_a(f)$ at $f = f_k$.

Lowpass-to-bandpass transformation, real-to-complex conversion, and frequency rescaling all enter into determining the sampling frequencies. (The first two items are optional, depending on the nature of the filter to be built.) All three would be combined together into a single builder and exercised by appropriate specification of builder parameters.

The method for evaluating $H_a(f_k)$ depends on its format. If $H_a(f)$ [or $H_a(s)$] is an analytical formula, the evaluator builder will be a routine that has knowledge of the formula. A different routine is needed for each formula to be included in the library. If $H_a(f)$ consists of tabular data, the evaluator-builder interpolates in the table. Guidelines for dependable interpolation are given in the text and are written into the code of an interpolating builder.

If the device data are provided as partitioned biquads, a separate digital filter will be built for each individual biquadratic section. To reduce the computing burden placed on the simulation processes, concatenate the sections within the builder to produce a single frequency response $H_d(k)$, consisting of all of the sections in cascade.

Time-Domain Builders

Only partitioned-biquad device data were examined for time-domain filter building. The same tasks encountered for frequency-domain builders—lowpass-to-bandpass transformation, real-to-complex conversion, frequency scaling, and A/D transformation—are to be accomplished. However, for time-domain builders these are all separate activities (with the exception of frequency rescaling). They are performed in separate builder routines and each generates its own distinctive output file. These actions cannot be combined together when building time-domain filters.

Standard Filters

Certain filter varieties are so well known and widely used that their characteristics are often included in a program's library as device-data files. Standard filters include Butterworth, Chebyshev, Bessel, and elliptic characteristics, which are specified by rational functions (partitioned biquads), and root-Nyquist, cosine-rolloff filters, which are specified by frequency-domain formula.

The cosine-rolloff formulas are delivered directly to a frequency-domain builder that constructs the digital frequency response $H_d(k)$. The formulas (or tabulated pole–zero values) for a selected rational filter are contained in a routine specialized for the filter type, a routine that generates coefficients of partitioned biquads for the regular time- or frequency-domain builders used with partitioned biquads.

Although Butterworth, Chebyshev, Bessel, and elliptic filters are often employed, they are not especially good choices for data-communications systems. Better filters can be devised.

17.6.3 STÆDT Processes

Conditioning Builders

BQCV: biquad real-to-complex conversion

BQST: biquad coefficients for standard filters; Butterworth, Chebychev, and Bessel

FRPC: frequency-response preconditioning

NLCD: nonlinear amplifier conditioner

Parameter-File Builders

BQAD: biquad A–D transformation (parameters for IIR time-domain filter)

BQFR: biquad A–D transformation (parameters for frequency-domain filter)

FRNP: frequency response interpolation

FRNQ: coefficients for root-Nyquist frequency response

LNCFLTR: prepare filter for linear convolution via FFT and overlap-and-add

SALH: curve fit to Saleh model

SPLN: prepare spline table

WNDW: window

18

GRAPHICAL DISPLAYS

Results of simulations generally have to be graphed if they are to be interpreted readily. Graphics details vary greatly among simulation programs. This chapter is devoted to the principles of displays and forgoes the details.

Simulation displays resemble the displays obtained from laboratory instruments, such as oscilloscopes, spectrum analyzers, and network analyzers. Displays are produced for time-domain waveforms, frequency-domain Fourier transforms of waveforms, time-domain eye patterns, frequency-domain spectra, time-domain impulse or step responses of filters, frequency-domain frequency responses, signal-state diagrams (Q versus I plots), constellation plots (strobed Q versus I plots), error-ratio plots versus SNR, and histograms. These are all discussed in the following pages.

Displays are commonly employed in either of two ways, at the discretion of the user:

1 Displays are generated as part of a main simulation run.
2 Data from the main simulation are collected and stored. Then displays are produced from the stored data as a post-run activity.

Although displays may seem to be a rather mundane topic, a number of important modeling issues do arise and are treated in the text.

18.1 TYPES OF PLOTS

Example displays are presented in the paragraphs and figures that follow. The examples illustrate the variety of plots that arise in a simulation and provide a concrete basis for the discussions. Most simulation programs can deliver the kinds of plots

TYPES OF PLOTS □ **499**

shown here. Data for these particular examples were generated in STÆDT, and most plots were produced with a commercial graphics program.

18.1.1 Time-Domain Plots

Signal Waveforms

Figure 18.1 shows a plot of a signal waveform, appearing much as if it were captured on a single-sweep oscilloscope. Several features are worth noting:

- Signals can be complex; a user selects either the real or imaginary part for plotting, or plots both parts.
- The figure consists of just one part of a much longer record. Signal records typically are quite long, and it is commonly necessary to display just a short portion to be able to see the details.
- It is difficult or impossible to determine the quality of the signal waveform (e.g., the amount of intersymbol interference) from an ordinary waveform plot, particularly if the signal has narrow bandwidth.
- The signal waveform is a one-shot display; there is no need to synchronize the display to the signal.
- All data consist of discrete points. The graphing routines connect adjacent

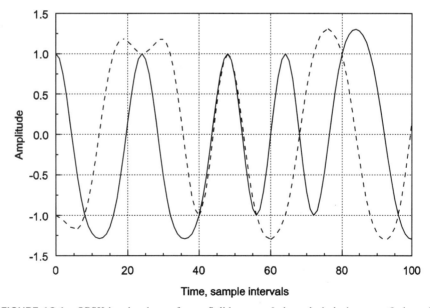

FIGURE 18.1 QPSK baseband waveforms. Solid curves, *I* channel; dashed curves, *Q* channel.

Filter Responses

The time-domain response of a filter—typically, response to a unit sample (the "impulse" response) or to a unit step—is often needed. Examples of impulse and step responses of the same filter are shown in Figs. 18.2 and 18.3. The examples have real time-domain responses, but a complex filter can have a complex response in the time domain. In that event, both the real and imaginary parts of the response would have to be plotted.

If the filter is simulated in the time domain (see Secs. 6.1.2 and 6.1.3), its impulse response is obtained by driving its input with a unit sample, and its step response is obtained by driving it with a unit step function. If the filter is simulated in the frequency domain, the impulse response $h(n)$ is simply the inverse Fourier transform of the frequency response $H(k)$.

To obtain the step response of a frequency-domain filter $H(k)$, first generate a time-domain unit step $u(n)$ and then Fourier transform it to $U(k)$ in the frequency domain. Next, form the product $U(k)H(k)$ and subject that to the inverse Fourier transform to get the time-domain step response. (In reality, the frequency-domain operations impose a periodicity upon the time-domain waveform, which therefore must be a rectangular pulse rather than a step of semi-infinite duration. The width of the pulse should exceed the duration of the filter's transient response.)

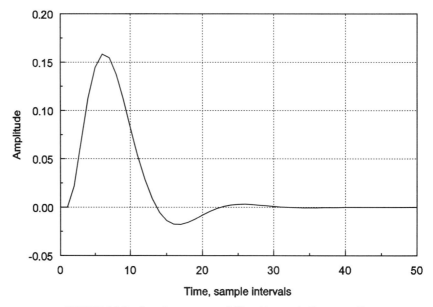

FIGURE 18.2 Impulse response of filter (three-pole Butterworth).

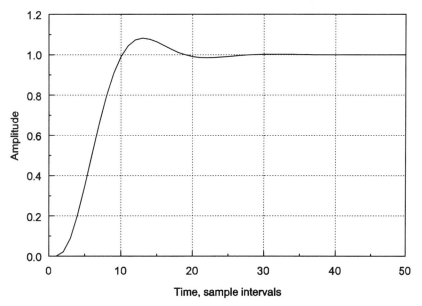

FIGURE 18.3 Step response of filter (same filter as in Fig. 18.2).

A plot of time-domain response is useful for determining rise times, fall times, ringing, and so on. Also, the tails of the impulse response can be examined for Nyquist behavior—nulls spaced by the symbol interval. Impulse response for an infinite impulse response (IIR) filter has to be truncated to a finite impulse response (FIR). A plot of the impulse response provides qualitative guidance for deciding upon a truncation point. Inspection of a printout of the impulse response or a measurement of truncated energy may be needed to come to a better decision.

Eye Patterns

Some of the limitations of signal-waveform displays can be overcome by cutting up the long waveform into an eye pattern. An eye display consists of a large number of short segments of the signal waveform overlaid upon one another. Figure 18.4 is an eye pattern of the real part of the same signal shown earlier in Fig. 18.1.

Separate eyes are needed for the real and imaginary parts of a complex signal. Eye patterns are meaningful only if the I and Q channels are independent. (Figure 18.4 was obtained from a screen dump since the STÆDT command for eye diagrams does not have provision for export of eye-formatted data.)

All signal segments in an eye have the same duration, typically one to two symbol intervals. Each segment must be synchronized to the signal pulses; it is convenient if the center of each segment coincides with the symbol strobe instant. If the segment length exceeds one symbol interval (as in Fig. 18.4), adjoining segments overlap. That is, the waveform toward the right portion of one segment will be repeated in the waveform toward the left portion of the next segment.

502 □ GRAPHICAL DISPLAYS

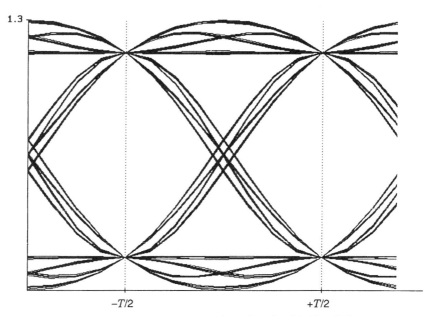

FIGURE 18.4 Eye pattern. (*I* channel of signal in Fig. 18.1).

An eye pattern is generated on an oscilloscope by synchronizing the scope to the symbol clock, adjusting the sweep width to approximately one or two symbol intervals, and displaying the repeated overlaid sweeps of the signal waveform. A simulated eye is generated by the same operations in software.

An eye pattern permits a large number of symbol intervals to be displayed in great detail because of the overlaying of multiple segments. The eye opening gives an experienced observer numerous clues to the quality of the waveform. Intersymbol interference shows up (in an otherwise noiseless display) as a spread of signal values at the strobe instant. A need for critical strobe-timing adjustment shows up as a narrow horizontal opening of the eye. A need for enhanced dynamic range of a transmitter shows up as peaking of the eye pattern away from the maximum eye opening. These features are illustrated in Fig. 18.4.

18.1.2 Frequency-Domain Plots

Signals in the Frequency Domain

A waveform in the time domain and its Fourier transform in the frequency domain contain the same information, but different aspects of the signal are emphasized in the two domains. The Fourier transforms of most signals, real or complex, will be complex. Both the real and imaginary parts (or magnitude and phase) have to be plotted for a complete presentation. Frequency-domain representations of signals are extremely valuable for computation purposes, but the power spectrum usually is the frequency-domain aspect of greatest interest for display. The power spectrum is considered separately below.

Frequency Response of Filters

Filters commonly are characterized by their frequency responses (Sec. 6.1.1), even when simulated in the time domain. The frequency response of any nontrivial realizable filter is complex even if the time response is real. A display could show the real and imaginary parts of the response (plotted versus frequency), but magnitude and phase displays are more customary. Figure 18.5 shows an example of the latter.

Internally to the computer program, the frequency-domain filter information is most likely to be kept in rectangular coordinates (see Sec. 6.1.1), so conversion to polar coordinates is needed for the display. Some programs (including STÆDT) incorporate the conversion routines within the display commands, but others require that the conversion be performed separately before the data are presented to the display commands.

Magnitude might be plotted on a linear ordinate (as delivered from a rectangular-to-polar conversion), or it might be plotted on a logarithmic ordinate (decibels). The latter requires a logarithmic conversion after the polar conversion. In some applications of filters, it is commonplace to display a logarithmic frequency abscissa, but communications applications are usually best served with a linear abscissa.

Phase is customarily plotted in degrees, although the computer internally works in radians. Also, the polar conversion delivers phase in the range $(-\pi, \pi]$; phase values outside this range are wrapped into it. Therefore, a plot of phase versus frequency is

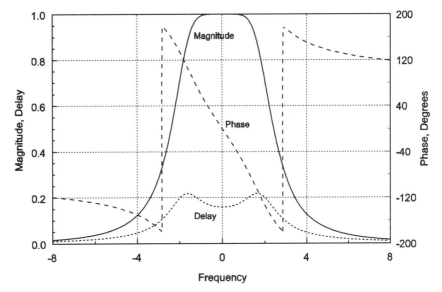

FIGURE 18.5 Frequency response of filter (same filter as in Figs. 18.2 and 18.3). Frequency and delay scales normalized to 3-dB bandwidth.

likely to have display discontinuities induced by the polar conversion, as exemplified by Fig. 18.5. A realizable filter has true phase discontinuities (as opposed to display discontinuities) only at frequencies of transmission nulls.

Sometimes a display of delay (derivative of phase with respect to frequency) is preferred to a phase display; a delay plot has been included in Fig. 18.5. A model for computing delay from phase is described in Sec. 18.3.1.

Power Spectrum

Represent a signal block in the frequency domain as $\{Z(k)\} = \{X(k) + jY(k)\}$. One definition of power spectrum is simply the magnitude squared $\{|Z(k)|^2\} = \{X^2(k)+Y^2(k)\}$. This is the *periodogram*—an unsophisticated representation, but easy to produce. There is a vast literature on more-effective spectral methods [18.1, 18.2], but they are not considered in this book. In Secs. 18.2.2 and 18.3.2 we delve into some features of periodogram displays.

The power spectrum is always real (and nonnegative), so only the magnitude squared is to be plotted. It may be displayed on a linear ordinate (ordinate proportional to power or energy) or on a decibel scale such as in Fig. 18.6.

18.1.3 Other Plots

Signal Space Diagrams

Consider a time-domain waveform $z(n) = x(n) + jy(n)$ to be a vector whose amplitude and angle in x–y space evolve along the third axis of time index n. A waveform

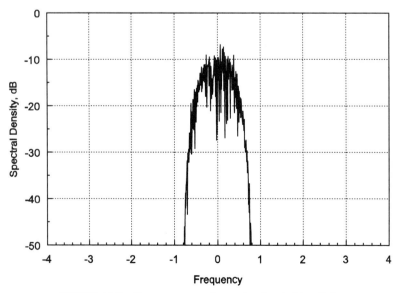

FIGURE 18.6 Power spectrum (same signal as in Fig. 18.1).

display is simply a plot of the projections of $z(n)$ onto perpendicular planes that intersect in the time axis. Now consider a different view of the same signal, sighting along the n axis instead of perpendicular to it. A *signal space diagram* [18.3] is a plot of the signal vector projected onto a plane that is normal to the time axis; it is a plot of y versus x or, equivalently, Q versus I. An example for a QPSK signal is shown in Fig. 18.7.

Space diagrams are most meaningful for two-dimensional (complex) signals. (A one-dimensional space diagram is only a straight line.) Space diagrams are useful for visualizing the behavior of complicated modulation schemes and for observing the effects of various types of distortion. They are generated on an oscilloscope by connecting the I channel of the signal to the horizontal input of the scope and the Q channel to the vertical input. They are displayed in a simulation program by plotting the imaginary component of a signal block against the real component.

In a space diagram for a simulated analog signal, there have to be enough sample points to define the waveforms adequately; the plotting will connect adjacent sample points to give a continuous display. As an alternative option, the display can take just the symbol-strobe points as data, one complex point per symbol, and plot those points without connecting lines. The result is a signal constellation plot. An example is shown in Fig. 18.8 for the same QPSK signal that was shown in Fig. 18.7. The spread of points and any other distortions of the constellation are indicators of the quality of signal processing and transmission. (The QPSK signal from which Fig. 18.8 was produced was perfect Nyquist, so all samples fall exactly at the four ideal constellation points $\pm 1 \pm j1$.)

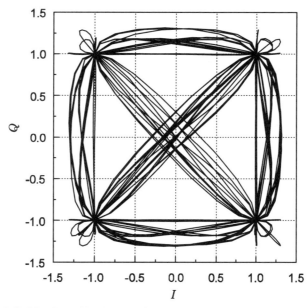

FIGURE 18.7 Signal space diagram (same signal as in Fig. 18.1).

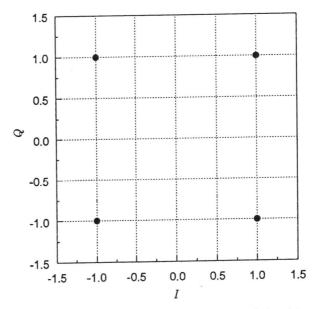

FIGURE 18.8 Constellation plot (strobed version of Fig. 18.7).

Error-Ratio Plots

Communications systems are evaluated in terms of their error performance—the likelihood of committing an error in transmission of information. The performance is typically plotted in a graph of either *error ratio* (a measured frequency of occurrence of errors) or *error probability* (a statistical prediction of errors) versus signal-to-noise ratio (SNR). An example is shown in Fig. 18.9.

If errors in the bits are to be evaluated, the performance measure is usually described as the *bit-error ratio* (BER), even in cases when *probability* is the correct term. Alternatively, error ratios or probabilities for symbols or words or any other unit of information might be evaluated instead of BER. The signal-to-noise ratio is usually defined as E_b/N_0 or E_s/N_0, where E_b is the energy per bit, E_s the energy per symbol, and N_0 the spectral density of additive white Gaussian noise. Other definitions of SNR might be employed where appropriate.

Almost invariably, the error ratio (or probability) will be displayed on a logarithmic ordinate, because of the large range typically encountered. It is also common practice to display SNR on a logarithmic (decibel) abscissa, but a linear abscissa could be used. Repeated evaluations are often performed for differing conditions in the simulated system, and a different error-ratio curve arises for each condition. It is convenient to be able to plot multiple curves on a single graph to facilitate comparisons among the results. Also, "theoretical" results exist for some signal formats (see Appendix 13B), and it is often convenient to be able to plot the theoretical curve on the same graph as the simulated results.

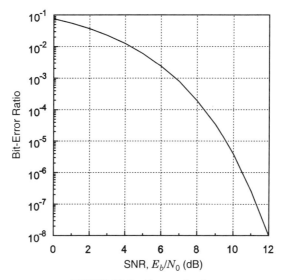

FIGURE 18.9 Error ratio plot.

Histograms

A histogram is a discrete approximation to a probability density function. It helps in visualizing the scatter of a quantity about a nominal value. The operations for generating a real histogram can be regarded as a quantizer (see Sec. 15.1), followed by a bank of counters. Each input sample is quantized to one of L bins. Each bin is assigned its own counter. For each sample that falls in any one bin, the counter for that bin is incremented by one count. The histogram is the accumulated count of each bin plotted versus the bin amplitude (or bin number).

18.2 UNITS AND SCALES

The point has been made repeatedly in earlier chapters that the computations of a simulation are performed on sequences of samples that carry no intrinsic time or frequency meaning. Exactly the same computations would be performed on a given system configuration, no matter what data rate or carrier frequency was involved. Think of the computations as operating in normalized time and frequency. Furthermore, the amplitudes of simulated signals are also dimensionless and frequently normalized to numerical values quite different from those of physical signals.

You may want to display simulation results in a denormalized time or frequency scale that is pertinent to a specific application. Or you may wish to select between the symbol interval and the sample interval as the basis for normalization of a display. Also, you may wish to denormalize the amplitudes for the displays.

What units and scales should be assigned to the displayed quantities when the computation results have no units? How are the axis division labels to be specified? Those questions, plus a few additional matters, will be the subjects of this section. Only time- and frequency-domain displays are considered here. The same issues do not arise for signal-space diagrams or error-ratio plots.

18.2.1 Abscissas

Simulation output data typically consist of sequences or blocks of complex samples representing a waveform or spectrum. Each sample bears an implied index according to its position in the block, but it is unlikely that an index or time instant or frequency value will be stored for each point. (For instance, STÆDT does not store explicit indexes in its data blocks; the memory consumption would be unacceptable.)

Abscissa Division Labels

There is a tacit assumption that points are uniformly spaced in time or frequency. However the block itself does not carry information on the point spacing or range; the abscissa values of the data blocks are undefined. Labels for abscissa divisions to be shown in displays have to be specified as side information. The needed side information might be supplied in the main simulation, or it may be deferred to the display processes. STÆDT accepts abscissa side information only in the display processes. Abscissa labeling is completely defined by specification of the number of points in the block (known to the program, not supplied by the user) and user specification of either (1) an origin and the spacing between points, or (2) the abscissas of the first and last points.

Any abscissas whatever could be specified (irrespective of any time or frequency scales that may have been assumed during the simulation), subject only to the constraints of uniform spacing of data samples and the limitations of the graphics program. Observe that the issue here is solely that of specifying the numerical labels to be assigned to the divisions of the abscissa. No matter what labels are assigned, the plotted graph itself remains entirely unchanged.

Abscissa Units

The labels applied to the abscissa divisions are dimensionless numbers; any units attached to the numbers reside in the mind of the user and not in the computer. The same time-domain graph with the same abscissa labels could represent time in nanoseconds or in hours, according to the whims of the observer. Definition of units might be conveyed in the axis title, or other side information.

Rather than assign specialized time or frequency units to a graph, you may prefer to keep the abscissa labels normalized: to the sample or symbol interval for time-domain plots or to the sample or symbol rate for frequency-domain plots. In STÆDT the default labels on the abscissas indicate sample number in the data block. Any other labeling requires user intervention.

Abscissa Wrapping

An N-element data block in either domain is likely to have an implied index running from 0 to $N-1$. This is the convention adopted in STÆDT and probably in most other simulation programs. That convention serves very well for time-domain blocks; a point occurring later in time will have an index that is higher than any point occurring earlier in time. Also, a time scale originating at zero is intuitively comfortable.

The same indexing from 0 to $N-1$ applies also in frequency-domain blocks. But a frequency-domain representation is periodic, so the block is regarded as circular (refer to Sec. 5.2.2). Therefore, it is not necessarily true that a point with higher index will have a higher frequency than that of a point with lower index (see Sec. 5.2.3 for a counterexample).

Two different interpretations might apply to the frequency-domain index:

1. The indexing is monotonic; a higher index implies a higher frequency. Data points will be plotted in the order in which they appear in the block and labels will be assigned to the abscissa divisions as described above.
2. The indexing is wrapped. Most commonly, the wrapping associates positive frequencies with index numbers 1 to $IP[(N-1)/2]$, zero frequency to index number 0, and negative frequencies to index numbers $IP[(N+1)/2]$ to $N-1$ (see Fig. 5.3). Just subtract N to find an unwrapped value for index numbers $IP[(N+1)/2]$ to $N-1$. (*Notation:* $IP[x]$ means the integer part of x.)

If the signal spectrum is concentrated about zero frequency, viewers of a plot will prefer to see zero frequency at the center of the plot. If zero frequency were placed at one end of the abscissa, the spectrum would be split, with one half at the left side of the plot and the other half at the right side. To place zero frequency at the center of the display requires that the block be unwrapped before plotting. The user is burdened least if unwrapping is built into the display process as an option.

Associating zero frequency with index zero is tantamount to fixing an origin. Then the label assignments for the abscissa divisions are completely defined by just one other specification—either the frequency spacing between points or the frequency at one end of the plot. Zero frequency is a natural origin with unique properties for frequency-domain quantities. No similar natural origin exists in the time domain.

18.2.2 Ordinates

Scaling and Units

Signals can take on a variety of amplitudes in their passage through a simulation. The particular numerical values that arise may be perfectly satisfactory for the purposes of computation but awkward for display and human interpretation. To display results with agreeable numbers for the ordinates, just rescale the data before plotting them.

Any simulation program will have facilities for scaling. Often, the amplitude of a signal will not be known accurately prior to a simulation, so the rescaling factor cannot be set beforehand. Instead, the experimental amplitude will have to be measured after the data have been collected and the rescaling factor calculated from the measurements. Simulation programs are well equipped to perform such calibration actions at the direction of the user.

Signal representations in the computer are dimensionless. Any units to be attributed to the graph ordinates exist only as side information supplied by the user; they are totally unrelated to the computations.

Magnitudes of frequency-domain quantities may be plotted on logarithmic scales, thereby requiring that a logarithmic conversion be applied to the simulation data. Strictly speaking, the logarithm should be taken of the ratio of the data quantity to a reference level. Specification of the reference is a scaling matter, as discussed above.

Power Spectrum Ordinates

If the amplitude of a signal can be rescaled at will, the units to be ascribed to the power spectrum are more than a little nebulous. Nonetheless, suppose that the signal amplitudes are scaled to meaningful values: How should the ordinates of the power spectrum be interpreted? There are various answers, depending upon the characteristics attributed to the signal data.

If the N-point signal block is denoted $\{z(n)\}$, its discrete Fourier transform is denoted $\{Z(k)\}$ (see Sec. 5.2.2). A simulation program might offer $\{|Z(k)|^2\}$ as the "power spectrum." This simplistic form is not the power spectrum, not rigorously. Note that the simple form does not even obey Parseval's rule (Sec. 5.4.1), which states that

$$\sum_{n=0}^{N-1} |z(n)|^2 = \frac{1}{N} \sum_{k=0}^{N-1} |Z(k)|^2 \qquad (18.1)$$

For rigor, the definition of spectral density should at least conform to Parseval's rule.

The interpretation of signal data will influence the definition of spectral density. Three distinct cases are examined below.

Finite-Duration Signal. Let the signal be represented as a finite block $\{z(n)\}$ of N points, spaced t_s seconds apart, and assumed to have value zero at all times outside the block. Such a signal is characterized in terms of its energy (Sec. 5.4.1):

$$E_N = t_s \sum_{n=0}^{N-1} |z(n)|^2 = \frac{t_s}{N} \sum_{k=0}^{N-1} |Z(k)|^2 \qquad (18.2)$$

from which it is logical to define the

$$\text{energy spectral density} = \frac{t_s}{N} |Z(k)|^2 \qquad (18.3)$$

If units of volts are imputed to $z(n)$, the energy spectral density has units of (volts^2-s)/(frequency increment), where the frequency increment is $1/Nt_s$. The presence of an undefined t_s in the definition (18.3) is awkward, but this is the correct definition for the stated representation of the signal.

Stationary Signal. Regard the signal as infinite in duration, with stationary (or cyclostationary) statistics. It then has infinite energy and has to be described in terms of its power. Only one segment of one realization of the signal is available, in the form of an N-point block $\{z(n)\}$ of samples, spaced at the sample interval t_s: that is, the same data as those available for calculating energy spectral density.

To obtain an estimate of the power spectral density, assume that the available signal block is typical of all such signal blocks: specifically, that all such blocks have the same energy. Then the estimated average power is simply the energy of the block divided by the duration

$$\hat{P}_N = \frac{E_N}{Nt_s} = \frac{1}{N}\sum_{n=0}^{N-1}|z(n)|^2 = \frac{1}{N^2}\sum_{k=0}^{N-1}|Z(k)|^2 \qquad (18.4)$$

from which a reasonable definition is

$$\text{power spectral density} = \frac{1}{N^2}|Z(k)|^2 \qquad (18.5)$$

This definition has dimensions of volts^2/(frequency increment), if $z(n)$ has units of volts. Dependence upon t_s has been eliminated from (18.5).

Periodic Signal. Assume that the signal is strictly periodic on the time interval Nt_s, allowing it to be represented as the sum of exponentials

$$z(n) = \sum_{i=0}^{N-1} a_i e^{j2\pi in/N} \qquad (18.6)$$

where a_i is the complex amplitude of the ith term and i/Nt_s is its frequency. Applying standard methods, the sum of squared magnitudes of the signal is found to be

$$\sum_{n=0}^{N-1}|z(n)|^2 = N\sum_{i=0}^{N-1}|a_i|^2 \qquad (18.7)$$

The mean-squared signal value (the power) is found by dividing the sum of squares by N:

$$\text{mean square} = \frac{1}{N} \sum_{n=0}^{N-1} |z(n)|^2 = \sum_{i=0}^{N-1} |a_i|^2 \quad (18.8)$$

Therefore, the power (mean square) in the ith discrete component is $|a_i|^2$.

Now let's relate a_i to $Z(k)$. Multiply both sides of (18.6) by $\exp(-j2\pi nk/N)$ and sum over n to obtain

$$\sum_{n=0}^{N-1} z(n) e^{-j2\pi nk/N} = Z(k) = \sum_{n=0}^{N-1} e^{-j2\pi nk/N} \sum_{i=0}^{N-1} a_i e^{j2\pi in/N}$$

$$= \sum_{i=0}^{N-1} a_i \sum_{n=0}^{N-1} e^{j2\pi n(i-k)/N} \quad (18.9)$$

The inner sum evaluates as

$$\sum_{n=0}^{N-1} e^{j2\pi n(i-k)/N} = \begin{cases} N, & i = k \\ 0, & i \neq k \end{cases} \quad (18.10)$$

from which is found the relation

$$a_k = \frac{Z(k)}{N} \quad (18.11)$$

Therefore, the power in the kth component is

$$P(k) = \frac{1}{N^2} |Z(k)|^2 \quad (18.12)$$

This has exactly the same formal definition as (18.5) for a stationary signal, but the units are power itself (actually, volts^2), not power per frequency increment.

Practicalities. Rather than becoming entangled in exacting interpretations of the character of the data, you may prefer instead simply to display $\{|Z(k)|^2\}$, or $\{20 \log |Z(k)|\}$. Just remember then that the display is qualitative and that the numerical labels on the ordinate are not truly the spectral density.

If you want to display the *power* spectral density and the signal $z(n)$ is scaled so that the result is meaningful, incorporate the factor of $1/N^2$ as in the definitions of (18.5) or (18.12). In STÆDT, the factor $1/N^2$ is best introduced by scaling $\{Z(k)\}$ by $1/N$ before submitting it to the display routine.

18.2.3 Plot Windows

Graphics programs commonly provide for user-specified *plot windows* that establish the limits of the coordinate axes. These limits are specified by minimum and maximum values on each axis. The window may be smaller than the range of the data, so that only part of the data will appear in the plot. This feature is helpful when small segments of the data have to be examined in detail. Changing the limits in effect moves the small window around on the larger space occupied by the data.

Conversely, the window can be specified larger than the range of the data; the program omits any plotting in a window region that is outside the data range. An oversized window might be employed as a means of obtaining simple round numbers for the axis division labels. Be aware that the window could be specified to lie entirely outside the data range, in which case no plot whatever would appear on the graph.

Many programs furnish an auto-window option, whereby the window limits are automatically set to encompass the entire data range. This feature is convenient for an initial test plot that determines the full span of data values and provides information to the user as an aid in specifying modified window limits. However, if the data span is excessive, an autoscaled window might provide a very strange-looking plot in which the data characteristics of interest are compressed to invisibility on a nearly flat line. As an example, STÆDT plots nulls in a dB spectrum at -10,000 dB, so any useful details in the range −100 to +100 dB will be squeezed out of sight.

18.3 DATA CONDITIONING

Data frequently will have to be preconditioned into a particular format before they can be plotted. Such processes as coordinate conversions, scaling, logarithmic conversions, magnitude computations, and Fourier transformations have all been indicated above. These all are either simple processes that need no further consideration here or have been covered adequately in earlier chapters. This section looks into three more-complicated conditioning processes that may be needed at times: computation of delay of a filter, smoothing of power spectrum plots, and windowing of data blocks.

18.3.1 Computation of Delay

The *group delay* of a continuous-frequency filter is defined formally as

$$\tau(f) = -\frac{1}{2\pi} \frac{d\phi(f)}{df} \qquad (18.13)$$

where $\phi(f)$ is the phase shift of the filter as a function of the frequency f. If the filter has a frequency response $H(f) = I(f) + jQ(f)$, where $I(f)$ and $Q(f)$ are the real and imaginary parts, respectively, of $H(f)$, the phase is defined as

$$\phi(f) = \tan^{-1} \frac{Q(f)}{I(f)} \qquad (18.14)$$

When evaluating a filter's characteristics, it may be more useful to inspect a plot of $\tau(f)$ instead of $\phi(f)$. A simulation program ought to be able to display delay as well as phase.

Two obstacles arise in evaluating (18.13) in a simulation: (1) the phase that is available is $\theta(f) = \phi(f)$ mod-2π, and (2) frequency is available only at discrete points, thereby impeding the differentiation operation of (18.13). Methods for circumventing these obstacles are presented below.

Avoidance of Phase

Because $\theta(f)$ is reduced modulo-2π, it has discontinuities wherever $\phi(f)$ passes through an odd multiple of π. Also, $\phi(f)$ itself has a phase discontinuity at any frequency where $H(f)$ has a null. Phase discontinuities would disrupt the computation of the derivative in (18.13), causing spurious irregularities in the computed delay.

To avoid the phase discontinuities, it is better to work with the rectangular components I and Q and never compute the phase at all. To that end, substitute (18.14) into (18.13) and perform the formal differentiation to obtain an alternative definition of delay as

$$\tau(f) = -\frac{1}{2\pi} \frac{1}{I^2 + Q^2} \left(I \frac{dQ}{df} - Q \frac{dI}{df} \right) \qquad (18.15)$$

The functions I and Q are always differentiable in any realizable filter and never have discontinuities. Therefore, delay can be defined without difficulties, even when phase shows discontinuities.

Frequency Sampling

A digital simulation does not provide a frequency-continuous $H(f)$; instead, the filter is represented as $H(k) = I(k) + jQ(k)$ at frequencies $k \Delta f, 0 \leq k < N$, where N is the number of complex points in the data block that specifies $H(k)$. The frequency increment is $\Delta f = 1/Nt_s$, and t_s is the sampling interval.

The continuous-frequency derivatives for (18.15) cannot be computed exactly from discrete-frequency data; an approximation is required. One simple approximation to the derivative is

$$I'(k) = \left. \frac{dI(f)}{df} \right|_{f = k\Delta f} \simeq \frac{I(k+1) - I(k-1)}{2\Delta f} \qquad (18.16)$$

and similarly for $Q'(k)$. With this approximation, the delay is computed as

$$\tau(k) = -\frac{1}{4\pi \Delta f} \frac{I(k)[Q(k+1) - Q(k-1)] - Q(k)[I(k+1) - I(k-1)]}{I^2(k) + Q^2(k)} \qquad (18.17)$$

The presence of $\Delta f = 1/Nt_s$ in (18.17) is a nuisance. Either a value could be assigned to t_s, or the delay could be displayed normalized to t_s. However, the symbol interval $T = M_s t_s$ is a more pertinent time division than the sampling interval t_s in a data communications system. By substituting NT/M_s for $1/\Delta f$ in (18.17), the delay can be expressed as normalized to T. The user specifies M_s, while N is known to the computer.

Delay cannot be computed at the endpoints of the block with the formula of (18.17) because I and Q values outside the block would be required for the calculations. A different formula is needed for the endpoints.

18.3.2 Spectrum Conditioning

Two operations are applied to data regularly to improve the characteristics of their spectrum displays: *spectrum smoothing* and *data windowing*. Data windowing is completely unrelated to the plot windowing considered above.

Smoothing

The spectrum plot of Fig. 18.6 is significantly ragged; it shows considerable apparent noise, even though no noise was included in the simulation that generated the plot. This noisy appearance is inherent to spectra of signals with random data under any but exceptional conditions. Periodograms are well known to exhibit such random behavior [1.9, Chap. 11].

The raggedness is distracting, particularly to persons not accustomed to the phenomenon. An uninitiated observer could easily be diverted into time-consuming examination of a meaningless artifact. More fundamentally, the apparent noise is a feature of the individual data realizations and is not a property of the spectrum of the ensemble from which the realizations are drawn. Therefore, the apparent noise is misleading, even though the plot is an exact representation of the spectrum of the particular realization.

For these reasons it is desirable to produce smoother spectral displays. Enormous efforts have been devoted to obtaining less-noisy spectra [18.1, 18.2]. The more-sophisticated methods avoid periodograms and Fourier transforms entirely; they are capable of much better results than can be obtained with periodograms. Nonetheless, because of the great utility of the fast Fourier transform, the periodogram is a simple and convenient spectral representation and is offered by most simulation programs. Nonperiodogram methods are not considered further here, despite their superior performance.

How can the periodogram be improved? It does no good to include more points in the periodogram; all that accomplishes by itself is to make the apparent fluctuations have a finer frequency scale. Instead, the periodogram has to be *smoothed* in some fashion. In the remainder of this section we consider methods of smoothing. The best-known method is due to Welch [18.4], wherein the data record is first broken up into multiple shorter (perhaps overlapping) segments. A periodogram is computed for each segment and the periodograms of the segments are averaged to produce a result that is appreciably smoother than the individual periodograms.

The averaged periodogram has coarser resolution than would a nonaveraged periodogram over the entire data span. Loss of resolution is inherent to any method for smoothing periodograms. Degraded resolution shows up as smearing of sharp features of a spectrum, such as narrow peaks and nulls, and of discrete-line components. In addition to the inherent smearing caused by smoothing, any segmenting method necessarily generates fewer than N points (the number of signal samples) in its spectrum. If the segments each have J signal samples, only J points will appear in the averaged periodogram.

A simpler approach to smoothing, adequate for most graphs, is to recompute each point in a periodogram as the average of some number of nearby points to obtain

$$\overline{S}(k) = \frac{1}{2L+1} \sum_{i=-L}^{L} |Z(k+i)|^2 \qquad (18.18)$$

where $\overline{S}(k)$ is the averaged periodogram. It is the central average of the kth point and the $2L$ points within $\pm L$ to either side of the kth point. Examples of averaged periodograms are compared in Fig. 18.6 ($L = 0$) and Fig. 18.10 ($L = 8$). The smoothing effect is evident; the loss of resolution can be seen with close inspection.

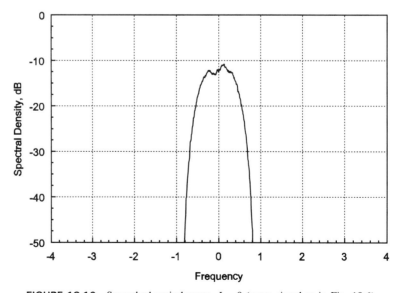

FIGURE 18.10 Smoothed periodogram, $L = 8$ (same signal as in Fig. 18.6).

Formula (18.18) unmodified can generate only $N - 2L$ points in $\overline{S}(k)$; the calculations would require points outside of the data block for k within L points of the ends of the block. One could be satisfied with $N - 2L$ points or could devise alternative smoothing rules for the $2L$ points at the block ends.

Weighting Windows

Spectral analysis is greatly concerned with the phenomenon of *spectral leakage*, in which spurious sidelobes of a strong component of the data overlap and mask the main lobe of a weak component. Refer to Figs. 10.1 to 10.4 for a vivid illustration of leakage. To reduce the leakage sidelobes, it is common practice to apply a tapered *weighting window* to the data sequence. Reference [18.5] provides an introduction to windows.

Instead of computing the Fourier transform of the signal sequence $\{z(n)\}$ to obtain a spectral estimate, the signal is first multiplied by a weighting window sequence $w(n)$ to form $\{u(n)\} = \{z(n)w(n)\}$.* The Fourier transform of the product is denoted $\{U(k)\}$ and the spectral estimate (before any averaging) is $\{|U(k)|^2\}$.

Large reductions of spurious sidelobes—major reductions of leakage—can be achieved with proper choice of window functions $w(n)$, at the cost of coarser resolution. An example of the effect of windowing is shown in Fig. 18.11, where a Hann (cosine-squared) weighting was applied to the signal from which the spectrum of Fig. 10.4 was produced. The improvement due to windowing is dramatically obvious.

---

*An erstwhile unwindowed sequence really is windowed by a rectangular function $w(n) = 1$, $n = 0$ to $N - 1$.

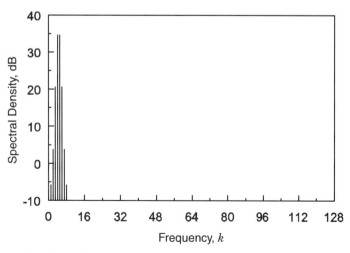

FIGURE 18.11 Spectrum of windowed signal (compare Fig. 10.4).

518 □ GRAPHICAL DISPLAYS

Windowing is well covered in the literature and is not discussed further here. The necessary multiplication of $z(n)$ by $w(n)$ is readily handled by elementary signal processes that are offered in all simulation programs. After the product is formed, $\{u(n)\}$ is processed in exactly the manner described earlier for an unwindowed sequence $\{z(n)\}$.

Leakage-Free Simulation

Although windowing is crucial in spectral estimation for experimental data, it does not play quite the same role in all simulations. In fact, windowing is superfluous for truly circular signal blocks that close around the ends without discontinuities in the signal or its derivatives; there will be no leakage at all for such signals.

To see how leakage can be avoided without windowing, consider an N-point signal block $\{z(n)\}$ that is a complex exponential (as in Figs. 10.1 to 10.4)

$$z(n) = e^{j 2\pi n f_s t_s} \tag{18.19}$$

where f_s is the signal frequency and t_s is the sample interval. To comply with the sampling theorem, restrict the signal frequency so that $|f_s t_s| < 0.5$.

Take the discrete Fourier transform of $z(n)$ to obtain

$$Z(k) = \sum_{n=0}^{N-1} e^{j 2\pi n f_s t_s} e^{-j 2\pi n k/N}$$

$$= \sum_{n=0}^{N-1} [e^{j 2\pi (f_s t_s - k/N)}]^n \tag{18.20}$$

The last line of (18.20) emphasizes the fact that $Z(k)$ of this signal is expressed as the sum of a finite geometric progression, and as such has the closed form

$$Z(k) = \frac{1 - e^{j 2\pi (N f_s t_s - k)}}{1 - e^{j 2\pi (f_s t_s - k/N)}} \tag{18.21}$$

After some manipulations, the magnitude is found as

$$|Z(k)| = \frac{\sin \pi (N f_s t_s - k)}{\sin \pi (f_s t_s - k/N)} \tag{18.22}$$

Case 1: Leakage-Free. Let $N f_s t_s$ be an integer m, where $0 \le m < N$. Then the magnitude of the Fourier transform evaluates as

$$|Z(k)| = \begin{cases} N, & k = m \\ 0, & k \neq m \end{cases} \qquad (18.23)$$

For this case there is no leakage whatever. All of the spectral contribution of $z(n)$ comes at its correct frequency $k/N = f_s t_s$. This condition applies to Figs. 10.1, 10.2, 18.6, and 18.10.

Case 2: Extensive Leakage. Suppose that $Nf_s t_s$ is not an integer (as in Figs. 10.3 and 10.4). Then the magnitude of the Fourier transform is not equal to zero at any value of k. In other words, the single-frequency signal of (18.19) contributes to the spectrum at all frequencies. This is what is meant by spectral leakage.

Any bandlimited, circularly continuous signal block can be resolved exactly as the sum of no more than N weighted components of the form (18.19), each with $Nf_s t_s$ products being integers from 0 to $N - 1$. Therefore, the periodogram (i.e., $|Z(k)|^2$) of a bandlimited circularly continuous signal is entirely free of leakage, without any need for windowing.

To appreciate further why a linear-block signal does have spectral leakage, consider a continuous-time bandlimited signal of infinite extent. Truncate that signal to a finite interval of duration Nt_s. The truncated signal is no longer bandlimited, by virtue of its having been truncated. Now replicate the truncated signal *ad infinitum* to create a periodic signal with period Nt_s. (Of course, only one cycle of the periodic signal is actually contained in the computer; periodization is implied in the subsequent discrete Fourier transformation used in computing the periodogram.) This periodic signal is not bandlimited either; in general, all harmonics of $1/Nt_s$ will be present.

When the signal is sampled, the higher-frequency harmonics are aliased into all frequencies in the lowest Nyquist interval (Sec. 5.1.3) and show up as leakage in the spectrum. The conclusion to be drawn is that better spectra can be obtained from circular blocks than from linear blocks. If spectra have to be computed from linear blocks, data windowing may be essential for control of leakage.

18.4 KEY POINTS

Graphical displays of simulation data are needed for ready interpretation of results. In this chapter we have presented some of the principal issues of displays.

18.4.1 Types of Plots

Simulation displays resemble the displays produced by laboratory instruments such as oscilloscopes, spectrum analyzers, and network analyzers. The following displays are valuable in simulation programs:

- Waveforms (time domain)
- Eye patterns (time domain)

- Frequency responses of filters (frequency domain)
- Power spectra (frequency domain)
- Signal space diagrams
- Constellations (sampled signal space diagrams)
- Error-ratio plots
- General x–y plots
- Histograms

18.4.2 Units and Scales

Simulations are performed with dimensionless times, frequencies, and amplitudes. However, results might better be displayed with scales and units that are meaningful to the system being simulated. Scales and units can be specified when the displays are being prepared; they do not enter into the simulations themselves.

Abscissas

Data blocks for time- or frequency-domain data are likely to be stored without explicit abscissas; an implied index attaches to a data point solely because of its position within the block. Unless the user chooses otherwise, results would be plotted with the sample indexes (numbers from 0 to $N - 1$) as the abscissa labels.

Division Labels. Other numbers can be user specified as the abscissa labels. These numbers are furnished to the plot routine as side information (not in the data block). The numbers specified affect only the labels, not the plot itself.

Units. The division labels are dimensionless numbers, with no units attached. Units (time or frequency) to be attributed to the abscissa are best conveyed as side information in the axis title. When simulating communications systems, it may be most convenient to normalize the units to the symbol interval or rate.

Wrapping. Frequency-domain blocks commonly are stored in wrapped format, with zero frequency at $k = 0$, positive frequencies at $k = 1$ to $IP[(N - 1)/2]$, and negative frequencies wrapped over to $k = IP[(N + 1)/2]$ to $N - 1$. An unwrapping option is needed so that zero frequency can be located at the center of the plot if so desired.

Ordinates

Amplitudes in a simulation usually are dimensionless numbers, unrelated to the physical system being simulated. To convert to meaningful numbers, rescale the data before plotting. Rescaled numbers are still dimensionless. Provide units as side information in the axis title.

Power Spectrum Ordinates. The *periodogram* used to represent the signal's spectral density might appear with any of three different scalings:

1. $|Z(k)|^2$
2. $t_s |Z(k)|^2 / N$
3. $|Z(k)|^2 / N^2$

Form 1 is not correctly scaled, so the numbers on its ordinates are meaningful only qualitatively. However, it is simple and convenient and is likely to be the form offered in many simulation programs. Form 2 is correct for energy spectral density of a finite-duration signal. The presence of the undefined sample interval t_s in the definition is rigorously necessary but is a practical nuisance. Form 3 is correct for the power spectral density of a stationary or periodic signal of infinite duration. Rather than one of the forms listed, the square root or the logarithm (dB) of any of them might be displayed instead.

Plot Windows

Most graphics programs offer user-specified plot windows for plotting just a portion of the data or for arranging the plot limits so that the data occupy only a portion of the graph. An auto-window facility adjusts the plot limits to encompass all the data.

18.4.3 Data Conditioning

Delay

Sometimes the group delay (derivative of phase) of a filter is of greater interest than the phase itself. To avoid problems with phase discontinuities, delay should be computed from the rectangular-coordinate representation of the frequency response, not from the phase.

Spectrum Processing

Smoothing. A periodogram of a random signal has a noisy appearance, even in the absence of noise. It is advisable to smooth the spectrum display to prevent distractions from the apparent noise. Smoothing necessarily impairs resolution, so a trade-off is required. A simple method of smoothing is to average over multiple nearby points of the unsmoothed periodogram.

Weighting Windows. Spectra computed from finite data records suffer from spectral leakage. It is common practice to weight the data block $z(n)$ by a window function $w(n)$ before computing the spectrum. A well-chosen window can reduce leakage greatly.

Leakage. A bandlimited signal in a circular block will not have any spectral leakage; it does not have to be windowed before its spectrum is computed. A signal in a linear block almost always will have spectral leakage. A window is needed if the leakage is excessive.

18.4.4 STÆDT Processes

Plot Routines

    PLER: error ratio plots
    PLEY: eye diagrams
    PLFD: frequency-domain plots
    PLHS: histograms
    PLTD: time-domain plots

Print Routines

    PRER: error ratio tables
    PRFD: frequency-domain blocks
    PRHS: histograms
    PRTD: time-domain blocks

Conditioning Routines

    GAIN: amplitude scaling
    WNDW: generates window functions

Also, PLFD, PLTD, PRFD, and PRTD include provisions for data conditioning.

19

PROGRAMMING ISSUES

This book has been written for communications engineers who use simulation programs as tools for analysis and design of communications systems. For that reason the emphasis in the book has been concentrated heavily on modeling of the systems and has given only passing attention to programming. Indeed, the programming details should be hidden for the most part; an engineering user does not want to be encumbered by programming.

Nonetheless, programming cannot be hidden completely. As a minimum, a user needs an understanding of the main programming issues to be able to apply a simulation program most effectively. Also, some knowledge of programming issues is invaluable when trying to choose one program from among several candidates. This chapter is a brief, simplified account of some programming issues that are important to a user. An account for programmers would be much more extensive.*

19.1 SIMULATION LANGUAGE

Even though it may not be evident, a user employs a *simulation language* to access most simulation programs, even those programs based on a graphical interface. In this section we describe how that comes about and the relation of the simulation language to a conventional programming language.

19.1.1 Evolution of a Simulation Language

To appreciate why a simulation language is used, let's start with a crude, impractical simulation scheme, and work up logically, step by step, to a simulation language.

---

*For the flavor of a computer-science approach and a list of other references, consult [19.1].

Ordinary Programs

A naive simulation could be written with a standard programming language in a one-piece program that models the entire system to be considered. That approach should be rejected out of hand before it begins, because it is so inflexible, so difficult to modify, and so antithetical to good programming practice, which insists that programs be modular.

Communications systems are very well suited to modular descriptions. Rather than simulating a system as one unbroken piece of code, it is much more logical to write each process and operation as a separate subroutine, or function, or macro, and to combine these modular pieces into a program to simulate a particular system. This approach allows the modular pieces to be written separately and kept in a library. Those pieces needed for a particular simulation are called up from the library.

If the simulation were carried out solely in a standard programming language, a *main program* or *driver program* would invoke the individual pieces needed. Each driver program would be specialized to its particular simulation and would have to be tailored to the configuration of each new system model; drivers could not be general-purpose pieces kept in the library.

An individualized driver program might be written as a series of calls on subroutines or functions in the library. This approach is far more flexible than attempting a monolithic, unmodular program. However, you would soon discover that the driver has to include much more than calls on the library routines. There are numerous housekeeping tasks to attend to, such as memory management; data passing; declarations of variables, arrays, and functions; file management; compiler instructions; print and plot formatting; and a host of other petty details. The simulation model would become submerged in these necessary housekeeping tasks.

Simulation Language

To reduce the labor of housekeeping, you probably would try to standardize the formats and methods of each of the regular tasks. In that way it would not be necessary to reinvent each task anew each time it was encountered. Once the housekeeping tasks were standardized, it would become possible to write housekeeping routines to perform some or all of the tasks, thereby reducing the labor of preparing a driver program. The housekeeping routines could be kept in the library with the signal processes.

The final step in arriving at a simulation language is to hide most or all of the housekeeping details so that the user need not be aware of them. One way of accomplishing this end is to construct a general-purpose *executive* program to direct the execution of the simulation. The executive is told the library processes that are to be executed, the order of execution, and a specification of the data flow among the processes. (Order of execution and data connectivity are important issues to be pursued below.)

The executive itself is a relatively simple program that manages the execution of the processes specified to it in a *list*. Upon reaching each process in a list, the executive decides which housekeeping functions are needed, activates the needed housekeeping functions, and directs the process to carry out its tasks. When the process is finished, the executive decides what to do with the resulting data, calls the housekeeping functions that handle the data, and goes on to the next process.

A user would not see the housekeeping functions or the executive. Only the simulation processes themselves would be visible in the list (to be prepared by the user). In effect, the names (or other identification) of the processes in the list serve as *commands* in a high-level simulation language.

For STÆDT we have given the name *command list* to the list of processes submitted to the executive; that is the terminology used in this chapter. Other programs may employ different nomenclature. Details of powerful simulation programs might differ enormously, but they are all based upon command lists whose commands, and the rules for using them, constitute a specialized simulation language.

Benefits of Simulation Language

A well-crafted simulation facility places many tools at your disposal. A simulation is prepared by selecting the available tools and putting them together. Once that has been accomplished without error, the simulation can be run expeditiously. A good simulation facility includes capabilities for displays, debugging, documentation, error checking, and so on.

Learning the rules and capabilities of a powerful simulation program demands substantial time and effort. One might sometimes be tempted to forgo the learning effort and just write a simple program from scratch. Our experience is that no time is saved, even for very simple models, at least partly because the extensive infrastructure of a good simulation program is not available in a homebrew program. Furthermore, models may start out appearing simple, but they have a nasty habit of growing more complicated. A versatile simulation program is far more convenient when that happens to you.

19.1.2 Relation to Programming Language

From the user's standpoint, the simulation language software could be written in just about any conventional programming language. In fact, different parts of the software could be written in different programming languages. Unless the user is going to write additional components to be added to the simulation software, the programming language is a matter that concerns only programmers. Simulation languages have been written in FORTRAN, C, Pascal, and compiled BASIC,* among others. User interfaces (see Sec. 19.2) have been written in these languages and in LISP.

19.1.3 High Level Versus Low Level

An inexperienced user wants the simulation program to present itself in as simple and transparent manner as possible; learning effort should be minimized. Ideally, an

---

*But not interpreted BASIC, which is much too slow for serious applications.

ordinary communications block diagram should translate directly into the computer model, with minimal effort on the part of the user. As many details as possible should be hidden. In this ideal, the simulation modules are necessarily rather complicated internally; each accomplishes multiple tasks in the simulation. Complicated modules are associated with a *high-level* simulation facility.

A need for thorough understanding of each module becomes more apparent as a user gains experience. For that reason, veteran simulationists often prefer that each module be specialized and simple—they want a *low-level* simulation program in which each module is easily understood and where model details are visible.

A high-level program undoubtedly demands less skill of the user if it fits the problem at hand. The user is relieved from much of the modeling burden, which is taken up by the programmer instead. A low-level program is much easier to adapt to different models, though. Furthermore, a programmer can never anticipate all of the high-level modules that may be needed for all applications; low-level provisions are necessary for flexibility.

A practical facility combines high- and low-level features. Some high-level modules that will be used often (filters are a prime example) are included so that they do not have to be reinvented over and over, but other modules are low-level, to allow scope for versatility. STÆDT tends toward the low-level pole, partly for the reasons stated, partly to ease the programming burden, and partly to expose the program workings to students of simulation.

Many simulation programs (including STÆDT) also resolve the conflict by providing an ample selection of low-level modules, plus a facility for the user to combine them into higher-level modules. These latter need be constructed only once, and then can be reused over and over (see Sec. 19.2.6).

19.2 ARCHITECTURE AND RULES

From the user's standpoint, the *user's interface*, *data connectivity*, and *order of execution* are vital features of a simulation program. Also important are *data and command types*, *parameter entry*, and *composite commands*.

19.2.1 User's Interface

The user has an interface with the program when preparing a command list, when preparing characteristics of simulation processes, and when observing results. Ordinarily, not much interaction takes place while the simulation is actually running. This section concentrates on the interface for preparing a command list. A command list is prepared by the user with the aid of the interface program—an editor of some sort—and is submitted (after translation or compilation) to another, usually distinct program (the simulation engine) for execution.

Text Editor Interfaces

When preparing a driver program (a progenitor of a command list, written in a programming language), one has to make do with whatever facilities are available in the environment of the programming language—perhaps nothing more than a simple editor. Early simulation programs incorporating simulation languages carried over this archetype into their interfaces for preparing a command list. A crude interface of this kind is serviceable but requires that the user thoroughly learn and remember all the commands of the language. Whenever any uncertainty arises, the user is compelled to return to the reference manuals. A friendlier interface is more inviting.

Menu Interfaces

The STÆDT interface presents on-screen menus as an improvement over unaided preparation of a command list. A scrollable name list, composed of the names and short descriptions of all commands in the STÆDT library, is shown to the user. Items may be selected freely from the name list for inclusion in the command list, using a mouse, or cursor keys, or typing abbreviated names. The on-screen name list substantially reduces the need for remembering or looking up commands.

Graphics Interfaces

A graphical block-diagram interface—included in several larger simulation programs—provides even greater support. Instead of constructing a text command list, the user constructs a block diagram on the computer screen. The text editors of the earlier methods are replaced by a graphics editor. Each block is assigned the name of a command, and data connections between blocks are shown by directed line segments. This scheme is enhanced even more if the command names for blocks can be selected from a name list, as in the menu scheme.

With a block-diagram interface, the user is no longer required to convert the system model—which invariably will be in block-diagram form—into a command list. It is only necessary to partition the system elements into pieces that can be modeled by processes (commands) in the simulation library. As its greatest benefit, a block diagram reveals system structure at a glance. It is far easier to grasp the essence of a system from a block diagram than from any command list.

But the block diagram is purely a display. The computer stores the essential information on the simulated system in a textual representation—most likely in a list. That list of essentials may or may not be in human-readable form and it may be hidden from the user, but it is clearly a command list. One part of the simulation facility produces an on-screen block diagram from the information in the list, while another part runs a simulation from the same list. It is for these reasons that we regard graphics block-diagram interfaces to be overlays for simulation languages, despite the reliance of graphics upon visual imagery instead of the written word. Their main simulation properties (other than the graphical displays) can be explained in the same terms as text-based simulation languages.

19.2.2 Parameter Entry

From the simulation point of view, commands have a multitude of different kinds of information to be conveyed. As inputs there might be signals, characteristics of the signal processes in a command (e.g., filter coefficients), initial conditions, or logical commands. Outputs might include signals, logical flags, error alarms, or displays. From the programming point of view, the data are transferred in one of two ways: either by specifying the actual numerical or string values or by specifying memory locations where the values are to be read or written. These specifications are called the *parameters* of the command and are one of the features essential to a workable program.

Standard programming languages typically list parameters in the *arguments* of the calls to subroutines. Arguments are usually enclosed in parentheses and are separated by commas. Early simulation languages adopted a similar style for specifying parameters. Each line in a command list had its command, followed by the command's parameters enclosed in parentheses. Each command, of course, had different parameters, and each command's parameters had to be supplied in exactly the correct order. Any error would cause the entire simulation to be wrong. No one can memorize all the parameters and their order for all commands in an extensive library, so preparing a command list demanded frequent reference to the program manuals.

More-recent simulation programs (including STÆDT) have eliminated this burden on the user. Instead, they show parameters for each command in an on-screen table. Each parameter has a brief description on-screen, and the user types in the information necessary to specify the parameter. The tables are easy to read. There is no need to remember parameter order; there is no need to remember simple definitions of parameters. Parameters are readily changed by typing new specifications into the applicable tables. Reference manuals need be consulted only for more-intricate matters that cannot be displayed in compact tables.

19.2.3 Order of Execution

Commands have to be executed in the proper order. Results would be meaningless if a command were to be executed before its correct inputs had been generated in previous commands.

A text-based command list is executed top to bottom, one command at a time, normally in the strict order in which commands are located in the list. Jump commands evoke departures from the strict order of listing when jumps are needed. The order of execution in a user-prepared, text-based command list is determined entirely by the user, who must take care that the order is correct.

A block diagram constructed in a graphical interface does not necessarily have an obvious or unique order of execution. The user prepares the block diagram but does not determine the order in which the blocks are to be exercised. To impose order, the simulation facility has a *scheduler* that analyzes the block diagram, decides on a correct order (there can be more than one), and arranges the hidden text-based command list accordingly. This scheduler relieves the user from the burden of maintaining order and is an additional benefit associated with block-diagram interfaces.

Command lists are often executed repeatedly in one simulation; the objective is to run enough samples to attain statistical reliability. Repeated executions are easily arranged in a text-based command list by incorporating conventional program-loop commands, such as FOR...NEXT or DO...WHILE. However, a block diagram is a static display; it ordinarily has no provisions to show repetitions or any other timing. Other arrangements for iterations are needed in the scheduler of a block-diagram interface; some flexibility and control of ordering could be lost.

19.2.4 Data Connectivity

Each command must know where to find its inputs and where to deliver its outputs. Output from one command is input to another. Every transmission simulation program needs a method for establishing data connections between commands.

Flow Patterns

Data flow between commands frequently takes a very different pattern from the order of execution of commands. Data do not necessarily flow from one command to the next in strict order of execution. Data-flow jumps are much more prevalent than execution-order jumps.

This distinction is familiar to programmers but may be overlooked by communications engineers. Engineers can easily be misled from their experience with real communications systems, wherein many operations take place simultaneously in parallel. Most physical operations require input to be present if they are to generate meaningful output—a property that seems to equate data flow and order of execution.

But a computer performs only one operation at a time; it simulates a system in serial fashion, not parallel. A serial computer is much more constrained with regard to data flow and order of execution than a physical system with elements running in parallel. A simulationist needs to keep the distinction in mind.

Connection Scheme

Many programs assign a name (or other identification) to each output from a module. The name refers to a memory location where the module is to write its output data. Each name is specified as a parameter of its module. (A *module* consists of a command, plus parameter specifications, plus output memory, plus state memory, if any.)

Connectivity is established by specifying the previously assigned name of an output location as an input parameter in a module accepting input. When the module needs an input, it simply reads it from the designated memory location. A user may not be required to name outputs explicitly if a block-diagram interface is employed. Connections are shown by directed lines drawn from one block to another on the screen. These connections are represented in the hidden command list by program-assigned identifiers. Then each input in a module is referenced to an output identifier, just as in a text-based command list, with the difference that the block-diagram program, not the user, assigns the identifiers.

Some block-diagram programs require (or allow) the user to name each signal output, even though connectivity is established graphically. These names are helpful in documentation of simulation results, since they identify the modules in which particular results originated.

19.2.5 Types

The concept of *data types* is central to programming. Different types of data have different properties and have to be processed differently. Types are decided by the programmer; a user has to work with the programmer's choices. Awareness of types helps a user understand why a program operates in one way rather than another. In this section we examine some of the type choices made for STÆDT as examples to illustrate the topic; other programs incorporate different choices.

Data Signals and Control Signals

STÆDT has defined two types of signals, as described at length in Chapter 14. A *data signal* is treated as complex; it can be processed in block format or single point; the values of its components are single-precision floating-point numbers. A *control signal* is always real; it can only be processed one point at a time, never in blocks. Its values are double-precision floating-point numbers. Each type of signal requires its own special signal-processing commands; a command intended for a data signal is incompatible with a control signal, and vice versa.

Why have these signals been split off as two different types? Why couldn't they have been combined as just "signals," with no distinction between them? Why should the user have to deal with the distinction? A short answer to these questions is that the signal processes applied to the two different types of signals are very different from each another. For example, a phase rotation makes sense as an operation on a complex data signal but has no meaning at all on a real control signal. For another example, integration is an operation very commonly applied to a control signal but nearly inconceivable for a data signal. In our judgment, processes applied to the two kinds of signals seemed so very divergent that the signals had to be treated formally as two separate types.

But why must the types be visible to the user? Why couldn't they be hidden, with the program guarding against misapplication? Indeed, some programs might hide the type distinctions. However, in STÆDT we wanted such issues to remain visible for instructional purposes and for the user to remain in close contact with the simulation model. As a matter of fact, though, STÆDT also has some hidden types within its code. Some complex data structures were deemed to be of no concern to the engineer-user and so were not made visible.

Real and Complex Data Signals

Another example from STÆDT is found in data signals, which are treated as just one type. A STÆDT data signal is always considered to be complex; all data-signal processes and all data-signal memory assignments make provision for complex signals. A real signal is handled as if it were complex but just happens to have a zero imaginary part.

In many applications, all data signals of interest will be real. Some users are in occupations in which they never see a complex signal, only real. With the STÆDT arrangement, half the complex memory assigned for storage of a real signal is wasted, and complex processes on real signals entail two to four times as many operations as needed for the corresponding real processes. These are penalties for treating real and complex signals as a single type instead of as two separate types.

On the other hand, there would also be penalties for separating into two types. Separate commands would have to be provided for the two types, or existing commands would have to be modified to recognize the type before acting on the signal. Either way, the program would increase in size and some additional user intervention would be required; as a minimum, the type of each data signal probably would have to be declared. The run-time wastage of a single complex type was deemed to be overshadowed by the extra programming and user effort that would have been required from two data types, complex and real.

Block and Single-Point Formats

STÆDT handles data signals in both block and single-point formats. These formats may be regarded as separate data types that are processed by the same commands. Many other programs offer only one format or the other, not both. All signal operations could be performed in single-point format but with a speed penalty in situations where large blocks were feasible. All operations could be performed in block format, with the restriction that each block could have only one element when simulating a feedback loop—a terribly inefficient restriction.

STÆDT offers both formats so that a user can experience the differences between them and so the more efficient format can be employed when desired.

Basis for Types

Decisions on types depend strongly upon programmers' judgments, both objective and subjective. Different programmers will make different choices.

19.2.6 Composite Commands

A hierarchical structure in a command list affords welcome assistance to a user. The overall model to simulate is broken up into convenient smaller pieces; then each piece is developed separately as a conglomeration of available commands. Once an individual piece has been checked out thoroughly and the user no longer wants to see its details, the piece should be encapsulated into what appears as a single module. The overall simulation then will be arranged as an interconnection of the encapsulated modules, each hiding details that would becloud the larger problem being investigated but which are essential for correct behavior of the individual modules.

All serious simulation programs have provisions for encapsulation of multiple commands into modules. In STÆDT, ordinary commands are called *primitive com-*

mands (a terminology employed by many others) and the encapsulated modules are called *composite commands* (our own terminology). A composite command, once it has been prepared, can be used just like any primitive command; it is virtually indistinguishable from a primitive. A composite is listed in the menus employed for preparing command lists, it accepts parameters in the same manner as a primitive, and it can be incorporated as a command in another composite if that is desired.

Composites afford the ability to construct a high-level simulation from low-level tools. The person constructing the simulation has to be well skilled and knowledgeable, but the model can then be employed by others with lesser skills. Composites also reduce the need for programming new commands when novel processes have to be implemented. A composite capability, plus a good foundation of primitives, allow a user to develop a wide variety of models without resorting to conventional programming.

19.3 INTERPRETERS VERSUS COMPILERS

Simulation programs can execute a command list in a variety of ways. The command list is said to be either *interpreted* or *compiled*, with some variations possible within each category. The choice between interpretation or compilation is made by the program's author; a user of the program has to accept whatever choice has been made. Thus the distinction between the two approaches takes on only secondary importance to the user.

19.3.1 Interpreters

Complete Interpreters

A "complete" interpreter (our nonstandard terminology) consists of all the commands of the simulation library, the executive routines, housekeeping routines, and anything else needed to make the program run. A complete interpreter is a compiled and linked stand-alone executable program.

The interpreter takes the user-prepared command list as input data. It analyzes each line of the command list one at a time. It then initiates immediate execution of the command and any other routines that are called for by the line. It then goes on to the next line. Thus each line of a command list tells an interpreter which part of itself to execute next. The same interpreter program is used for any command list that could be written.

Each command and each routine will have been written as a separate piece of source code, and each will have been compiled individually into separate pieces of object code. All of the pieces of object code are then linked to form the stand-alone interpreter. New commands are added to such an interpreter by (1) writing source code for the new command, (2) compiling the command's source code into an object-code module, and (3) relinking the entire interpreter into a new executable program that includes the new command. STÆDT is a complete interpreter.

Relinkable Interpreters

A complete interpreter contains all commands of the library, even though only a relatively small number of the commands are ever used in any one simulation. The unused commands are wasteful of memory, and operation may be slowed down by their presence. A "relinking" interpreter (nonstandard terminology) is one that contains only those commands that are required by the particular command list; it would occupy less memory than a complete interpreter and might run faster.

The components of a relinkable interpreter would be kept in the library as individual pieces of object code—a separate piece for each command, each housekeeping operation, and for the executive. An additional preprocessor (nonexistent with a complete interpreter) would be needed to analyze each command list submitted for execution; the preprocessor would decide which object-code pieces were needed to run the command list. Those pieces, and only those pieces, would be linked into an executable interpreter, which then runs its particular command list in exactly the same manner as a complete interpreter. A new relinkable interpreter has to be linked for every new command list, or for any alteration to an existing command list that requires additional commands.

19.3.2 Compilers

If a simulation is to be compiled, assume that all commands, housekeeping routines, and maybe the executive (if it even exists) are kept in the library in source code format, written in the programming language. A user-prepared command list is submitted to a compiler as input data. The compiler scans the command list to determine which library pieces will be needed for the simulation. The compiler takes source code for each indicated command, for each needed housekeeping operation, plus any run-time library operations that are required, and compiles them into a single piece of object code. This object code is then linked into an executable program that runs the simulation without further reliance upon the command list. That is, the command list has been incorporated into the run-time program, which consists of a complete set of machine language instructions to the computer.

Besides using only those commands needed for the particular simulation, a compiler can look at an entire program as a whole. The compiler is thereby able to make more-efficient arrangements than is possible in an interpreter, which is constrained to look at just one command in isolation at a time.

Code Generators

Commands can be written in source code as subroutines, as functions, or as macros—fragments of source code to be incorporated in-line into a larger program. Instead of submitting all the many pieces separately to the compiler, a *code generator* can be used to combine all the source components into a single piece of source code, which in turn is submitted to the compiler. Including a code generator in the procedure offers even better efficiencies than a compiler alone. Code generators might also produce source code that can be compiled to run on an entirely different machine, such as a DSP chip. That prospect is examined further below.

19.4 AUXILIARY FEATURES

Previous chapters were devoted to modeling of signals and signal processes. Except for a couple of chapters on builder routines and graphical displays, no attention has been given to subsidiary matters. Large simulation programs tend to include numerous features that accomplish auxiliary tasks other than simulation as such. It is the auxiliary features that help, in part, to make the program large. In this section we give a concise description of some of the features provided in existing programs. Implementation is not discussed.

Run Documentation

A nontrivial simulation needs to be well documented if it is to be of more than passing value. A user may have to be able to understand the simulation and its results many months after it has been run. Persons other than the user may have to interpret the simulation and its results without the help of the user. There is a widespread need for documentation that is thorough and self-explanatory, even to viewers who may not be adept in the ways of simulation.

On the other hand, most engineers despise the clerical burden of documentation—it is a chronic occupational syndrome, not easily remedied. Many engineers will skimp or omit the documentation, given half a chance. Inadequately documented simulations become difficult to interpret and lose much of their value as the simulationist gradually forgets why and how things were done the way they were done.

In an effort to counteract this common failing and to relieve a conscientious user of some clerical labor, several programs have provisions for automatic documentation of those items amenable to automatic recording. Any program will preserve the command list (either in text format or in a block-diagram display) and the user-entered parameters. Also, the program should be able to preserve any arbitrary data designated for saving by the command list.

Some programs go further. They automatically save all data entered or produced in the simulation and provide formatted printout (or other display) of selected data. Formatted display would include the data values, of course, but also such items as data type, numerical precision, module of origin, and a unique identifier for each simulation run. There is a substantial difference between a simple printout of data values and a formatted printout with copious side information. STÆDT has various simple documentation features. Good automatic documentation can tell much of the what and how of a simulation, but users' written records are irreplaceable for explaining the why.

Data Base

A large program, heavily exploited by multiple users, quickly generates enormous amounts of data. There are command libraries, utilities, command lists, parameter tables, module characteristics, and vast quantities of results. If this conglomeration were unmanaged, any wanted piece of information would soon be irretrievable from amid growing chaos.

To try to maintain order, large simulation programs place all files into a database system. Each file is classified according to its role, a run identifier (if applicable), command-list identifier, date, originator or user identity, project identifier, or any other characteristics that distinguish it from innumerable other files that are very similar. The searching and reporting capabilities of a powerful data base program greatly ease the task of retrieving information. Furthermore, the data base takes on the burden of storing the data effectively in the first place, thereby relieving the user of that load.

In a small program (such as STÆDT) intended primarily for single users, the files do not proliferate nearly so rapidly, so a sophisticated data base is not as necessary. (Indeed, the simulation program would not be small if it also included a powerful data base.) In the absence of a data base, responsibility for maintaining order is placed on the user. We urge that careful thought be given to the matter if anything more than casual simulations are to be performed. As a minimum, we recommend that each simulation be given its own subdirectory.

Filter Design

It is generally accepted that filter building (preparation of run-specific filter coefficients from established characteristics of a filter) needs to be part of the simulation suite (see Chapters 6 and 17). Filter design itself—establishing the desired characteristics in the first place—is a nontrivial task that is usually considered to be separate from the simulator.

Nonetheless, an active simulationist is continually called upon to provide filter characteristics. These might be designed in a completely independent program and then imported into the simulation program. (Importation is eased if data formats of the two programs are compatible.) That is the only option if the simulation program lacks a filter-design facility.

The need for filter design is so pervasive that some of the largest simulation programs offer a filter-design option in the form of a separate program with data formats and user interface that are compatible with the simulator. It can be a great convenience to have the filter designer and the simulator available in a common environment. A good filter design program is itself quite sizable and so is not likely to be bundled with a small simulation program.

Data Import and Export

Other programs (such as spreadsheets, graphics programs, editors, word processors, mathematics and statistics programs, etc.) might be better equipped than the simulation program to prepare module characteristics or to analyze or display results. That is a strong reason for writing data files in a standardized format compatible with the most useful outside programs. STÆDT has provision for accepting and delivering ASCII-format files.

Sometimes one would like to capture or generate a physical analog signal in conjunction with the simulation program. That would allow the simulation program to analyze an actual signal, not an academic model thereof; or the program could generate a signal that might be difficult to describe by other methods. An analog signal is captured with the aid of a sampler and analog-to-digital convertor; an analog signal is generated with the aid of a digital-to-analog convertor. The analog signals could be brought into the simulator if the driver software needed by the convertors were made part of the simulation facility.

Digital Synthesis

For a digitally implemented communications element, it is often said that the simulation is the design, meaning that the digital design is the same as the simulation model. That statement is not necessarily quite correct, but it can be made exact for specific hardware devices.

To have a one-to-one correspondence with particular hardware, such as a DSP microprocessor, an application-specific integrated circuit (ASIC), or a particular chip set of specialized communications processors, it is necessary to have a simulation command library that faithfully models the actions of the target hardware. Commands would model algorithms of the DSP microprocessor, library cells of the ASIC, or individual chips of the chip set.

Command lists are prepared with these commands, and simulations are performed just as with any command library. Once the simulations have proven satisfactory, the information needed for hardware implementation resides in the command list. A translation facility provides output in a form that is useful to other design tools pertinent to the particular hardware. A DSP microprocessor accepts source or assembly code, an ASIC is designed with the aid of a high-level design language, and the chip set interconnections could be specified from a net list.

To demand *bit-true* performance of the system simulator (modeling of the logic and number system of the target application) might encompass too many levels of the simulation hierarchy (see Sec. 1.2.1) and thereby inflict excessive complexity upon the program. As an alternative, one or more intermediate language steps might be interposed between the highest-level program and the hardware design or code. A simulation from a high level in the hierarchy hands part of its command list over to another program that works lower down in the hierarchy, a program able to handle fine details. Extensive work on interconnecting different levels is in progress at this writing. STÆDT has none of these multilevel capabilities; a much larger program is needed to support them.

Graphics Editor

Since graphical output is far more informative than screens or pages full of numbers, most of the results of a simulation will be presented to the user in graphical form. Graphical displays need customizing with respect to titles, scales, ticks, grids, notations, sizes, plot style, and other features. A versatile easy-to-use graphics editor would be welcome to any user.

Help

Even a small simulation program needs extensive reference material. A large program has multiple volumes of manuals. Trying to look up isolated facts in big manuals is a painful task, not relished by anyone. Well-organized on-screen help is an attractive adjunct in a large simulation facility. Note that the help capability is part of the user interface, not the simulation program itself.

Large and Small Programs

To be useful, even a small simulation program needs to have a sufficient number of core commands in its library. A large program might have more library commands and so would be more versatile, but there does not exist an infinite number of worthwhile commands; one would not expect a large command library to contain vastly more signal processes than those in an adequate small library. On a given machine, one would not expect a large program necessarily to run any faster (if faster at all) than a small program.

A large program may have a more convenient command-list interface (graphical block diagram instead of text-based) and better presentation graphics. But these additions do not have to grow enormously large either. The major difference in size and capability between a "small" program and a "large" program lies in the auxiliary features that are included. A really small program has almost none, whereas a really large program has many features.

19.5 KEY POINTS

A simulation is described by a *command list* prepared by the user. In most advanced simulation programs, the command list consists of statements in a high-level *simulation language*. Each statement represents a procedure of greater or lesser complexity that has been written in a conventional programming language. In a high-level simulation language, typical commands accomplish complicated tasks, leaving less preparation burden to the user. In a low-level program, typical commands tend to be simple and readily understood, but the user has to arrange any complicated interactions. Complicated established models are easier to work with in a high-level language, but novel models are easier to develop in a low-level language. A high-level program is attractive to a novice user, but one gradually comes to prefer low level as one's experience grows.

The user prepares a command list with the aid of a *user interface*, which might be a text editor for a text-based command list or a graphical editor for an on-screen block-diagram. Although the user may see only a block-diagram display, a textual command list (perhaps not human-readable) will be prepared nonetheless. The interface is usually a separate program, logically distinct from the run-time simulation program. Data needed by commands are provided through user-specified *parameters*. Modern interfaces arrange for entry of parameters via informative screen-displayed tables.

A text-based command list is executed by the simulation program in the *order* in which commands appear in the list. Jump commands are employed when necessary to alter the strict top-to-bottom order. A graphical block diagram does not have a unique, inherent order of execution. A *scheduler* is needed to establish a correct order from the diagram.

Data *connections* between modules have to be designated. Data flow often differs substantially from order of execution. Connections may be made by assigning a name to each data output from each module. A name identifies a unique memory location to which the data are to be written. Another module that is to use the stored data as input is supplied the assigned name to identify the memory location from which to read the data. In a text-based interface, the user specifies an identifying name as a parameter of the module. In a block-diagram interface, the program may specify the identifiers from user-drawn connecting lines.

Various *types* of signals and processes appear within a simulation program. Types are chosen as seen fit by the programmer. Types may be hidden from the user. If not hidden, the user has to take account of the different types.

Composite commands provide a hierarchical structure to a program. They permit close attention to details in the construction of a composite, and then they hide the details when they are encapsulated. Composites afford many of the potential benefits of both high- and low-level features in the same program.

A simulation may be executed, one command at a time, by an *interpreter* that works on any command list; or each command list may be *compiled* into its own unique executable program. The program author chooses between interpreter or compiler; a user accepts whatever has been established in the program.

Large programs tend to have numerous *auxiliary features* beyond those needed for the core simulation activities. More than anything else, it is these auxiliary features that determine the overall size of a program.

REFERENCES

Chapter 1

1.1. J. M. Wozencraft and I. M. Jacobs, *Principles of Communication Engineering*. New York: Wiley, 1965.

1.2. J. G. Proakis, *Digital Communications*, 2nd ed. New York: McGraw-Hill, 1989.

1.3. R. W. Lucky, J. Salz, and E. J. Weldon, *Principles of Data Communication*. New York: McGraw-Hill, 1968.

1.4. E. A. Lee and D. G. Messerschmitt, *Digital Communication*. Boston: Kluwer, 1988.

1.5. S. Benedetto, E. Biglieri, and V. Castellani, *Digital Transmission Theory*. Englewood Cliffs, NJ: Prentice Hall, 1987.

1.6. R. D. Gitlin, J. F. Hayes, and S. B. Weinstein, *Data Communications Principles*. New York: Plenum Press, 1992.

1.7. M. K. Simon, S. M. Hinedi, and W. C. Lindsey, *Digital Communications Techniques*. Englewood Cliffs, NJ: Prentice Hall, 1995.

1.8. J. A. C. Bingham, *The Theory and Practice of Modem Design*. New York: Wiley, 1988.

1.9. A. V. Oppenheim and R. W. Schafer, *Digital Signal Processing*. Englewood Cliffs, NJ: Prentice Hall, 1975.

1.10. A. V. Oppenheim and R. W. Schafer, *Discrete-Time Signal Processing*. Englewood Cliffs, NJ: Prentice Hall, 1990 (update of [1.9]).

1.11. L. R. Rabiner and B. Gold, *Theory and Application of Digital Signal Processing*. Englewood Cliffs, NJ: Prentice Hall, 1975.

1.12. J. G. Proakis and D. G. Manolakis, *Introduction to Digital Signal Processing*. New York: Macmillan, 1988.

1.13. M. C. Jeruchim, P. Balaban, and K. S. Shanmugan, *Simulation of Communication Systems*. New York: Plenum Press, 1992.

1.14. Special Issue on Computer-Aided Modeling, Analysis, and Design of Communication Systems, *IEEE J. Sel. Areas Comm.*, pp. 8–203, Jan. 1984.

1.15. Special Issue on Computer-Aided Modeling, Analysis, and Design of Communication Systems II, *IEEE J. Sel. Areas Comm.*, pp. 5–125, Jan. 1988.

Chapter 2

2.1. A. Papoulis, *The Fourier Integral and Its Applications*. New York: McGraw-Hill, 1962.

2.2. R. N. Bracewell, *The Fourier Transform and Its Applications*. 2nd ed. New York: McGraw-Hill, 1986.

2.3. K. Knopp, *Theory of Functions*. New York: Dover, 1945.

2.4. E. Goursat, *Theory of Functions of a Complex Variable*. New York: Dover, 1959.

2.5. S. A. Gronemeyer and A. L. McBride, "MSK and Offset QPSK Modulation," *IEEE Trans. Comm.*, vol. COM-24, pp. 809–819, Aug. 1976.

2.6. H. Nyquist, "Certain Topics in Telegraph Transmission Theory," *Trans. AIEE*, vol. 47, pp. 617–644, Apr. 1928.

2.7. W. R. Bennett and J. R. Davey, *Data Transmission*. New York: McGraw-Hill, 1965.

2.8. R. W. Lucky, J. Salz, and E. J. Weldon, Jr., *Principles of Data Communication*. New York: McGraw-Hill, 1968.

2.9. H. S. Black, *Modulation Theory*. New York: Van Nostrand, 1953.

2.10. S. O. Rice, "Envelopes of Narrow-Band Signals," *Proc. IEEE*, vol. 70, pp. 692–699, July 1982.

2.11. L. E. Franks, "Complex Envelope Representation of Signals," Appendix 7.A in K. Feher (ed.), *Digital Communications: Satellite–Earth Station Engineering*. Englewood Cliffs, NJ: Prentice Hall, 1983.

2.12. L. E. Franks, *Signal Theory*. Englewood Cliffs, NJ: Prentice Hall, 1969, Chap. 4.

2.13. A. Lender, "The Duobinary Technique for High-Speed Data Transmission," *AIEE Trans. Comm. Electron.*, vol. 82, pp. 214–218, May 1963.

2.14. E. R. Kretzmer, "Generalization of a Technique for Binary Data Communication," *IEEE Trans. Comm.*, vol. COM-14, pp. 67–68, Feb. 1966.

2.15. P. Kabal and S. Pasupathy, "Partial-Response Signaling," *IEEE Trans. Comm.*, vol. COM-23, pp. 921–934, Sept. 1975.

2.16. H. Kobayashi, "A Survey of Coding Schemes for Transmission or Recording of Digital Data," *IEEE Trans. Comm.*, vol. COM-19, pp. 1087–1100, Dec. 1971.

2.17. J. B. Anderson, T. Aulin, and C.-E. Sundberg, *Digital Phase Modulation*. New York: Plenum Press, 1986.

2.18. J. K. Omura and D. Jackson, "Cutoff Rates for Channels Using Bandwidth Efficient Modulations," *Conf. Rec. Nat'l Telecomm. Conf.*, NTC'80, vol. 1, paper 14.1.

2.19. R. deBuda, "Coherent Demodulation of Frequency-Shift Keying with Low Deviation Ratio," *IEEE Trans. Comm.*, vol. COM-20, pp. 429–435, June 1972.

2.20. W. R. Bennett, "Statistics of Regenerative Digital Transmission," *Bell Syst. Tech. J.*, vol. 37, pp. 1501–1542, Nov. 1958.

2.21. F. M. Gardner, "Self Noise in Synchronizers," *IEEE Trans. Comm.*, vol. COM-28, pp. 1159–1163, Aug. 1980.

2.22. A. Papoulis, *Probability, Random Variables, and Stochastic Processes*. New York: McGraw-Hill, 1965.

2.23. W. B. Davenport, Jr., and W. L. Root, *An Introduction to the Theory of Random Signals and Noise*. New York: McGraw-Hill, 1958.

2.24. A. D. Spaulding and D. Middleton, "Optimum Reception in an Impulsive Interference Environment," *IEEE Trans. Comm.*, vol. COM-25, pp. 910–934, Sept. 1977.

2.25. J. R. Barry and E. A. Lee, "Performance of Coherent Optical Receivers," *Proc. IEEE*, vol. 78, pp. 1369–1394, Aug. 1990.

2.26. W. P. Robins, *Phase Noise in Signal Sources*. London: Peter Peregrinus, 1982.

2.27. G. W. Wornell, "Wavelet-Based Representations for the $1/f$ Family of Fractal Processes," *Proc. IEEE*, vol. 81, pp. 1428–1450, Oct. 1993.

Chapter 3

3.1. A. Papoulis, *Signal Analysis*. New York: McGraw-Hill, 1977, Chap. 4.

3.2. M. F. Gardner and J. L. Barnes, *Transients in Linear Systems*. New York: Wiley, 1942, Sec. VIII.8.

3.3. T. Kailath, *Linear Systems*. Englewood Cliffs, NJ: Prentice Hall, 1980.

Chapter 4

4.1. G. D. Forney, Jr., "Maximum Likelihood Sequence Estimation of Digital Sequences in the Presence of Inter-symbol Interference," *IEEE Trans. Inf. Theory*, vol. IT-18, pp. 363–378, May 1972.

4.2. A. J. Viterbi, "Error Bounds for Convolutional Codes and an Asymptotically Optimum Decoding Algorithm," *IEEE Trans. Inf. Theory*, vol. IT-13, pp. 260–269, Apr. 1967.

4.3. G. D. Forney, Jr., "The Viterbi Algorithm," *Proc. IEEE*, vol. 61, pp. 268–278, Mar. 1973.

Chapter 5

5.1. J. L. Brown, Jr., "First-Order Sampling of Bandpass Signals: A New Approach," *IEEE Trans. Inf. Theory*, vol. IT-26, pp. 613–615, Sept. 1980.

5.2. A. J. Jerri, "The Shannon Sampling Theorem—Its Various Extensions and Applications: A Tutorial Review," *Proc. IEEE*, vol. 65, pp. 1565–1596, Nov. 1977.

5.3. J. B. Thomas and B. Liu, "Error Problems in Sampling Representations," *1964 IEEE Int. Conf. Rec.*, vol. 12, part 5, pp. 269–277.

5.4. J. L. Brown, Jr., "On Mean-Square Aliasing Error in the Cardinal Series Expansion of Random Processes," *IEEE Trans. Inf. Theory*, vol. IT-24, pp. 254–256, Mar. 1978.

5.5. A. Papoulis, "Error Analysis in Sampling Theory," *Proc. IEEE*, vol. 54, pp. 947–955, July 1966.

5.6. D. C. Stickler, "An Upper Bound on Aliasing Error," *Proc. IEEE*, vol. 55, pp. 418–419, Mar. 1967.

5.7. M. Abramowitz and I. A. Stegun (eds.), *Handbook of Mathematical Functions*, Natl. Bur. Std. Appl. Math. Ser. 55. Washington, DC: U.S. Government Printing Office, June 1964.

5.8. G. E. Valley, Jr., and H. Wallman, *Vacuum Tube Amplifiers*, MIT Radiat. Lab. Ser., vol. 18. New York: McGraw-Hill, 1948.

Chapter 6

6.1. A. Antoniou, *Digital Filters: Analysis and Design*. New York: McGraw-Hill, 1979.

6.2. F. M. Gardner, "A Transformation for Digital Simulation of Analog Filters," *IEEE Trans. Comm.*, vol. COM-34, pp. 676–680, July 1986.

6.3. C. K. Campbell, "Applications of Surface Acoustic and Shallow Bulk Acoustic Wave Devices," *Proc. IEEE*, vol. 77, pp. 1453–1483, Oct. 1989.

Chapter 7

7.1. S. W. Golomb, *Shift Register Sequences*. San Francisco: Holden-Day, 1967; rev. ed., Laguna Hills, CA: Aegean Park Press, 1982.

7.2. N. Zierler, "Linear Recurring Sequences," *J. Soc. Ind. Appl. Math.*, vol. 7, pp. 31–48, 1959.

7.3. F. J. MacWilliams and N. J. A. Sloane, "Pseudo-random Sequences and Arrays," *Proc. IEEE*, vol. 64, pp. 1715–1729, Dec. 1976.

7.4. M. K. Simon, J. K. Omura, R. A. Scholtz, and B. K. Levitt, *Spread Spectrum Communications*, Vol. I. New York: Computer Science Press, 1985, Chap. 5.

7.5. W. W. Peterson and E. J. Weldon, *Error-Correcting Codes*. Cambridge, MA: MIT Press, 1972.

7.6. W. A. Gardner and C.-K. Chen, "On the Spectrum of Pseudo-noise," *Proc. IEEE*, vol. 74, pp. 608–609, Apr. 1986.

7.7. E. L. Pinto and J. C. Brandao, "A Way of Efficiently Using Computer Simulated Digital Signals to Evaluate Error Probability," *Conf. Rec., Globecom'86*, vol. 1, paper 4.5, Dec. 1986.

7.8. E. L. Pinto and J. C. Brandao, "On the Efficient Use of Computer Simulated Digital Signals to Evaluate Performance Parameters," *IEEE J. Sel. Areas Comm.*, vol. 6, pp. 52–57, Jan. 1988.

7.9. M. Kavehrad and P. Balaban, "Digital Radio Transmission Modeling and Simulation," *Conf. Rec., Globecom'89*, vol. 2, paper 35.4, Nov. 1989.

7.10. D. V. Sarwate and M. B. Pursley, "Crosscorrelation Properties of Pseudorandom and Related Sequences," *Proc. IEEE*, vol. 68, pp. 593–619, May 1980.

7.11. J. R. Carson, "Notes on the Theory of Modulation," *Proc. IRE*, vol. 10, pp. 57–64, Feb. 1922.

7.12. A. A. M. Saleh, "Frequency-Independent and Frequency-Dependent Nonlinear Models of TWT Amplifiers," *IEEE Trans. Comm.*, vol. COM-29, pp. 1715–1720, Nov. 1981.

7.13. W. H. Press, B. P. Flannery, S. A. Teukolsky, and W. T. Vetterling, *Numerical Recipes*. New York: Cambridge University Press, 1986, Chap. 14.

7.14. L. Erup and R. A. Harris, "On Numerical Optimization of Communications System Design," *IEEE J. Sel. Areas Comm.*, vol. SAC-6, pp. 106–125, Jan. 1988.

7.15. R. E. Crochiere and L. R. Rabiner, *Multirate Digital Signal Processing*. Englewood Cliffs, NJ: Prentice Hall, 1983.

Chapter 8

8.1. B. Goldberg (ed.), *Communications Channels: Characterization and Behavior*, IEEE Press Repr. Ser. New York: IEEE Press, 1976.

8.2. H. Cravis and T. V. Crater, "Engineering of T1 Carrier System Repeatered Lines," *Bell Syst. Tech. J.*, vol. 42, pp. 431–486, Mar. 1963.

8.3. A. S. Rosenbaum, "Binary PSK Error Probabilities with Multiple Cochannel Interferences," *IEEE Trans. Comm. Tech.*, vol. COM-18, pp. 241–253, June 1970 (reprinted in [8.4]).

8.4. P. Stavroulakis (ed.), *Interference Analysis of Communications Systems*, IEEE Press Repr. Ser. New York: IEEE Press, 1980.

8.5. P. Mertz, "Model of Impulsive Noise for Data Transmission," *IRE Trans. Comm. Syst.*, vol. CS-9, pp. 130–137, June 1961.

8.6. J. W. Modestino and K. R. Mathis, "Interactive Simulation of Digital Communication Systems," *IEEE J. Sel. Areas Comm.*, vol. SAC-2, pp. 51–76, Jan. 1984.

8.7. S. D. Personick, P. Balaban, J. H. Bobsin, and P. R. Kumar, "A Detailed Comparison of Four Approaches to the Calculation of the Sensitivity of Optical Fiber System Receivers," *IEEE Trans. Comm.*, vol. COM-25, pp. 541–548, May 1977.

8.8. D. G. Duff, "Computer-Aided Design of Digital Lightwave Systems," *IEEE J. Sel. Areas Comm.*, vol. SAC-2, pp. 171–185, Jan. 1984.

8.9. L. Bellato, B. Polacco, S. G. Pupolin, and M. Tamburello, "Problems in the Computer Simulation of a Fiber Optic Digital Transmission System," *Conf. Rec., Int. Conf. Comm., ICC'79*, vol. 2, paper 28.

8.10. D. Datta and R. Gangopadhyay, "Simulation Studies on Nonlinear Bit Synchronizers in APD-Based Optical Receivers," *IEEE Trans. Comm.*, vol. COM-35, pp. 909–917, Sept. 1987.

8.11. A. F. Elrefaie, J. K. Townsend, M. B. Romeiser, and K. S. Shanmugan, "Computer Simulation of Digital Lightwave Links," *IEEE J. Sel. Areas Comm.*, vol. SAC-6, pp. 94–105, Jan. 1988.

8.12. J. K. Townsend and K. S. Shanmugan, "On Improving the Computational Efficiency of Digital Lightwave Link Simulation," *IEEE Trans. Comm.*, vol. COM-38, pp. 2040–2048, Nov. 1990.

8.13. G. Ascheid, "On the Generation of WMC-Distributed Random Numbers," *IEEE Trans. Comm.*, vol. COM-38, pp. 2117–2118, Dec. 1990.

8.14. F. Jorgensen, *Handbook of Magnetic Recording*, 3rd ed. Blue Ridge Summit, PA: TAB Books, 1988.

8.15. J. C. Mallinson, "Maximum Signal-to-Noise Ratio of a Tape Recorder," *IEEE Trans. Magn.*, vol. MAG-5, pp. 182–186, Sept. 1969 (reprinted in [8.18]).

8.16. L. Thurlings, "Statistical Analysis of Signal and Noise in Magnetic Recording," *IEEE Trans. Magn.*, vol. MAG-16, pp. 507–513, May 1980 (reprinted in [8.18]).

8.17. K. Tarumi and Y. Noro, "A Theoretical Analysis of Modulation Noise in DC Erased Noise in Magnetic Recording," *Appl. Phys. A*, vol. 28, pp. 235–240, Aug. 1982 (reprinted in [8.18]).

8.18. R. M. White (ed.), *Introduction to Magnetic Recording*, IEEE Press Repr. Ser. New York: IEEE Press, 1985.

8.19. S. U. H. Qureshi, "Adaptive Equalization," *Proc. IEEE*, vol. 73, pp. 1349–1387, Sept. 1985.

8.20. K. Davies, *Ionospheric Radio Propagation*, Natl. Bur. Std. Monogr. 80. Washington, DC: U.S. Government Printing Office, 1965.

8.21. W. C. Jakes, Jr. (ed.), *Microwave Mobile Communications*. New York: Wiley, 1974.

8.22. J. H. Painter, S. C. Gupta, and L. R. Wilson, "Multipath Modeling for Aeronautical Communications," *IEEE Trans. Comm.*, vol. COM-21, pp. 658–662, May 1973.

8.23. H. C. Salwen, "Characteristics of Satellite-to-Aircraft Links," *Conf. Rec., IEEE Int. Conf. Comm., ICC'71*, pp. 29-14 to 29-18.

8.24. I. L. Lebow, K. L. Jordan, and P. R. Drouilhet, "Satellite Communications to Mobile Platforms," *Proc. IEEE*, vol. 59, pp. 139–159, Feb. 1971.

8.25. W. D. Rummler, R. P. Coutts, and M. Liniger, "Multipath Fading Channel Models for Microwave Digital Radio," *IEEE Comm. Mag.*, Nov. 1986 (revised and reprinted in [8.26]).

8.26. L. J. Greenstein and M. Shafi (eds.), *Microwave Digital Radio*, IEEE Press Repr. Ser. New York: IEEE Press, 1988.

8.27. P. A. Bello, "Characterization of Randomly Time-Variant Linear Channels," *IEEE Trans. Comm. Syst.*, vol. CS-11, pp. 360–393, Dec. 1963 (reprinted in [8.1]).

8.28. C. C. Watterson, J. R. Juroshek, and W. D. Bensema, "Experimental Confirmation of an HF Channel Model," *IEEE Trans. Comm. Tech.*, vol. COM-18, pp. 792–803, Dec. 1970 (reprinted in [8.1]).

8.29. B. Goldberg, R. L. Heyd, and D. Pochmerski, "Stored Ionosphere," *Conf. Rec., IEEE Int. Conf. Comm., ICC'65*, pp. 619–622 (reprinted in [8.1]).

8.30. J. J. Bussgang, E. H. Getchell, and B. Goldberg, "VHF Channel Simulation," *IEEE EASCON'74 Rec.*, pp. 562–564 (reprinted in [8.1]).

8.31. L. Ehrman, L. B. Bates, J. F. Eschle, and J. M. Kates, "Real-Time Software Simulation of the HF Radio Channel," *IEEE Trans. Comm.*, vol. COM-30, pp. 1809–1817, Aug. 1982.

8.32. R. G. McKay, M. Shafi, and C. J. Carlisle, "Trellis-Coded Modulation on Digital Microwave Radio Systems: Simulations for Multipath Fading Channels," *Conf. Rec., Globecom'88*, vol. 1, paper 8.1.

8.33. M. Kavehrad and P. Balaban, "Digital Radio Transmission Modeling and Simulation," *Conf. Rec., Globecom'89*, vol. 2, paper 35.4.

8.34. A. B. Johnson, "Simulation of Digital Transmission over Mobile Channels at 300 Kb/s," *IEEE Trans. Comm.*, vol. COM-39, pp. 619–627, Apr. 1991.

8.35. K. Brayer (ed.), *Data Communications via Fading Channels*, IEEE Press Repr. Ser. New York: IEEE Press, 1975.

8.36. C. E. Cook, F. W. Ellersick, L. B. Milstein, and D. L. Schilling (eds.), *Spread-Spectrum Communications*, IEEE Press Repr. Ser. New York: IEEE Press, 1983.

8.37. N. Abramson (ed.), *Multiple Access Communications*, IEEE Press Repr. Ser. New York: IEEE Press, 1993.

8.38. M. K. Simon, J. K. Omura, R. A. Scholtz, and B. K. Levitt, *Spread Spectrum Communications*. New York: Computer Science Press, 1985.

Chapter 9

9.1. D. E. Knuth, *The Art of Computer Programming*, Vol. 2; *Seminumerical Algorithms*, 2nd ed. Reading, MA: Addison-Wesley, 1981, Chap. 3.

9.2. R. F. W. Coates, G. J. Janacek, and K. V. Lever, "Monte Carlo Simulation and Random Number Generation," *IEEE J. Sel. Areas Comm.*, vol. 6, pp. 58–66, Jan. 1988.

9.3. N. J. Kasdin, "Discrete Simulation of Colored Noise and Stochastic Processes and $1/f^\alpha$ Power Law Noise Generation," *Proc. IEEE*, vol. 83, pp. 802–827, May 1995.

9.4. G. W. Wornell, "Wavelet-Based Representations for the $1/f$ Family of Fractal Processes," *Proc. IEEE*, vol. 81, pp. 1428–1450, Oct. 1993.

9.5. G. A. Tsihrintzis and C. L. Nikias, "Performance of Optimum and Suboptimum Receivers in the Presence of Impulsive Noise Modeled as an Alpha-Stable Process," *IEEE Trans. Comm.*, vol. COM-43, pp. 904–914, Feb./Mar./Apr. 1995.

Chapter 10

10.1. E. H. Armstrong, "A Method of Reducing Disturbances in Radio Signaling by a System of Frequency Modulation," *Proc. IRE*, vol. 24, pp. 689–740, May 1936.

10.2. K. K. Clarke and D. T. Hess, *Communication Circuits: Analysis and Design*. Reading, MA: Addison-Wesley, 1971, Chap. 11.

10.3. F. M. Gardner, *Phaselock Techniques*. New York: Wiley, 1979, Chap. 9.

10.4. F. J. Harris, "On the Use of Windows for Harmonic Analysis with the Discrete Fourier Transform," *Proc. IEEE*, vol. 66, pp. 51–83, Jan. 1978.

10.5. J. W. Adams, "A Subsequence Approach to Interpolation Using the FFT," *IEEE Trans. Circ. Syst.*, vol. CAS-34, pp. 568–570, May 1987.

10.6. R. E. Crochiere and L. R. Rabiner, *Multirate Digital Signal Processing*. Englewood Cliffs, NJ: Prentice Hall, 1983.

10.7. F. M. Gardner, "Interpolation in Digital Modems. Part I: Fundamentals," *IEEE Trans. Comm.*, vol. COM-41, pp. 501–507, Mar. 1993.

10.8. L. Erup, F. M. Gardner, and R. A. Harris, "Interpolation in Digital Modems. Part II: Implementation and Performance," *IEEE Trans. Comm.*, vol. COM-41, pp. 998–1008, June 1993.

10.9. R. W. Schafer and L. R. Rabiner, "A Digital Signal Processing Approach to Interpolation," *Proc. IEEE*, vol. 61, pp. 692–702, June 1973.

Chapter 11

11.1. R. A. Scholtz, "Frame Synchronization Techniques," *IEEE Trans. Comm.*, vol. COM-28, pp. 1204–1212, Aug. 1980.

11.2. W. H. Press, B. P. Flannery, S. A. Teukolsky, and W. T. Vetterling, *Numerical Recipes: The Art of Scientific Computing*. New York: Cambridge University Press, 1986, Chap. 10.

11.3. G. Ascheid and H. Meyr, "Maximum Likelihood Detection and Synchronization by Parallel Digital Signal Processing," *Conf. Rec., Globecom'84*, vol. 2, paper 32.2.

11.4. L. E. Franks, "Synchronization Subsystems: Analysis and Design," in *Digital Commu-*

nications: Satellite/Earth Station Engineering, K. Feher (ed.). Englewood Cliffs, NJ: Prentice Hall, 1981.

11.5. Special Issue on Synchronization, *IEEE Trans. Comm.*, vol. COM-28, Aug. 1980.

11.6. L. E. Franks, "Carrier and Bit Synchronization in Data Communications: A Tutorial Review," *IEEE Trans. Comm.*, vol. COM-28, pp. 1107–1120, Aug. 1980.

11.7. W. C. Lindsey and M. K. Simon, *Telecommunication Systems Engineering*. Englewood Cliffs, NJ: Prentice Hall, 1973.

11.8. F. M. Gardner, *Phaselock Techniques*, 2nd ed. New York: Wiley, 1979, Chap. 11.

11.9. E. A. Lee and D. G. Messerschmitt, *Digital Communication*. Norwell, MA: Kluwer Academic Publishers, 1988, Chaps. 13–15.

11.10. J. A. C. Bingham, *Theory and Practice of Modem Design*. New York: Wiley, 1988, Chaps. 6 and 7.

11.11. J. J. Stiffler, *Theory of Synchronous Communications*. Englewood Cliffs, NJ: Prentice Hall, 1971.

11.12. J. J. Spilker, Jr., *Digital Communications by Satellite*. Englewood Cliffs, NJ: Prentice Hall, 1977, Chaps. 12 and 14.

11.13. V. K. Bhargava, D. Haccoun, R. Matyas, and P. P. Nuspl, *Digital Communications by Satellite*. New York: Wiley, 1981, Chap. 5.

11.14. J. B. Anderson, T. Aulin, and C.-E. Sundberg, *Digital Phase Modulation*. New York: Plenum Press, 1986, Chap 9.

11.15. J. P. Costas, "Synchronous Communications," *Proc. IRE*, vol. 44, pp. 1713–1718, Dec. 1956.

11.16. W. C. Lindsey and M. K. Simon, "Data-Aided Carrier Tracking Loops," *IEEE Trans. Comm.*, vol. COM-19, pp. 157–169, Apr. 1971.

11.17. W. C. Lindsey and M. K. Simon, "Carrier Synchronization and Detection of Polyphase Signals," *IEEE Trans. Comm.*, vol. COM-20, pp. 441–454, June 1972.

11.18. G. L. Hedin, J. K. Holmes, W. C. Lindsey, and K. T. Woo, "Theory of False Lock in Costas Loops," *IEEE Trans. Comm.*, vol. COM-26, pp. 1–11, Jan. 1978 (reprinted in [11.64]).

11.19. M. K. Simon and J. G. Smith, "Carrier Synchronization and Detection of QASK Signal Sets," *IEEE Trans. Comm.*, vol. COM-22, pp. 98–106, Feb. 1974.

11.20. A. Leclert and P. Vandamme, "Universal Carrier Recovery Loop for QASK and PSK Signal Sets," *IEEE Trans. Comm.*, vol. COM-31, pp. 130–136, Jan. 1983.

11.21. D. N. Godard, "Self-Recovering Equalization and Carrier Tracking in Two-Dimensional Data Communication Systems," *IEEE Trans. Comm.*, vol. COM-28, pp. 1867–1875, Nov. 1980.

11.22. S. Moridi and H. Sari, "Analysis of Four Decision-Feedback Carrier Recovery Loops in the Presence of Intersymbol Interference," *IEEE Trans. Comm.*, vol. COM-33, pp. 543–550, June 1985.

11.23. L. C. Palmer and S. A. Klein, "Phase Slipping in Phaselocked Loop Configurations That Track Biphase or Quadriphase Modulated Carriers," *IEEE Trans. Comm.*, vol. COM-20, pp. 984–991, Oct. 1972.

11.24. A. J. Viterbi and A. M. Viterbi, "Nonlinear Estimation of PSK-Modulated Carrier Phase with Application to Burst Digital Transmission," *IEEE Trans. Inf. Theory*, vol. IT-29, pp. 543–551, July 1983.

11.25. M. K. Simon and J. G. Smith, "Offset Quadrature Communications with Decision-Feedback Carrier Synchronization," *IEEE Trans. Comm.*, vol. COM-22, pp. 1576–1584, Oct. 1974.

11.26. W. R. Braun and W. C. Lindsey, "Carrier Synchronization Techniques for Unbalanced QPSK Signals," *IEEE Trans. Comm.*, vol. COM-26, pp. 1325–1341, Sept. 1978.

11.27. J. R. Lesh, "Costas Loop Tracking of Unbalanced QPSK Signals," *Conf. Rec., IEEE Int. Conf. Comm., ICC'78*, paper 16.2.

11.28. P. Y. Kam, "Maximum Likelihood Carrier Phase Recovery for Linear Suppressed-Carrier Digital Data Modulations," *IEEE Trans. Comm.*, vol. COM-34, pp. 522–527, June 1986.

11.29. W. R. Bennett, "Statistics of Regenerative Digital Transmission," *Bell Syst. Tech. J.*, vol. 37, pp. 1501–1542, Nov. 1958.

11.30. B. R. Saltzberg, "Timing Recovery for Synchronous Data Transmission," *Bell Syst. Tech. J.*, vol. 46, pp. 593–622, Mar. 1967.

11.31. E. Roza, "Analysis of Phase-Locked Timing Extraction Circuits for Pulse Code Transmission," *IEEE Trans. Comm.*, vol. COM-22, pp. 1236–1249, Sept. 1974.

11.32. R. D. Gitlin and J. Salz, "Timing Recovery in PAM Systems," *Bell Syst. Tech. J.*, vol. 50, pp. 1645–1669, May/June 1971.

11.33. U. Mengali, "A Self Bit Synchronizer Matched to the Signal Shape," *IEEE Trans. Aerosp. Electron. Syst.*, vol. AES-7, pp. 686–693, July 1971.

11.34. M. K. Simon, "Nonlinear Analysis of an Absolute Value Type of an Early–Late Gate Bit Synchronizer," *IEEE Trans. Comm.*, vol. COM-18, pp. 589–597, Oct. 1970.

11.35. M. K. Simon, "Optimization of the Performance of a Digital-Data-Transition Tracking Loop," *IEEE Trans. Comm.*, vol. COM-18, pp. 686–690, Oct. 1970.

11.36. L. E. Franks and J. P. Bubrouski, "Statistical Properties of Timing Jitter in a PAM Timing Recovery Scheme," *IEEE Trans. Comm.*, vol. COM-22, pp. 913–920, July 1974.

11.37. F. M. Gardner, "Self-Noise in Synchronizers," *IEEE Trans. Comm.*, vol. COM-28, pp. 1159–1163, Aug. 1980.

11.38. N. A. D'Andrea and U. Mengali, "A Simulation Study of Clock Recovery in QPSK and 9QPRS Systems," *IEEE Trans. Comm.*, vol. COM-33, pp. 1139–1142, Oct. 1985.

11.39. D. Godard, "Passband Timing Recovery in an All-Digital Modem Receiver," *IEEE Trans. Comm.*, vol. COM-26, pp. 517–522, May 1978.

11.40. K. H. Mueller and M. Müller, "Timing Recovery in Digital Synchronous Data Receivers," *IEEE Trans. Comm.*, vol. COM-24, pp. 516–531, May 1976.

11.41. F. M. Gardner, "A BPSK/QPSK Timing-Error Detector for Sampled Receivers," *IEEE Trans. Comm.*, vol. COM-34, pp. 423–429, May 1986.

11.42. M. Oerder and H. Meyr, "Digital Filter and Square Timing Recovery," *IEEE Trans. Comm.*, vol. COM-36, pp. 605–612, May 1988.

11.43. C.-P. J. Tzeng, D. A. Hodges, and D. G. Messerschmitt, "Timing Recovery in Digital Subscriber Loops Using Baudrate Sampling," *Conf. Rec., IEEE Int. Conf. Comm., ICC'85*, vol. 3, paper 37.6.

11.44. M. Moeneclaey, "The Influence of Four Types of Symbol Synchronizers on the Error Probability of a PAM Receiver," *IEEE Trans. Comm.*, vol. COM-32, pp. 1186–1190, Nov. 1984.

11.45. M. Moeneclaey, "A Comparison of Two Types of Symbol Synchronizers for Which Self Noise Is Absent," *IEEE Trans. Comm.*, vol. COM-31, pp. 329–334, Mar. 1983.

11.46. H. Kobayashi, "Simultaneous Adaptive Estimation and Decision Algorithm for Carrier Modulated Data Transmission Systems," *IEEE Trans. Comm.*, vol. COM-19, pp. 268–280, June 1971.

11.47. M. H. Meyers and L. E. Franks, "Joint Carrier Phase and Symbol Timing for PAM Systems," *IEEE Trans. Comm.*, vol. COM-28, pp. 1121–1129, Aug. 1980.

11.48. D. D. Falconer, "Jointly Adaptive Equalization and Carrier Recovery in Two-Dimensional Digital Communication Systems," *Bell Syst. Tech. J.*, vol. 55, pp. 317–334, Mar. 1976.

11.49. U. Mengali, "Synchronization of QAM Signals in the Presence of ISI," *IEEE Trans. Aerosp. Electron. Syst.*, vol. AES-12, pp. 556–560, Sept. 1976.

11.50. U. Mengali, "Joint Phase and Timing Acquisition in Data Transmission," *IEEE Trans. Comm.*, vol. COM-25, pp. 1174–1185, Oct. 1977.

11.51. R. W. D. Booth, "An Illustration of the MAP Estimation Method for Deriving Closed-Loop Phase Tracking Topologies: The MSK Signal Structure," *IEEE Trans. Comm.*, vol. COM-28, pp. 1137–1142, Aug. 1980.

11.52. F. D. Natali, "AFC Tracking Algorithms," *IEEE Trans. Comm.*, vol. COM-32, pp. 935–947, Aug. 1984 (reprinted in [11.64]).

11.53. F. M. Gardner, "Properties of Frequency Difference Detectors," *IEEE Trans. Comm.*, vol. COM-33, pp. 131–138, Feb. 1985 (reprinted in expanded form in [11.64]).

11.54. C. F. Schaeffer, "The Zero-Beat Method of Frequency Discrimination," *Proc. IRE*, vol. 30, pp. 365–367, Aug. 1942.

11.55. D. Richman, "Color Carrier Reference Phase Synchronization Accuracy in NTSC Color Television," *Proc. IRE*, vol. 42, pp. 106–133, Jan. 1954.

11.56. J. H. Park, Jr., "An FM Detector for Low S/N," *IEEE Trans. Comm.*, vol. COM-18, pp. 110–118, Apr. 1970.

11.57. D. G. Messerschmitt, "Frequency Detectors for PLL Acquisition in Timing and Carrier Recovery," *IEEE Trans. Comm.*, vol. COM-27, pp. 1288–1295, Sept. 1979.

11.58. T. Alberty and V. Hespelt, "A New Jitter Free Frequency Error Detector," *IEEE Trans. Comm.*, vol. COM-37, pp. 159–163, Feb. 1989.

11.59. A. N. D'Andrea and U. Mengali, "Performance of a Quadricorrelator Driven by Modulated Signals," *IEEE Trans. Comm.*, vol. COM-38, pp. 1952–1957, Nov. 1990.

11.60. H. Sari and S. Moridi, "New Phase and Frequency Detectors for Carrier Recovery in PSK and QAM Systems," *IEEE Trans. Comm.*, vol. COM-36, pp. 1035–1043, Sept. 1988.

11.61. M. K. Simon and D. Divsalar, "Doppler Corrected Differential Detection of MPSK," *IEEE Trans. Comm.*, vol. COM-37, pp. 99–109, Feb. 1989.

11.62. S. Bellini, C. Molinari, and G. Tartara, "Digital Frequency Estimation in Burst Mode QPSK Transmission," *IEEE Trans. Comm.*, vol. COM-38, pp. 959–961, July 1990.

11.63. E. Biglieri, F. Abrishamkar, and Y.-C. Jou, "Doppler Frequency Shift Estimation for Differentially Coherent CPM," *IEEE Trans. Comm.*, vol. COM-38, pp. 1659–1663, Oct. 1990.

11.64. W. C. Lindsey and C. M. Chie, *Phase-Locked Loops*, reprint volume. New York: IEEE Press, 1986.

11.65. J. M. Tribolet, "A New Phase Unwrapping Algorithm," *IEEE Trans. Acoust. Speech Signal Process.*, vol. ASSP-25, pp. 170–177, Apr. 1977.

11.66. K. P. Zimmerman, "On Frequency-Domain and Time-Domain Phase Unwrapping," *Proc. IEEE*, vol. 75, pp. 519–520, Apr. 1987.

11.67. H. al-Nashi, "Phase Unwrapping of Digital Signals," *IEEE Trans. Acoust. Speech Signal Process.*, vol. ASSP-37, pp. 1693–1702, Nov. 1989.

11.68. J. B. Thomas, *Statistical Communication Theory*. New York: Wiley, 1969, Chap. 6.

11.69. M. Abramowitz and I. A. Stegun (eds.), *Handbook of Mathematical Functions*, Natl. Bur. Std. Appl. Math. Ser. 55. Washington, DC: U.S. Government Printing Office, 1964, Chap. 6.

11.70. I. S. Gradshteyn and I. M. Ryzhik, *Table of Integrals, Series, and Products*, 4th ed. New York: Academic Press, 1965, Chap. 1.

11.71. B. E. Paden, "A Matched Nonlinearity for Phase Estimation of a PSK-Modulated Carrier," *IEEE Trans. Inf. Theory*, vol. IT-32, pp. 419–422, May 1986.

11.72. M. Luise and R. Reggiannini, "A Fast Carrier Frequency Estimation Algorithm for Burst-Mode M-PSK Satellite Transmissions," *Proc. 3rd Int. Workshop on DSP Techniques*, pp. 421–434, European Space Agency: ESA WPP-038, Sept. 1992.

Chapter 12

12.1. A. J. Viterbi, "Convolutional Codes and Their Performance in Communication Systems," *IEEE Trans. Comm. Tech.*, vol. COM-19, pp. 751–772, Oct. 1971.

12.2. G. D. Forney, Jr., "Maximum Likelihood Sequence Estimation of Digital Sequences in the Presence of Intersymbol Interference," *IEEE Trans. Inf. Theory*, vol. IT-18, pp. 363–378, May 1972.

12.3. A. J. Viterbi and J. K. Omura, *Principles of Digital Communication and Coding*. New York: McGraw-Hill, 1979, Chap. 4.

12.4. S. Stein and J. J. Jones, *Modern Communication Principles*. New York: McGraw-Hill, 1967, Sec. 10-2.

Chapter 13

13.1. M. C. Jeruchim, "Techniques for Estimating the Bit Error Rate in the Simulation of Digital Communication Systems," *IEEE J. Sel. Areas Comm.*, vol. SAC-2, pp. 153–170, Jan. 1984.

13.2. M. D. Knowles and A. I. Drukarev, "Bit Error Rate Estimation for Channels with Memory," *IEEE Trans. Comm.*, vol. COM-36, pp. 767–769, June 1988.

13.3. P. Balaban, "Statistical Evaluation of the Error Rate of the Fiber Guide Repeater Using Importance Sampling," *Bell Syst. Tech. J.*, vol. 55, pp. 745–766, July 1976.

13.4. K. S. Shanmugam and P. Balaban, "A Modified Monte-Carlo Simulation Technique for the Evaluation of Error Rate in Digital Communications Systems," *IEEE Trans. Comm.*, vol. COM-28, pp. 1916–1924, Nov. 1980.

13.5. B. R. Davis, "An Improved Importance Sampling Method for Digital Communication System Simulations," *IEEE Trans. Comm.*, vol. COM-34, pp. 715–719, July 1986.

13.6. P. M. Hahn and M. C. Jeruchim, "Developments in the Theory and Application of Importance Sampling," *IEEE Trans. Comm.*, vol. COM-35, pp. 706–714, July 1987.

13.7. D. Lu and K. Yao, "Improved Importance Sampling Technique for Efficient Simulation of Digital Communication Systems," *IEEE J. Sel. Areas Comm.*, vol. SAC-6, pp. 67–75, Jan. 1988.

13.8. G. Orsak and B. Aazhang, "On the Theory of Importance Sampling Applied to the Analysis of Detection Systems," *IEEE Trans. Comm.*, vol. COM-37, pp. 332–339, Apr. 1989.

13.9. M. C. Jeruchim et al., "An Experimental Investigation of Conventional and Efficient Importance Sampling," *IEEE Trans. Comm.*, vol. COM-37, pp. 578–587, June 1989.

13.10. J. S. Sadowsky and J. A. Bucklew, "On Large Deviations Theory and Asymptotically Efficient Monte Carlo Estimation," *IEEE Trans. Inf. Theory*, vol. IT-36, pp. 579–588, May 1990.

13.11. R. J. Wolfe, M. C. Jeruchim, and P. M. Hahn, "On Optimum and Suboptimum Biasing Procedures for Importance Sampling in Communication Simulation," *IEEE Trans. Comm.*, vol. COM-38, pp. 639–647, May 1990.

13.12. G. C. Orsak and B. Aazhang, "Constrained Solutions in Importance Sampling via Robust Statistics," *IEEE Trans. Inf. Theory*, vol. IT-37, pp. 307–316, Mar. 1991.

13.13. S. A. Rhodes and L. C. Palmer, "Computer Simulation of a Digital Satellite Communications System Utilizing TDMA and Coherent Quadriphase Signaling," *Conf. Rec., IEEE Int. Conf. Comm.*, pp. 34.19–34.24, June 1972.

13.14. D. Hedderly and L. Lundquist, "Computer Simulation of a Digital Satellite Communications Link," *IEEE Trans. Comm.*, vol. COM-21, pp. 321–325, April 1973.

13.15. M. Abramowitz and L. A. Stegun, *Handbook of Mathematical Functions*, Natl. Bur. Std. Appl. Math. Ser. 55. Washington, DC: U.S. Government Printing Office, 1964.

13.16. V. Castellani, M. Elia, L. LoPresti, and M. Pent, "Performance Analysis of a DCPSK Up-Link for a Regenerative Satellite Repeater," *IEEE J. Sel. Areas Comm.*, SAC-1, pp. 63–73, Jan. 1983.

13.17. A. Morello and M. Pent, "Semianalytic BER Evaluation Method for FSK Demodulation with Discriminator and Post-detection Filter," *Conf. Rec., IEEE Int. Conf. Comm.*, ICC'88, vol. 3, paper 51.5.

13.18. M. Pent, L. LoPresti, G. D'Aria, and G. De Luca, "Semianalytic BER Evaluation by Simulation for Noisy Nonlinear Bandpass Channels," *IEEE J. Sel. Areas Comm.*, vol. SAC-6, pp. 34–41, Jan. 1988 (comments: R. Giubilei, vol. SAC-7, pp. 167–169, Jan. 1989).

13.19. N. C. Beaulieu, "A Simple Series for Personal Computer Computation of the Error Function ($Q \cdot$)," *IEEE Trans. Comm.*, vol. COM-37, pp. 989–991, Sept. 1989.

13.20. P. O. Boerjesson and C. E. Sundberg, "Simple Approximations of the Error Function $Q(x)$ for Communications Applications," *IEEE Trans. Comm.*, vol. COM-27, pp. 639–643, Mar. 1979.

13.21. A. J. Viterbi, *Principles of Coherent Communication*. New York: McGraw-Hill, 1966, Chap. 4.

13.22. H. Meyr and G. Ascheid, *Synchronization in Digital Communications*. New York: Wiley, 1990, Chap. 6.

13.23. S. O. Rice, "Noise in FM Receivers," Chap. 25 in *Time Series Analysis*, M. Rosenblatt (ed.). New York: Wiley, 1963.

13.24. N. B. Mandayam and B. Aazhang, "Importance Sampling for Analysis of Direct Detection Optical Communication Systems," *IEEE Trans. Comm.*, vol. COM-43, pp. 229–239, Feb./Mar./Apr. 1995.

13.25. K. B. Letaif, "Performance Analysis of Digital Lightwave Systems Using Efficient Computer Simulation Techniques," *IEEE Trans. Comm.*, vol. COM-43, pp. 240–251, Feb./Mar./Apr. 1995.

Chapter 15

15.1. N. S. Jayant and P. Noll, *Digital Coding of Waveforms*. Englewood Cliffs, NJ: Prentice Hall, 1984, Chap. 4.

15.2. K. W. Cattermole, *Principles of Pulse Code Modulation*. Sevenoaks, Kent, England: Ilife, 1969, Chap. 3.

15.3. A. Antoniu, *Digital Filters: Analysis and Design*. New York: McGraw-Hill, 1979, Chap. 11.

15.4. T. A. C. M. Claasen, W. F. G. Mecklenbräuker, and J. B. H. Peek, "Second-Order Digital Filter with Only One Magnitude-Truncation Quantizer and Having Practically No Limit Cycles," *Electron. Lett.*, vol. 9, pp. 531–532, Nov. 1, 1973 (reprinted in *Selected Papers in Digital Signal Processing, II*, New York: IEEE Press, 1976).

15.5. R. A. Pepe and J. D. Rogers, "Simulation of Fixed-Point Operations with High-Level Languages," *IEEE Trans. Acoust. Speech Signal Process.*, vol. ASSP-35, pp. 116–118, Jan. 1987.

15.6. P. M. Ebert, J. E. Mazo, and M. G. Taylor, "Overflow Oscillations in Digital Filters," *Bell Syst. Tech. J.*, vol. 48, pp. 2999–3020, Nov. 1969.

Chapter 16

16.1. D. G. Messerschmitt, "A Tool for Structured Functional Simulation," *IEEE J. Sel. Areas Comm.*, vol. SAC-2, pp. 137–147, Jan. 1984.

16.2. J. W. Modestino and K. R. Matis, "Interactive Simulation of Digital Communication Systems," *IEEE J. Sel. Areas Comm.*, vol. SAC-2, pp. 51–76, Jan. 1984.

16.3. D. G. Messerschmitt, "Structured Interconnection of Simulation Programs," *Conf. Rec. Globecom'84*, vol. 3, paper 24.1.

16.4. E. A. Lee and D. G. Messerschmitt, "Synchronous Data Flow," *Proc. IEEE*, vol. 75, pp. 1235–1245, Sept. 1987.

16.5. R. E. Crochiere and L. R. Rabiner, *Multirate Digital Signal Processing*. Englewood Cliffs, NJ: Prentice Hall, 1983.

Chapter 17

17.1. L. Erup, *Unified Interchange Format for Computer Data (Standard File Format)*, ESTEC Report ESA STR-225. Noordwijk, The Netherlands: European Space Agency, Jan. 1988.

17.2. L. Ljung and T. Söderstrom, *Theory and Practice of Recursive Identification*. Cambridge, MA: MIT Press, 1983.

17.3. L. Ljung, *System Identification Toolbox: User's Guide*. Natick, MA: Mathworks, 1986.

17.4. L. Ljung, *System Identification: Theory for the User*. Englewood Cliffs, NJ: Prentice Hall, 1987.

17.5. T. Söderstrom and P. Stoica, *System Identification*. Hemel Hempstead, Hertfordshire, England: Prentice Hall, International, 1989.

17.6. J. Schoukens and R. Pintelon, *Identification of Linear Systems*. Elmsford, NY: Pergamon Press, 1991.

17.7. W. H. Press, B. P. Flannery, S. A. Teukolsky, and W. T. Vetterling, *Numerical Recipes*. New York: Cambridge University Press, 1986.

17.8. A. A. M. Saleh, "Frequency-Independent and Frequency-Dependent Nonlinear Models of TWT Amplifiers," *IEEE Trans. Comm.*, vol. COM-29, pp. 1715–1719, Nov. 1981.

17.9. A. Antoniu, *Digital Filters: Analysis and Design*. New York: McGraw-Hill, 1979, Chap. 5.

17.10. J. D. Rhodes, *Theory of Electrical Filters*. New York: Wiley, 1976, Chap. 2.

17.11. W. C. Yengst, *Procedures of Modern Network Synthesis*. New York: Macmillan, 1964, Chap. 14.

17.12. L. Weinberg, *Network Analysis and Synthesis*. New York: McGraw-Hill, 1962, Chap. 11.

17.13. R. W. Hamming, *Numerical Methods for Scientists and Engineers*. New York: McGraw-Hill, 1962, Chap. 20.

17.14. F. B. Hildebrand, *Introduction to Numerical Analysis*. New York: McGraw-Hill, 1956, Chap. 9.

Chapter 18

18.1. D. G. Childers (ed.), *Modern Spectrum Analysis*. New York: IEEE Press, 1978.

18.2. S. B. Kesler (ed.), *Modern Spectrum Analysis, II*. New York: IEEE Press, 1986.

18.3. J. R. Davey, "Digital Data Signal Space Diagrams," *Bell Syst. Tech. J.*, vol. 43, pp. 2973–2984, Nov. 1964.

18.4. P. D. Welch, "The Use of the Fast Fourier Transform for the Estimation of Power Spectra: A Method Based on Time Averaging Over Short, Modified Periodograms," *IEEE Trans. Audio Electroacoust.*, vol. AU-15, pp. 70–73, June 1967 (reprinted in [18.1]).

18.5. F. J. Harris, "On the Use of Windows for Harmonic Analysis with the Discrete Fourier Transform," *Proc. IEEE*, vol. 66, pp. 51–83, Jan. 1978 (reprinted in [18.2]).

Chapter 19

19.1. E. A. Lee and T. M. Parks, "Dataflow Process Networks," *Proc. IEEE*, vol. 83, pp. 773–799, May 1995.

INDEX

Additive white Gaussian noise, *see* AWGN
AGC (automatic gain control), 227, 248
Aliasing, 126–134
   energy, 128–129, 153–157
   error, 128
      magnitude, 129–130, 157–159
   examples, 130–134
   filter poles and zeros, 191–192
   power, 128–129
Alignment, *see* Synchronization
Alphabet, 10, 27, 41
Ambiguity, 346, 360, 375
   resolution, 209–210
Amplifiers, 226–235, 239–240. *See also* Nonlinear amplifiers
   gain scaling, 226–227
   linear, 226–227
   nonlinear, 227–235, 239–240
Analytic signal, *see also* Complex envelope; Signal
   definition, 38
   periodogram, 60
Attenuators, *see* Amplifiers, linear
Averages, 49
AWGN (additive white Gaussian noise), 13, 70
   additive property, 62
   complex envelope of, 62
   generation of, 263–266
   linear invariance, 62
   probability density (pdf), 61
   scaling, 264–266, 271–273
   spectral density, 61–62
   variance, 62

Bandwidth, definitions, 178
BER (bit error ratio), 356, 358. *See also* Error probability; Performance evaluation; Symbol detection, error
Biquads, *see* Filters, biquadratic sections
Bit error ratio, *see* BER
Bit grouping, 206–207
Bit-true operations, 536
Block editing:
   extracting and inserting, 462
   zero padding, 462
Blocks, 151, 196
   circular, 142, 143–144, 168
   concatenated, 142, 145
   linear 142, 145, 167
BPSK (binary phase shift keying), 40
Buffers, 455–457
   dynamic, 456
   fill tags, 456–457
   flexible, 455
   multiple, 457
   rules, 456–457
   single-element, 457
Builder routines, 465–497
   conditioning, 468–470
      analytical functions, 477
      conjugate extension, 477
      delay integration, 474–477
      frequency response of filters, 473–477
      input data (filters), 473
      nonlinear amplifiers, 470–473
      nonlinear functions, 470–471
      tasks, 471–473

Builder routines (*Continued*)
    transfer functions, 473–474
    units and formats, 475
  device-data files, 466–468, 483
  filters, 481–497
    A/D transformation, 493
    categories, 483–484
    constituents, 481–484
    formula evaluation, 489–490
    frequency determination, 485–487
    frequency domain, 485–490, 496
    frequency-domain interpolation, 487–489
    frequency indexing, 487
    frequency limits, 490
    Nyquist filters, 484–485
    partitioned biquads, 489–491
    procedure, 491–492
    rational transfer functions, 484
    tasks, 485–487
    time and frequency scales, 492, 493–495
    time domain, 490–494, 496–497
    user parameters, 486–487
  nonlinear functions, 478–481, 495
    artifacts of interpolation, 480
    fitting, 478–479
    interpolation, 479–481
    overview, 478
  originator, 468
  overview, 466–468
  parameter files, 466–468, 482–483
  user, 468

Calibration:
  phantom noise, 376–379, 402–406
  signal input to nonlinear amplifier, 233–234
  synchronization, 297–299
  SNR, 378–379, 381–382
Carrier, 11
Channels:
  memory, 204
  time varying, 256–257
Coherent demodulation, *see* Demodulation, coherent
Compilers, 533–534
Complex envelope, *see also* Analytic signal; Signal
  bandlimiting, 38
  in carrier regeneration, 312–313
  CPM, 45
  definition, 37, 39
  periodogram, 60
  polar format, 38
  power, 55–56
  rectangular format, 37
  spectrum, 57–58
Composite commands, 531–532
Constellations, 41–42, 69, 114, 206
Continuous phase modulation, *see* CPM
Control signals, 227, 418–423
  accumulators, 422
  configuration, 418
  initial values, 421–422
  integrators, 420–421
  loop filters, 420–422
  lowpass filters, 420–421
  numerical precision, 422
  properties, 419–420
  state memory, 421–422
Convolution, 76–77, 166, 195
Coordinate system, 164–166, 196
  polar, 38, 165
  rectangular, 37, 165–166
Correlation:
  auto, 56–57
  cross, 56–57
  of cyclostationary signal, 56
  ensemble, 56–57, 69
  of signals, 56–57
  for synchronization, 298–299, 334–338
  time-averaged, 56–57
CPM (continuous phase modulation), 44–47, 224–226
  complex envelope, 45
  data memory, 46
  definition, 44–45
  frequency pulse, 45, 225
  full response, 45–46
  modulation index, 45
  MSK, *see* MSK
  multi-h, 45
  partial response, 46
  phase, 45, 225
  phase accumulation, 226
  phase pulse, 45
  pulse shaping, 225–226
Crosstalk, *see* Interference
Cycle slips, 360–361, 400–403
Cyclostationarity, 48, 56, 69

Data base, 535
Data entry, *see* Entry routines
Data formats, 141–145
  block, *see* Blocks
  elements, 141
  impact on scheduling, 142

points, 141
samples, 141
single point, *see* Single point data
Data import and export, 535–536
Data types, 530–531
Decimation, *see* Resampling
Delay in simulation, 443
Delay integration:
  rules, 474–476
  treatment of nulls, 476–477
Delay, 284–286, 294–295
  block ends, 286
  continuous adjustment, 285–286
  discrete, 284–285
  dual of frequency shift, 285–286
  by interpolation, 293
Demodulation, 14, 99–104, 246–247
  coherence, 100–104, 121
  coherent, 14, 100–101
  I-Q, 99–100, 102, 246–247
  noncoherent, 14, 103–104
Detection of symbols, *see* Symbol detection
Device data files, *see* Builder routines
DFT (Discrete Fourier transform), 136–138, 150–151
  circularity, 137
  definition, 136–137
  inverse, 137
  periodicity, 137
Differential decoding, 102
Differential demodulation, *see* Differential detection
Differential detection, 14, 101–102, 350–353, 354
  CPM, 353
  multisymbol, 352–353
  pi/M-PSK, 352
  purpose, 350–351
  rules, 351–352
Differential encoding, 101, 209–213, 351
  block size, 211
  examples, 241–243
  with Gray encoding, 212
  rules, 210–211
  square constellations, 241–243
  state, 211
Differentially coherent demodulation, *see* Differential detection
Digital implementation, 249
Digital operations, 8
Digital synthesis, 536
Digitization, 124–135, 150
Dimensions, 124

Discrete Fourier transform, *see* DFT
Displays:
  abscissa labels, 508
  abscissa units, 508
  abscissa wrapping, 509
  constellations, 505–506
  delay of filter, 513–515
  error ratio, 506–507
  eye patterns, 501–502
  Fourier transforms of signals, 502–503
  frequency domain, 502–504
  frequency response of filters, 503–504
  histograms, 507
  leakage-free spectra, 518–519
  ordinate scales and units, 509–510
  power-spectrum ordinates, 510–512
  signal space, 504–505
  spectrum, 504
  spectrum conditioning, 515–519, 521–522
    smoothing, 515–517
    weighting windows, 517–519
  transient response of filters, 500–501
  types, 498–507, 519–520
  units and scales, 507–512, 520–521
  waveforms, 499
Distortion, 5, 126, 177, 193
  AM-AM conversion, 228
  AM-PM conversion, 228
Diversity, 258
Documentation, 534
Domain:
  frequency, 23
  time, 23
Doppler shift, 13, 140, 256
Downsampling, *see* Resampling

Echo cancellers, 14
Energy, 47–49, 145–150, 152–153
  of analog pulse trains, 147–148
  of continuous and sampled signals, 146
  data-averaged, 148
  per bit, 49
  relations between analog and digital signals, 147, 159–161
  of truncated pulse trains, 148
Entry routines:
  file formats, 469
  tools, 469–470
Equalizers, 14
Error counting, *see* Performance evaluation, Monte Carlo
Error probability. *See also* BER; Performance evaluation; Symbol detection, error

Error probability (*Continued*)
  evaluation of probability integral Q(X), 394–395
  theoretical, 374, 388–394
Excess bandwidth, 29
Expectation, 48

Fading, 257–258
Fast Fourier transform, *see* FFT
Feedback, 144, 249
Feedback loops, 409–418, 423
  block failure, 410–411
  indexing, 412
  inherent delay, 412–415
  initialization, 412–415
  multiple, 415–418
  open for simulation, 411–412
  quantized, 429, 430
  simulation constraints, 409–410
  single-point format, 410–411
FFT (Fast Fourier transform), 137
Fill tags, *see* Buffers
Filters, 11, 71–95, 162–198
  A/D transformation, 174–195
    aliasing of poles and zeros, 191–192
    bilinear, 191
    dc gain, 192
    frequency domain, 175–184
    frequency warping, 193
    impulse invariant, 187–188
    matched-z, 191
    stability, 191
    step invariant, 188–190
    time domain, 185–195
  adaptive, 14, 168
  analog, 71–95
  balanced and unbalanced, 172
  biquadratic sections, 74, 87–88, 90–95, 170, 489–490
  causal, 76
  complex, 72, 76–86, 88–89, 170–173
    bandpass, 78–82
    decomposition to real, 72, 78–79, 88
    envelope, 77–78, 81
  complex conversion, 82–86, 89, 90–95, 178–181, 186
    frequency domain, 83–85, 178–181
    time domain, 86, 186
  convolution, 76–77, 166, 195
  difference equation, 163, 168–170, 195
  differential equation, 71–72, 87
  digital simulation, 162–198
  envelope, 89
  FIR (finite impulse response), 166, 194–195
  frequency domain, 163–166, 195
    cascading sections, 184
    circularity, 164
    computation, 164
    coordinates of data, 164–166
    frequency window, 184
    linear convolution, 173–174
    overlap and add/save, 174
  frequency response, 74–75
  IIR (infinite impulse response), 166
  IIR truncation to FIR, 194, 406–408
  impulse response, 75–77, 87, 163, 195
  LP-BP (lowpass to bandpass) transformation, 176–178, 186
  matched, *see* Matched filters
  noncausal, 86
  Nyquist, *see* Nyquist
  preparation, 176, 196–197
  pseudobandpass, 179–181
  pulse shaping, *see* Pulse shaping filters
  real, 76
  signal transmission through bandpass, 79–82
  simulation algorithms, 163–174
  transfer function:
    nonrational, 75
    partitioning, 74
    poles, 73–74
    rational, 72–75
    relation to difference equation, 169–170
    zeros, 73–74
  transfer function, 72–75, 87, 163
  transversal, 166
  unit-sample response, 163
  $x/\sin(x)$ compensation, 107
Folding frequency, 126
Fourier transform. *See also* DFT; FFT
  continuous, 20, 23, 28, 30–31, 45, 47, 50, 51, 57, 74–76, 79–81, 98, 126, 136
Frequency discriminator, 102–103
Frequency increment, 138–139
Frequency modulator, 99. *See also* Frequency shifters
Frequency response, *see* Filters, frequency response
Frequency scaling, 181–183, 193–194, 493–495
Frequency shift keying, *see* FSK; MFSK
Frequency shifters, 98, 276–284, 293–294. *See also* NCO
  analytic signal, 284
  complex signals, 277
  frequency domain, 278–282
  interpolation, 279

leakage hazard, 279–282
phase accumulation, 277–278
real signals, 282–284
time domain, 277–278
Frequency span, 138–139
Frequency translation, 246
Frequency warping, 193
Frequency/time relations, 138
Frequency/time scales, 138–139
FSK (frequency shift keying), 34, 44, 223–224

Gain control, *see* AGC
Graphical displays, *see* Displays

Hard decisions, 115
Hilbert transform, 39, 43, 172

Importance sampling, 356, 363–364
Impulse noise, *see* Noise, impulse
Impulse response, *see* Filters, impulse response
Impulse trains, *see* Pulse shaping, impulse trains
Indexes:
    frequency, 136–137, 138
    time, 138
Integrate and dump, 309–310
Interference, 13, 252–253
    adjacent channel, 252
    cochannel, 252
Interpolation, 220, 247, 286–293, 295. See also Resampling
    basepoints, 287
    characteristics, 288
    for clock extraction, 324
    computation, 291–292
    control, 307–309, 324, 451–452
    cubic, 289–290
    downsampling, 292
    filter, 287
    fractional interval, 287
    frequency domain, 487–489
        amplitude, 487–488
        phase, 488
        treatment of nulls, 488–489
    ideal, 288
    inverse, 343–344
    Lagrange, 289
    linear, 289
    model, 287–288
    out-of-range points, 480–481
    polynomial, 288–290, 479–480
    for resampling, 447, 451–452
    simulation, 290–291

splines, 480
state memory, 291
strobes, 306–309
timing, 292
upsampling, 292
variable delay, 293
Interpreters and compilers, 532–534
Intersymbol interference, *see* ISI
ISI (intersymbol interference), 21, 29, 67

Jamming, 13. See also Interference

Laplace transform, 72–74
    relation to Fourier transform, 74–75

M-ary frequency shift keying, *see* MFSK
Matched filters, 109–112, 122
    colored noise formulation, 112
    examples, 111–112
    impulse response, 110
    properties, 110–111
    signal-to-noise ratio, 109
    sufficient statistic, 110–111
    transfer function, 110
    white noise formulation, 109–110
Message sources, 8, 200–205
MFSK (M-ary frequency shift keying), 104
Minimum shift keying, *see* MSK
MLSE (maximum likelihood sequence estimation), 112–113
Models:
    analog filters, 162–198
    processes, 124–161
    signals, 124–161
Modulation, 11
    amplitude (AM), 34, 97, 120
    angle, 34, 97–98 120
    complex envelope, 97
    envelope variation, 35
    formats, 34–35, 67–68
    frequency (FM), 35. See also Frequency shifters
    index of CPM, 45
    linear, 35, 97
    nonlinear, 35
    phase (PM), 35
    phase, *see* Phase shifters
    pulse amplitude, *see* PAM
    quadrature amplitude, *see* QAM
    residual carrier, 35–36
    spectrum, 35
Modulators, 220–226, 237–239. See also Nonlinear modulation

Modulators (*Continued*)
  complex envelope for linear modulation, 220
  real signal, 221
Modules, 2, 140–141, 151
  parameters, 141
  signal processes, 141
  state memory, 141
Monte Carlo method, 251, 356–364. *See also*
    Performance evaluation, Monte Carlo
MPSK (*M*-ary phase shift keying), 42–43, 101
MSK (minimum shift keying), 23, 46–47
  as CPFSK, 46
  as CPM, 46
  as OQPSK, 46
  symbol interval, 46–47
Multipath, 257–258

NCO (number controlled oscillator), 221–222,
    224, 226, 277–278, 330–331
Noise, 12, 61–66
  additive, 12
  additive white Gaussian, *see* AWGN
  colored, 252, 266–267
  complex envelope of AWGN, 269–273
  impulse, 65, 253–254
  in magnetic media, 255
  non-Gaussian, 267
  periodograms, 63–64
  phase, 13, 65–66, 267
  quantum, 13, 65
  shot, 254–255
  uniformly distributed, 261–263
  white, 61
Nonlinear amplifiers, 227–235
  backoff, 233
  curve fitting, 231–232
  data conditioning, 232–233
  functions, 229–232
  input loading, 233
  interpolation, 229–231
  models, 227–229
  signal level adjustment, 233–234
  spectral spreading, 234–235
  spline interpolation, 231
Nonlinear modulation:
  CPM, 224–226
  FSK, 223–224
  PPM, 222–223
Nonlinear processes, 116–119, 123
  envelope nonlinearity, 117–119
  frequency doubler, 118
  hard limiter, 118–119
  memoryless, 116–117

  square-law device, 118
  superposition, 116–117
  zero memory, 116–117
Number controlled oscillator, *see* NCO
Number representation, 440–441, 444
Nyquist:
  cosine rolloff, 30–31
  criterion, 28
  filters, 27–31
  frequency, 127
  properties, 28–31
  shaping, 27–31, 67
  spectrum, 28–30

Overflow, 441–443, 445

PAM model, 21, 27, 40–43, 66–67, 104–109
PAM-QAM link:
  complex envelope, 104–106
  impulse assumption, 106–107
  model, 104–106, 121–122
  strobes, 106
  transmit filters, 106–107
Parameter files, *see* Builder routines
Parameters, *see* Modules, parameters
Parseval's rule for sample sequences, 146
Partial response, 43, 46
Performance evaluation, 250–251. *See also*
    Symbol detection
  configuration, 250–251
  delay and phase alignment, 381, 395–402
  error counting, 251
  Monte Carlo method, 251, 380–381
    BERC meters, 361
    confidence limits and levels, 361–362
    configuration, 357–361
    cycle slips, 360–361
    delay and phase alignment, 358–361
    error counting, 358
    importance sampling, 363–364
    independence of errors, 362–363
    phase ambiguity, 360
    statistics, 361–363
    symbol comparison, 358
  QA (quasianalytic) method, 251–252, 364–381
    ambiguity resolution, 375
    configuration, 366–367
    decision formulas, 369–374
    decision regions, 367–369
    error probability, 367–374
    noise calibration, *see* Phantom noise
    noise-free errors, 373
    non PAM-QAM formats, 375–376

overview, 364–367
  probability of error, 373–374, 382–388
  restrictions, 365–366
  time invariance, 379–380
  signal and noise calibration, 381–382
Periodogram, 50, 69, 510–512
  of analytic signal, 60
  of bandpass signal, 59
  of complex envelope, 60
  examples, 52–53
  noise, 63–64
PFM (pulse frequency modulation), 34
Phantom noise, 251, 376–379, 402–406
  calibration:
    noise transmission, 376–377, 402–405
    procedure, 377–378
    signals, 405–406
    SNR specification, 378–379
    statistics, 404–405
Phase accumulation, 98. *See also* NCO
  for CPM, 226
  for FSK, 224
  for real modulated signals, 221–222
Phase detectors, 325–327
Phase modulator, 98–99. *See also* Phase shifters
Phase noise, *see* Noise, phase
Phase shifters, 274–276, 293
  analytic signal, 284
  complex signals, 274–275
  computation, 275
  frequency domain, 275
  real signals, 276, 284
Phase unwrapping, 313–314
pi/4-QPSK (4-8PSK), 241
pi/M-PSK, 352
Plots, *see* Displays
Power, 47–49 145–150, 152–153
  average, 49
  estimation, 51–52, 54–55, 69, 149
  in complex envelope, 55–56
  time-averaged, 69
PPM (pulse position modulation), 31–32, 222–223
Primitive commands, 531–532
Process parameter files, *see* Builder routines
Processes, 2. *See also* Modules, signal processes
Pseudorandom sequences, *see* Shift register sequences
Pulse:
  bandlimited, 20–21, 23, 66
  causal, 23
  complex, 43
  cosine rolloff, 67
  crowding, 21
  energy, 47–49, 69
  examples, 21–23
  noncausal, 23
  nulls, 28, 31, 67
  Nyquist, *see* Nyquist
  overlap, 28, 216
  partial response, 43
  raised cosine, 132–133
  rectangular, 130–132
  sinc, 23, 29, 31
  spectrum, 23–27, 49
  tails, 28, 67
  time-limited, 20–21, 66
Pulse amplitude modulation, *see* PAM
Pulse frequency modulation, *see* PFM
Pulse position modulation, *see* PPM
Pulse shaping, 213–220, 237
  for CPM, 225–226
  impulse trains, 214
  square pulses, 215, 217
  time base, 213–215
Pulse shaping filters, 215–220
  amplitude of impulse response, 220
  circular block, 217, 218
  DFT, 218
  FIR, 215–217
  frequency domain, 218–219
  IIR, 217–218
  initial transient, 217
  initial values, 217
  rational transfer function, 219
  relation to interpolation, 220
  state variables, 216–217

QAM (quadrature amplitude modulation):
  4QAM, 40
  balanced, 39
  constellations, 41–42
  definition, 36–37
  examples, 40–43
  format:
    polar, 38
    rectangular, 36–38
  M-ary (MQAM), 41–42
  M-ary phase shift keying (MPSK), 42–43
  offset (OQAM), 39
  properties, 68–69
  staggered (OQAM), 39
  unbalanced, 39
QPSK (quadrature phase shift keying), 40
  offset (OQPSK), 40
  staggered, 40

Quadrature amplitude modulation, *see* QAM
Quantization, 125, 426–440, 444
  A/D convertors, 437
  axis translations, 428
  dead zone, 429, 430
  dither, 430
  error, 429
  feedback, 429, 430
  of filter coefficients, 438
  input scaling, 428
  levels, 428
  midriser, 430–432
  midstep, 429–430
  models, 433–435
  noisy inputs, 432–433
  offset, 432
  polarity, 429
  rectification, 429
  rounding, 429
  saturation, 427–428
  scales, 428
  sign-magnitude, 432
  of signals, 438–440
  specification of parameters, 435–437
  truncation, 432
  uniform, 426–428
  unipolar, 432
Quantizing noise, 429
Quantum noise, *see* Noise, shot
Quasianalytic (QA) method, 251–252, 356, 364–380. *See also* Performance evaluation, QA

Reblocking, *see* Resampling, reblocking
Receiver processes, 245–250
Representative sequences, 203–205, 235–236
Resampling, 292, 446–464. *See also* Interpolation
  block mode, 447–452, 463
  buffers, 455–457, 464
  bypass, 453–454, 464
  clocked bypass, 454
  conditional jumps, 454
  decimators, 460
  definitions, 447–448
  downsampling, 292, 460–461
  flag-mediated, 453–455
  hold, 461
  integer ratios, 447
  interpolation, 447, 451–452, 462
  irrational ratios, 448
  memory management, 449–450
  number representation, 448–449
  rational ratios, 447–448
  reblocking, 450–451
    circular blocks, 450, 452
    elastic buffer, 450
    scheduling, 450
  scheduling, 453
    internal, 455
    mixed, 457–458
    parallel operations, 459–460
    traffic jams, 458–459
  single-point mode, 452–460, 463
  sum and dump, 460–461
  upsampling, 292, 461
Rise time, 132

Sample-rate conversion, *see* Resampling
Samples, signal reconstruction from, 126, 127
Sampling, 125–126
  density, 130
  interval, 125
  rate, 130, 133
  samples per symbol, 130, 133
  sequence, 134–135
Scales:
  frequency, 138–139
  time, 138–139
Sequence detection, *see* MLSE
Shift register sequences, 201–205, 235
  binary, 204
  completion, 203
  M-ary, 204–205
  m-sequences, 201
  maximal-length, 201, 205
  pseudorandom properties, 201–203
  representative, 203–205, 235–236
Shot noise, *see* Noise, shot
Sideband regrowth, 234–235, 315
Signal, 11, 19–54
  analytic, 38, 98
  baseband, 11, 19–34, 47–53, 66–67
  carrier, 98
  complex envelope, 97–98
  coordinates:
    polar, 38
    rectangular, 36–38
  I-Q, 37, 97
  inphase-quadrature, *see* Signal, I-Q
  modulated, 11
  passband, 11, 34–47, 67–69
  periodogram, *see* Periodogram
  power, 47–49, 54–56
  processes, 96–123
  pulses, 20–34
  sampled, 134

spectrum, 49–53
waveform, 19–20
Signal formats, *see* Data formats
Signal-to-noise ratio, *see* SNR
Simulation:
   capabilities, 2
   data connectivity, 529–530
   duration, 124
   hierarchy, 7
   limits, 5
   modules, *see* Modules
   order of execution, 528–529
   role of, 1
Simulation language, 523–526
   benefits, 525
   evolution, 523–525
   high *vs.* low level, 525–526
   relation to programming language, 525
Single point data, 144–145, 151, 167
SNR (signal-to-noise ratio), 271–273, 378–379
Soft decisions, 115
Spectral leakage, 279–282, 517–519
Spectral spreading, 234–235, 315
Spectrum:
   of complex envelope, 57–58, 60
   conditioning, 515–519, 521–522
   cosine rolloff, 30–31
   of ensemble, 50–51, 57, 69
   estimation, 51–53, 58–60, 69
   excess bandwidth, 29
   of finite-duration signals, 58–60
   Nyquist, 28–30
   periodogram, *see* Periodogram
   rolloff, 29
   of simulated signals, 51–53
   smoothing, 517–519
   time-averaged, 57
   windows, 515–517
Spread spectrum processes, 258–259
SSB (single sideband), 43
Statistical scatter, 5, 55, 58, 69, 124, 147, 148, 149, 204, 205
Steady state, 143, 144
Strobes, 27, 106, 113–115, 247
   integrate and dump, 309–310
   by interpolation, 306–309
   by sample selection, 306, 309
Symbol, 10
   complex, 10, 42
   examples, 10
   interval, 20, 66, 130, 139
   samples per, 130
   sequence, 19

Symbol decisions, *see* Symbol detection
Symbol demapping, 15, 248, 348, 354
Symbol detection, 14–15, 122–123, 248, 346–354
   binary, 348
   decision boundaries, 113–115
   decision regions, 113,
   decision rules, 113
   differential, *see* Differential detection
   errors, 15
   hard, 14
   largest-of, 354
   minimum distance, 347–348, 354
      rule, 347
      tables, 347–348
   MPSK, 349
   multiamplitude, 349
   multiple-strobe, 353
   PAM-QAM decisions, 346–350
   quantizers, 348–350, 354
   slicers, *see* Symbol detection, quantizers
   soft, 14
   symbol-by-symbol, 27, 112–113
   threshold, 349
Symbol mapping, 10, 205–213, 236–237
   combinatorial, 207
   for nonlinear modulations, 213
   for square constellations, 208–209
   Gray encoding, 207
   rules, 207
   tables, 207–209, 243, 347–348
Symbol strobes, *see* Strobes
Synchronization, 249, 296–344
   calibration, 297–299
   carrier, 14
   clock, 14
   by correlation, 298–299, 334–338
      frequency-domain method, 337–338
      search, 336–337
   hardwire, 297–298
   search, 298
   timing, 14
Synchronizers:
   carrier extractors, 320–322
   carrier regenerators, 310–315, 338–342
      absolute value, 340–342
      carrier wave, 311
      complex envelope, 312–313
      fourth law, 339–340
      M-th law, 314
      phase jumps, 313–314
      phase unwrapping, 313–314
      sampling rate, 315
      signal representation, 338–339

Synchronizers (*Continued*)
    simulation, 311–312
    square law, 310–314
  categories, 300
  clock extractors, 322–324
  clock regenerators, 315–319
    location of zero crossings, 318–319, 343–344
    raw clock, 317–318
    square law, 317–318
  configurations, 300–301
  controllers, 303–304, 307–310
  correctors, 303, 305–307
  Costas detector, 326
  data aided (DA), 302
  decision directed (DD), 302
  error detectors, 303, 325–331
  extractors, 300–301, 304, 319–324
  feedback, 300–301, 303–304
  feedforward, 300–301, 304
  frequency-error detectors, 325
  integrate and dump, 309–310
  loop filters, 303
  non-data aided (NDA), 302
  phase error detectors, 325–327
  recovery, 299–304
  regenerators, 300–301, 304, 310–319
  timing correction, 305–307
  timing-error detectors, 327–331
    decision-directed, 327–329
    maximum likelihood based, 327–328
    non-data aided, 329
    sequential edge, 329–331
    two-sample, 328–329

Tape flutter, 256–257
Time base, *see* Pulse shaping, time base
Time delay, *see* Delay
Time duration, 138
Time increment, 138
Time scale, 135
Time-base variation, 13
Time/frequency relations, 138, 152
Time/frequency scales, 138–139
Timing fluctuations, 13
Timing recovery, *see* Synchronization, timing
Transducers, 13
Transfer function, *see* Filters, transfer function
Transients, 143, 144
Transmission:
  disturbances in medium, 12
  links, 6–15
  synchronous, 10, 14, 20, 66
Transmitter processes, 200
Truncation error, 134
TWTA (traveling wave tube amplifier), 232

Unit sample, 107, 166
Upsampling, *see* Resampling
User interface, 526–528
  graphical block diagram, 527
  menu based, 527
  parameter entry, 528
  text based, 527
Viterbi detector, 112. *See also* MLSE
VSB (vestigial sideband), 43

Waveform generator, 10. *See also* Signal, waveform
Windows, 517–518

z-Transform, 135